COMPUTATIONAL SCIENCE AND ENGINEERING

PROCEEDINGS OF THE INTERNATIONAL CONFERENCE ON COMPUTATIONAL SCIENCE AND ENGINEERING (ICCSE2016), BELIAGHATA, KOLKATA, INDIA, 4–6 OCTOBER 2016

Computational Science and Engineering

Editors

Arpan Deyasi
Department of Electronics and Communication Engineering, RCC Institute of Information Technology, Kolkata, West Bengal, India

Soumen Mukherjee
Department of Computer Application, RCC Institute of Information Technology, Kolkata, West Bengal, India

Pampa Debnath
Department of Electronics and Communication Engineering, RCC Institute of Information Technology, Kolkata, West Bengal, India

Arup Kumar Bhattacharjee
Department of Computer Application, RCC Institute of Information Technology, Kolkata, West Bengal, India

CRC Press is an imprint of the
Taylor & Francis Group, an **informa** business

A BALKEMA BOOK

CRC Press/Balkema is an imprint of the Taylor & Francis Group, an informa business

Typeset by V Publishing Solutions Pvt Ltd., Chennai, India
Printed and bound in the UK and the US

Published by: CRC Press/Balkema
P.O. Box 11320, 2301 EH Leiden, The Netherlands
e-mail: Pub.NL@taylorandfrancis.com
www.crcpress.com – www.taylorandfrancis.com

ISBN: 978-1-138-02983-5 (Hbk)
ISBN: 978-1-315-37502-1 (eBook)

Computational Science and Engineering – Deyasi et al. (Eds)
© 2017 Taylor & Francis Group, ISBN 978-1-138-02983-5

Table of contents

Computational Science and Engineering – Deyasi et al. (Eds)
© 2017 Taylor & Francis Group, ISBN 978-1-138-02983-5

Preface

The growth of a Nation may be indexed by the degree and rate of progress in scientific research and technological development which also scales the socio-economical and geo-political stature in global context. A perfect blend of innovation and invention helps the sustainable augmentation of a Country through indigenous development, which not only contributes to industrial recreation, but also akin to social pursuit. Market-driven economy in twenty-first century not only wants innovative ideas, but also manifestation of them for productive outcome in consumer-based society. In this context, the Vision of Mission of an Engineering Institution should focus on development and nurturing of the igniting minds which may honour the Nation by providing professional engineering solutions mingled with professional ethics and code of conduct towards day-to-day societal and environmental complex problems. The essence of progress may really be smelled when the fruit of technology can be reached at the farthest points of Nation where civilization just begins to spread its wings. It is the sole aim for organizing an International Conference where podium is set in the environment of pluralistic participatory democracy for the budding theoreticians and experimentalists with a dream to serve Society and Mankind.

The International Conference on Computational Science and Engineering (ICCSE 2016) organized by Department of Electronics and Communication Engineering and Department of Computer Application of RCC Institute of Information Technology covers a few tracks of research which are growing at exponential pace and henceforth affecting our livelihood in a variety of essential means. Present market of consumer electronics is dependent on VLSI & MEMS, Distributed System & ICT, Medical Instrument based System, Image Processing & System Security, Embedded System, Soft Computing, Microwave Communication, Heterojunction Devices & Circuits, Wireless Communication & Computer Vision and Quantum & Photonic Devices. To create an atmosphere for sharing knowledge in these domains, the conference is organized where research papers are chosen with the vision of providing out-of-the-box thoughts among the present day education system. The purpose of engineering education is outcome-oriented instead of input-output based system, and double-blind review process is introduced with authenticity checking (using iThenticate software in collaboration with NIT Durgapur) to acknowledge the originality and novelty of research contributions. Keynote lectures and invited talks are arranged by connoisseurs of the domains of Device (Prof. (Dr.) Subir Kumar Sarkar, Jadavpur University; Prof. (Dr.) Siddheswar Maikap, Thin Film Nano Technology Laboratory and Biomedical Engineering Research Center, Department of Electronic Engineering, Chang Gung University, Taiwan); Electromagnetics (Prof. (Dr.) Arokiaswami Alphones, School of Electrical and Electronic Engineering, Nanyang Technological University, Singapore), Communication (Prof. (Dr.) Santi Prasad Maity, IIEST Shibpur) and Computer Science (Prof. (Dr.) Tandra Pal, NIT Durgapur; Mr. Prabuddha Samaddar, Cognizant Technology Solutions) to make awareness on the immense area of opportunities for research. A blend of academic experience, foster collaborations across industry and academia, and evaluation of emerging technologies are the extract of this conference; and to make icing on the cake, discussion forums are offered at the end of each technical session where ideas are rejuvenated for the coming days.

The conference is organized in the year which is world-widely celebrated for the 150th year of Maxwell's Equation, 100th of year of Gravitational Wave and also of Stimulated Emission, 100th year of first Transcontinental Conversation, 50th of year of Optical Fibre, 25th year of GSM network and also of Carbon Nanotube, and 25th year of 3D Photonic Crystal. This memorizing consequence is highlighted in the opening day of the programme, and to make the event successful, it is technically co-sponsored by IEEE Electron Device Society (Kolkata Chapter) and The Institution of Engineers (INDIA), and financially sponsored by TEQIP-II. The respected session chairs are a few of invited speakers and also the respected contributors in the different tracks of the conference (Prof. (Dr.) Partha Pratim Sarkar, Kalyani University; Prof. (Dr.) Angsuman Sarkar, Kalyani Govt Engineering College; Prof. (Dr.) Rajarshi Gupta, University of Calcutta, Prof. (Dr.) Siddhartha Bhattacharyya, RCCIIT; Prof. (Dr.) Pramatha Nath Basu, RCCIIT; Prof. (Dr.) Minakshi Banerjee, RCCIIT; Prof. (Dr.) Tirtha Sankar Das, RCCIIT).

Presenters of the 51.1% selected papers on the multifaceted aspects on computing, communication and device engineering in the three-day conference (4th–6th October 2016) has benefitted by the knowledge of Dignitaries (Prof. (Dr.) Sparshamani Chatterjee, Chairman, RCCIIT; Prof. (Dr.) Arup Kumar Bhaumik, Principal, RCCIIT; all the members of RCCIT Society, Governing Body & Advisory Committee) and Session Chairs. Contribution of TEQIP-II Coordinator Prof. Biswanath Chakraborty may not be expressed by using phrases only, and organizing committee members put their great effort to make the event different from other conferences in terms of eminence and excellence. Editors convey their regard to all the people who are associated with making the event memorisable.

Arpan Deyasi
Soumen Mukherjee
Pampa Debnath
Arup Kumar Bhattacharjee

Technical session I: VLSI and MEMS

Computational Science and Engineering – Deyasi et al. (Eds)
ISBN 978-1-138-02983-5

Design of quantum cost efficient MOD-8 synchronous UP/DOWN counter using reversible logic gate

H. Maity & A. Biswas
ECE Department, NSHM Knowledge Campus Durgapur, Durgapur, India

A.K. Bhattacharjee
ECE Department, National Institute Technology Durgapur, Durgapur, India

ABSTRACT: The Reversible logic synthesis techniques will be a necessary part of the long-term future of computing due to its low power dissipating characteristic. Today, reversible logic circuits have considerable attention in improving some field like nanotechnology, quantum computing etc. In this paper we proposed the design of MOD-8 synchronous up/down counter with reduced number of quantum cost to implement it using existing reversible logic gates. The quantum cost, number of gates, constant inputs and garbage outputs are respectively are 35, 13, 9 and 12 for the proposed work.

1 INTRODUCTION

Landauer (Landauer 1961), shown that due to irreversible computation, each bit of information loss create KTln2 joules of energy, where K is the Boltzmann constant and T is the absolute temperature at which the process is performed. Further Bennett, showed that one can avoid KTln2 joules of energy dissipation from the circuit if input can be extracted from output and it would be possible if and only if reversible gates are used (Bennett 1973). Research is going on in the field of reversible logic synthesis and excellent research work has been carried out in the area of reversible combinational logic. However, there is not much more work in the area of sequential circuit. A counter is a sequential circuit that capable of counting the numbers. This paper proposes the design of MOD-8 synchronous up/down counter with reduced number of quantum cost, constant inputs and less number of gates to implement it using existing reversible gates.

2 BASIC CONCEPTS

2.1 *Reversible logic function*

It is an equal number of inputs and output logic function in which there is a one-to-one correspondence between the input and the output. The input vector can be uniquely determined from the output vector. This avoids the loss of information which is the root cause of power dissipation in irreversible logic circuits.

2.2 *Garbage output*

A garbage output is an output that is needed to change an irreversible gate to a reversible one and are not used to the input to the other gates.

2.3 *Quantum cost*

All reversible 1×1 and 2×2 gates have unity quantum cost. As every reversible gate is a combination of 1×1 or 2×2 quantum gate, so the quantum cost of a reversible gate is the numbers of quantum gate (Thapliyal et al. 2010).

2.4 *Reversible gate*

It is a gate with equal number of inputs and outputs. If the input vector of a reversible gate is denoted by $I_V = (I_1, I_2, I_3, \ldots, I_K)$, the output vector can be represented as $O_V = (O_1, O_2, O_3, \ldots, O_K)$. A reversible gate can be represented as $K \times K$ in which the number of input and output is K. There are several reversible gate, some of them discuss below.

2.4.1 *Reversible NOT gate*

The reversible NOT gate is a simplest 1×1 gate (Garipelly et al. 2013), Fig. 1 shows the reversible NOT gate with zero quantum cost.

2.4.2 *Feynman Gate (FG)*

The Feynman gate (FG) having 2 inputs and 2 outputs. A, B are the inputs and P, Q are the outputs where ($P = A$, $Q = A \oplus B$) (Feynman 1985). The quantum cost of FG is 1. Fig. 2 shows the block diagrams of FG.

Figure 1. Diagram of reversible NOT gate.

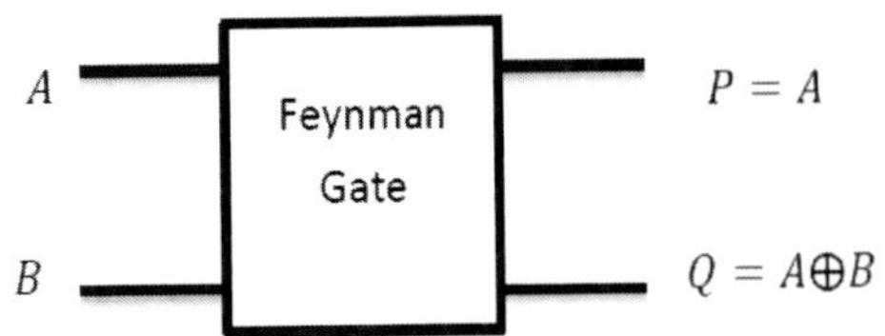

Figure 2. 2 × 2 Feynman gate.

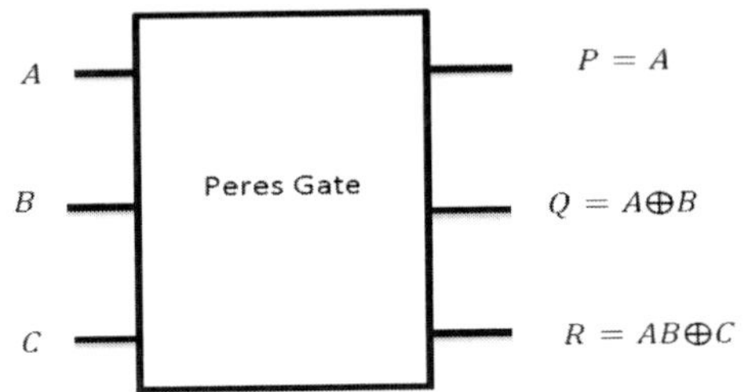

Figure 3. 3 X 3 Peres gate.

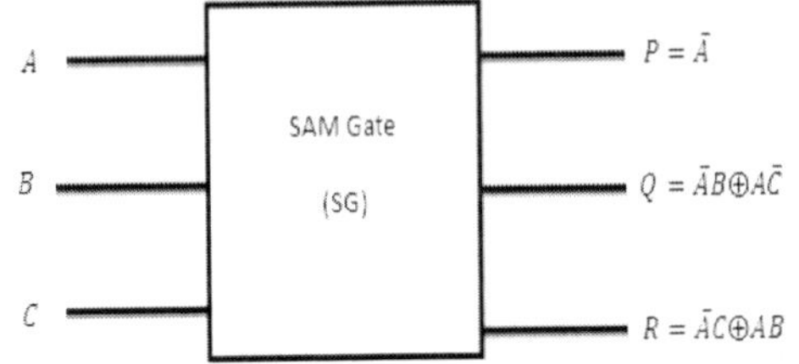

Figure 4. 3 X 3 Block diagram of SAM gate.

2.4.3 *Peres Gate (PG)*

A Peres gate (Peres 1985) having 3 inputs and 3 outputs. A, B, C are the inputs and P, Q, R are the outputs where P = A, Q = A ⊕ B, R = A · B ⊕ C. Fig. 3 shows the Peres gate with quantum cost of 4.

2.4.4 *SAM Gate(SG)*

A SAM gate (Mamun et al. 2012, 2013, 2104) is a (3 × 3) reversible gate having the quantum cost 4. Figure 4 shows the SAM gate and Fig. 5 shows the OR and AND operation of SAM gate. The quantum cost of SAM gate is 4.

2.5 *Clocked T Flip-flop*

The characteristic equation of the clocked T flip-flop (Thapliyal et al. 2007, 2010 and Rice 2008) is Q = (T.CLK) ⊕ Q. The clocked T flip-flop is designed by using a Peres gate and a Feynman gate as shown in Fig. 6.

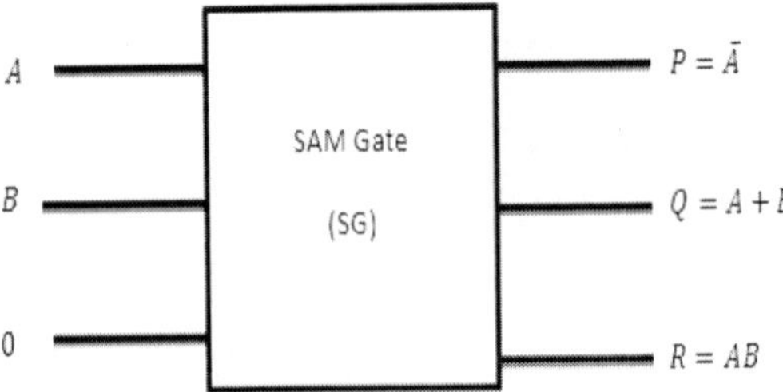

Figure 5. Functional diagram of SAM gate.

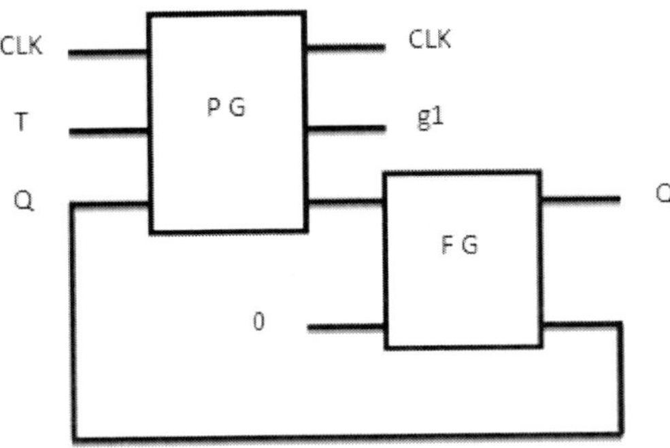

Figure 6. Flip-Flop using Peres and Feynman gate.

3 PROPOSED WORK

The Block diagram of 3-bit Up/Down counter is shown in Fig. 7.

When C = 1 in Fig. 7, the counter will work in the Up mode, i.e. it counts from state Q0Q1Q2 = 000 to state Q0Q1Q2 = 111, when C = 0, it will work in the down mode, i.e. it counts from stateQ0Q1Q2 = 111 to state Q0Q1Q2 = 000. The table-1 shows the excitation table of proposed work.

Derived equations from excitation table using K-Map for T_2, T_1, T_0 are as follows:

$$T_2 = C'Q1'Q0' + CQ1Q0 \tag{1}$$

$$T_1 = C'Q0' + CQ0 \tag{2}$$

$$T_0 = 1 \tag{3}$$

Implementation of these derived equations 1, 2, 3 are shown in our proposed design in Fig. 8 of MOD-8 UP/DOWN Synchronous Counter using reversible NOT gate, Feynman gate, Peres gate and SAM gate.

Table 1. Excitation table for MOD-8 UP/DOWN Counter.

Control Input	Present State			Next State			Excitation Inputs		
C	q_2	q_1	q_0	Q_2	Q_1	Q_0	T_2	T_1	T_0
0	0	0	0	1	1	1	1	1	1
0	0	0	1	0	0	0	0	0	1
0	0	1	0	0	0	1	0	1	1
0	0	1	1	0	1	0	0	0	1
0	1	0	0	0	1	1	1	1	1
0	1	0	1	1	0	0	0	0	1
0	1	1	0	1	0	1	0	1	1
0	1	1	1	1	1	0	0	0	1
1	0	0	0	0	0	1	0	0	1
1	0	0	1	0	1	0	0	1	1
1	0	1	0	0	1	1	0	0	1
1	0	1	1	1	0	0	1	1	1
1	1	0	0	1	0	1	0	0	1
1	1	0	1	1	1	0	0	1	1
1	1	1	0	1	1	1	0	0	1
1	1	1	1	0	0	0	1	1	1

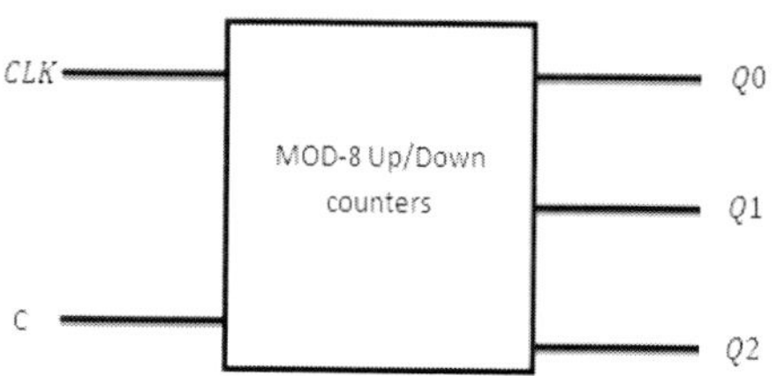

Figure 7. Block diagram of 3-bit Up/Down counter.

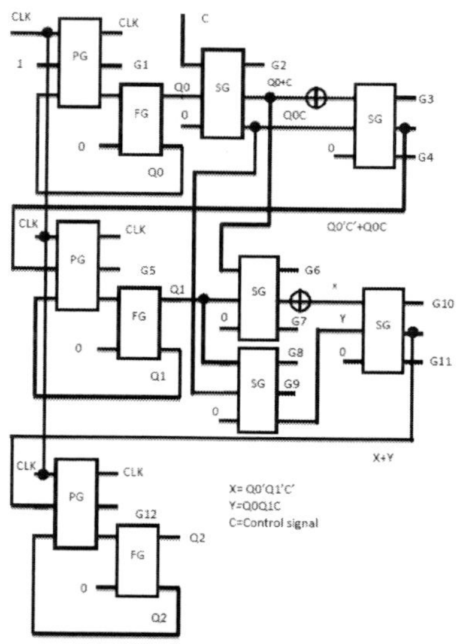

Figure 8. Proposed design of MOD-8 UP/DOWN synchronous counter.

Table 2. Comparison result of MOD-8 Up/Down synchronous counter.

Parameters	Existing design (Singh et al. 2014)	Existing design (Dey et al. 2015)	Proposed design
Quantum cost	53	50	35
No of gates	17	14	13
No. of constant input	13	12	9
Garbage outputs	12	12	12

4 RESULT AND DISCUSSION

In the implementation of MOD-8 UP/DOWN Synchronous counter we use three Peres gate having QC = 4, three Feynman gate having QC = 1, two NOT gate having zero quantum cost and five SAM gate having QC = 4. Table-2 shows the comparison result of proposed work.

5 CONCLUSION

In this paper we have presented the basic concepts of reversible gates. This paper proposed the quantum cost efficient MOD-8 synchronous up/down counter. The proposed designs are better in terms of quantum cost, number of gates, number of constant inputs and garbage outputs. The proposed MOD-8 synchronous up/down counter has great impact in quantum computing. The proposed synchronous counter can be used in reversible ALU, processor, nanotechnology, low power circuit design etc.

REFERENCES

Bennett. C.H. (1973). Logical Reversibility of Computation. *IBM J. Research and Development*. 17:525–532.

Dey A, Singh S. K. and Maity H. (2015). Improved Design of MOD-8 Synchronous UP/DOWN Counter Using Reversible Gate. *International Journal for Scientific Research and Development*. 3:787–790.

Feynman R. (1985). Quantum Mechanical Computers. *Optics News*. 11:11–20.

Garipelly. R, Kiran Madhu P. and Sontosh Kumar A. (2013). A review on reversible logic gates and their implementation. *International journal of emerging technology and advanced engineering*. 3:417–423.

Landauer, R. (1961). Irreversibility and heat generation in the computing process. *IBM J. Research and Development*. 5:183–191.

Mamun Md. Selim Al. and Hossain Syed Monowar (2012). Design of Reversible Random Access Memory. *International Journal of Computer Applications*. 56:18–23

Mamun Md. Selim Al. and Menvill David (2013). Quantum Cost Optimization for Reversible Sequential Circuit. *International Journal of Advanced Computer Science and Applications.* 4:15–21.

Mamun Md. Selim Al. and Karmaker B. K. (2014). Design of Reversible Counter. *International Journal of Advanced Computer Science and Applications.* 5:124–128

Peres A. (1985). Reversible Logic and Quantum Computers. *Physical Review A.* 32:3266–3276.

Rice J. E. (2008). An introduction to reversible latches. *The Computer journal.* 51:700–709.

Singh S. K, Maity H. and Dey A. (2014). A Novel Design of MOD-8 Synchronous UP/DOWN Counter Using Reversible Gate. *International Journal Of Scientific Research And Education.* 2:1968–1976.

Thapliyal H. and Vinod A. P. (2007). Design of reversible sequential elements with feasibility of transistor implementation. *In Proc. IEEE Intl. Symposium On Circuit and System.* 625–628.

Thapliyal. H. and Nagarajan R. (2010). Design of Reversible Sequential Circuits Optimizing Quantum Cost, Delay, and Garbage Outputs. *ACM Journal on Emerging Technologies in Computer Systems.* 6:1–14.

Computational Science and Engineering – Deyasi et al. (Eds)
© 2017 Taylor & Francis Group, ISBN 978-1-138-02983-5

Modeling of MEMS based capacitive micromachined ultrasonic transducer array element at static voltage

Reshmi Maity & N.P. Maity
Mizoram University (A Central University), Aizawl, India

Swapnil Vaish
National Institute of Technology, Kuruskhetra, India

ABSTRACT: Capacitive Micromachined Ultrasonic Transducers (CMUTs) are used to receive and transmit ultrasonic waves. The device is constructed from many small, in the order of microns, circular membranes, which are connected in parallel. In this paper, static voltage analysis of CMUT has been modeled and described. The membrane displacement with respect to various parameters such as radius, membrane thickness, gap thickness and applied static bias has been analyzed. By consideration of Mason's analysis, the equation of motion of membrane under tension has been discussed. The graphical results support the theoretical modeling of the CMUT cell.

1 INTRODUCTION

Ultrasonic waves, is a band above 20 kHz. It continues up to some MHz range, finally, at around 1 GHz, goes over into the hypersonic regime. Most of the applications described in this paper take place in the range of 1 to 100 MHz, corresponding to wavelength in a typical solid of approximately 1 mm to 10 micro meters, where average sound velocity is about 5000 m/s. In air, the sound velocity is about 500 m/s, with wavelengths of the order of 3 mm to 30 micro meters for the above frequency range. A transducer is a device that converts one form of energy to another. An ultrasonic transducer is a device that converts energy into ultrasound, or sound waves above the normal range of human hearing. The term is more appropriate to be used to refer to piezoelectric transducers that convert electrical energy into sound. Piezoelectric crystals have the property of changing size when a voltage is applied, thus applying an alternating current across them, causes to oscillate at very high frequencies, thus producing very high frequency sound waves. Since piezoelectric crystals generate a voltage when force is applied to them, the same crystal can be used as an ultrasonic detector. Some systems use separate transmitter and receiver components while others combine both in a single piezoelectric transceiver. Non-piezoelectric principles are also used in construction of ultrasound transmitters. Magnetostrictive materials slightly change size when exposed to a magnetic field; such materials can be used to make transducers. CMUT mainly consists of three layers and it is having one fixed electrode and one free electrode, which fulfills its capacitive action. It is fabricated using surface micromachining technique that is why it is called micromachined. A capacitor microphone uses a thin plate which moves in response to ultrasound waves; changes in the electric field around the plate convert sound signals to electric currents, which can be amplified CMUT membrane behavior during a high-voltage excitation [1]. Measurements were performed with a homemade interferometer system. Experimental results in air and fluid (here oil) are discussed. CMUTs with a perforated membrane have been fabricated and characterized in air [2].

Two types of CMUT device have been fabricated having perforation ratio (area ratio of holes = AR) of 10% and 20%, and analyzed about electrical and mechanical characteristics. An analytical CMUT model was proposed [3–6] that couples three physical domains using plate vibration and radiation theories. The model accounts for the spatial distribution of the normal deflection over the CMUT cell membrane. Apart from that, in the same year, the collapse mode of operation of CMUTs was shown to be a very effective way to achieve high output pressures [7]. However, no accurate analytical or equivalent circuit model exists for understanding the mechanics and limits of the collapse mode.

The design, development, fabrication and characterization of a 12-MHz ultrasound probe for medical imaging; based on a CMUT array was presented [8]. The CMUT array is micro fabricated and packed using a novel fabrication concept specifically conceived for imaging transducer arrays. The performance of the developed probe is optimized by including analog front-end reception electronics. Characterization and imaging results are used to assess the performance of CMUTs with respect to conventional piezoelectric transducers. In fact, because of its many applications, it gives the researcher more interesting and hopes for innovating something. These occur in a very broad range of disciplines, covering engineering, physics, chemistry, biology, medicine, food industry, oceanography, seismology, etc. based on two unique features. Ultrasonic waves travel slowly, about 100,000 times slower than electromagnetic waves. This provides a way to display information in time, create variable delay, etc. Ultrasonic waves can easily penetrate opaque materials, whereas many other types or radiation such as visible light cannot. Since ultrasonic waves sources are inexpensive, sensitive, and reliable, this provides a highly desirable way to probe and image the interior of opaque objects.

CMUT technology was first introduced in 1990s. It is a desired combination of the micro-fabrication technology and electrostatic principle. The principle of operation of electrostatic transducers is based on electrostatic transducer of a thin metalized membrane suspended over a back plate. After two decades of development, a piezoelectric transducer is replaced by CMUT. A CMUT consists of a silicon or silicon nitride membrane and substrate using micromachining technology which is currently employed in Micro-Electro-Mechanical System (MEMS). CMUTs have much lower mechanical impedance than the piezoelectric transducers which offers better coupling with the medium and higher bandwidth.

2 THEORETICAL MODELING

A single CMUT is designed layer by layer. The whole structure lies on silicon substrate, whose top is highly doped. Vibrating silicon nitride membrane is supported by silicon dioxide stands. A metal (aluminium or gold) inside the membrane (whose position may vary) forms the top electrode. This thin poly-silicon or silicon-nitride membrane is separated by a small vacuum cavity from the substrate to prevent short circuits between the electrodes when membrane collapses. The gap that is formed inside the structure may or may not be sealed depending on the application.

The basic structure of single CMUT cell is shown in Fig. 1. When a voltage is applied between the top electrode and bottom electrode, regardless of the polarity, the membrane will deflect towards the substrate due to the attractive electrostatic forces. As the voltage is increased, the slope of the voltage-deflection curve also increases, that is displacement of membrane from an initial horizontal condition. This shows the increase of sensitivity with the applied voltage.

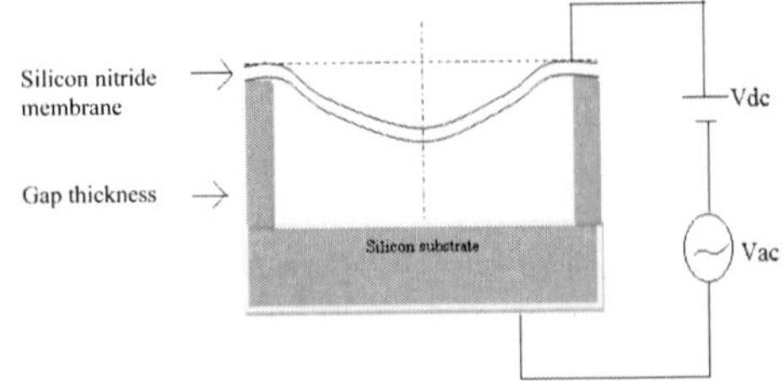

Figure 1. A CMUT cell.

This electrostatic force is resisted by a mechanical restoring force due to membrane stiffness (tension or residual stress). During transmission an alternating bias is applied with the static bias whose magnitude is very less comparable to applied static bias, and due this applied alternating voltage a small resistive force will generate in opposite to the membrane displacement due to static bias. Membrane will start fluctuating, and vibration of membrane generates an ultrasound. In reception, the acoustic pressure on the membrane results in a change in the capacitance between the electrodes, thus an alternating current signal can be detected.

By consideration of Mason's analysis, [9] the equation of motion of membrane under tension can be expressed by (1).

$$\frac{(Y_m+T_m)t_m^{\ 3}}{12(1-\sigma_m^{\ 2})}\nabla^4 w - T_m\nabla^2 w - P + t_m\rho_m\frac{\partial^2 w}{\partial t^2} = 0 \tag{1}$$

Where t_m is the thickness of the membrane, σ_m is the value of poison's ratio (ratio of lateral concentration to the longitudinal expansion of material), w is displacement normal to the plane of the membrane, T_m is the tension on membrane, P is the transverse pressure on the membrane and ρ_m is density of the material. We have considered a very thin stretched membrane actuated by a constant pressure over the surface [9]. As $t_m^{\ 3} \to 0$ and $\nabla^2 w = \left[d^2w/dr^2 + (1/r)(dw/dr)\right]$. Where, r is the radial position measured from the center. Then eq. (1) becomes,

$$T_m\left(\frac{d^2w}{dr^2}+\frac{1}{r}\frac{dw}{dr}\right) - \rho_m t_m\frac{\partial^2 w}{\partial t^2} + P = 0 \tag{2}$$

Under static bias condition (2) becomes,

$$\left[r^2\frac{d^2w}{dr^2}+r\frac{dw}{dr}\right] = -\left[\frac{P}{T_m}r^2\right] \tag{3}$$

Let us consider $r = e^Z$, then the complementary function $= (C_1 + C_2 z)$. Where, C_1 and C_2 are constants, whose values will be derived by the boundary conditions. The particular integral is $-[P/4T_m][e^{2Z}]$. So the general solution of above equation will be,

$$w(z) = (C_1 + C_2 z) - \frac{P}{4T_m} e^{2z} \tag{4}$$

Now 'w' will be maximum at $r = 0$ so, $C_2 = 0$. And $w(r)$ will be zero at $r = a_m$ so,

$$w(r) = \frac{P}{4T_m}(a_m^2 - r^2) \tag{5}$$

$$w(0) = \frac{P}{4T_m}(a_m^2 - 0) \tag{6}$$

3 RESULTS AND DISCUSSION

The different fabrication parameters value is selected from Haller and Khuri-Yakub [10]. This paper present the theory of operation, the fabrication technique used and the characterization of the device and the device performance is compared with the theory. The parameters value is given in Table 1, which are used for the theoretical calculations. When the applied static voltage across the membrane increases, the force on membrane increases nonlinearly shown in Fig. 2. The membrane displacement at center starts increasing from its initial zero displacement. When the applied electrostatic potential will increase then there will be a strong electrostatic force generated between the top and bottom electrodes and it will attract the top membrane towards substrate. This represents the membrane displacement at the centre.

In Fig. 3, we find that when the space between the membrane and the substrate is the minimum, then due to applied static bias voltage the displacement goes toward maximum but when space increases corresponding displacement decreases. An electrostatic force of attraction is always dependent inversely to the electrode separation. When the electrode space (or point charge) start increasing the columbic force of attraction between the top and bottom electrode is decreasing and the membrane displacement is decreasing. Fig. 4 represents the variation of membrane at center due to applied static bias with respect to radius of membrane. Displacement of membrane increases when the radius of circular membrane increases. This is happened due to decrement of the membrane tension. Tension is inversely related with the length of string, so when the length of the membrane is start increasing corresponding to that, the tension (or residual stress) inside the membrane starts decreasing. When the tension will be reduced, the membrane displacement due to same applied static bias will start increasing.

Table 1. Parameter specification.

Sl. No	Parameters	Values
1	a_m	26×10^{-6} m
2	ϵ_g	8.854×10^{-22} F/m^2
3	t_g	0.50×10^{-6} m
4	t_m	7500×10^{-10} m
5	ρ_m	2816 kg/m^2
6	V_{dc}	$40V$
7	Pr	280×10^6 N/m^2
8	ϵ_m	4.17×10^{-11} F/m^2

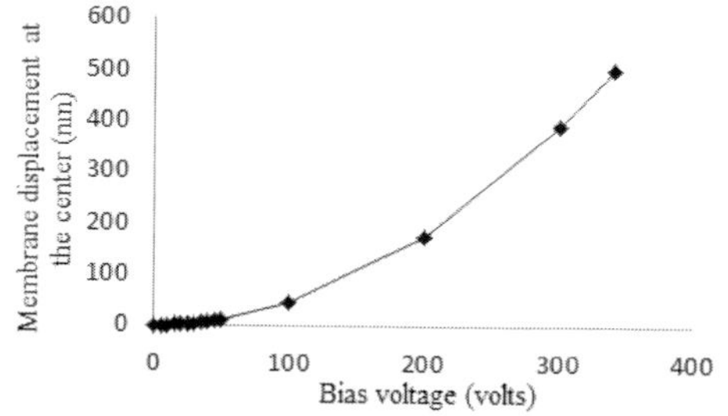

Figure 2. Membrane displacement behavior with respect to applied DC bias.

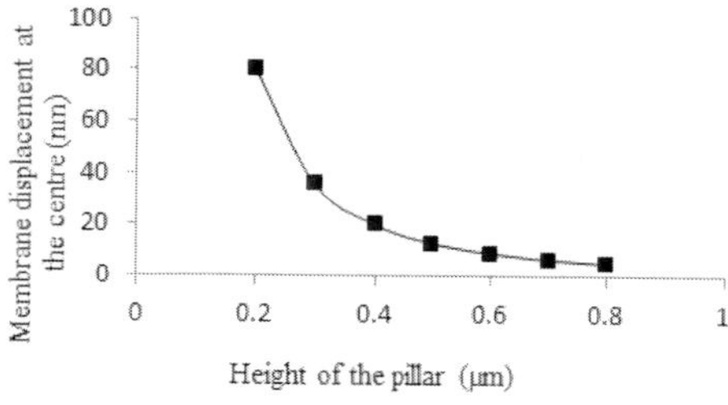

Figure 3. Displacement variation of membrane at center with respect to gap thickness.

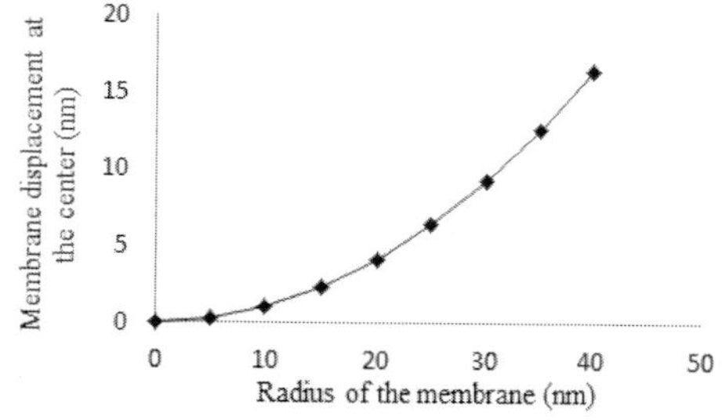

Figure 4. Variation of membrane at center with respect to radius of membrane.

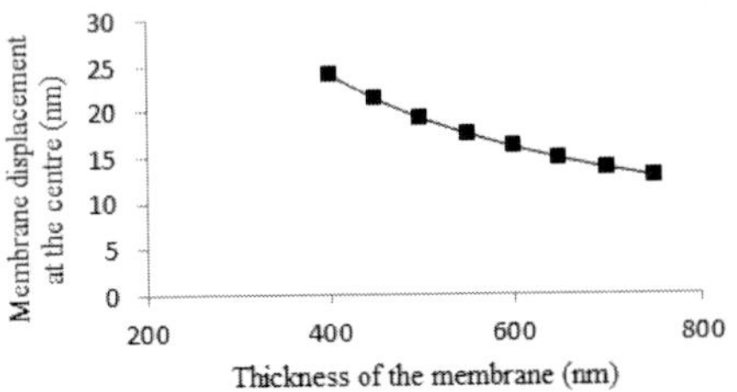

Figure 5. Variation of membrane at center with respect to radius of membrane.

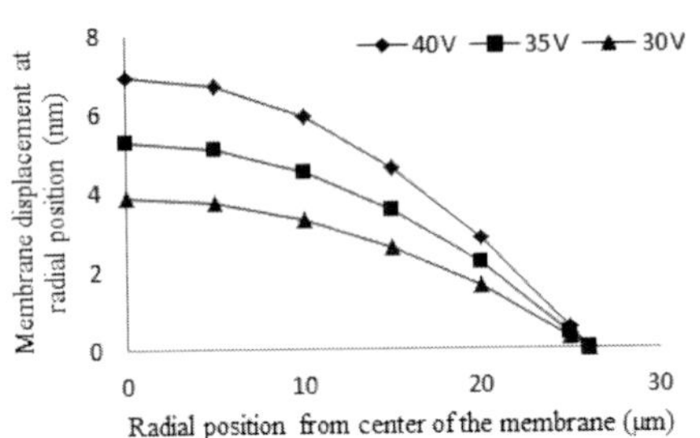

Figure 6. Variation of membrane at center with respect to radius of membrane.

Fig. 5 represents that, as we are increasing the membrane thickness at same bias and physical condition, the displacement of membrane start decreasing. This is because due to increment in membrane thickness, the membrane stiffness (toughness) starts increasing which will strongly oppose the electrostatic force of attraction between the top and bottom electrode which will cause the decrement in membrane displacement at center.

Fig. 5 represents that, as we are increasing the membrane thickness at same bias and physical condition, the displacement of membrane start decreasing. This is because due to increment in membrane thickness, the membrane stiffness (toughness) starts increasing which will strongly oppose the electrostatic force of attraction between the top and bottom electrode which will cause the decrement in membrane displacement at center. Fig. 6 is obtained by analyzing the (3) and this explained that when a constant voltage is applied, the displacement will be maximized at center and moving toward edge of cell the displacement of membrane decreases and is zero at the end. When voltage is changed to a higher value corresponding displacement at center is maximized and it starts decreasing towards edge.

4 CONCLUSION

In this paper, a CMUT array element has been characterized. These results confirm the validity of the CMUT model as a circular membrane. The model can be used to predict further improvements in performance and, in fact, is a powerful tool for future device designs. CMUT is an attractive alternative to piezoelectric transducers. The ease of fabrication and electronics integration, wide bandwidth, and the large dynamic range of the CMUT make it a superior transducer choice for medical imaging.

ACKNOWLEDGMENT

The authors are highly indebted to University Grant Commission (UGC), Ministry of Human Research Development (MHRD), Govt. of India for supporting this technical work.

REFERENCES

Cha, B., Lee, S., Kanashima, T., Okuyama, M. & Tanaka, T. 2010. Influences of perforation ratio in characteristics of capacitive micromachined ultrasonic transducers in air. *Sensor and Actuators A: Physical* 171: 191–198.

Haller, M., & Khuri-Yakub, B. 1996. A surface micromachined electrostatic ultrasonic air transducer. *IEEE transactions on Ultrasonics, Ferroelectrics and Frequency Control* 43: 1–6.

Maity, R., Maity, N.P., Thapa, R.K. & Baishya, S. 2014. Impedance Response Behavior of Capacitive Micromachined Ultrasonic Transducers (CMUTs). *International Conference on Advances in Engineering & Technology, October 2014.* Jamshedpur: India.

Maity, R., Maity, N.P., Thapa, R.K. & Baishya, S. 2015. Analysis of Frequency Response Behaviour of Capacitive Micromachined Ultrasonic Transducers. *Journal of Computational & Theoretical Nanoscience* 12 (10): 3492–3494.

Maity, R., Maity, N.P., Thapa, R.K. & Baishya, S. 2015. Analytical Characterization and Simulation of a 2-D Capacitive Micromachined Ultrasonic Transducer Array Element. *Journal of Computational & Theoretical Nanoscience* 12 (10): 3692–3696.

Mason, W. 1942. Electromechanical Transducers and Wave Filters. New York: D. Van Nostrand.

Olcum, S., Yalcin, Y., Bozkurt, A., Koymen, H. & Atalar, A. 2011. An Equivalent Circuit Model for Transmitting Capacitive Micromachined Ultrasonic Transducers in Collapse Mode. *IEEE Transactions on Ultrasonic's, Ferroelectrics, and Frequency Control* 58: 1468–1477.

Savoia, A., Caliano, G. & Pappalardo, M. 2012. A CMUT Probe for Medical Ultrasonography: From Microfabrication to System Integration. *IEEE Transactions on Ultrasonic's, Ferroelectrics, and Frequency Control* 59: 1127–1138.

Senegond, N., Teston, F., Royer, D., Meynie, C. & Certon, D. 2009. High Voltage Time Domain Response of CMUT Membrane. *Physics Proceedia* 3: 1011–1016.

You, W., Cretu, E. & Rohling, R. 2011. Analytical Modeling of CMUTs in Coupled Electro-Mechano-Acoustic Domains Using Plate Vibration Theory. *IEEE Sensors Journal* 11: 2159–2168.

Computational Science and Engineering – Deyasi et al. (Eds)
© 2017 Taylor & Francis Group, ISBN 978-1-138-02983-5

Three-dimensional finite element method of capacitive micromachined ultrasonic transducer: A simulation study

Reshmi Maity & N.P. Maity
Department of Electronics and Communication Engineering, Mizoram University (A Central University), Aizawl, India

ABSTRACT: In this paper Finite Element Method (FEM) simulation has been studied for Capacitive Micromachined Ultrasonic Transducers (CMUTs). This work presents a comprehensive simulation study for the design and analysis of CMUTs for the structural element circular using FEM simulation tools. 3-D modeling is carried out by commercially available PZFLEX SolidWorks simulation tool. Although complexity of the problem increases, 3-D modeling results are more accurate as expected. Strain analysis is carried out to validate the performance reliability of the modeled structure. The study also includes result of the peak displacement.

1 INTRODUCTION

Capacitive Micromachined ultrasonic transducers constitute the present generation of ultrasound imaging transducers used in biomedical applications and non-destructive evaluation tests. It journey as a commercial application has started since early 2000 [1] and emerged as a promising alternative for piezoelectric ultrasound transducer. Different element geometries have been proposed to suit for a given application [2]-[6]. Desirable performance characteristics for CMUTs include high membrane displacement, low dc bias voltages, reliability, easiness in fabrication, and more importantly integration with CMOS processes. This work presents a comprehensive simulation study for the design and analysis of CMUTs for the structural element circular using FEM simulation tools. The study also includes result of the peak displacement. In this study, the future design requirement is considered as the benchmark reference. Strain analysis is carried out to validate the performance reliability of the modeled structure. Model for the practical membrane shape circular is reported. 3-D modeling is carried out by PZFLEX SolidWorks simulation tool. Although complexity of the problem increases, 3-D modeling results are more accurate as expected. Even though a 3-D model needs more time and processing power to simulate, performance of different element geometries relies on 3-D modeling [7]–[9].

Figure 1. Flow chart to perform simulation in SolidWorks.

2 THREE DIMENSIONAL FINITE ELEMENT SIMULATION MODEL

PZFLEX presents the initial release of the SolidWorks modeling interface. This interface is designed in building models with complex geometry. As SolidWorks tends to build models in arbitrary space and the origin is not necessarily assigned to a convenient point in the model, it is needed to translate the model to a known reference point. This will allow easy access to the node and elemental locations generated within the model. A flowchart highlighting the important steps in the simulation is shown in Fig. 1.

Meshing is the most important step in modeling the structure. Meshing builds a grid for the model. The SolidWorks mesher has direct access to the materials used within the model; it generates the correct element size. For the frequency of interest for the model and the number of elements per wavelength, the mesher calculates the element size required and displays the number of elements that will be created in each direction. The default condition for any boundary is free. Except the side where the pressure load is applied, the other sides are kept fixed. Pressure load is applied to the model by selecting the appropriate side and is applied normal to the selected surface. The radius of the membrane, area and pressure loads needs to be calculated before designing. The transparency slider on the interface can be used to examine inside the model.

3 RESULTS AND DISCUSSION

Advances in the silicon micromachining techniques have enabled the fabrication of thin membranes if silicon nitride (0.1 μm–2 μm) over very small gaps (0.05 μm–1 μm) [10]–[13]. The membrane thickness and diameter considered in this simulation are the mostly referred in the literature in the regime of medical imaging as 750 nm and 50 μm respectively [14]–[17]. The modeled structure for a circular membrane CMUT is shown in Fig. 2. The structural properties of the same are shown in Table 1. The loads pressure and force is applied uniformly towards the membrane as in Table 2.

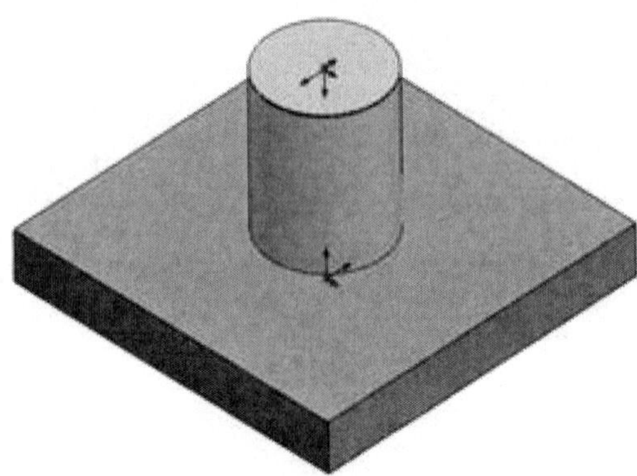

Figure 2. Model name: circle assembly.

The pressure load is applied according to the expression,

$$P = \left(e_{eff} V_{dc} V_{ac}\right) / d_{eff}^2 + (1/2)\left\{e_{eff} / d_{eff}^2\right\} V_{dc}^2 \quad (1)$$

Giving a value of 8638.176 N/m² for a static bias of 40 V and a signal of 100 mV pressure

Table 1. 3-D model physical and material properties.

Document name and reference	Treated as	Volumetric properties	Material properties
Boss-extrude1	Solid body	Mass: 1.17386×10^{10} kg Volume: 4.42965×10^{14} m³ Density: 2650 kg/m³ Weight: 1.15038×10^{9} N	Name: Silicon dioxide Model type: Linear Elastic Isotropic Elastic modulus: 3.1×10^{11} N/m² Poisson's ratio: 0.2
Boss-extrude1	Solid body	Mass: 5.20842×10^{12} kg Volume: 1.59279×10^{15} m³ Density: 3270 kg/m³ Weight: 5.10425×10^{11} N	Name: Silicon nitride Model type: Linear Elastic Isotropic Elastic modulus: 3.2×10^{11} N/m² Poisson's ratio: 0.263
Boss-extrude1	Solid body	Mass: 1.0485×10^{9} kg Volume: 4.5×10^{13} m³ Density: 2330 kg/ m³ Weight: 1.02753×10^{8} N	Name: Silicon Model type: Linear Elastic Isotropic Elastic modulus: 1.69×10^{11} N/m² Poisson's ratio: 0.3

Table 2. Loads and fixtures and contact.

Fixture name	Fixture image	Fixture details	
Fixed-1		Entities: Type:	6 face(s) Fixed Geometry
Load name Force-1	Load image	Load details Entities: Type: Value: Unit:	 1 face(s) Apply normal force 1.83361×10^{-5} N
Pressure-1		Entities: Type: Value: Units:	1 face(s) Normal to selected face 8638.18 N/m^2
Contact Global contact	Contact image	Contact properties Type: Components: Options:	 Bonded 1 component(s) Compatible mesh

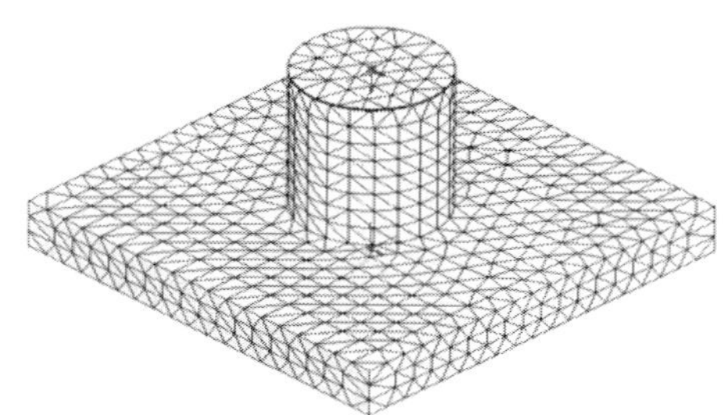

Figure 3. A meshed structure of a circular CMUT cell.

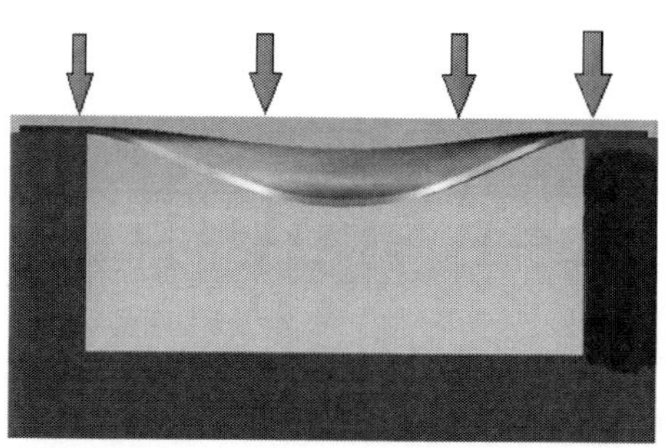

Figure 4. Inner view of a 3-D CMUT cell with circular membrane having maximum displacement at the centre of the membrane.

Figure 5. Strain limit description of a circular CMUT biased at 40 V.

Figure 6. Inner view of a 3-D CMUT.

load has been calculated. The force is evaluated as 18.336 μN. All faces of the models are fixed except the face 1 (membrane). A meshed structure of a circular CMUT with 12831 nodes and 7646 elements is shown in Fig. 3. After applying the loads the membrane starts deforming with the maximum displacement at the centre of the membrane as in Fig. 4, with peak displacement of 5.392 nm.

This large displacement corresponds to a strain of 3.03456×10^{-6} as in Fig. 5. The fracture strain for silicon nitride is 3.6×10^{-2} [15]. Thus the CMUT is operated well below the fracture limit and fatigue would not be a significant problem. The inner view of the circular CMUT cell is seen in Fig. 6.

4 CONCLUSION

Different membrane geometries have been proposed to suit better for a given application. However to arrive at a full proof structure, array and parameter design optimization technique more research is needed. This work analysed circular CMUT elements through FEM simulations and attempted to arrive at a conclusion for choosing the best peak displacement. 3-D modeling is carried out by SolidWorks simulation tool.

ACKNOWLEDGMENT

The authors are highly indebted to University Grant Commission (UGC), Ministry of Human Research Development (MHRD), Govt. of India for supporting this technical work.

REFERENCES

Ahrens, O., Hohlfeld, D., Buhrdorf, A., Glitza, O., & Binder, J. 2000. A New Class of Capacitive Micromachined Ultrasonic transducers. *Proc. IEEE Ultrasonics Symposium*: 939–942.

Cetin, A.M., & Bayram, B. 2013. Diamond-Based Capacitive Micromachined Ultrasonic Transducers in Immersion. *IEEE Trans on Ultrasonics, Ferroelectrics, and Frequency Control* 60(2): 414–420.

Ergun, A.S., Huang, Y., Zhuang, X., Oralkan, O., Yaralıoglu, G., & Khuri Yakub, B. 2005. Capacitive Micro-machined Ultrasonic Transducers: Fabrication Technology. *IEEE Trans on Ultrasonics, Ferroelectrics, and Frequency Control* 52(12): 2242–2258.

Haller, M., & Khuri-Yakub, B. 1994. A Surface Micromachined Electrostatic Ultrasonic Air Transducer. *Proc. Ultrasonics Symposium*: 1241–1244.

Huang, Y., Ergun, A., Haeggstrom, E., & Khuri-Yakub, B. 2002. Fabrication of Capacitive Micromachined Ultrasonic Transducers (CMUTs) Using Wafer Bonding Technology for Low Frequency (10 kHz-150 kHz) Sonar Applications. *Proc. IEEE OCEANS*: 2322–2327.

Jin, X., Oralkan, O., Degertekin, F., & Khuri-Yakub, B. 2001. Characterization of One-Dimensional Capacitive Micromachined Ultrasonic Immersion Transducer Arrays. *IEEE Trans on Ultrasonics, Ferroelectrics, and Frequency Control* 48(3): 750–760.

Kaltenbacher, M., Landes, H., Niederer, K., & Lerch, R. 1999. 3-D simulation of controlled micromachined capacitive ultrasound transducers. *Proc. IEEE Ultrasonic Symposium*: 1155–1158.

Kaltenbacher, M., Landes, H., Reitzinger, S., & Peipp, R. 2002. 3-D simulation of electrostatic mechanical transducers using algebraic multigrid. *IEEE Trans. Magn.* 38(2): 985–988.

Maity, R., Maity, N.P., Thapa, R.K. & Baishya, S. 2014. Impedance Response Behavior of Capacitive Micromachined Ultrasonic Transducers (CMUTs). *International Conference on Advances in Engineering & Technology, October 2014*. Jamshedpur: India.

Maity, R., Maity, N.P., Thapa, R.K. & Baishya, S. 2015. Analysis of Frequency Response Behaviour of Capacitive Micromachined Ultrasonic Transducers. *Journal of Computational & Theoretical Nanoscience* 12 (10): 3492–3494.

Maity, R., Maity, N.P., Thapa, R.K. & Baishya, S. 2015. Analytical Characterization and Simulation of a 2-D Capacitive Micromachined Ultrasonic Transducer Array Element. *Journal of Computational & Theoretical Nanoscience* 12 (10): 3692–3696.

Oralkan, O., Ergun, A.S., Cheng, C.H., Johnson, J.A., Karaman, M., Lee T.H., & Khuri-Yakub, B. 2003. Volumetric Ultrasound Imaging Using 2-D CMUT Arrays. *IEEE Trans on Ultrasonics, Ferroelectrics, and Frequency Control* 50(11): 1581–1594.

Oralkan, O., Ergun, A.S., Johnson, J.A., Karaman, M., Demirci, U., Kaviani, K., Lee T.H., & Khuri-Yakub, B. 2002. Capacitive micromachined ultrasonic transducers: Next-generation arrays for acoustic imaging?. *IEEE Trans on Ultrasonics Ferroelectrics and Frequency Control* 49(11): 1596–1610.

Park, K., Lee, H., Kupnik, M., & Khuri-Yakub, B. 2011. Fabrication of Capacitive Micromachined Ultrasonic Transducers via Local Oxidation and Direct Wafer Bonding. *Journal of Microelectromechanical Systems* 20(1): 95–103.

Tsuji, Y., Kupnik, M., & Khuri-Yakub. B. 2010. Low Temperature Process for CMUT Fabrication with Wafer Bonding Technique. *Proc. IEEE Ultrasonics Symposium*: 551–554.

Zhelezina, E., Kaltenbacher, M., & Lerch, R. 2002. Numerical simulation of acoustic wave propagation by a time and space adaptive finite element method. *Proc. IEEE Ultrasonic Symposium*: 1213–1216.

Zhuang, X., Wygant, I., Lin, D., Kupnik, M., Oralkan, O., & Khuri-Yakub, B. 2009. Wafer-bonded 2-D CMUT arrays incorporating through-wafer trench-isolated interconnects with a supporting frame. *IEEE Trans on Ultrasonics, Ferroelectrics, and Frequency Control* 56(1): 182–192.

Computational Science and Engineering – Deyasi et al. (Eds)
© 2017 Taylor & Francis Group, ISBN 978-1-138-02983-5

Effect of intrinsic and extrinsic device parameter variations on Schmitt trigger circuit

Subhasish Banerjee
ECE Department, MCKVIE, Howrah, India

Eshita Sarkar
VLSI Design, ECE Department, MCKVIE, Howrah, India

Sagar Mukherjee
ECE Department, MCKVIE, Howrah, India

ABSTRACT: Regenerative circuits are most popular sequential MOS logic circuits. One of the main regenerative circuit types, the bistable circuits which is most widely used. Schmitt triggers are bistable networks that are used in analog and digital circuit as wave shaping circuit to solve the noise problem. The approach of our work is based on the effect of source-to-substrate bias voltage for both nMOS and pMOS and the body effect coefficient on the conventional CMOS Schmitt trigger circuit. In this approach one can observed that how the triggering points of Schmitt trigger that is upper and lower threshold points UTP and LTP are being control with this device parameter. Simulation has been done on Tanner EDA tool. TSPICE simulation results of the circuit confirm the effectiveness of the approach. Typically, 1 μm CMOS technology has been used to implement Schmitt trigger circuit and the simulation results are presented.

Keywords: Schmitt trigger, UTP, LTP

1 INTRODUCTION

Example of regenerative circuit is Schmitt trigger circuit, is very useful for VLSI circuit design (Kang *et al.*, 2003). Schmitt trigger output state depends on input state and changes only as input level crosses a preset threshold level. Schmitt trigger is having a voltage transfer characteristic which is similar to inverter, but with two different logic threshold voltages known as Upper Threshold Point (UTP) for increasing signal and Lower Threshold Point (LTP) for decreasing input signals. The circuit can be exploit for the determination of low-to-high and high-to-low switching events in the presence of noise with the help of this distinct property (Baker 2010, Uyemura 2002, Kumar *et al.*, 2012, Ramesh *et al.*, 2015). This work is mainly focus on how the two different threshold voltages or triggering points of Schmitt trigger circuit are being controlled using physical parameter like source-to-substrate voltage V_{sb}. As for zero substrate bias and non zero substrate bias voltages directly influencing the threshold voltage of a MOS structure.

The most general expression of threshold voltage VT is:

$$V_T = V_{T0} + \gamma \cdot (\sqrt{|-2\varphi + Vsb|} - \sqrt{|2\varphi|}) \quad (1)$$

where the parameter

$$\gamma = \frac{\sqrt{2q \cdot N_A \cdot \epsilon_{Si}}}{C_{ox}} \quad (2)$$

γ is known as *substrate–bias* coefficient *or body-effect* coefficient. Where V_{T0} = Threshold voltage for zero V_{sb}

V_T = Threshold voltage for non-zero V_{sb}
2φ = Surface inversion potential
N_A = Doping concentration
ϵ_{Si} = Silicon permittivity
Cox = Oxide thickness.

From the above two equations it has been observed that changing the device parameter V_{sb} and γ affecting the threshold voltage of a MOS structure.

2 CIRCUIT DESCRIPTION

2.1 *Conventional Schmitt trigger*

In this section, basic design issues and constraints of Schmitt trigger which covers the overall circuit description will be discussed.

The conventional Schmitt trigger is the combination of three pMOS transistor and three nMOS transistor, (i.e. upper pMOS and lower nMOS) and considered as low for each other, the lower two nMOS transistor can be considered as series connection (Kang *et al.*, 2003, Baker 2010, Uyemura 2002).

The Schmitt circuit has an inverter like voltage transfer characteristics but with two different threshold voltages for increasing and decreasing input signals and two extra transistors are using for hysteresis (Kumar *et al.*, 2012).With this special property this circuit can be applicable for the detection of low-to-high and high-to-low switching of output in the presence of noise. In Fig. 1 the two transistors M2 and M4 having higher threshold voltage than M1 and M5 due to non-zero substrate bias and so that the output V0 switches to high-to-low and low-to-high only when M1 or M4 become ON.

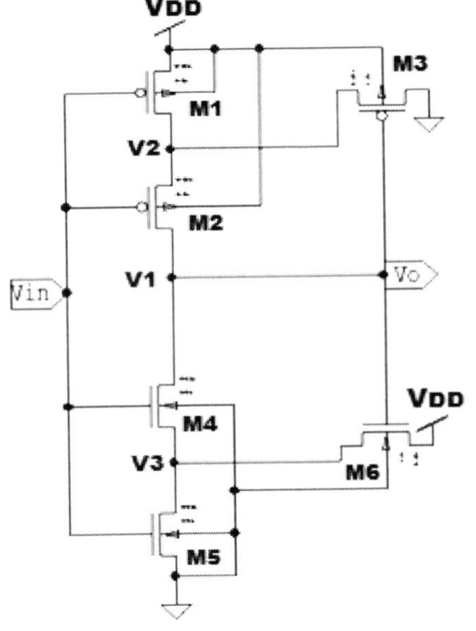

Figure 1. Schematic of conventional Schmitt trigger.

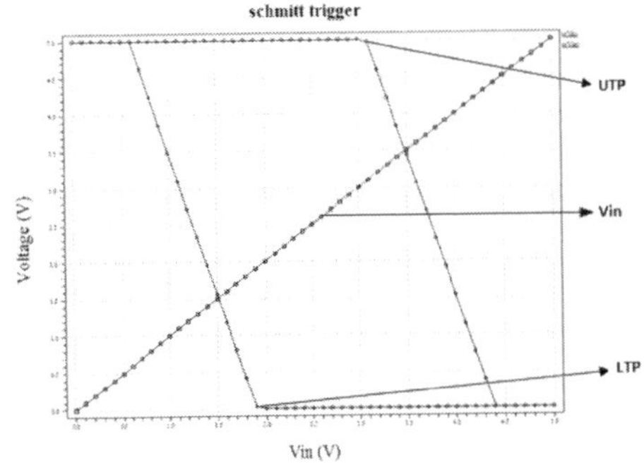

Figure 2. The transfer curve.

Now with the help of another two transistors in Fig. 1 M3 and M6 the circuit is providing hysteresis.

At zero input voltage (Vin = 0) M1 and M2 are turned on and at the same time, M4 and M5 are turned off and output is at logic high.

Now M5 will be on when the input voltage nearly equal with to threshold voltage of M5 transistor, while M4 remains OFF and M6 will be on so for that reason output will be high, so M5 Try to decreases the node voltage V3while M6 try to increase this node to voltage VDD-VT, so transistor M4 sustain the output to become HIGH logic level.

Now when the input voltage is nearly equal with the threshold voltage of M4 then only output turns into LOW logic level, so with the effect of this the switching point shifted to higher voltage referred as VIH. Similar in case when input voltage swing from higher logic level to lower logic level then pMOS's are more effective then nMOS's throughout the performance of the circuit and then the switching point at output is shifted to some lower voltage consider as VIL (Kumar *et al.*, 2012, Ramesh *et. al.*, 2015, Khurana *et al.*, 2015).

Fig. 2 shows the SPICE simulation results of circuit shown in Fig. 1 for both increasing and decreasing input voltages. The expected circuit behaviour and the two switching thresholds upper threshold and lower threshold points are clearly seen in the simulation results (Rashid *et al.*, 2013).

3 SIMULATION RESULT

In this work conventional Schmitt trigger circuit is simulated in Tanner EDA tool using level 2 PTM model. Table 1 and 2 shows the summary of Simulations results. These results sight that the triggering points of conventional Schmitt trigger has been controlled by the change in source-to-substrate bias

Table 1. Variation of source-to-substrate voltage.

pMOS substrate bias (V)	nMOS substrate bias (V)	Upper threshold point (V)	Lower threshold point (V)
3.5	1.5	2.8	2.2
3.6	1.4	2.9	2.1
3.9	1.1	3	2
4.2	0.8	3.1	1.9
4.3	0.7	3.1	1.9
4.4	0.6	3.1	1.9
4.5	0.5	3.1	1.9
4.6	0.4	3.1	1.9
4.7	0.3	3.1	1.9
4.8	0.2	3.1	1.9
4.9	0.1	3.1	1.9

Table 2. Variation of body effect coefficient.

$\gamma(V^{1/2})$	Upper threshold point (V)	Lower threshold point (V)
0.35	3	1.9
2.35	3.2	1.8
3.35	3.3	1.7
4.35	3.4	1.6

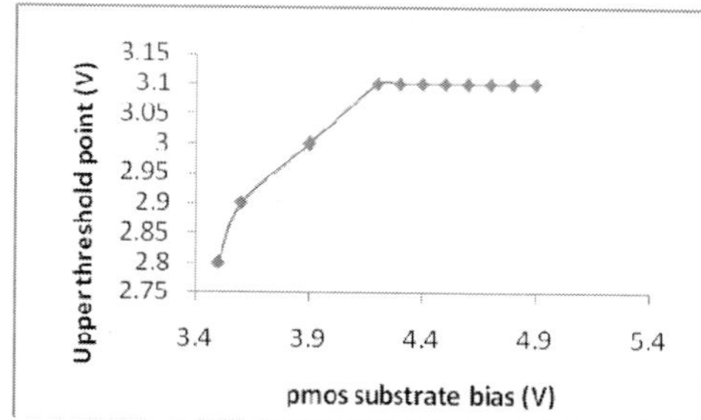

Figure 3. Variation of UTP with pmos substrate bias.

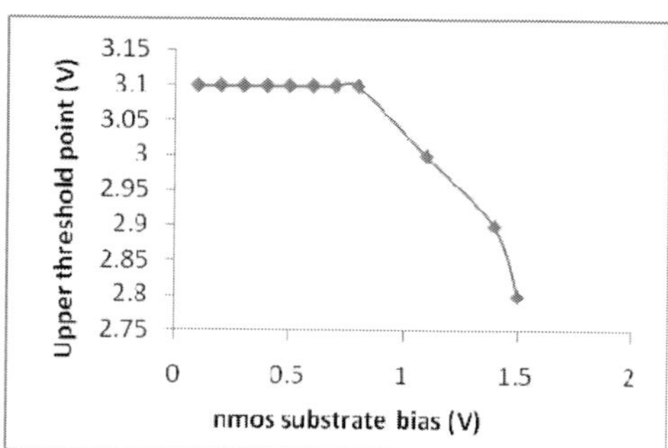

Figure 4. Variation of UTP with nmos substrate bias.

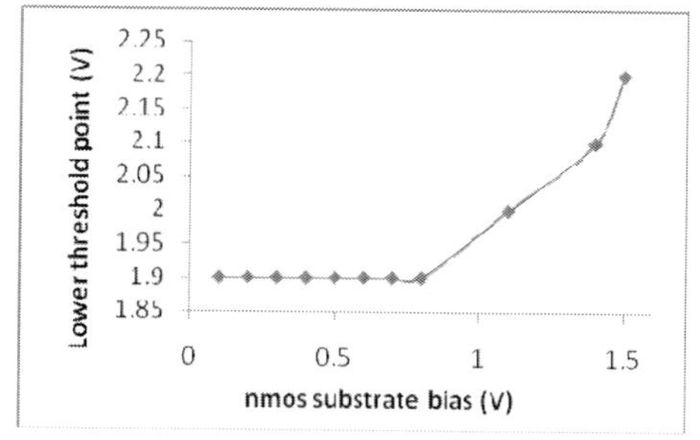

Figure 5. Variation of LTP with pmos substrate bias.

voltage for pMOS and nMOS, and due to body effect coefficient. Hence different outputs of triggering points of Schmitt trigger based on the variation of intrinsic and extrinsic device parameter has been obtained in Figs. 3, 4, 5, 6, 7 and 8.

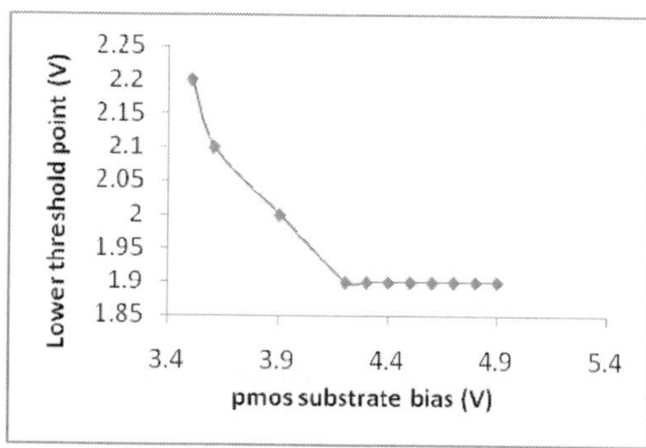

Figure 6. Variation of LTP with nmos substrate bias.

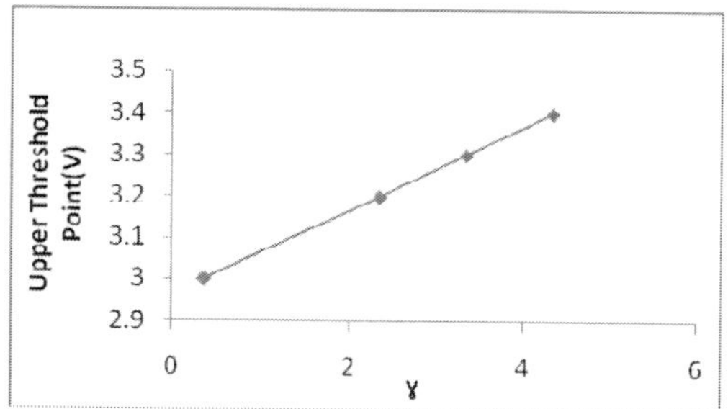

Figure 7. Variation of UTP with γ.

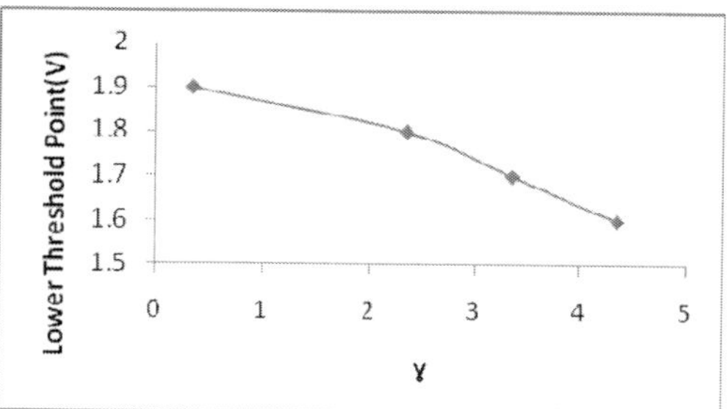

Figure 8. Variation of LTP with γ.

3.1 *Effect of variation of source-to-substrate voltage (V_{sb}) on triggering points of Schmitt trigger*

The threshold voltage of a MOS can be controlled by applying voltage at the back gate. The voltage difference between the bulk and source (V_{sb}) changes the depletion layer width and also causes change in voltage across the oxide due to the change of the depletion region charge. The impact of physical parameters which characterize the MOS structure on the threshold voltage value (Kang *et al.*, 2003, Baker 2010). In this section the effect of variation of γ& V_{sb} is observed.

For the V_{sb} range of 3.5 V to 4.9 V (for pMOS) and 1.5 V to 0.1 V (for nMOS) the UTP and LTP changes are in the range of 2.8V to 3.1 V and 2.2 V to 1.9 V respectively.

3.2 *Effect of variation of substrate bias (or body-effect) coefficient (γ) on triggering points of Schmitt trigger*

The change in the threshold voltage of a MOSFET, because of the voltage difference between body and source is called body effect. 'γ' is called the substrate-bias or the body effect coefficient is the electrical parameter of MOS device. The substrate bias coefficient (γ) is positive for nMOS and negative for pMOS. The body effect is sometimes called the "back-gate effect" (Kang *et al.*, 2003, Uyemura 2002).

This work describes the effect of (γ) on the triggering points of Schmitt trigger. For the (γ) range of 0.35 $V^{1/2}$ to 4.35 $V^{1/2}$ the UTP and LTP changes are in the range of 3 V to 3.4 V and 1.9 V to 1.6 V respectively.

4 CONCLUSION

In this work the triggering points of Schmitt trigger that is the Upper Threshold Point (UTP) and Lower Threshold Point (LTP) were modified by using the variation of electrical parameters such as source-to-substrate voltage (V_{sb}) and γ. For the V_{sb} range of 3.5 V to 4.9 V and 1.5 V to 0.1 V the UTP and LTP changes are in the range of 2.8V to 3.1 V and 2.2 V to 1.9 V respectively. For the (γ) range of 0.35 $V^{1/2}$ to 4.35 $V^{1/2}$ the UTP and LTP changes are in the range of 3 V to 3.4 V and 1.9 V to 1.6 V respectively. This analysis gives the overview that how the triggering points of Schmitt trigger has been controlled based on the effect of intrinsic and extrinsic device parameter of MOS device.

ACKNOWLEDGEMENTS

This work was supported by the TANNER EDA Tool design system.

REFERENCES

Baker, R. Jacob. (2010). *CMOS*: Circuit Design, Layout & Simulation; Vol.1 Wiley-IEEE press, Third Edition.

Kang, Sung-Mo. & Leblebici, Yusuf. (2003). CMOS digital integrated circuits: analysis and design; Tata Mcgraw-Hill, Third Edition.

Kumar, M., Kaur, P., Thapar, S. (2012). Design of CMOS Schmitt Trigger; IJEIT, Vol.2(1).

Khurana, A., Saxena, A., Jain, N. (2015). Evaluation of Low Power Schmitt Trigger for Communicaion System; IRJET, Vol.2(5).

Ramesh, S. S., Gandhe, S. T., Dhulekark, P. A. (2015). Design of CMOS Schmitt Trigger; IETE 46th Mid Term Submission, MTS.

Rashid, H., Mamun, Md., Amin, Md. S., Husain, H. (2013). Design of Low Voltage Schmitt Trigger in 0.18um CMOS Process With Tunable Hysteresis; Modern Applied Science, 7(4).

Uyemura, J. P. (2002). Circuit Design for CMOS VLSI; Wiley.

Computational Science and Engineering – Deyasi et al. (Eds)
© 2017 Taylor & Francis Group, ISBN 978-1-138-02983-5

A comparative study of reversible circuits using QDCA and formulation of new universal reversible gate

S. Ghosal & K. Chakraborty
Acharya Prafulla Chandra College, Kolkata, West Bengal, India

Barnik Mandal
B.P. Poddar Institute of Management and Technology, Kolkata, West Bengal, India

ABSTRACT: The quantum dot cellular automata is one of the various nanostructure based paradigms that have been proposed to overcome the limits being faced by the existing CMOS technology. The prevalent building block for QDCA has been the majority voter which is not functionally complete as it lacks conditional inversion. In an earlier work we have proposed a new architecture for a NAND/NOR based universal gate and shown it to be more compact as compared to the earlier proposed AOI and NNI structures. Landauer's limit places a lower bound on energy dissipation in conventional circuits which may be overcome by reversible computing. In this paper we have realized several standard reversible gates using our structure and shown those to be much more compact compared to the prevalent Majority voter based design. We have also proposed a new Universal reversible gate and made a comparative study with the popular PERES gate.

1 INTRODUCTION

In recent years, much research has been devoted to the possibility of shifting from the traditional field-effect transistor system (ITRS roadmap) to a nanostructure-based paradigm of encoding binary information (Lent et. al., 1993, Lent et. al., 1997, Feynman 1985, Grover 1998). This would provide much larger device densities of 10^{11} to 10^{12} devices/cm^2 and clock frequencies several orders of higher magnitude. Of the several technologies which have been proposed to replace the semiconductor-transistor approach the Quantum Dot Cellular Automata (QDCA) is one of the most promising (Lent et. al., 1993, Lent et. al., 1997). In this system, the binary information is represented by an array of cells, each of which contain four quantum dots containing two excess electrons, where the binary information is now encoded in the two possible polarization states of the two electrons on the dots.

It is expected that minimization of power density will be the determining factor in the choice of any future computing paradigm. The question of minimal energy requirements in computation is connected to a larger discussion of entropy and its statistical interpretation. Irreversible logic gates, which are the building blocks of Classical computing machines are dissipative by nature as loss of each bit of information is accompanied by a loss of thermal energy equal to KTln2 (Landauer 1961). In 1973 Charles Bennett (IBM research) showed that use of Reversible logic computation can prevent dissipation of this KTln2 joules of energy.(Bennett 1973). Hence Reversible logic design naturally gets priority to design combinational as well as sequential circuit. So far the conventional MOSFET switches dissipate an energy considerably higher than this lower limit, however as we move towards single electron devices and quantum computing, reaching this lower bound becomes a plausible limit. Thus we need to seriously consider reversible logic design as a means of going below the KTln2 barrier.

The rest of the paper is organized as follows. In Section II we introduce QCA, including the majority voter and our new NAND-NOR based structure (DPNNI). In Section III a brief overview of reversible functions is provided. In Section IV we present, implementations of various conventional and reversible structures using the DPNNI and their relative size compared to implementations using the majority vote. Section V introduces our newly proposed universal reversible gate and concludes with implementations of various standard reversible gates using our structure and a comparative study with PERES gate. All simulations have been done using the QCAD simulator (Walus 2002).

2 PHYSICAL STRUCTURE

The QCA cell was first introduced by Prof. C. S. Lent at the University of Notre Dame (Lent et. al., 1993,

Lent et. al., 1997). A QCA cell is composed of four quantum-dots arranged in the form of a square, such that only two of them have an excess electron each. The electrostatic interactions (repulsion) between the charges, ensures that they occupy the diagonally opposite quantum-dots. Consequently only two stable configurations are possible corresponding to the two diagonals in a square. These two stable configurations correspond to the two lower energy states and are encoded as the binary values '0' and '1' as shown in Fig 1. The NULL state configuration depends on the physical implementation; for example in some cases two extra quantum-dots are placed in the center of the cell to be occupied during this state. Logic operations are performed by means of the Coulomb interactions and linear superposition of the electrical field between adjacent cells.

The basic logic operation is the majority voter (Lent et. al., 1993, Lent et. al., 1997, Tougawand et. al., 1994) of three inputs and the majority gate shown in Figure 2. It is described as the following logic function

MV (A, B, C) = AB + BC + CA

The majority voter structure cannot be used as a universal building block as it lacks the inversion property (Tougawand et. all 1994, Akers 1962, Zhang et. al., 2004). To overcome this various structures have been proposed like the And-OR-Inverter (AOI) and other structures (Momenzadeh et. al., 2005, Sen et. al., 2007, Das et. al., 2009). However while the majority voter suffers from the lack of an inverter, the AOI while offering the advantage of an inbuilt inverter has the drawback of unequal spacing of input and output binary wires- that is, in fixing the intercell spacing's. Such asymmetrical structures are not ideally suited for mass production. Moreover And-Or-Inverter is not a universal gate and all Boolean functions cannot be de-signed using only A-O-I gate. The larger space requirement and complexity of structure is another major drawback of AOI (Brayton 1984, Lawler 2011) compared to that of MV or NOT gates. The tile based structures although robust in functionality take considerably more space. Based on these shortcomings we proposed a new NAND/NOR based structure, the Dual Polarization NAND-NOR Inverter (DPNNI) (Ghosal et. al., 2013). It has been shown that this circuit can give a NAND/NOR representation as opposed to AND/OR gate. Moreover NOT gate can also be realized. More complex functions using the above structure have also been demonstrated.

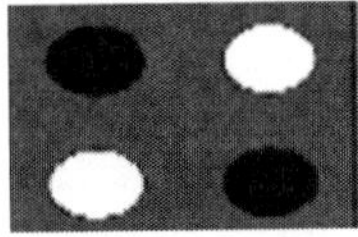

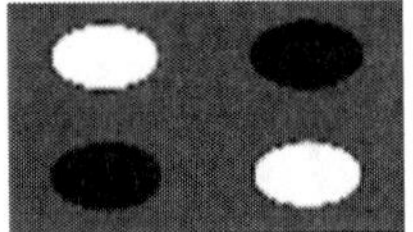

Figure 1. Representation of the two logic values of a QCA cell with four quantum-dots. Black filled circles represent occupied quantum-dots while white filled circles represent unoccupied quantum-dots.

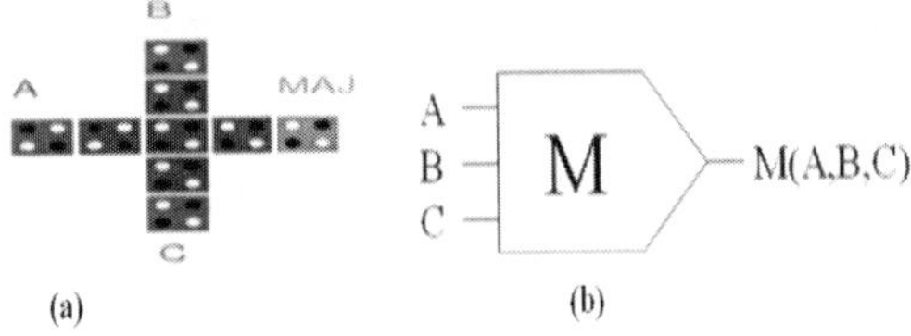

Figure 2. (a) QCA layout of a majority gate. (b) Logic symbol.

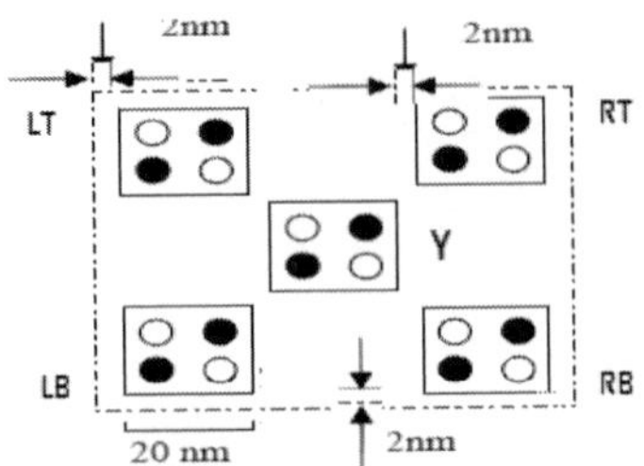

Figure 3. The dual polarity nand nor inverter structure.

The proposed structure consists of five cells (Fig. 3) with a central cell providing the output while any two of the remaining cells may act as polarizing cells while other two are used as input cells. The proposed structure has a logical relationship of the form

$$\mathbf{Y} = \overline{\mathbf{LT}}\,\overline{\mathbf{RT}}\left(\overline{\mathbf{LB}} + \overline{\mathbf{RB}}\right) + \overline{\mathbf{LB}}\,\overline{\mathbf{RB}}\left(\overline{\mathbf{LT}} + \overline{\mathbf{RT}}\right)$$

And $\overline{\mathbf{Y}} = \mathbf{LBRBRT} + \mathbf{LT} + \mathbf{RTLT}(\mathbf{LB} + \mathbf{RB})$

Thus for RT = 0; LB = 0, we get

$\mathbf{Y} = \overline{\mathbf{LT}} + \overline{\mathbf{RB}} = \overline{\mathbf{LTRB}}$ i.e. NAND Gate

And for RT = 1; RB = 1, we get

$\overline{Y} = \mathbf{LB} + \mathbf{LT}$ Or $\mathbf{Y} = \overline{LB + LT}$ i.e. NOR Gate

FOR LT = 1; RB = 0; RT = 0 and LB as the input cell we get the NOT GATE.

3 REVERSIBLE LOGIC

Logical functions lacking a single-valued inverse are typical of all computing machines implying loss of information or in the binary world "loss" of a bit. This logical irreversibility is associated with physical irreversibility and the laws of thermodynamics ensure that, the erasure of a single bit of information from the mechanical degrees of freedom of a system must be accompanied by the thermalization of an amount kT of energy per cycle [Landauer 1961].This Von Neumann-Landauer (VNL) limit places a lower bound on energy dissipation using conventional logic.In today's CMOS technology the energy dissipation is still several orders of magnitude larger than this theoretical minimum. However,with the advent of nanotechnology based architectures like QDCA and other paradigms the "kT" barrier looms as the single most significant obstacle to greater computer performance. In 1973 Charles Bennett [Bennett 1973] (IBM research) shows that in case of Reversible logic computation KTln2 joules energy will not dissipate. Hence Reversible logic design naturally gets priority to design combinational as well as sequential circuit [Kerntopf 1961].

We shall call a device logically irreversible if the output of a device does not uniquely define the inputs. Logical irreversibility, in turn implies physical irreversibility, and the latter is accompanied by dissipative effects. Function f(x1, x2,..., xn) of n Boolean variables is called reversible iff the number of outputs is equal to the number of inputs and any input pattern maps to a unique output pattern, i.e given its output, it is always possible to determine back its input. Logical reversibility implies conservation of information. A reversible logic gate is an n-input n-output logic device with one-to-one mapping. The simplest example of a reversible logic function is negation while logical conjunction or disjunction, display irreversible nature since there are several input configurations corresponding to the same output.

Landauer argued that transformation of an irreversible computation into a reversible one is possible by embedding the former into a larger computation where no information is lost, eg. by replicating every output in the input ('sources') and every input in the output ('sinks').This has lead to the development of various gates like;Toffoli – $(x, y, z) \rightarrow (x, y, z \oplus xy)$; Feynman – $(x, y) \rightarrow (x, x \oplus y)$; Peres $(x, y, z) \rightarrow (x, x \oplus y, xy \oplus c)$ and others [Feynman 1985, Toffoli 1980, Fredkin 1982, Peres 1985, Krishnaswamy et. al., 2008].

4 IMPLEMENTATION OF CONVENTIONAL AND REVERSIBLE GATES USING DPNNI

We had demonstrated the realization of several basic Boolean functions using our proposed DPNNI structure in our earlier paper (Ghosal et. al., 2013) and shown a minimum of 43% to a

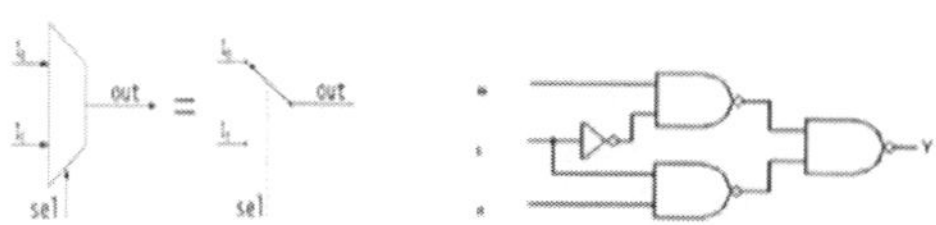

Figure 4. (a) 2:1 Mux switch; (b) Implementation using DPNNI.

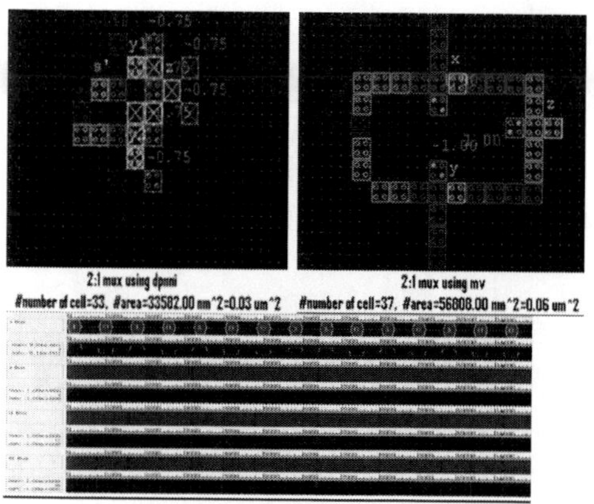

Figure 5. Comparative study of realization using (a) DPNNI and (b) Majority voter (c) Simulation result.

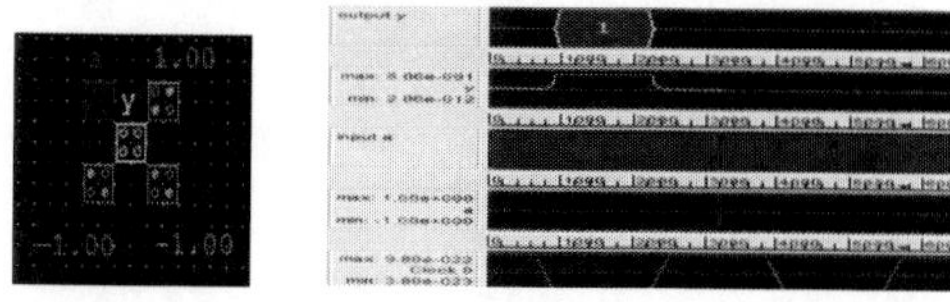

Figure 6. (a) PMOS equivalent structure; (b) Simulation output.

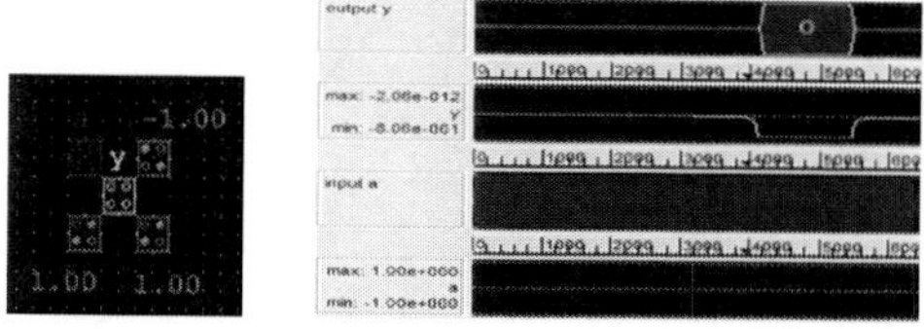

Figure 7. (a) NMOS equivalent structure; (b) Simulation output.

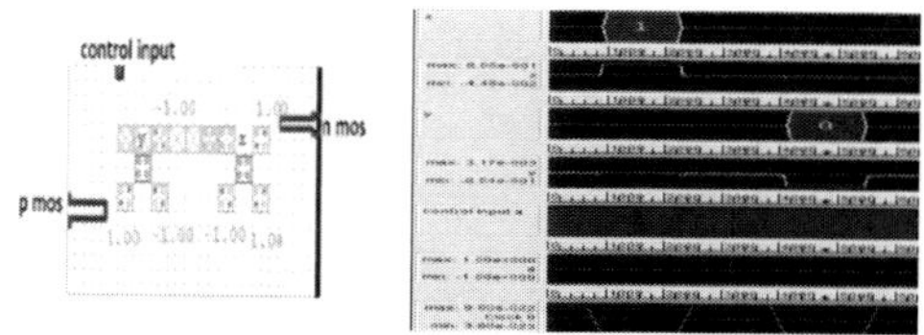

Figure 8. (a) Structure; (b) simulation output of controlled switch (transmission gate).

maximum of 176% reduction in size of circuits with this structure in comparison to that with the AOI and NNI gates. In this paper we have shown below the implementation of various combinational and reversible gates using our DPNNI structure and also made a relative comparison in size with the same structures implemented using the widely used majority voter structure.

4.1 *2:1 Multiplexer*

The multiplexer is a fundamental building block of any modern electronic circuit. We have implemented this structure using DPNNI and shown it to be more compact than that using MV.

4.2 *Transmission gate*

As we know, the transmission gate is realized as a combination of a PMOS and an NMOS transmission. Here we have simulated two controlled switches working on opposite stimuli similar to PMOS and NMOS structures which in tandem work as a transmission gate.

4.3 *Realization of reversible gates*

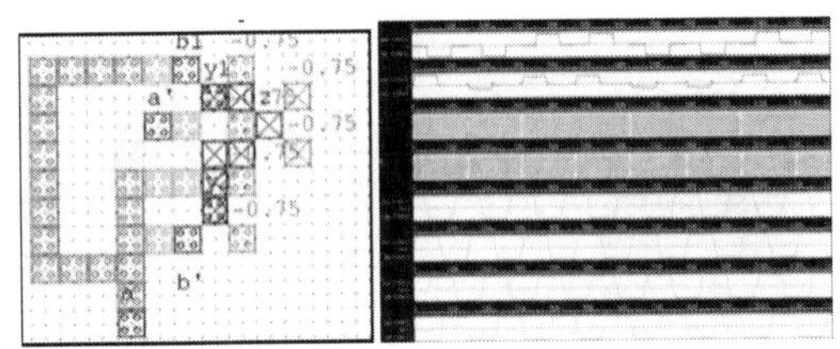

Figure 9. (a) Structure (b) simulation output of Feynman Gate.

Figure 10. (a) Structure (b) simulation output of Toffoli gate.

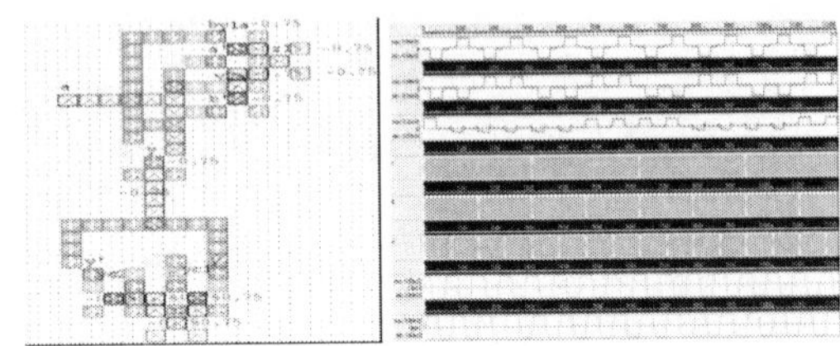

Figure 11. (a) Structure (b) simulation output of Peres gate.

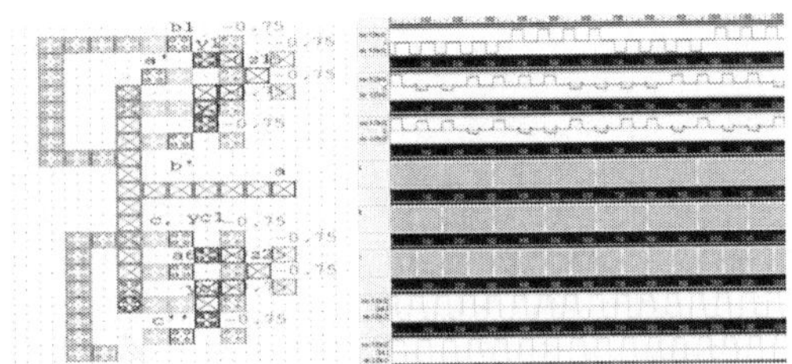

Figure 12. (a) Structure (b) simulation output of Double Feynman gate.

Table 1. Comparative study of area consumption in design of reversible gates using the majority voter and the DPNNI structure.

Gate	Design using MV		Design using DPNNI		
400 200 0 Dou... Toffoli Peres	No. of cells	Area (nm^2)	No. of cells	Area (nm^2)	% Reduction
Feynman	96	101618.00	49	51714.00	49.1%
Toffoli	103	134524.00	76	70446.00	47.6%
Peres	206	370512.00	139	160468.00	56.6%
Double Feynman	201	295816.00	126	98514.00	66.6%

5 PROPOSED REVERSIBLE GATE

In the field of low power design Reversible logic design has the potential to turn out to be the ideal design strategy for logic synthesis as it is not limited by the lower bound of KTln2 J. Most of the existing reversible gates (Feynman 1985, Toffoli 1980, Fredkin 1982, Peres 1985) were proposed in the field of quantum computation. Here we have proposed a new universal reversible logic gate which is suited for the QCA paradigm and can realize all basic Boolean functions using lesser number of gates than the existing ones. We have implemented classical logic gates namely NOT, AND and NOR gate using our proposed new Reversible structure named as Universal Reversible Module [URM] and given a comparative study with the popular PERES Reversible gate.

Next, we describe our proposed 3*3 URM gate. Here reversible logic gate is defined as input vector and output vector must be with one to one correspondence i.e., objective. The mapping function is given by

f:M => M. The circuit has three inputs and three outputs.

The input vector is given by –

Iv = (A, B, C) and output vector is given by

Ov = [P = A'B' + BC; Q = B; R = B'C' + AB]

and the truth vector is [5,4,2,6,1,0,3,7].

In our design, we use four clocking zones and input, output has been designed on same clocking zone, and we found 171 number of QCA cells were needed to implement proposed URM gate.

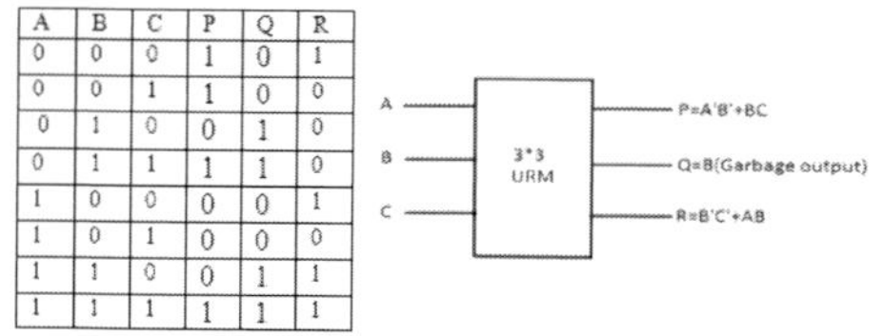

A	B	C	P	Q	R
0	0	0	1	0	1
0	0	1	1	0	0
0	1	0	0	1	0
0	1	1	1	1	0
1	0	0	0	0	1
1	0	1	0	0	0
1	1	0	0	1	1
1	1	1	1	1	1

Figure 13. (a) Truth table of proposed reversible URM Gate; (b) Block diagram of reversible 3*3 URM Gate.

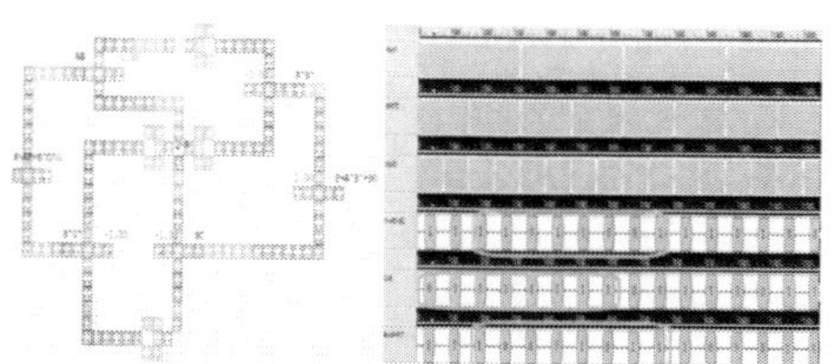

Figure 14. (a) Structure; (b) Simulation output of 3 × 3 reversible URM Gate.

From the Boolean expression of the ouputs we can summarise the possible ouputs for some basic input combinations as shown in the table below

A	B	C	P	Q	R
A	0	0	A' (NOT)	0	1
A	B	1	A' + B	B	AB (AND)
0	B	C	B' + C	B	(B + C)' (NOR)

In Fig 15 (a), we take B = 0, C = 0 that is why the ultimate outcome of the output will be NOT GATE, coming from P = A'.

In Fig 15 (b), we take C = 1, and A & B is considered as input for which we obtain R = AB (AND GATE).

In Fig 15 (c) at 3*3 URM GATE considering A = 0, B & C as input we get R = (B + C)' (NOR GATE).We have shown next the implementation of these.

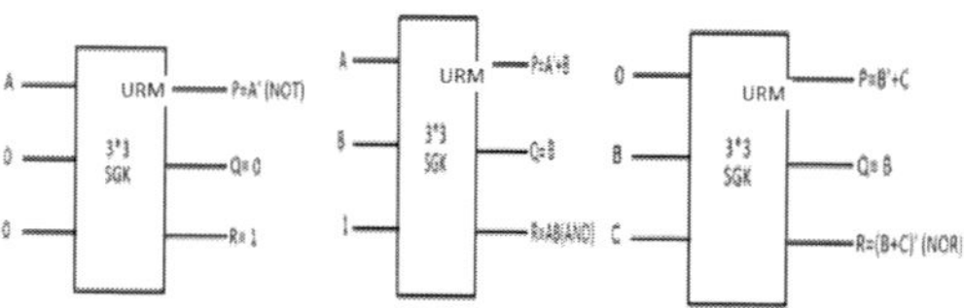

Figure 15. (a) NOT gate, (b) AND gate, (c) NOR gate.

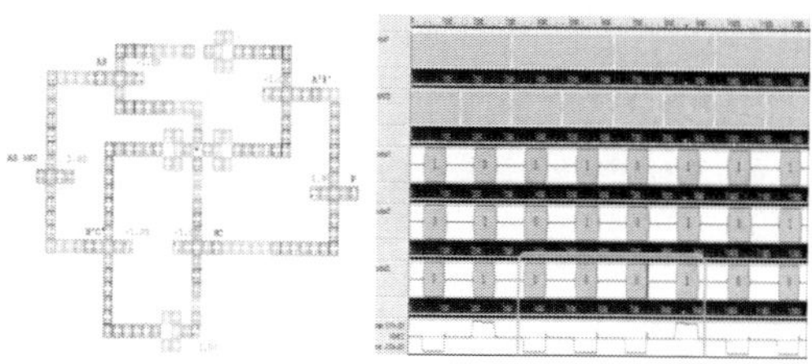

Figure 16. (a) Structure; (b) simulation output of NOT gate using URM.

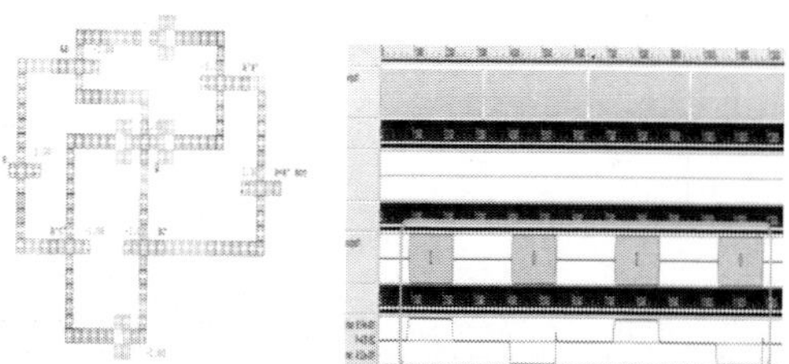

Figure 17. (a) Structure; (b) simulation output of AND gate using URM.

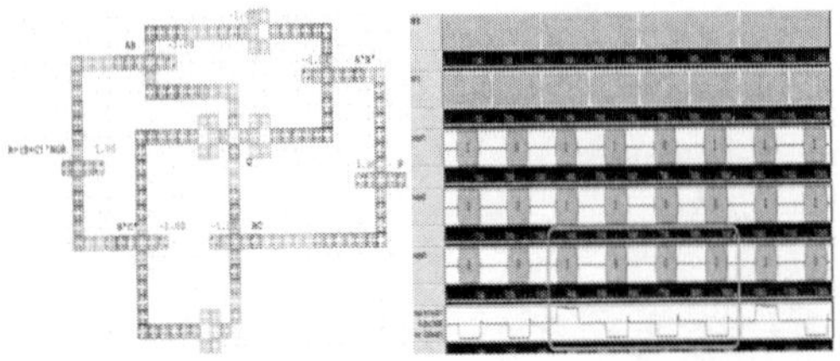

Figure 18. (a) Structure; (b) simulation output of NOR gate using URM.

Thus using different input we can get the basic gates and further cascading the structures we may proceed in such a way to implement more complicated Boolean functions. We have highlighted below the structural block for some other standard Boolean functions.

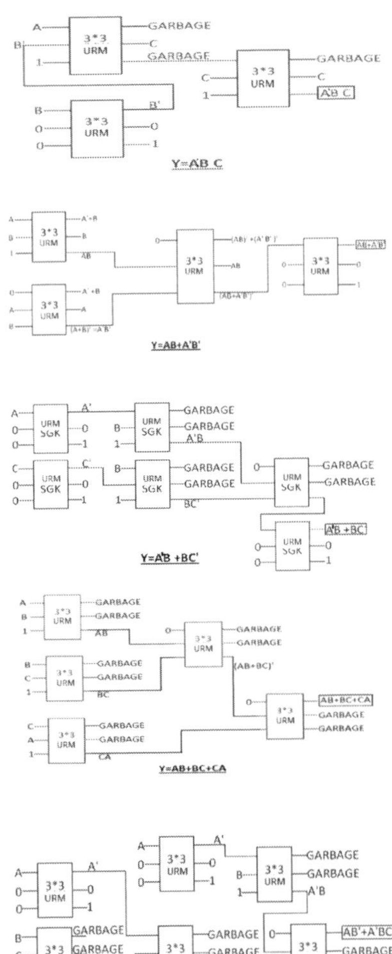

Table 2. Comparative chart with PERES gate.

Some standard boolean functions	PERES gate	URM gate
A'BC	2	3
AB	1	1
A'B + BC'	5	6
AB' + A'BC	8	6
A	1	1
AB + BC + CA	11	5
AB + A'B'	3	4
Total no. of gates	31	26
Average no. of gates	4.429	3.714

From the above table we can get an estimate of the relative efficiency of this structure with respect to a popular reversible gate, the PERES gate. The improvement factor may be estimated as

$$\frac{4.429-3.714}{40429}(100)=16.14\%$$

6 CONCLUSION

In this paper we have reiterated the inevitability of opting for reversible circuits as we adopt nanotechnology based paradigms as the next level of computational technology. In this context, QDCA as one of the most promising prospects in this regard has been studied. We have, as an extension of our earlier work, shown that the DPNNI structure is superior to the prevalent majority voter structure in the implementation of both conventional and reversible circuits. We have shown a more compact realization of a 2:1 multiplexer and perhaps for the first time implementation of a transmission gate like structure using QDCA. As is well known transmission gate is a very useful building block of CMOS digital integrated circuits and implementation of a similar controlled switch structure in the QDCA paradigm will be immensely beneficial for any future co-simulation of QDCA and MOS hybrid circuits. Besides its compactness as shown, the advantage of using a NAND based circuit is that the existing logic minimization rules are NAND/NOR based hence the DPNNI offers an inherent advantage as its basic building block is a NAND gate. Moreover a NAND actually propagates an improved error probability [Krishnaswamy et. all 2008, BoMarr et. all 2010] due to its logic properties through its natural computation. We have also proposed a new universal reversible logic gate URM and made a comparative study (with respect to size only) with the PERES gate. We hope to make a more detailed study with respect to garbage minimization and other aspects like fault tolerance of this structure and give a detailed comparison with other standard reversible logic gates like Feynman, Tofolli etc. in our future work.

REFERENCES

A. Peres, 1985, Reversible logic and quantum computers", Physical Review: A, 32(6), 3266–3276.

Bennett C. H., 1973, Logical reversibility of computation, IBM J. Research and Development, 17,525–532

B. Sen, B.K. Sikdar, 2007, Characterisation of universal Nand-Nor-Inverter QCA gate, 11thIEEE VLSI Design and Test Symposium,Kolkata, 8–11

C. S. Lent, P. D. Taugaw, W. Porod, G. H. Bernstein, 1993, Quantum cellular automata, Nanotechnology, 4, 49–57

C. S. Lent, P. D. Tougaw, 1997, A device architecture for computing with quantum dots, Proc. IEEE, 85(4), 541–557

International Technology Roadmap for Semiconductors (ITRS). [Online].Available: http://www.itrs.net

E. Lawler, 2011, An Approach to Multilevel Boolean Minimization, JACM, 283–295.

R. Feynman, Quantum mechanical computers. 1985, Optics News, 11, 11–20

K. Walus, 2002, ATIPS Laboratory QCA Designer Homepage. ATIPS Laboratory, Univ. Calgary, Calgary, Canada. [Online]. Available: http://www.atips.ca/projects/qcadesigner

K. Das, D. De, 2009, A Novel Approach for AOI gate design for QCA, International conference on Computers and Devices for Communication

L. K. Grover, 1998, A Framework for Fast Quantum Mechanical Algorithms, Proc. Symp. On Theory of Computing

R. Landauer, 1961, Irreversibility and heat generation in the computing process, IBM J. Research and Development, 5(3),183–191

M. Momenzadeh, Huang. J, Tahoori, M.B. Lombardi, 2005, Characterization, test, and logic synthesis of And-Or-Inverter (AOI) gate design for QCA implementation, IEEE Trans.Computer-Aided Design of Integrated Circuits and Systems, 24, 1881–1893

P. Kerntopf, 2000, A Comparison of Logical Efficiency of Reversible and Conventional Gates, 261–269.

P.D. Tougawand, C. S. Lent, 1994, Logical devices implemented using quantum cellular automata, J. Appl. Phys., 75(3), 1818–1825

R. Zhang, K. Walus, W. Wang, G. A. Jullien, 2004, A method of majority logic reduction for quantum cellular automata, IEEE Trans. Nanotechnol., 3(4), 443–450

R. K. Brayton, C. McMullen, G. D. Hachtel, A. Sangiovanni-Vincentelli, 1984, Logic Minimization Algorithms for VLSI Synthesis, Kluwer Academic Publishers, Norwell, MA

S. B. Akers, 1962, Synthesis of combinational logic using three-input majority gates, Proc. 3rd Annu. Symp. Switching Circuit Theory Logical Des. Logic minimization techniques, 149–157

S. Ghosal, D. Biswas, 2013, Study and Defect Characterization of a Universal QCA Gate, IJCA, 74(15)

S. Krishnaswamy, G. F. Viamontes, I. L. Markov, J. P. Hayes, 2008, Probabilistic Transfer Matrices in Symbolic Reliability Analysis of Logic Circuits, ACM Transactions on Design Automation of Electronic Systems, 13(1), Article 8.

T. Toffoli, 1980, Reversible computing, Tech Memo MIT/LCS/TM-151, MIT Lab for Computer Science

Computational Science and Engineering – Deyasi et al. (Eds)
© 2017 Taylor & Francis Group, ISBN 978-1-138-02983-5

Realization of hybrid single electron transistor based low power circuits in 22 nm technology

Tahesin Samira Delwar & Sourav Biswas
Department of Electrical and Electronics Engineering, University of Science and Technology Chittagong, Bangladesh

Anindya Jana
Department of Electronics and Communication Engineering, Sree Vidyanikethan Engineering College (Autonomous), Tirupati, India

ABSTRACT: Continuous downscaling in MOSFET channel length for achieving highly dense IC has led the researchers to face some performance related issues due to short channel effects. Performance improvement of device is still possible by changing the device material or by modifying the principle of device operation. For this tenacity, CMOS technology, in near future, going to share its space in present IC technology with basically new aspirant of nanotechnology: Single Electron Transistor (SET). Despite the fact that the current driving abilities and power dissipation of SETs and CMOS are quite opposite to each other, but co-integration of SET with CMOS can convey some new functionalities, which are beneficial over pure CMOS technology. The aim of this paper is to find some better solutions of basic digital circuits in 22 nm technology using this novel concept. LPBSIM4.6.1.0 and MIB models are used to prove the feasibility of the circuits in 22 nm Technology. All the simulations are done in TCAD.

1 INTRODUCTION

Research on semiconductor technology trends reduction in dimension & power consumption in the recent decades. The main aim of this research is to reduce the power consumption by minimizing the dimensions of the devices without any undesirable impact either on speed or on the active surface area of the circuits. However, according to the ITRS roadmap the miniaturization of CMOS dimension beyond a certain limit is not possible (Dennard, 1974) as it faces some problems, collectively known as Short Channel Effects (Yu, 2008, Ono, 2000). To triumph over this problem, researchers have introduced single electron transistor, which operates by controlling the individual electrons through a certain barrier. Using the SET technology, manipulation on the flowing of individual electrons is possible with better scaling potential & reduced power consumption than CMOS circuit. Due to higher packing density and greater scaling potentiality, SET has attracted the attention of the researchers. But SET exhibits some problems too; background charge effects, low input impedance, high output impedance and low current gain when CMOS exhibits sufficient current gain. To surmount all the problems and to obtain high performance device in this new era of nanotechnology, scientists introduced Hybrid SET-CMOS, a co-integrated novel device having benefits of both the CMOS and SET. It has been observed that the transistor have higher operational speed and lower power consumptions. It is also observed that the gains of the Hybrid SET-CMOS based circuits are greater than the conventional CMOS based circuits. Previously some basic circuits were implemented using this hybrid concept (Jana, 2013). In this work some digital circuits have been designed and simulated using hybrid SET-CMOS.

The operations of the proposed 22 nm circuits are verified in Tanner environment, which establishes the novelty and feasibility of the circuits.

2 SINGLE ELECTRON TRANSISTOR AND ITS HYBRIDIZATION WITH CMOS

Single Electron Transistor (SET) is such a device that controls the tunneling of individual electrons to amplify current for switching. The tunnel junction is the fundamental factor of SET which can be obtained by positioning both tunnel junctions in series. The two tunneling junctions form a "Quantum dot or Coulomb island". The operation is based on the principles of quantum mechanics.

Schematic illustration of a tunnel junction is presented in Fig-1.

The schematic and equivalent circuit of SET is depicted in Figure 2.

The gate terminal is connected to a node in the middle of the two tunnel-junctions. Due to its high thickness, capacitor acts like a third tunnel junction so that no electrons can pass through it. Therefore the role of the capacitor is basically setting of the charge of the electrons on the coulomb island. The simple constructions of SETs are almost identical as that of MOSFETs. But, SETs encompasses of tunneling junctions by replacement of pn-junction whether a quantum dot switches the tunnel section of the MOSFETs. A central island is necessary in the source to drain tunneling. Charging energy EC must be superior to the thermal energy and the tunneling resistance R_T should be greater than the quantum resistance while tunneling. Conditions of the single-electron event can be expressed as $EC = e^2/2C_\Sigma >> K_BT$ and $R_T >> h/e^2$ where C_Σ symbolizes the island capacitance with respect to the ground, K_B expresses the Boltzmann's constant, h is Planck's constant and T is temperature. SETs might have a optional 2nd gate, interlinked to the island that can be useful for controlling the phase shift in the time of coulomb oscillation. Due to some problems of SET like background charge effects, low input impedance, high output impedance and low current gain scientists have gone for the hybridization of SET with CMOS. CMOS can provide high gain. Previously researchers have implemented different circuits using the hybrid SET-CMOS technology (Jana, 2013). In this paper some more circuits are reflected using this technology in 22 nm. All the simulations are done using TCAD.

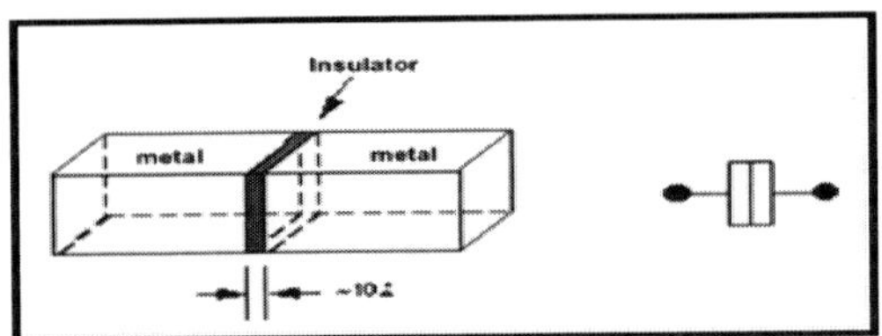

Figure 1. Schematic illustration & symbol of tunnel junction [4].

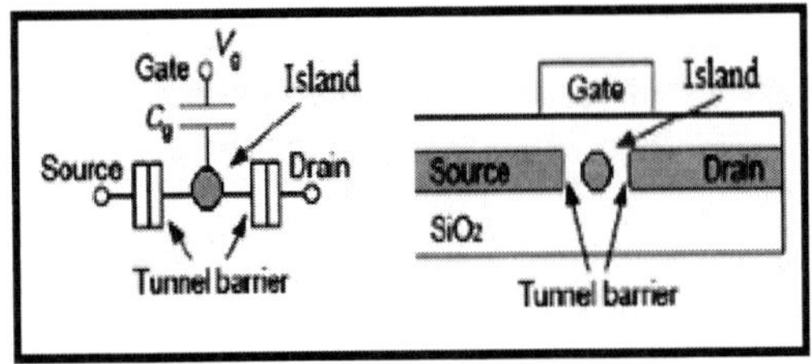

Figure 2. Schematic illustration & equivalent circuit of SET [4].

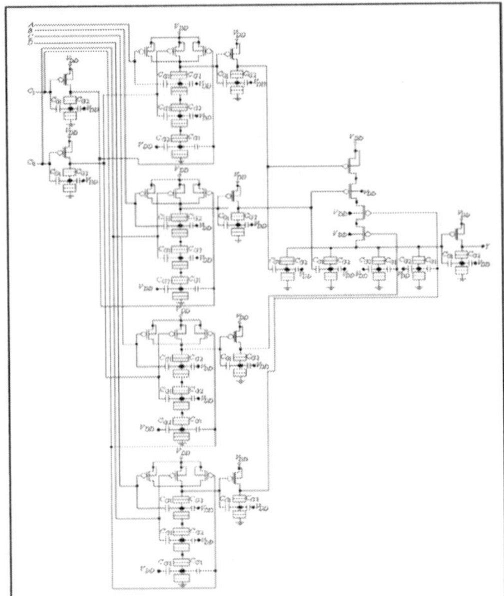
Figure 3. Hybrid SET-CMOS based multiplexer circuit in 22 nm technology.

3 HYBRID SET-CMOS BASED CIRCUITS IN 22 NM

3.1 *Hybrid SET-CMOS based multiplexer circuit in 22 nm*

Figure 3 provide some basic multiplexer circuit designed in 22 nm technology in Tanner environment. This circuit has been verified and proves the feasibility of this model in 22 nm.

3.2 *Hybrid SET-CMOS based 2 to 4 decoder circuit in 22 nm technology*

Figure 4 reflects the hybrid SET-CMOS based 2:4 decoder circuits in 22 nm technology.

4 RESULTS & DISCUSSIONS

All the circuits are simulated with the help of MIB compact model, which is described by Analog Hardware Description Language (AHDL) for SET and LPBSIM4.6.1.0 model for CMOS in Tanner

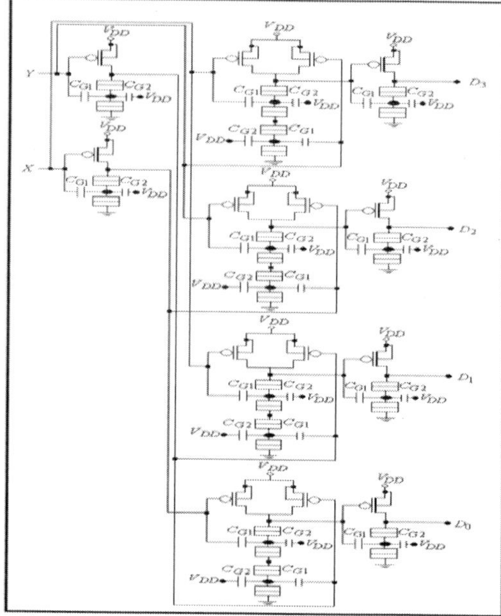

Figure 4. Hybrid SET-CMOS based 2 to 4 decoder circuit in 22 nm technology.

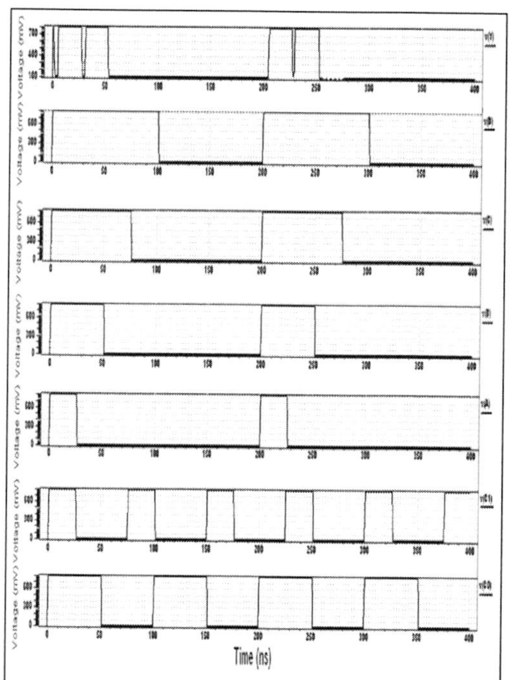

Figure 5. Output of the Hybrid SET-CMOS based multiplexer circuit in 22 nm technology.

Table 1. Parameters used for simulation.

Device	Parameters	Voltage level
SET	$R_{TD} = R_{TS} = 1M\Omega$, $C_{TD} = C_{TS} = 0.16aF$, $C_{G1} = 0.274aF$, $C_{G2} = 0.125aF$, $C_L = 0.01fF$, $V_c = 0.02V$	Logic 0 = 0V Logic1 = 0.8V $V_{DD} = 0.8V$
PMOS	$V_{TH} = -220$ mV, W/L = 33 nm/22 nm and default values of BSIM4.6.1 model for other parameters	

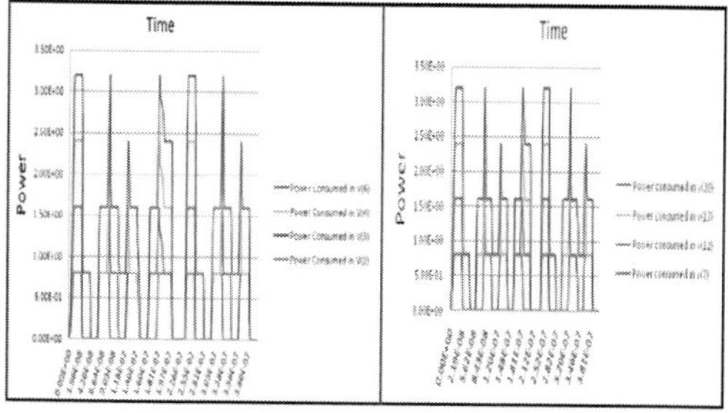

Figure 6. Comparison of consumed power in CMOS based Multiplexer and Hybrid SET-CMOS based multiplexer in 22 nm technology.

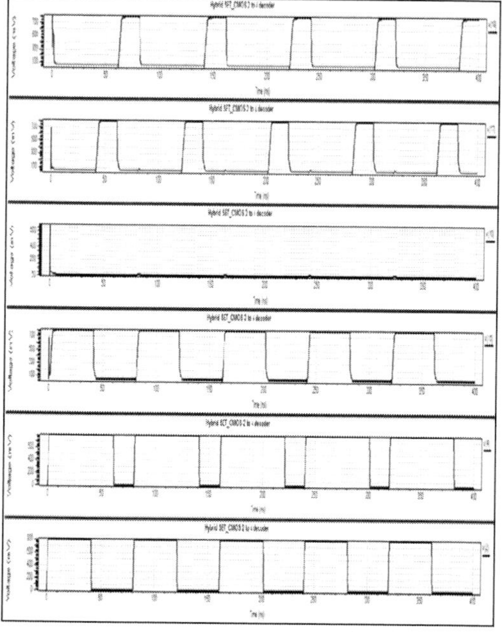

Figure 7. Output of the Hybrid SET-CMOS based 2 to 4 decoder circuit in 22 nm technology.

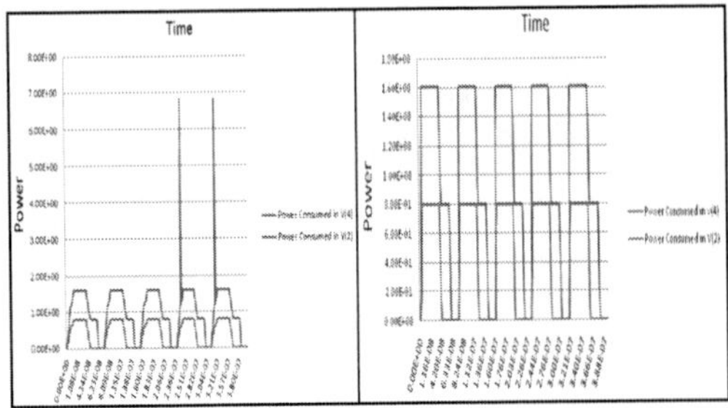

Figure 8. Comparison of consumed power in CMOS based 2 to 4 decoder and its Hybrid model in 22 nm technology.

Table 2. Comparative analysis of average & maximum power consumption.

Sl. no.	Name of the circuit	Average power consumed in CMOS model (watt)	Average power consumed in 22 nm hybrid model (watt)
1.	4-To-1 MUX	5.350260e-010	2.621596e-009
2.	2-To-4 Decoder	1.862519e-009	1.61210e-008

environment. All the used values of the parameters for the simulation are operable in room temperature for both SET [11, 12] & CMOS, specified in Table 1.

Figure 5 to Figure 8 shows the outcomes of the proposed circuits.

The following table 2 reflects the comparative Analysis of Average & Maximum Power Consumption.

5 CONCLUSION

In this paper, some novel circuits in 22 nm technology have been represented. After comparing the power consumption of both the circuits *i.e.* the conventional CMOS based circuits and hybrid circuits it has been found that hybrid model based circuits exhibits lesser power consumption than conventional CMOS based circuits. It has been found that these novel circuits are faster in operation than conventional circuits. Hence it also proves the feasibility of the circuits by showing satisfactory graphical outputs.

REFERENCES

Dennard R. et al. "Design of ion-implanted MOSFET's with very small physical dimensions," IEEE Journal of Solid-State Circuits, Volume 9, No. 5, pp. 256–268, 1974.

Jana Anindya et.al. Microelectronics Reliability, Elsevier, Volume 53, Issue 4, pp. 592–599, 2013.

Ono Yukinori et.al., Applied Physics. Letters, Volume 76, pp. 3121–3130, 2000.

Yu B. et. al. "Scaling of Nanowire Transistors," IEEE Transactions on Electron Devices, Volume 55, No. 11, pp. 2846–2858, 2008.

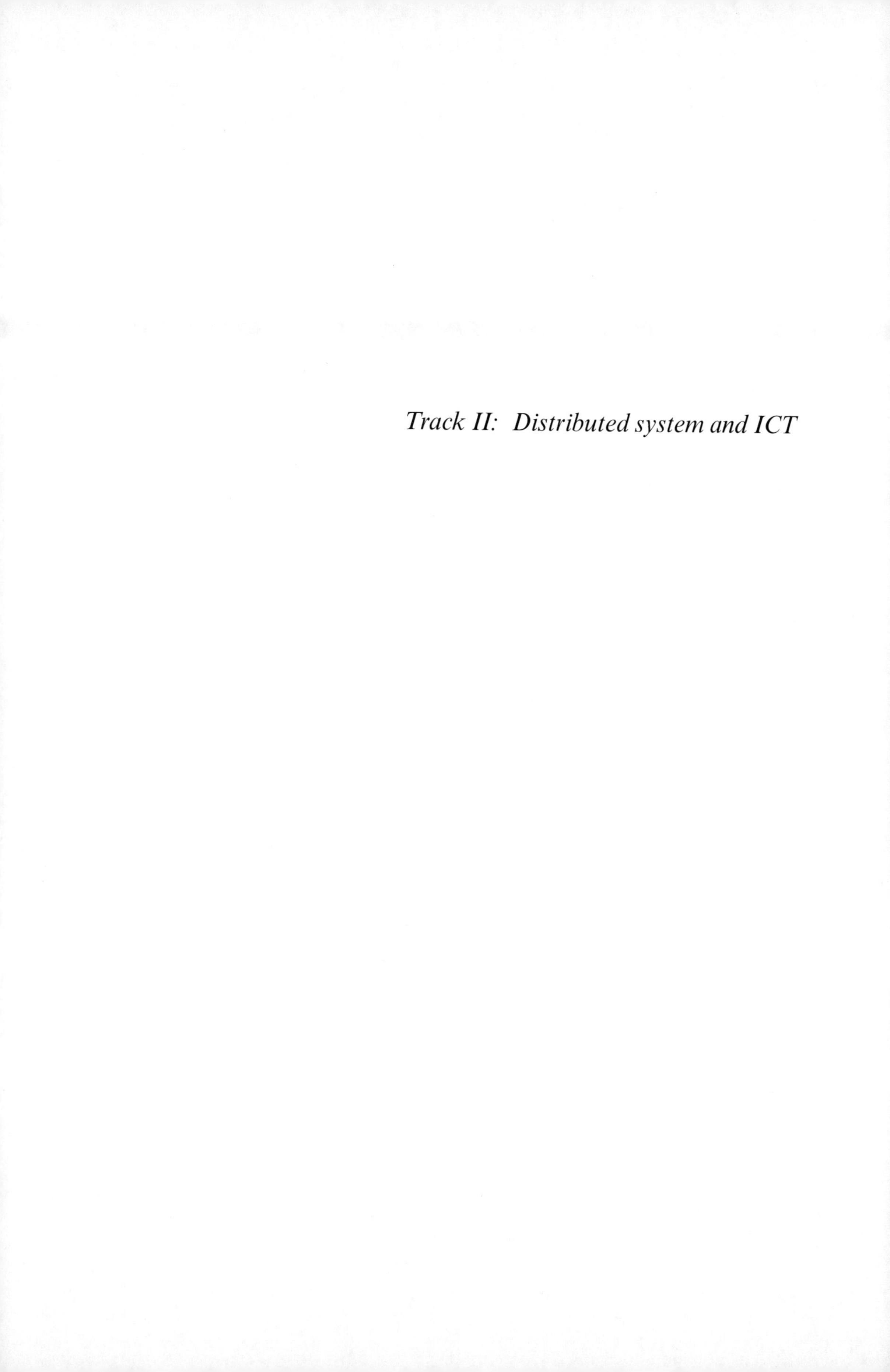

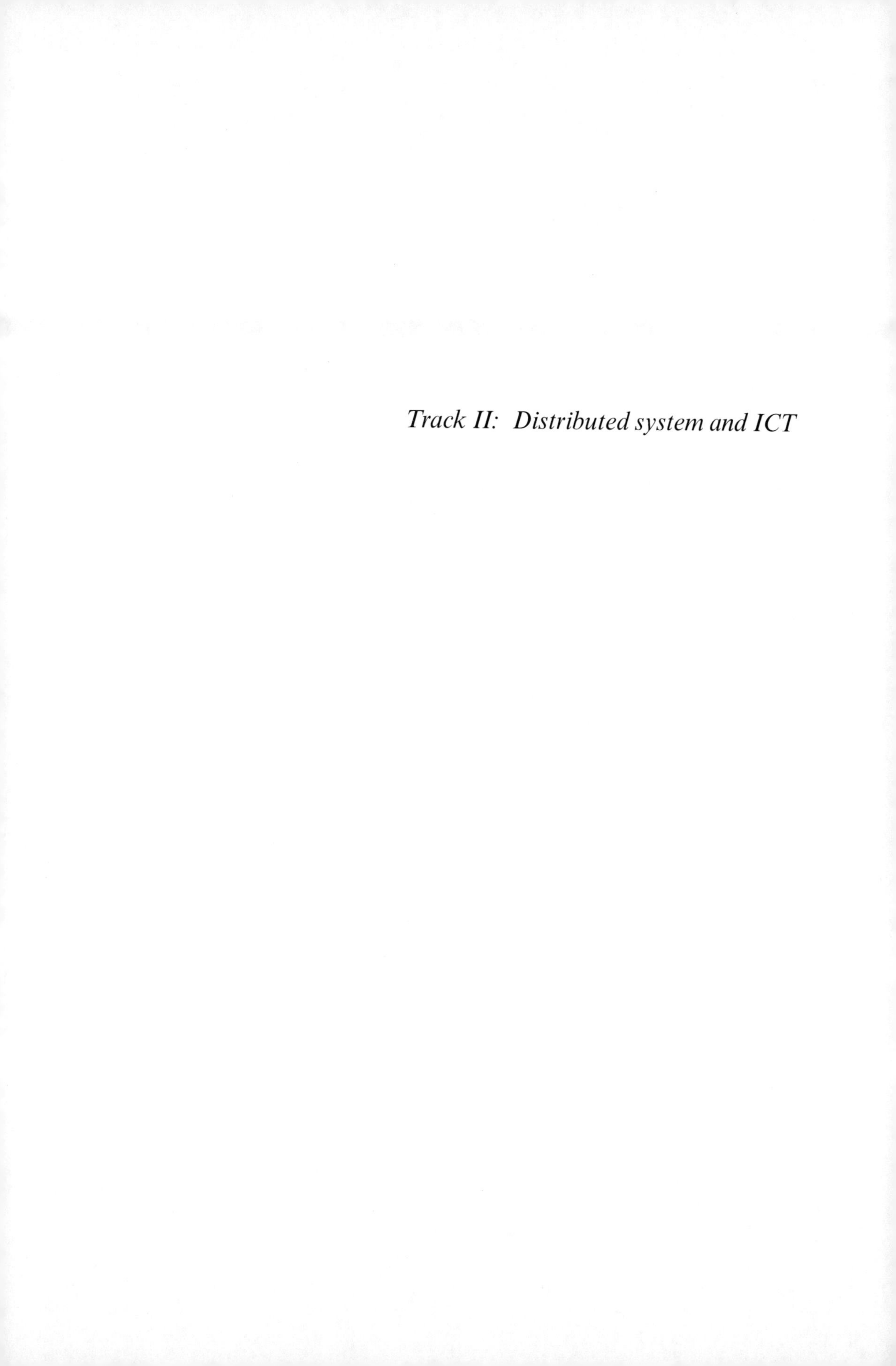

Track II: Distributed system and ICT

Computational Science and Engineering – Deyasi et al. (Eds)
© 2017 Taylor & Francis Group, ISBN 978-1-138-02983-5

An algorithmic approach for detecting students' learning style in e-learning

Ankita Podder & Rajeev Chatterjee
National Institute of Technical Teachers' Training and Research, Kolkata, India

ABSTRACT: In e-learning, many a times different tools, techniques, strategies may be incorporated to increase the performance a learner. This may also increase efficiency in due course of a particular learner. In case, the knowledge dissemination is efficiently done, e-learning may become successful. In order to have successful e-learning model, individual learning style may plays an important and crucial role but is not the ultimate reason. In this proposed research work, the authors of this paper propose an approach for detecting individual learning style. This is done based on Visual, Auditory, Read/Write, and Kinesthetic (VARK) model. Based on the recommendation of the VARK model an algorithm is proposed in this research paper.

1 INTRODUCTION

In the process of E-learning we re-integrate our former approach for better understanding of learning. According to Benjamin S. Bloom (CBloom, 1968) there are three domain of learning. They are cognitive domain: deals with intellectual capabilities i.e. knowledge; affective domain: concern with feelings, emotions, and attitudes and psychomotor domain: understanding with manual and physical skills. In e-learning both learners and learning style are from diverse field, by creating a requirement to manage both learning preferences and subject matter so that the purpose of learning may be efficiently fulfilled. Different learner perceive in a different way so individual learner has different styles of learning and how they preferred to learn is very important. Our proposed approach is based on Visual Learner, Auditory Learner, Read/Write Learner, and Kinesthetic Learner (VARK) learning style model. Each learner may fall in one or more category or may not fall in any category with respect to their behavior during the process of learning.

The paper is organized as follows. In section 2, we discuss Review of the Existing Works. In section 3, we briefly explain Our Proposed Design Approach. In section 4, Results and comparisons are given. In section 5, Conclusion and Future Scope are given.

2 REVIEWS ON EXISTING WORK

Goyal, M., Yadav, D., Tripathi, A. (Goyal, 2015) focuses on knowledge extraction using McCarthy model. Author proposed a method for evaluating learner's learning style using fuzzy approach. It calculates the mixed trait of learning style of a learner. After getting this result learner could be suggested to get proper type of learning content with respect to their learning style.

Rong, W.J., Min, Y.S. (Rong, 2005) focus on the effectiveness of e-learning by using learning style and flow experience that facilitates learner performance and a better learning experience. They use demographic survey, Kolb learning style scale, pre-test and post-test of content, learning satisfaction, introspection questionnaire, and flow experience scale for this purpose.

Chuang, H.M., Shen, C.C. (Chuang, 2008) describe relationship between learning path, learning style and E-learning performance.

Huiting, H., Prasad, P.W.C., Alsadoon, A., Bajaj, K.S. (Huiting, 2015) investigate influences of learning style in e-learning environment at universities. They proposed a framework of the e-learning environment for higher education student.

Mironova, O., Amitan, I., Vilipold, J., Saar, M., Ruutmann, T. (Mironova, 2013) divides students in a two study groups i.e. (a) Reference group (Students learn through provided learning material and practical assignment according to the traditional study); (b) Test group (Students learn through provided learning material and practical assignment according to learning styles) and suggests test group perform better than reference group.

Hamtini, T., Ateia, H. (Hamtini, 2015) have proposed dynamic technique for found learning style using VAK learning style model. The technique based on literature based approach.

Deborah, L.J., Sathiyaseelan, R., Audithan, S., Vijayakumar, P. (Deborah, 2015) have proposed a marvellous technique for detecting learning style. It accurately categorizes learners using fuzzy approach based on some parameters: (a) Number of mouse movement in the y axis, (b) ratio of document length to the time span of a page, (c) ration of image area to document length and scroll distance, (d) number of visits to a document. Performance evaluation is calculating by experimental setup, Evaluation Snapshot, and results and interface. Learners provide their complete information through a web interface. On successfully, collecting and recording the information of individuals the metric analyzer analyze learning style using different metrics.

X = No of mouse movement in the y axis.

Y = Ration of document length to the time spent on a page.

Z = Ration of image area to the document length and scroll distance.

3 PROPOSED DESIGN APPROACH

3.1 *A general overview of our proposed approach*

In a learning environment different people perceive in learning various way based on their personal behavior and independent of each other. Each perspective learner defines the learning style differently. Learning style may be defined as how learner likes to improve their quality of knowledge. There are more than one model available to describe learning style; where each model based on different learning theory that may be used by different learner. In order to fetch learning style of a particular learner we use literature based approach that may in turn improve the quality of learning of a particular learner. In our study we adopted Neil Fleming's learning style model known as VARK model to determine in which learning style they belong to—Visual, Auditory, Read/Write, Kinesthetic. The VARK model is shown in Fig. 1.

3.2 *Propose learning process*

Learning process may be defined as the steps which should be following in the process of learning. It consist different constraints. In our proposed research

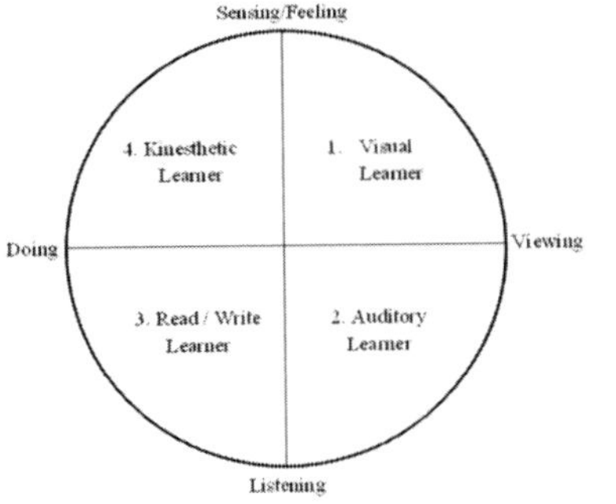

Figure 1. The four major learning style in VARK model.

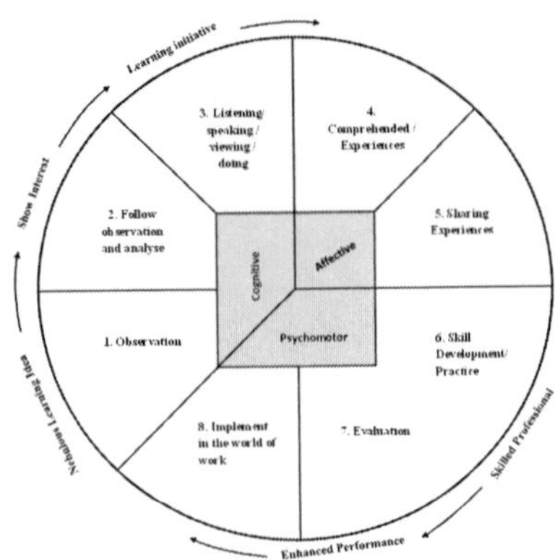

Figure 2. Cycle of learning.

work a learning process is proposed where different factor present in learning environment are considered which may affect learning process as a whole. The proposed learning process is referred using Cycle of learning shown in Fig. 2. The proposed learning process is referred using Cycle of learning shown in Fig. 2.The proposed learning process is referred using Cycle of learning shown in Fig. 2.

The proposed learning process is referred using Cycle of learning shown in Fig. 2.

3.3 *Methodology*

The proposed methodology has been divided into five steps; first, choose VARK model as a learning style model; second, generate two set of survey question with respect to VARK model; third, collect data from survey; fourth, apply algorithm to analyze data, finally conclude from survey result. In our learning environment we consider learning style with respect to conventional learning as well as online learning approach as some student learns best when they are in conventional mode and other have better learning when they learn online. The conventional learning is outside the scope of this research.

3.3.1 *An algorithmic approach for selecting learning style of individual learner*

GQ[ ]: Questions
S[ ][ ]: Answers given by respective students
RA[ ][ ]: Right answers
VQ[i]: Visual learner questions
AQ[j]: Auditory learner questions
RQ[k]: Read/Write learner questions
KQ[l]: Kinesthetic learner questions
i = j = k = l
Q[i….j]∈ V // defines the visual type questions
Q[p…q]∈ A// defines the auditory type questions
Q[s….t]∈ R// defines the read-write type questions
Q[y….z]∈ K// defines the kinesthetic type questions
Q[i….j]+Q[p….q]+Q[s….t]+Q[y….z] = Total no. of questions in the system

```
VC = Visual answer count
AC = Auditory answer count
RC = Read/Write answer count
AC = Kinesthetic answer count
Procedure Correct_Answer_Prediction
 Begin
for(i = 0; i<n; i++)
   {for (j = 0; j<n; j++)
     {if ( RA [0][j] = = S[i][j]) /* Answers given by
     the students are compared with right answers
     set*/
       S[i][j] = 1 ; // for right answer 1 is assigned
       Else
        S[i][j] = 0; // otherwise 0 is assigned
     }
   }
 End
Procedure Learning Style Count
 Begin
  Loop
      For each student
/* the respective answer counts are calculated agan
      ist each right answer given by the student*/
     {for each question Q ∈ Q[i....j]
           If (S[i][j] = = 1) Count VC
     for each question Q ∈ Q[p....q]
          If (S[i][j] = = 1) Count AC
     for each question Q ∈ Q[s....t]
          If (S[i][j] = = 1) Count RC
     for each question Q ∈ Q[y....z]
          If (S[i][j] = = 1) Count KC
     FindLearningStyle();
     }
  End Loop
 End.
Procedure Find Learning Style
 Begin
   If (VC>AC && VC>RC && VC>KC)
      Student is visual learner;
   Else If (AC>VC && AC>RC && AC>KC)
     Student is auditory learner;
   Else If (RC>AC && RC>RC && RC>KC)
     Student is read/write learner;
   Else If (KC>AC && KC>RC && KC>KC)
     Student is kinesthetic learner;
       Else
     Can't be determined
End
```

3.4 *Results and comparisons*

We made two surveys first based on aptitude questions and second based on behavioral questions. Same participants attended both surveys. After completion of survey responses are recorded for analyzing the result. Using proposed algorithm we analyze responses mathematically.

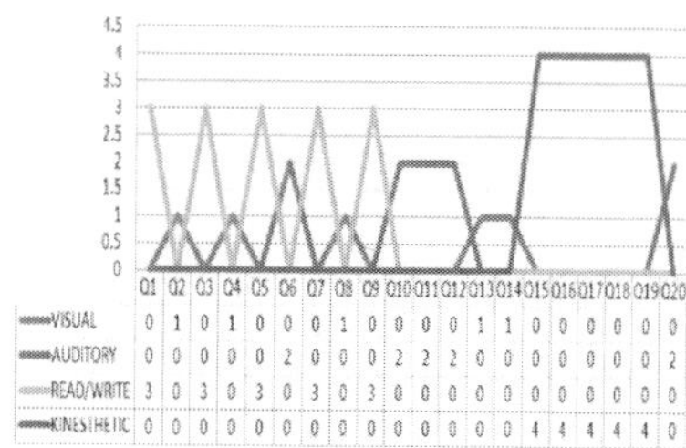

	Q1	Q2	Q3	Q4	Q5	Q6	Q7	Q8	Q9	Q10	Q11	Q12	Q13	Q14	Q15	Q16	Q17	Q18	Q19	Q20
VISUAL	0	1	0	1	0	0	0	1	0	0	0	0	1	1	0	0	0	0	0	0
AUDITORY	0	0	0	0	0	2	0	0	0	2	2	2	0	0	0	0	0	0	0	2
READ/WRITE	3	0	3	0	3	0	3	0	3	0	0	0	0	0	0	0	0	0	0	0
KINESTHETIC	0	0	0	0	0	0	0	0	0	0	0	0	0	0	4	4	4	4	4	0

Figure 3. Analysis of result of first survey.

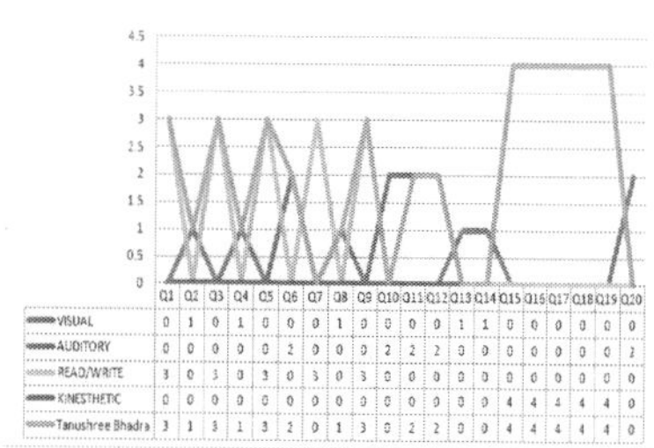

	Q1	Q2	Q3	Q4	Q5	Q6	Q7	Q8	Q9	Q10	Q11	Q12	Q13	Q14	Q15	Q16	Q17	Q18	Q19	Q20
VISUAL	0	1	0	1	0	0	0	1	0	0	0	0	1	1	0	0	0	0	0	0
AUDITORY	0	0	0	0	0	2	0	0	0	2	2	2	0	0	0	0	0	0	0	2
READ/WRITE	3	0	3	0	3	0	3	0	3	0	0	0	0	0	0	0	0	0	0	0
KINESTHETIC	0	0	0	0	0	0	0	0	0	0	0	0	0	0	4	4	4	4	4	0
Tanushree Bhadra	3	1	3	1	3	2	0	1	3	0	2	2	0	0	4	4	4	4	4	0

Figure 4. Analysis of result of individual student doing first survey.

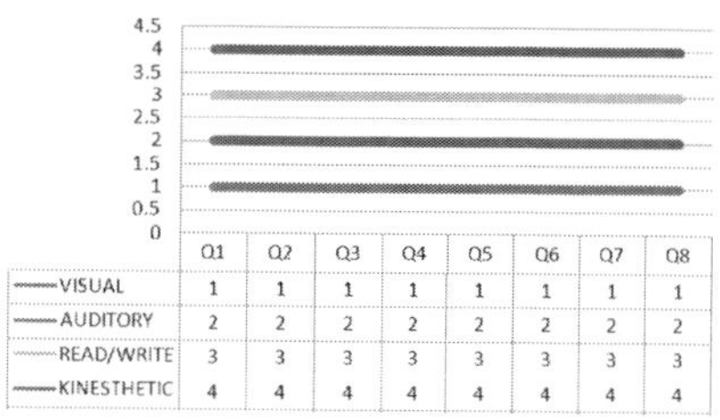

	Q1	Q2	Q3	Q4	Q5	Q6	Q7	Q8
VISUAL	1	1	1	1	1	1	1	1
AUDITORY	2	2	2	2	2	2	2	2
READ/WRITE	3	3	3	3	3	3	3	3
KINESTHETIC	4	4	4	4	4	4	4	4

Figure 5. Analysis of result of second survey.

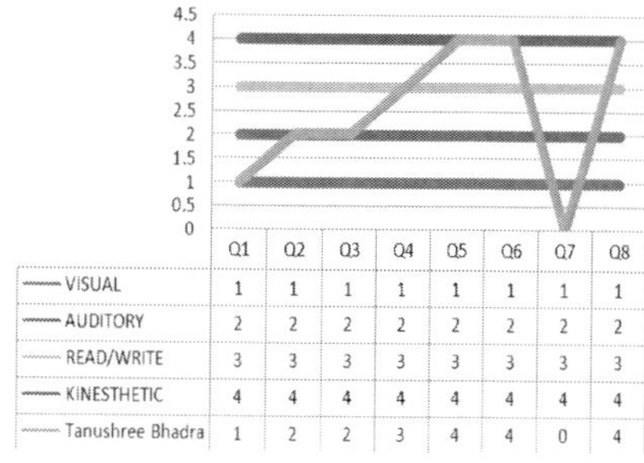

	Q1	Q2	Q3	Q4	Q5	Q6	Q7	Q8
VISUAL	1	1	1	1	1	1	1	1
AUDITORY	2	2	2	2	2	2	2	2
READ/WRITE	3	3	3	3	3	3	3	3
KINESTHETIC	4	4	4	4	4	4	4	4
Tanushree Bhadra	1	2	2	3	4	4	0	4

Figure 6. Analysis of result of individual student doing second survey.

The study shows various types of learners which are mainly divided into four groups i.e. visual, auditory, read/write, kinesthetic learner. The survey is organized with our institutes students, 5 questions have been given in every segment all total 20 questions to signify the learning style as well as what kind

Table 1. Comparison of proposed work with existing research works.

Sl. no.	Parameter	Existing research	Proposed work	Remarks
1	Individual learning style	Learning style static in nature.	Learning style variable in nature.	Better for continuous learning.
2	Design principle	Based on literature based approach	Based on algorithmic approach.	This approach enhances performance.
3	Learner satisfaction	Based on Assessment Instrument	Based on Survey Result.	Flexible design enhances satisfaction.
4	Framework	Standard design framework generate relationship among learning path, style, and performance	Flexible enough to support various e-learning tools and generate relationship among different learner expectations.	Support ubiquitous learning that enhances learning.

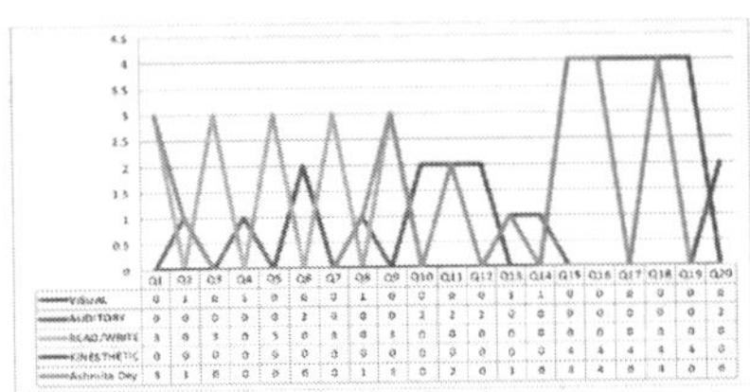

Figure 7. Analysis of first survey which constitute no learning style.

of learner on the basis of aptitude questions and behavioral questions. For right answers we assign 1 to each question else assign zero. For example, a student, answered total 14 no. of questions form survey one (5 from kinesthetic, 2 from auditory, 3 from read/write, 4 from visual) and its clear that he/she answered maximum no of questions from kinesthetic segment's questions then that learner must be in a category of kinesthetic segment i.e. kinesthetic type of learner. When same student participate in a second survey on applying our algorithm we got same result i.e. student is a kinesthetic learner. When we trying to illustrated those segments of students with a graph the result look like below.

In certain cases we may not be able to differentiate learners according to a particular learning style. They share composite learning style.

We have proposed an algorithm for finding learning style of a particular learner in e-learning system. The proposed idea may generate a path of success for better performance in both synchronous and asynchronous e-learning mode.

4 CONCLUSION AND FUTURE SCOPE

Since learning style is important in e-learning we may concluded that if we want more efficient performance of a learner in e-learning system, then we can guide them according to their learning style. In future, we may propose a design that may enlarge the performance of a learner with respect to learning style and deliver content appropriate with the learning style.

REFERENCES

CBloom, B. S.: Learning for Mastery. Instruction and Curriculum. Regional Education Laboratory for the Carolinas and Virginia, Topical Papers and Reprints, Number 1. Evaluation comment, 1(2), n2 (1968).

Chuang, H.M., Shen, C.C.: A study on the relationship among learning path, learning style, and e-learning performance. In Machine Learning and Cybernetics, 2008 International Conference, vol. 5, pp. 2481–2486 (July 2008).

Deborah, L.J., Sathiyaseelan, R., Audithan, S., Vijayakumar, P.: Fuzzy-logic based learning style prediction in e-learning using web interface information, Sadhana, 40(2), 379–394 (2015).

Goyal, M., Yadav, D., Tripathi, A.: Fuzzy approach to detect learning style using McCarthy model as a tool for e-learning system. In Emerging Trends and Technologies in Libraries and Information Services (ETTLIS), 2015 4th International Symposium, IEEE, pp. 295–300 (January 2015).

Hamtini, T., Ateia, H.: A proposed dynamic technique for detecting learning style using literature based approach. In Applied Electrical Engineering and Computing Technologies (AEECT), 2015 IEEE Jordan Conference. IEEE, pp. 1–6 (November 2015).

Huiting, H., Prasad, P.W.C., Alsadoon, A., Bajaj, K.S.: Influences of learning styles on learner satisfaction in E-learning environment. In Computing and Communication (IEMCON), 2015 International Conference and Workshop, IEEE, pp. 1–5 (October 2015).

Mironova, O., Amitan, I., Vilipold, J., Saar, M., Ruutmann, T.: Computer science e-courses for students with different learning styles. In Computer Science and Information Systems (FedCSIS), 2013 Federated Conference, pp. 735–738 (September 2013).

Rong, W.J., Min, Y.S.: The effects of learning style and flow experience on the effectiveness of e-learning. In Advanced Learning Technologies, 2005. ICALT 2005. Fifth IEEE International Conference, pp. 802–805 (July 2005).

Computational Science and Engineering – Deyasi et al. (Eds)
© 2017 Taylor & Francis Group, ISBN 978-1-138-02983-5

Participation measurement of individual member in an ICT-based collaborative work

Tanushree Bhadra & Rajeev Chatterjee
National Institute of Technical Teachers' Training and Research, Kolkata, India

ABSTRACT: Collaborating online with the help of information and communication technology reduces individual work load with respect to goal and may encourages better outcome, higher productivity, and better market response. Online collaboration incorporates the concept of appropriate measurement of group performance. Several techniques are designed in order to evaluate group performance but there may be a requirement to measure individual contribution of the participants. The valid and reliable measurement of participant's contribution motivates the member for better performance on the job. This paper proposes a technique for evaluating the contribution of individual member by using the concept of genetic algorithm.

1 INTRODUCTION

The revolution in Information and Communication Technology (ICT) brings out several possibilities that facilitate people from diverse environment to work together in a group. ICT serves this purpose by involving people in collaborative activities in order to perform certain task by sharing and cooperating different ideas among the workers. Working in a collaborative environment encourages people in e-learning where they can share their knowledge to others, solve problem together by discussing respective views and motivates other. ICT provides the flexibility of working collaboratively beyond the restrictions of time and geography and encourages utilization of wide knowledge repository. Being part of a team helps people to develop interpersonal skills like leadership, working with and motivating others.

Collaborating with others helps people to enhance self-awareness about their job by identifying their own strength and weakness. However working as a group involves the importance of fair assessment of individual team member as team success or failure lies on the contribution of individual team member and his/her performance. Many times the group members are reluctant about their own work and impose their job to other. This situation creates conflicts among the group members in collaborative environment which in terns reduce the working potential of the group. The valid and reliable measurement of participant's contribution motivates the member for better performance on the job. Researchers describe several approaches towards group assessment by assessing each member individually, by measuring overall group performance or by evaluating members by their peers. It is very important that a valid and transparent technique may be used to determine the individual contribution/performance in a group work. Therefore it is proposed that such transparent evaluation technique is very important and required and may be implemented.

The paper is organized as follows. In section 2, we discuss Review of the Existing Works. In section 3, we briefly explain Our Proposed Design Approach. In section 4, Results are given. In section 5, Conclusion and Future Scope are given.

2 REVIEWS ON EXISTING WORK

Mills, K.L (Mills, 2003 describes how Computer Supported Cooperative Work (CSCW) provides similar environment of working collaboratively in a group as a team and focuses on key dimensions, design areas and the features of CSCW.

Ludvigsen, S.R. and Morch, A.I. (Ludvigsen, 2010) highlight the basic concepts, common technologies, important perspectives of Computer Supported Collaborative Learning (CSCL) by defining two approaches of CSCL i.e. systemic and dialogical approaches.

Atkins, A.T. (Atkins, 2010) emphasizes three important aspects of collaborative project: assessing the project or task; technology to organize the project and technology to present the project.

Tran, V. N. and Latapie, H. M. (Tran, 2006) define several organizational models in order to structure objectives, strengths, weaknesses, and implementation requirements of globally distributed collaborative working environment.

Mota, D., de Carvalho, C. V., and Reis, L. P. (Mota, 2011) describe an Advanced Collaborative Educational Model (ACEM) that facilitates the collaborative working environment by assisting the educators on designing collaborative learning architecture.

Kuisma, R. (Kuisma, 1998) tries to find out an effective way to measure individual participation in a collaborative project and also analyzes that it is difficult to find individual member's contribution to the final product development.

Scott, E., van der Merwe, N., Smith D. (Scott, 2005) emphasize on peer evaluation and also describe important impacts of peer evaluation on student's knowledge and skill.

Hughes, R. L. and Jones, S. K. (Hughes, 2011) describe several ways to assess individual team member: Written Teamwork Tests, Comprehensive Assessment of Team Member Effectiveness (CATME), and Valid Assessment of Learning in Undergraduate Education (VALUE).

Hayes, J. H., Lethbridge, T. C., Port, D. P. (Hayes, 2003) focus on several issues related to the assessment of a collaborative group like the benefits of group projects, the difficulties of evaluating the participants, the scoring and grading criteria for the instructor.

Bastick, T. (Bastick, 1999) suggests the method of separating the assessment of final group product (assessed according to the standards) and assessing individual's contribution to the group product (assessed by peer-evaluation).

Lurie, S.J., Schultz, S.H. and Lamanna, G. (Lurie, 2011) introduce an instrument for measuring teamwork i.e. Practice Environment Checklist (PEC).

Wei, T., Xu, Y., Zhao, Y., Khanna, N., Gao, B., and Coady, Y. (Wei, 2015) propose a peer-to-peer architecture design among relayers and listeners named ThinkTogether, in computer supported collaborative work (CSCW) field that facilitates the network performance of CSCW application by creating and sharing content in almost real-time.

3 DESIGN APPROACH

3.1 *A general overview*

The collaborative working environment provides more productivity and performance over working as an individual. In collaboration people can apply wide range of knowledge and skill to perform certain task through sharing and discussion of different ideas. The most common problem with working in collaboration may be unequal participation of individuals. It is very often that all the members of a group may not perform equally. Working especially in a group, it may become very easy to avoid own work and leave it to others to complete. This in terms reduces productivity and also introduces conflicts and grievances within the group. However there must be a fair, valid and reliable method for assessing the contribution of each individual member of a particular group. Many a times a group is assessed on the basis of the performance of an individual member. It may also be accessed by peer evaluation or by giving same mark to all the members. All these methods have respective pros and cons. Therefore it is being proposed that there is a requirement to model the whole process of evaluation of a collaborative work on individual basis. In this paper, we propose a method that evaluates contribution of individual member towards group work with reliability by using the concept of genetic algorithm.

3.2 *Methodology*

In our proposed technique, individual contributions in a collaborative group work are evaluated on the basis of genetic algorithm. The overall procedure is depicted in the form of flow chart shown in the Figure 1. Here we use the roulette wheel selection method recursively for the purpose of analyzing the participation of members working collaboratively. The selection method works according to the fitness value calculated by using fitness function of each member. The fittest member is more probable to be selected according to the probability of selection. β can be depend upon many factors such as population size, quality of population, etc.

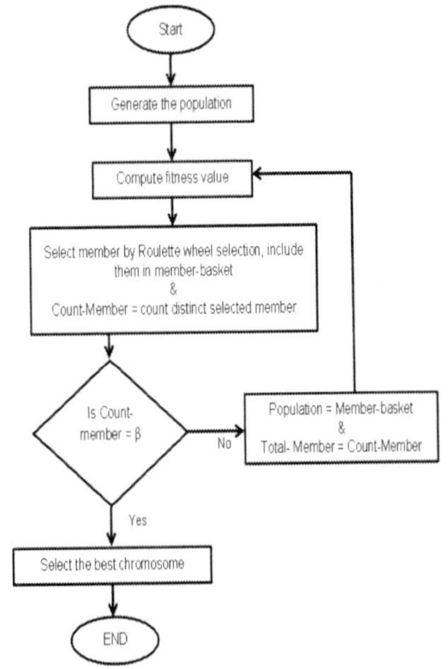

Figure 1. A Flow-chart of the proposed method using genetic algorithm.

3.3 *A genetic algorithmic approach towards contribution evaluation of individual participant*

```
/* Variable Declaration
total_member: Total no of members in collaborative group
member [ ][ ]: Array consists of the marks of individual group member for each parameter (for example parameters are like P, Q, R, S, T, U etc.)
member_fitness [ ]: Array consists of the fitness value of individual member
weight [ ]: Array consists of the weight of the parameter. Weight values are in chronological order with parameter (i.e. P, Q, R, S, T, U) accordingly.
Total_parameter: Total no of parameter to be considered. */

Procedure Create-Population
Begin
value: input
   /* value indicates the mark given to each member considering respective parameters */
  For i← 1 to total_member
   {
   For j← 1 to total_parameter
   {
          member[i][j] ←value
   }
    }
End.

Procedure Fitness ()
Begin
For i← 1 to total_member
  { sum_parameter = 0;
  For j← 1 to total_member
   {
       sum_parameter + = member[i][j]*weight[j];
   }
member_fitness[i] ←sum_parameter;
   }
End.

Procedure Collaborative-Work-Roulette-Wheel-Selection()
Begin
sum_fitness← 0
For i← 1 to total_member
  {
sum_parameter←sum_fitness+member–fitness[i];
  }
 For i← 1 to total_member
  {
r← random (0,sum_fitness)
  /* locate a point P */
j← 1;
total_fitness←0;
While(total_fitness< r )
 Do
  {
  total_fitness←total_fitness + member_fitness[j];
           j = j+1;
  }
  End-while;
  k=j-1;
 Select member (k) and include them in a member-basket
  }
  count_member=count distinct selected member in member-basket
  End.
```

4 RESULTS

In the proposed search work it was observed that different person has different perception of a same problem and hence their contributions are different. Therefore it is obvious to measure their proper contribution and in order to do so we generate some parameter based on which we try to measure the contribution of individual member. The parameter that we generally follow are Literature Survey, Data Collection from Literature Study, Identify and fix

Table 1. Parameters and their respective weight value.

SI. no.	Parameter	Description	Weight value
1	Literature survey (P)	No. of paper he/she read, no. of articles referred	2.60 (W1)
2	Data collection from literature study (Q)	How many useful data can be collected through the survey in order to solve the problem	2.34 (W2)
3	Identify and fix team-project related problem (R)	What are the different kinds of problems individual have faced and necessary solutions for those problem	2.84 (W3)
4	Target fulfilment (S)	Percentage of work completed within some time interval	4.24 (W4)
5	Quality measure (T)	Individual helps other team members by solving some of their modules	4.70 (W5)
6	Cooperate with other team members to accomplish job (U)	Job performed by individual must follow some qualitative parameters like standard methods, company norms	3.90 (W6)

team-project related problem, Target fulfillment and Co-operation with other team members to accomplish a work. We create survey to calculate weight value for each parameter. The parameters along with their weight values are given in table 1. The weight values are proposed from the survey are generated based of the result.

For example in a collaborative work there are 10 employees and instructor gave marks with respect to each parameter and from there we generate GA-population. In this population we applied our fitness function algorithm.

Here we use the fitness function as:

$$Fitness(X) = W1*P + W2*Q + W3*R + W4*S + W5*T + W6*U.$$

The data are shown in a Table 2.

In our problem we use the concept of roulette wheel selection that ensures the fittest member is more probable to be selected, where f(X) describes the fitness value of X. A typical example of roulette wheel is given in Figure 2.

Table 2. Population of the member along with respective fitness value.

		Parameters (Weight value)						
Sl no.	Member name	P (2.6)	Q (2.34)	R (2.84)	S (4.24)	T (4.7)	U (3.9)	Fitness value
1	Akash	8	7	4	6	4	4	108.38
2	Sourav	6	6	3	8	5	6	118.98
3	Mridul	7	6	5	7	7	8	140.06
4	Tanvi	5	7	3	4	5	3	90.06
5	Arjun	7	5	6	7	6	5	124.32
6	Rohit	8	6	6	8	8	7	148.20
7	Indu	2	3	4	5	6	4	88.58
8	Barnita	6	5	4	5	6	6	111.46
9	Partho	8	7	6	7	8	7	148.24
10	Mita	8	6	5	6	2	3	95.58

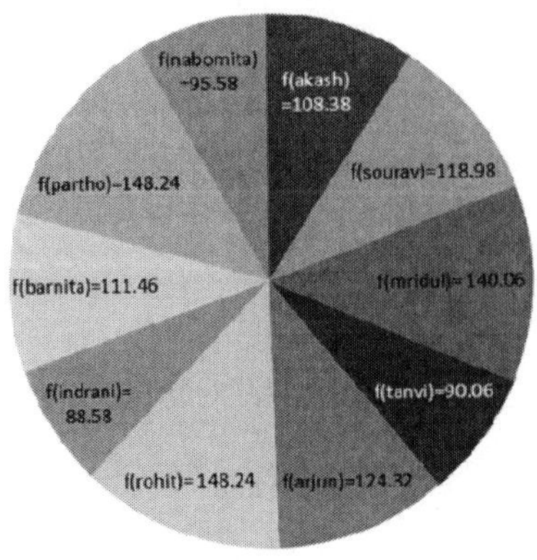

Figure 2. Roulette wheel.

5 CONCLUSIONS AND FUTURE SCOPES

Collaborative work helps in reducing work pressure of individuals and it facilitates better performance and productivity. It is important that individual member must be more responsive and devoted to their work in a collaborative group. If we want more reliable system then measuring the contribution of each member is important.

REFERENCES

Atkins, A.T.: Collaborating Online: Digital Strategies for Group Work. Writing Spaces: writingspaces,2010-books.google.com, pp. 235–248 (2010).

Bastick, T.: Reliable and Valid Measurement of Individual's Contributions to Group Work. (1999).

Hayes, J. H., Lethbridge, T. C., Port, D. P.: Evaluating Individual Contribution Toward Group Software Engineering Projects. IEEE Computer Society, In Proceedings of the 25th International Conference on Software Engineering, pp. 622–627 (May 2003).

Hughes, R. L., Jones, S. K.: Developing and Assessing College Student Teamwork Skills. New Directions for Institutional Research,2011(149), pp. 53–64 (2011).

Kuisma, R.: Assessing Individual Contribution to a Group Project. How and Why, Assessment Portfolio, Students' Grading-Evaluation of the Student Experience Project, 2. Assessment of University Students in Hong Kong, pp. 79–106 (1998).

Ludvigsen, S.R., Mørch, A.I.: Computer-Supported Collaborative Learning: Basic Concepts, Multiple Perspectives, and Emerging Trends. The International Encyclopedia of Education, 3rd Edition, edited by B. McGaw, P. Peterson and E. Baker, Elsevier (2010).

Lurie, S.J., Schultz, S.H., Lamanna, G.: Assessing Teamwork. Family medicine, vol.43, no.10, pp. 731–734 (2011).

Mills, K.L.: Computer-Supported Cooperative Work. Encyclopedia of Library and Information Science, DOI: 10.1081/E-ELIS 120008706, pp. 666–677 (2003).

Mota, D., de Carvalho, C. V., and Reis, L. P.: Fostering Collaborative Work between educators in higher education. In Systems, Man, and Cybernetics (SMC), 2011 IEEE International Conference, pp. 1286–1291 (2011, October).

Scott, E., van der Merwe, N., Smith D.: Peer Assessment: A Complementary Instrument to Recognise Individual Contributions in IS Student Group Projects. The Electronic Journal of Information Systems Evaluation, Vol. 8, no. 1, pp. 61–70 (2005).

Tran, V. N., Latapie, H. M.: Models for Structuring Teams and Work in globally Collaborative Projects. In Engineering Management Conference, 2006 IEEE International. IEEE, pp. 425–431 (17, Sep 2006).

Wei, T., Xu, Y., Zhao, Y., Khanna, N., Gao, B., and Coady, Y.: Exploring peer-to-peer infrastructure for computer supported collaborative work applications. In Communications, Computers and Signal Processing. 2015 IEEE Pacific Rim Conference (PACRIM), pp. 304–308 (2015, August).

Computational Science and Engineering – Deyasi et al. (Eds)
© 2017 Taylor & Francis Group, ISBN 978-1-138-02983-5

A channel trust based approach for congestion control in IoT

Moumita Poddar & Debdutta Pal
CIEM, Kolkata, India

ABSTRACT: Nowadays in the area of Internet of Things (Iot), congestion control is treated as a important research area. Congestion control is originated from the point of network bandwidth, node processing ability and server capacities. In IoT people and devices are progressively get connected over the network, hence the necessity of congestion control increases. A monitoring device in a IoT, the monitoring device notices different network features like flow, capacity of a link in network, number and size of distinct flows. In this paper we have presented literature review of some of existing congestion control mechanisms. A congestion control model has also been proposed, which uses the measure of channel trust for decision making.

1 INTRODUCTION

IoT (Atzori *et. al.*, 2010) is a new networking framework which integrate a large number of smart objects and keeping them in a tight and continuous interaction. These objects are attached with several kinds of sensors and actuators which are interconnected for interlinking between physical and cyber world. Such heterogeneous networks are enabled by different technologies such as embedded sensing and actuating, Radio Frequency Identification (RFID), wireless sensor networks, real-time and semantic web services, etc (Cao *et. al.*, 2008). On the one hand, billions of smart objects are absorbed in the environment for sensing, interacting, and cooperating with each other to enable efficient services thus the environment, the economy and the society are benefited as a whole. On the other hand, these smart objects are extremely diversed and heterogeneous in terms of resource capabilities, lifespan and communication technologies, thus the scenario become complicated.

In IoT, especially in sensor networks and wireless access networks, the parameters such as bandwidth (Chen *et. al.*, 2009, Chen *et. al.*, 2010), data storage capacity and battery power (Chen *et. al.*, 2010) of each node are considered as important resources. The throughput of the network depends on the amount of congestion. Congestion increases delay, packet loss probability of the network, and it plays an important role in a network. It is caused due to the following condition (i):

Σresource demand> usable resource (i)

Normally user demand, network resource, and network topology are the factors that tend to influence congestion.

The rest of the paper is organized as follows: Section 2 describes different types of congestion control mechanisms briefly. Section 3 shows the research gap and open issues in congestion control for IoT and section 4 shows the proposed work. Finally we conclude the paper in section 5.

2 REVIEW METHOD

In (Accettura *et. al.*, 2013), authors have deployed in amulti-hop Low-power and Lossy Networks (LLN) (Winter *et. al.*, 2012) and it uses IETFRPL routing protocol for Internet of things. This technique computes optimum multihop schedule for the whole network in a decentralized manner and control queue congestion. It also reduces overheads due to using very low amount of signaling messages. In (Ray *et. al.*, 2013) authors have proposed a technique for distributed congestion control PAISMD that allocates transmission rates to M2M (machine to machine)flows in proportion to their demands. It is based on AMID (Jain *et. al.*, 1987, Miller *et. al.*, 2008). This mechanism yields unstable throughput for increasing and decreasing the congestion window of the flow. In (Xu *et. al.*, 2008) authors have proposed modified TCP congestion control mechanism based on bandwidth estimation (Capone *et. al.*, 2004) and double AIMD (Additive-Increase Multiplicative-Decrease) (Chiu *et. al.*, 1989), named Switch-TCP. Main idea of this mechanism is that first, it uses ABSE filter mechanism to estimate available bandwidth. In (Leith *et. al.*, 2007) proposes an algorithm that can be used to solve the coexistence problem by choosing the probabilistic back off function. In (Budzisz *et. al.*, 2011) authors have used delay and packet loss based AIMD congestion control algorithm. It also uses Probabilistic Early Response TCP (PERT) (Bhandarkar *et. al.*, 2007). In PERT, switching between loss and delay based flow is possible. The deployment environment for this mechanism has similar characteristics with RED AQM (Que *et. al.*, 2008, Firoiu *et. al.*, 2000). In (Huang *et. al.*, 2014), authors have proposed improved Random Early Drop (IRED) algorithm where framework and queuing theory are used. This algorithm can calculate drop rate of the systems. Here active Queue Management (AQM) (Firoiu *et. al.*, 2000) mechanism is also used to avoid congestion. In contrast to standard RED (Bonald *et. al.*, 2000, Bauso *et. al.*, 2004, Floyd *et. al.*, 2011), here probability of packet dropping is reduced. IRED is not capable to consider packets coming from multiple sources. In HTHRED (Liu *et. al.*, 2001, Byers *et. al.*, 2002, Liu *et. al.*, 2007), authors have used hop-to-hop controlled hierarchical multicast congestion control mechanism in heterogeneous net hierarchical system work. It is a combination of RED and HTH. In this mechanism, each router determines its congestion status based on current network state and capable to takes appropriate measurement for increasing or decreasing transmission rate of packet. At the same time, number of packet discards is considered to ensure the fairness.

3 PROPOSED WORK

3.1 *Assumptions*

In the proposed work, following assumptions are taken to perform trust based congestion control

- The network consists of heterogeneous node.
- Data are catagorised into two types Medical data (MEDDATA) and Other Data (OTHDATA).
- Each node *i* has *n* number of equal sized queues for two types of sensed data.

Two states are considered to form a multiperiod transition probability matrix using a stochastic method. These two states are CONGESTION (C) and NO CONGESTION (NC).

3.2 *Description*

The proposed logic aims to detect whether channel is reliable for transmission of the packet from source to destination using the metrics: node level congestion and channel trust. A priority scheme is chosen to ascertain quicker response time in case of urgency. In the proposed model network layer interact with the MAC layer to perform congestion control function. Here, the application layer generates node data (if it is a source node) and the hop data (if it is not a source node) from the other nodes and the combined data traverse through the network layer. Fig.1 shows the block diagram of proposed model. Proposed work has been divided into three sub phases, Priority setter, node_level_congestion measure and compute_channel_trust.

PHASE 1: Priority Setter

In this phase before assigning the priority the rate of data generation by a specific node is calculated as the originating_rate (R^i_{gr}) where R^i_{gr} of node i is:

$$R^i_{gr} = \sum_{i=1}^{n} D_i / s \tag{1}$$

Here, D_i represents data originating at ith instance, for all i, $R^i_{gr} > 0$. This rate will be used for congestion control in the network.

In this phase a node with maximum number of link with the neighboring nodes in the network is designated as monitoring node and it assigns the priorities for heterogeneous traffic data. Each data queue has its own priority known as Inter_queue_rank (obj_{inter_rank}). MEDDATA are inserted in the queue having higher inter_queue_rank and OTHDATA are inserted in the queue having lower inter_queue_rank.

Scheduler decides the selection order of the data packets from the queues and manages the queues according to the Inter_queue_rank. In the queue Hop_data are given more priority than Node_data. Hop_data have already traversed some hop (s) hence loss of any Hop_data would cause more wastage of network resources than Node_data. The classifier assigns the priority between the Node_data and Hop_data by examining the source address in the packet header. This priority is known as Intra_queue_rank (obj_{intra_rank}).

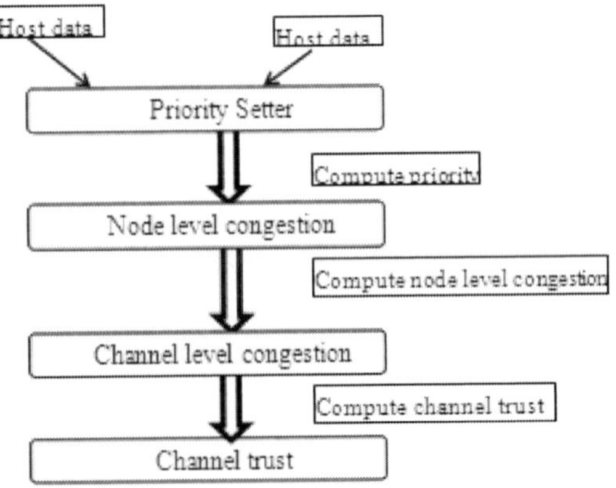

Figure 1. Block diagram of proposed model.

The queue has the following structure

Type	Data_ index	Status	Source_ addr	Des_ addr	Hop_ count

PHASE 2: Measure_Node_Level_Congestion

Node_level_congestion (C_t^i) for ith node at tth instance of time is computed as the ratio of Forwarding_rate (R_{fwd}^i) to Scheduling_rate (R_{sch}^i) of a node.

$$C_t^i = R_{fwd}^i / R_{sch}^i \tag{2}$$

Here i. if $\left(R_{fwd} / R_{sch} = 1\right)$ then scheduling rate and forwarding rate are same hence this status can be considered as low congestion.

ii. If $(R_{fwd} / R_{sch} > 1)$ then Scheduling_rate is less than Forwading_rate hence this status can be consider as low congestion.

iii. If $(R_{fwd} / R_{sch} < 1)$ then Scheduling_rate is greater than Forwading_rate hence this status can be consider as congestion.

In this phase Scheduling_rate and Forwarding_rate are used to determine node level congestion. The Scheduling_rate(R_{sch}^i) is defined as how many packets the scheduler schedules per unit time from the queues. The scheduler forwards the packets to the MAC layer from which the packets are delivered to the next node (i + 1) along the path towards the Monitoring Node (MN).

$$R_{sch}^i = queue_{len} / \mathrm{T} \tag{3}$$

Where $queue_{len}$ is the number of packets in the queue at ith instance of time and T is time period require to forward total number of packets.

On the other hand Forwarding_rate (R_{fwd}^i) is the rate at which packets are forwarded from MAC layer at ith instance of time to the next node.

$$R_{fwd}^i = 1 / \mathrm{t}_{\mathrm{pkt\_trans\_time}} \tag{4}$$

Packet_transmission_time ($\mathrm{t}_{\mathrm{pkt\_trans\_time}}$) is the time interval requires to successfully sending packet from the current node to the next hop node. It is calculated using Exponential Weighted Moving Average formula (EWMA). So every time $\mathrm{t}_{\mathrm{pkt\_trans\_time}}$ will be updated by

$$t^i_{pkt\_trans\_time} = \left(1 - \beta_{pkt_{conts}}\right) \times t_{pkt\_trans\_time} + \beta_{pkt\_conts} \times prompt\left(t_{pkt\_trans\_time}\right) \tag{5}$$

Here β_{pkt_conts} is a constant where $0 < \beta_{pkt\_conts} < 1$

PHASE 3: Compute_channel_trust

Channel_ trust (Ch^i_{trust}) is used to decide whether the selected channel is dependable for secure transmission of packets to the next node or not. Each time Channel_trust is updated by determining Multiperiod Transition Probability (MTP) using stochastic process. The Multiperiod Transition Probability Matrix (MTPM) will be –

$$\begin{bmatrix} P_{NC}^{NC} & P_{C}^{NC} \\ P_{NC}^{C} & P_{C}^{C} \end{bmatrix} \tag{6}$$

Here each and every conditional probability is the state transition probability of a node. The matrix diagonal depicts the state retention probabilities. The rows and columns of matrix imply change of state from No_Congestion (NC) to Congestion or vice versa. Finally the calculated Channel_trust gives the probability of the channel's retention of one state. The state transition probability is calculated using the following equations.

$$NCRat_{Psch}^{PFwd} = \int_1^n R_{fwd}\, dt / \int_1^n R_{sch}\, dt \tag{7}$$

$$CRat_{Psch}^{PFwd} = \int_1^n R_{sch} - \mathrm{R}_{\mathrm{fwd}}\, dt / \int_1^n R_{sch} dt \tag{8}$$

$$p_C^{NC} = (NCRat_{Psch}^{PFwd} + CRat_{Psch}^{PFwd}) / 2n \tag{9}$$

$$p_{NC}^{NC} = 1 - p_C^{NC} \tag{10}$$

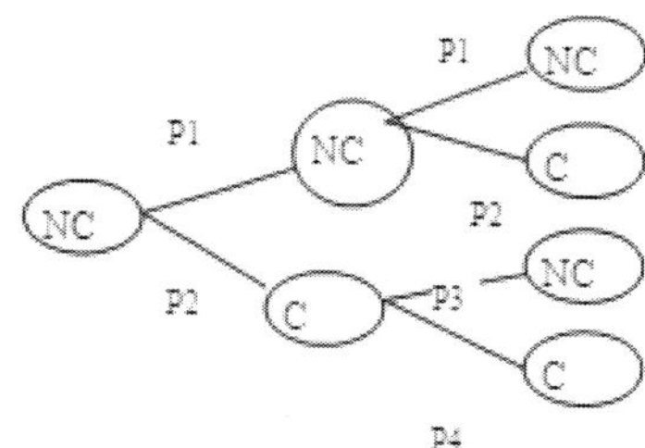

Channel_ trust(Ch^i_{trust}) P_{NC}^{NC} = p1 × p1 + p2 × p3

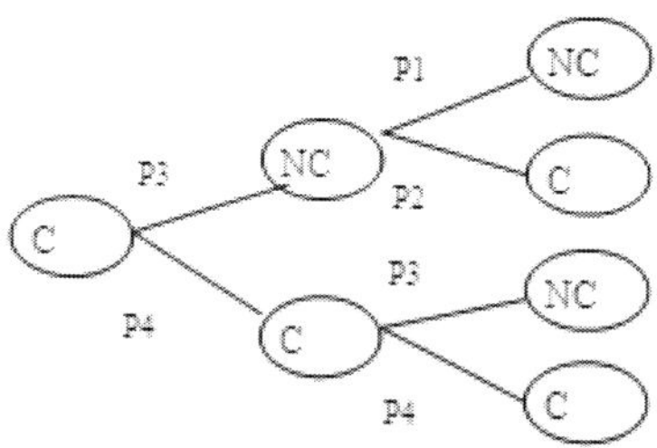

Channel_ trust(Ch^i_{trust}) P_C^C = p3 × p2 + p4 × p4

Figure 2. Channel_trust process for state NC and C.

Performance metric
i=1.... n time index
R^i_{gen}= Data rate generation
$obj_{inter\ queue\ rank}$= Inter queue rank
$obj_{intra\ queue\ rank}$= Intra queue rank
R^i_{sch}= Scheduling data rate in ith instance of time
R^i_{Fwd}= Forwarding data rate in ith instanceof time
$t_{pkt\ trans\ time}$=Packet transmission time
C= congestion state
NC= no congestion state
p_C^{NC}= Congestion to no congestion transition state probability
p_{NC}^{C}= No congestion to congestion transition state probability
p_C^{C}= Congestion to congestion transition state probability
p_{NC}^{NC}= No congestion to no congestion transition state probability.

Figure 3. Performance metric.

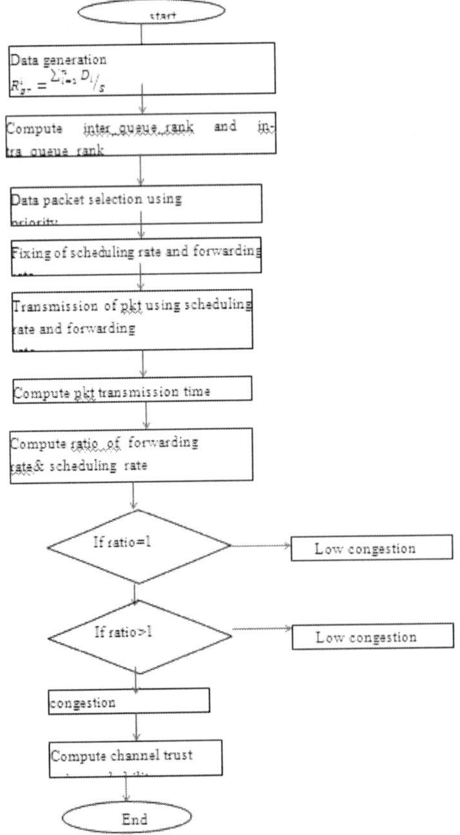

Figure 4. Flowchart.

In Eq. 7 $NCRat_{Psch}^{PFwd}$ implies the ratio of rate of packets forwarded in the time interval n to the rate of packets scheduled for forwarding in n time intervals when the channel is in No_Congestion State.

In Eq. 8 $CRat_{Psch}^{PFwd}$ indicates the ratio of rate of queued packet in the n time interval to the rate of packets scheduled for forwarding in n time intervals when the channel is in Congestion State.

Similarly remaining probabilities p_{NC}^{C}, p_{C}^{C} are computed as in Eq. 9 and Eq. 10 respectively. The Fig. 2 shows the process of determining the Channel_trust for different states after two periods.

The performance metric and flowchart of the proposed work is given in Fig. 3 and Fig. 4 respectively.

4 CONCLUSION & FUTURE WORK

In this paper we have reviewed some existing research papers on congestion control in IoT. In the study of recent works it is observed that now a day's increasing use of mobile devices increases network traffic. Thus the probability of occurrence of congestion becomes a serious problem in IoT. If this congestion is not controlled in efficient manner then communication delay, packet loss degrades the performance of network. It is required to develop an efficient and effective congestion control protocol that will reduce the chance of delayed communication and loss of packets in IoT environment. Thus a review of different congestion control protocols, comparing the various features is necessary to come up with new proposals for congestion control in IoT. The performance of congestion control protocols depend on various parameters. Thus this paper has come up with a conscientious survey and comparison of different classes of congestion control protocols. Our goal is to develop an algorithm that will efficiently perform congestion control in IoT and maintain an optimal data rate during packet transmission. We have identified certain parameters for efficient congestion control. Channel_trust is the key parameter which will be used to measure the channel reliability or trust. To compute channel_trust, we are working now on underlying channel capacity, queue length and the opinion from other nodes about a channel.

REFERENCES

Accettura, N., Palattella, M. E., Boggia, G., Grieco, L. A., Dohler, M. (2013) Decentralized Traffic Aware Scheduling for Multi-hop Low Power Lossy Network sin the Internet of Things, IEEE.

Atzori, L., Iera, A, Morabito, G. (2010) The internet of things: A survey, Computer Networks, 54(15), 2787–2805.

Bauso, D., Giarre, L., Neglia, G. (2004) Active Queue Management Stability in Multiple Bottleneck networks Control, Communications and Signal Processing, 369–372.

Bhandarkar, S., Reddy, A., Zhang, Y., Loguinov, D. (2007) Emulating AQM from end hosts, Comput. Commun. Rev., 37(4), 349–360.

Bonald, T., May, M„ Bolot, J. C. (2000) Analytic Evaluation of RED Performance, Proc. INFOCOM, 1(3), 1415–1424.

Budzisz, L., Stanojevic, R., Schlote, A., Baker, F., Shorten, R. (2011) On the Fair Coexistence of Loss- and Delay-Based TCP, IEEE/ACM transactions on networking, 19(6).

Byers, J. W., Horn, G., Luby, M. (2002) Congestion Control for Layered Multicast, IEEE Journal of Selected Areas in Communications 20, 1558–1570.

Cao, X., Chen, J., Zhang, Y., Sun, Y. (2008) Development of an integrated wireless sensor network micro-environmental monitoring system, Isa Transactions, 47, 247–255.

Capone, A., Fratta, L., Martignon, F. (2004) Bandwidth estimation schemes for TCP over wireless networks, IEEE Transactions on Mobile Computing, 3(2), 129–143.

Changbiao, X., Wei, S. (2008) New TCP Mechanism over Heterogeneous networks, International Conference on Embedded Software and Systems, pp. 303–307.

Chen, J., He, S., Sun, Y., Thulasiraman, P. (2009) Optimal flow control for utility-lifetime tradeoff in wireless sensor networks, Computer Networks, 53, 3031–3041.

Chen, J., Xu, W., He, S., Sun, Y., Thulasiraman, P., Shen, X. (2010) Utility-based asynchronous flow control algorithm for wireless sensor networks, IEEE Journal on Selected Areas in Communications, 28, 1116–1126.

Chiu, D. M., Jain, R. (1989) Analysis of the increase and decrease algorithms for congestion avoidance in computer networks, Computer Networks and ISDN Systems, 17(1), 1–14.

Firoiu, V., Borden, M. (2000) A Study of Active Queue Management for Congestion Control, Ieee infocom.

Floyd, S., Jacobson, V. (2014) Random early detection gateways for congestion avoidance, IEEE/ACM Transactions on, Networking, 1(4), 397–413.

Huang, J., Du, D., Duan, Q., Sun, Y., Yin, Y., Zhou, T., Zhang, Y. (2014) Modeling and Analysis on Congestion Control in Modeling and Analysis on Congestion Control, IEEE ICC-Ad-hoc and Sensor Networking Symposium, 434–439.

Jain, R., Ramakrishnan, K. K., Chiu, D. M. (1987) Congestion avoidance in computer networks with a connectionless network layer, Digital Equipment Corporation, Tech. Rep. DEC-TR-506.

Lam, R. K., Chen, K. C. (2013) Congestion Control for M2M Traffic with Heterogeneous Throughput Demands, IEEE WCNC, 1452–1457.

Leith, D., Heffner, J., Shorten, R., McCullagh, G. (2007) Delay-based AIMD congestion control, Proc. 5th PFLDnet, 1–6.

Liu, J., Li, B., Zhang, Y. (2001) A hybrid adaptation protocol for tcp-friendly layered multicast and its optimal rate allocation, In: Proceedings of IEEE INFOCOM, 1520–1528.

Liu, K., Cheng, Z., Zhao, Y. (2007) Multicast congestion control based on hop to hop, Computer Engineering, 33, 99–101.

Miller, K., Harks, T. (2008) Utility max-min fair congestion control with time-varying delays, Proceedings of IEEE INFOCOM, 331–335.

Palattella, M. R., Accettura, N., Dohler, M., Grieco, L. A., Boggia, G. (2012) Traffic Aware scheduling algorithm for reliable low power multi hopIEEE 802.15.4e Networks, 23rd IEEE international symposium on personal, indoor and mobile radio communictions

Que, D., Chen, Z., Chen, B. (2008) An Improvement Algorithm Based on RED and Its Performance Analysis, ICSP Proceedings.

Winter, T., Thubert, P., Brandt, A., Hui, J., Kelsey, R., Levis, P., Pister, K., Struik, R., Vasseur, J .P., Alexander, R. (2012) IPv6 Routing Protocol for Low-Power and Lossy Networks, IETF RFC 6550.

Computational Science and Engineering – Deyasi et al. (Eds)
© 2017 Taylor & Francis Group, ISBN 978-1-138-02983-5

Real time task scheduling in clustered homogeneous distributed systems

L. Datta
CEMK, West Bengal, India

ABSTRACT: A distributed system is a collection of autonomous processing nodes that appears to its user as a single processor system. Some of the nodes may be overloaded due to high rate of job arrivals while other nodes may remain idle. Real-time scheduler must be aware of deadlines. Meeting the deadlines of tasks in a distributed system depends crucially on dividing up work effectively among the computing nodes. So a way is needed to share load across all the computing nodes so that maximum tasks can meet their deadline. In centralized scheduling schemes, the scheduling decision is taken by a central server. So this scheme is not scalable. Fully distributed schemes are scalable, but they use local information. A hierarchical dynamic scheduling model is proposed in this paper, taking deadline of each task in consideration, where an ordinary node does not need to have a global system wide knowledge about the states of other nodes in the system. The proposed model is semi distributed as each cluster is represented by a cluster master.

1 INTRODUCTION

The objective of a proper scheduling protocol is to allocate tasks among the computing nodes, such that no processing element become neither overloaded nor idle, hence balance the load at the computers and increase the overall system performance. Task scheduling in distributed computing system can be local scheduling and global scheduling. Fundamental properties of a task are: arrival time and approximate execution time. A real-time task also specifies a deadline by which it must complete its execution.

The correctness of a real-time system depends not only on the system's outputs but also on the time at which these outputs are produced. The completion of a request after its deadline is considered to be of degraded value, and could even lead to a failure of the whole system. So, the real-time scheduler for distributed systems must consider the deadline of each task along with the load balancing issue while allocating tasks to processor. Real time tasks may be broadly classified into two categories. Periodic processes, which execute on a regular basis, aperiodic processes are viewed as being activated randomly, following for example a Poisson distribution. Real time schedulers can be categorized as either static or dynamic depending on the time at which necessary information for the successful completion of application is available. A scheduling algorithm is said to be static if it does not depend on the sequence of requests; otherwise it is dynamic (Baruah et al. 1991). Current allocation information is used to determine feasibility of application by dynamic scheduler. A scheduler is static and offline if all scheduling decisions are made prior to the running of the system. Hence this scheme is workable only if all the processes are effectively periodic.

To reduce the huge traffic in comparison to fully distributed system, a method that uses a two-level strategy for aperiodic task scheduling to meet the deadline constraint and to balance workload among the nodes, is proposed. The total system is grouped into clusters consisting of a subset of nodes in the system.

The work is organized as follows: The status of the considered domain is presented in section 2. Section 3 describes the system model and the responsibilities of each type of node. The 2-Level Task Scheduling (2 LTS) policy is discussed in Section 4. Section 5 analyses the communication cost. Simulation results of our experiments are presented in Section 6. Section 7 includes the conclusion part.

2 RELATED WORK

The theoretical foundation to all modern scheduling algorithms for real-time systems was provided (Liu et al. 1973) for hard real-time tasks executing on a single processor. According to the authors, that upper bound of processor utilization quickly drops to approximately 70% as the number of tasks increases. Then Liu and Layland suggested a new, deadline-driven scheduling algorithm, which assigns dynamic priorities to tasks according to their deadlines. Earliest Deadline First is too complex to be implemented in real-time operating system (Li et al. 2009). These algorithms were developed for uniprocessor system. They can be extended for centralized controlled distributed system. Then they would have to suffer from all the difficulties of centralized control. The RT-SADS (Atif et al. 1998) algorithm is designed for scheduling aperiodic, non-preemptable, independent, soft real-time tasks with deadlines on a set of identical processors with distributed memory architecture. RT-SADS self-adjusts the scheduling stage duration depending on processor load, task arrival rate and slack. Epoch scheduling (Karatza et al. 2001) is a special type of scheduling for distributed processor. In this scheme at the end of an epoch, the scheduler recalculates the priority of each task in the queue using Shortest Task First (STF) criteria. LLF (Mok et al. 1978) assigns priorities depending on the laxity. In (Arabnejad et al. 2014) authors present a novel list-based scheduling algorithm called Predict Earliest Finish Time (PEFT) for heterogeneous computing systems. In (Zeng et al. 2008) authors have presented a modified dynamic critical path algorithm (CBL) to find the earliest possible start time and the latest possible finish time of a task using the distributed nodes network structure. Tasks are sorted by the ascending order of their loads and the processors are sorted by the descending order of their current loads in (Zhang et al. 2012). Tasks are assigned to processor to make the loads assigned to each processor balanced as much as possible. Ant colony based task scheduling for real time operating systems is proposed in (Shah et al. 2010).

3 SYSTEM MODEL

In this paper a network is considered consisting of N homogeneous nodes P_1, P_2, ..., P_N connected by a communication network. Each node has the same computational power and local memory. Jobs are assumed to arrive at each node i according to Poisson rate λ. The service time of the jobs are exponentially distributed with mean of $1/\mu$. It is assumed that there is no precedence relationship among the jobs. Each node maintains a ready queue to store jobs which are assigned to the node but yet to be executed. The jobs are assumed to be non-preemptive and aperiodic. In the proposed model the whole system is divided into L clusters. Each cluster has a specific node designated as the Cluster Master (CM).

This proposed model is semi-distributed and decentralizes the load balancing process. It is scalable as it minimizes communication overhead.

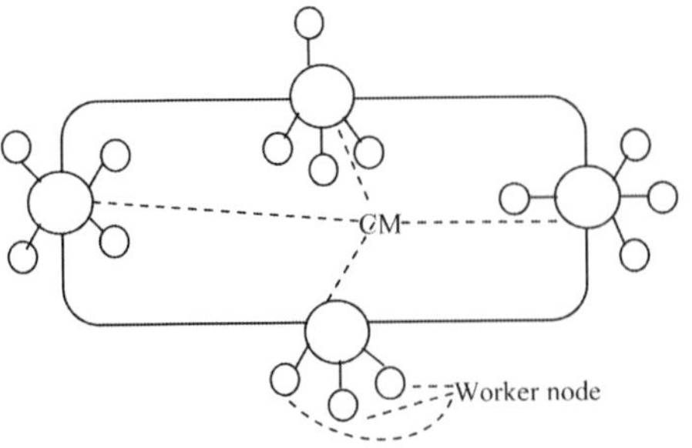

Figure 1. Clustered representation of a distributed network.

4 2 LEVEL TASK SCHEDULING (2 LTS)

Given a set of independent tasks submitted to the system. The algorithm makes an effort to assign the task to a computing node such that the deadline of the task can be met. Firstly, intra cluster task scheduling is tried. CM searches for an idle node in home cluster. On failure, CM searches for a node whose remaining workload is less than the slack time of new task. If such a node is found, the task is sent to that node. Otherwise, inter cluster task scheduling is required. The CM broadcasts a message to all other CMs. If the receiver CM is able to find a suitable node, it sends response to the initiating CM with its minimum load information. The initiating CM selects the cluster with least minimum load. The new task is transferred to the selected CM. If no response is received, the task is assigned to the least loaded node in home cluster and it misses deadline.

4.1 *Cluster master algorithm*

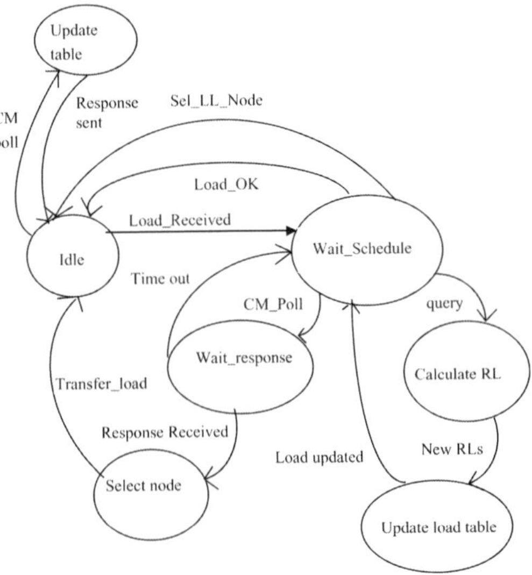

Figure 2. Cluster master state machine.

Process CM

1. Begin
2. While TRUE
3. Wait for Load_Receipt;
4. Send Load_query to all worker nodes in its cluster;
5. Receive RL from each worker node
6. Update load table
7. Search for idle node in home cluster
8. If no idle node found
9. Search for a node whose load is less than slack time
10. If such a node found
11. Send New_Load to selected node;
12. Else
13. Broadcast CM_Poll to all other CMs;
14. If response is received before time out
15. Select the response with least RL;
16. Send New_Load to selected CM;
17. Else
18. Send New_Load to least loaded node in home cluster
19. End

4.2 *Worker node algorithm*

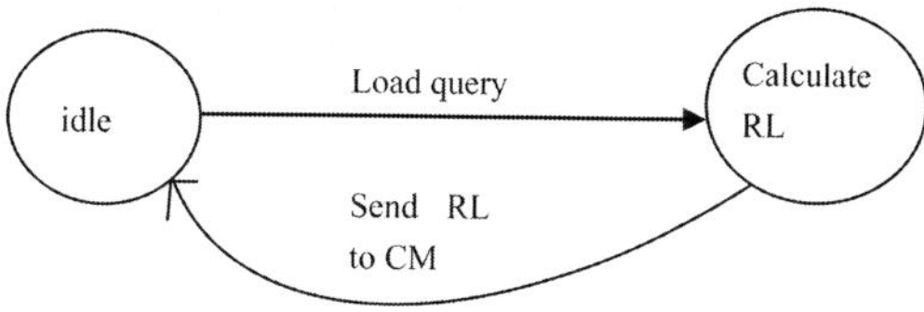

Figure 3. Worker node state machine.

Process Worker node

1. Begin
2. While TRUE
3. Wait for load query from CM.
4. Calculate RL
5. Send RL to CM
6. End

5 ANALYSIS

Let us assume k, m and d be the upper bounds on the number of clusters, nodes in a cluster in the distributed system, and the diameter of a cluster respectively. A node can communicate to its CM in maximum d steps.

Theorem 1. The total time to assign a task is between 2dT + L and (4d + k) T + L where T is the average message transfer time between adjacent nodes and L is the actual average load transfer time.

Proof: Sending query and receiving load from each node in home cluster requires 2d steps. So, total time to load transfer is 2dL. If Intra cluster task scheduling is not possible, CM poll to k – 1 CMs and their response require maximum k hops. Query and response in each cluster requires max 2d steps. So number of hops to load transfer is (2d + k + 2d) resulting in (4d + k) T + L time. If the diameter of cluster decreases this approach produces better result than (Erciyes et al. 2005) in inter cluster task scheduling.

Theorem 2. The total number of messages to assign a task for load balancing is between (2 m) to (2 km + 2(k – 1)).

Proof: Total number of messages in intra cluster scheduling is 2 m. In inter cluster load balancing k – 1 request messages are sent and at most k – 1 replies can be received. Each CM sends and receives replies from each worker node resulting (k – 1) * 2 m messages. So maximum number of messages is (2 m + 2(k – 1) + (k – 1) * 2 m). This is much less than required in (Chatterjee et al. 2015).

6 SIMULATION AND RESULT

Simulation experiments were conducted to evaluate the performance of the proposed 2 LTS policy. The experiments were performed by varying several performance parameters in the system namely the number of worker nodes and the number of jobs.

From figure 4, it can be seen that the proposed algorithm reduces percentage of deadline missed

Table 1. Parameter values.

Parameters	Values
Number of processors	10 ~ 40
Number of tasks	100 ~ 1000
Service time	Exponentially distributed with mean 20 ns
Job inter arrival time	Exponentially distributed with mean 2 ns

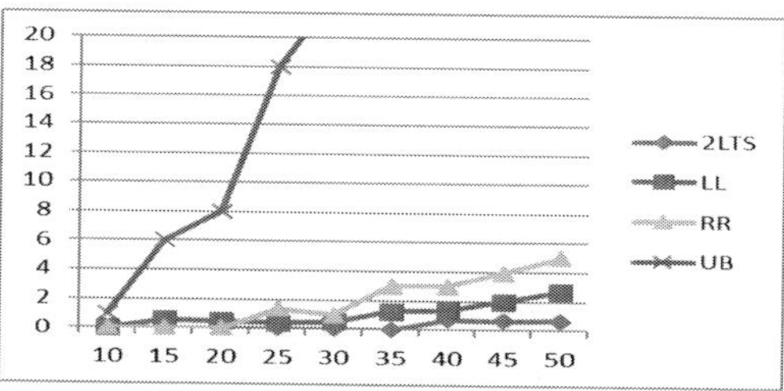

Figure 4. % deadline missed for varying avg. no of tasks per node.

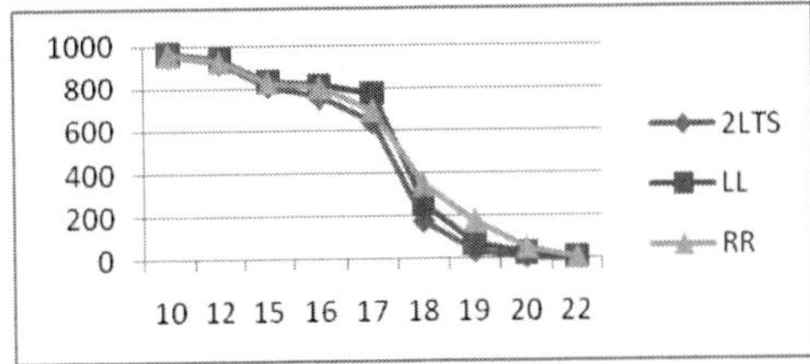

Figure 5. Deadline missed for 1000 tasks.

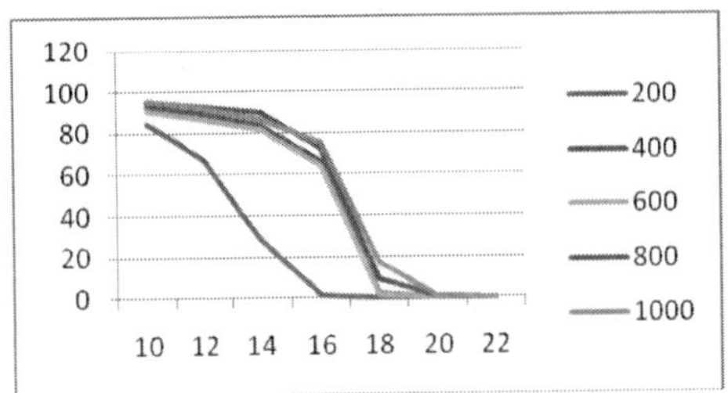

Figure 6. % of missed deadline for varying no of tasks.

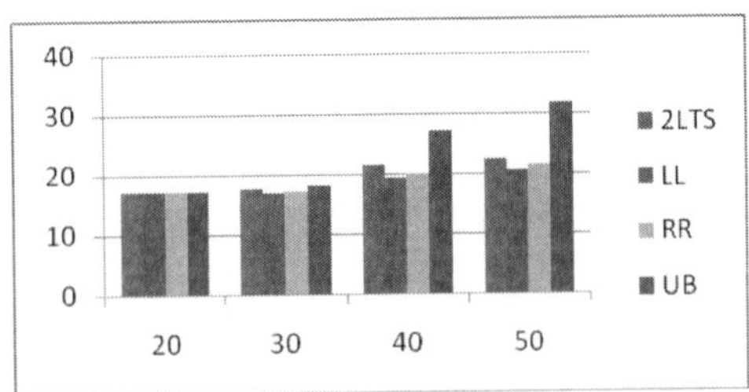

Figure 7. Average turnaround time for varying no of tasks per node.

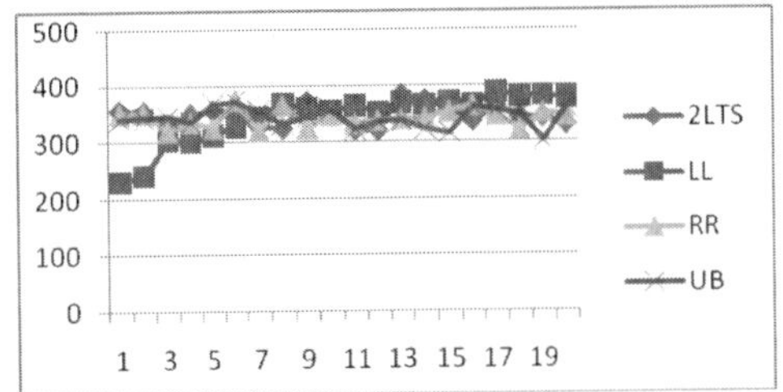

Figure 8. Average load on nodes.

for varying avg. no of tasks per node in compare to unbalanced system and the existing algorithms. For 1000 tasks in varying number of computing nodes the number of tasks missed deadline is represented in figure 5 for 3 different algorithms. It is found that 2 LTS performs better than the existing algorithms. From figure 6 it is observed that performance of the system, in terms of missing deadline, while scheduling is done according to 2 LTS algorithm is not varying much for medium and high load. Figure 7 shows the Average Turn-Around Time (ATAT) for varying no of tasks per node for 3 different algorithms and unbalanced system. In LL algorithm for assigning each new task huge number of message passing is needed, which takes some time and incur cost. That time is not considered while calculating the TAT of each job here. Even though, if is observed that 2 LTS's performance is same as LL. Figure 8 shows the load on each node of the system for varying average workload. It is observed that, while tasks are assigned using 2 LTS algorithm, load on each node is varying a little. So, the aim of balancing load for proper resource utilization is assured. From the graphs it can be inferred that if the nodes are heavily loaded, then the performance of the system is much better in terms of missing deadline than existing algorithms. So taking all variations in consideration it is seen that 2 LTS improves the overall system performance.

7 CONCLUSION

In this paper a semi-distributed task scheduling method for aperiodic tasks is proposed for clustered homogeneous distributed system. New task is assigned a processing node considering the present load status of each node and the slack time of the new task. The task assignment method is scalable and has low message and time complexities. The method of partitioning the system into clusters and the method of load transfer are not addressed. It is assumed that the tasks have no precedence. Cluster Master may fail and due to their important functionality in the proposed model, new masters should be elected. Also, a mechanism to exclude faulty nodes from a cluster and add a recovering or a new node to a cluster is needed. These procedures can be implemented using algorithms as in (Tulani et al. 2000). As a future work I will focus on heterogeneous system and tasks with precedence relations.

REFERENCES

Arabnejad, H. & Barbosa, J.G. 2014. List scheduling algorithm for heterogeneous systems by an optimistic cost table, *IEEE Transactions on Parallel and Distributed Systems*, Vol 25, Issue 3.

Atif, Y. & Hamidzadeh, B. 1998. A Scalable Scheduling Algorithm for Real-Time Distributed Systems, *Proceedings of the 18th International Conference on Distributed Computing Systems*, pp. 352–359.

Baruah, S., Rosier, L. & Varvel, D. 1991. Static and Dynamic Scheduling of Sporadic Tasks for Single-Processor Systems, *IEEE proceedings on Real Time Systems, Eiromicro*, Paris-Orsay.

Chatterjee, M. & Setua, S.K. 2015. A new clustered load balancing approach for distributed systems, *IEEE Conference on Computer, Communication, Control and Information Technology.*

Erciyes, K. & Payli, R.U. 2005. A Cluster-Based Dynamic Load Balancing Middleware Protocol for Grids, *Advances in Grid Computing—EGC*, LNCS 3470, 805–812, Springer-Verlag, Berlin.

Karatza, H.D. & Hilzer, R.C. 2001. Epoch Load Sharing in a Network of Workstations, *IEEE Simulation Symposium*, Proceedings. 34th Annual, 36–42.

Li, X., Jia, Z., Ma, L., Zhang, R. & Wang, H. 2009. Earliest deadline scheduling for continuous queries over data streams, *International Conferences on Embedded Software and Systems.*

Liu, C.L. & Layland, J.W. 1973. Scheduling Algorithms for Multiprogramming in a Hard Real-Time Environment, *Journal of the ACM*, vol. 20, no. 1, pp. 46–61.

Mok, A. & Dertouzos, M. 1978. Multiprocessor scheduling in a hard real-time environment, *7th Texas Conference on Computing Systems.*

Shah, A. & Kotecha, K. 2010. Scheduling Algorithm for Real-Time Operating Systems using ACO, *IEEE International Conference on Computational Intelligence and Communication Networks.*

Tunali, T., Erciyes, K. & Soysert, Z. 2000. A Hierarchical Fault-Tolerant Ring Protocol For A Distributed Real-Time System. *Special issue of Parallel and Distributed Computing Practices on Parallel and Distributed Real-Time Systems*, 2(1), 33–44.

Zeng, B., Wei, J. & Liu, H. 2008. Research of Optimal Task Scheduling for Distributed Real-time Embedded Systems, *IEEE-ICESS.*

Zhang, K., Qi, B., Jiang, Q. & Tang, L. 2012, Real-time periodic task scheduling considering load-balance in multiprocessor environment, *3rd IEEE International Conference on Network Infrastructure and Digital Content.*

Computational Science and Engineering – Deyasi et al. (Eds)
© 2017 Taylor & Francis Group, ISBN 978-1-138-02983-5

A smart strategy of the Automatic Generation Control (AGC)

MD.T. Hoque, A.K. Sinha, D. Sau & T. Halder
Electrical Engineering Department, Kalyani Government Engineering College, West Bengal, India

ABSTRACT: The paper presents on the improved scheme of the automatic generation control in terms of the permissible limit of the power frequency, one or multi-area control, load forecasting, supervisory control robust turbine and generator model tying up with a closed loop system. The consumers wait for good quality of un-interrupted power to capitalize monetary assets inviting huge battery backup storage systems and fast charging of the electrical vehicles during the off peak hour load as cost effective solutions of the power crisis for the green power class managements.

1 INTRODUCTION

The load of the power system changes persistently so the change of power frequency is a regular phenomenon from the generation to consumer areas. A chunk of electricity cannot be stored maintaining the definitive balance between generation and power utilization to control the load frequency. In order to adjust power frequency, the real and reactive power balance is controlled by the generators, Static Var Compensator (SVC) and controlled load shedding (Suvire, 2011).

The generic objective function and performance are adeptly adopted in the governor control scheme to accomplish the dynamic response of the system at which it holds the unrestrained generation outing together with load changes. It also presents the quite quantity of mechanical input to turbines to balance the electrical output of the one and the same generators in terms of the power frequency and angular speed of the turbine maintaining the constant time constant (Lazarewicz, 2010).

The styles of the parameters of the system are encountered within the units to power transfer on the supervisory control over dynamic governor responses (Miller, 2010).

Having the performance optimization of the governors on-line and within allowable droop. Characteristics will minimize the magnitude of frequency variation. The input parameters in the steady state speed drooping characteristics, governor dead-band, initial phase boiler pressure of the vapour units and head hydro power units. If there constantly is a demand response in market price of electricity, there will undeniably be proffers, awards and agreement, but the market will underplays the demand response at which the administrator response is over totally by the time domain getting secondary control signals in response to the disturbances of the system. The random generation between units should be achieved by the security and profitable objectives in connection with frequency to the standard value. The changes of the governor frequency (50 Hz) governor at the set-points of the units should be accomplished by the program values which are ludicrously agreed by the power market (Halder, 2013).

The base power programmed in association with the deployments is incredibly feasible of the power flow, power balance, outstanding regulations, responsive and non-spinning reserves. The AGC of the dead band, gains and frequency biasing parameters of the condition monitoring of the quality services is estimated by the performance parameters (Halder, 2014).

2 COMMON GOVERNOR & TURBINE OF THE GENERATOR SYSTEM

This integrating system has two main parts: GNSS Receiver Module and the Microcontroller and the block diagram of the system is shown in Figure 1.

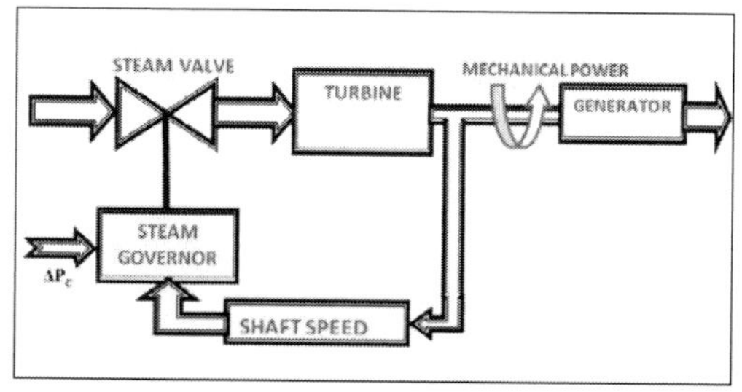

Figure 1. Common governor & turbine of the generator system.

GNSS module continuously transmits serial data through RS232 port in the form of sentences according to the NMEA (National Marine Electronic Association) standards. The information of the latitude, longitude, altitude, velocity and time are available in the \$GNRMC sentence of the NMEA message and the information of the satellites i.e. the information of the Elevation, Azimuth and CNO for each satellites are available in the \$GPGSV and \$GLGSV sentence of the NMEA message. Example NMEA message format is shown in the Figure 2. Proper NMEA message sentence can be chosen to be used by the µC as for processing and use as per the requirement.

The essential transfer function of the turbine model is considered in terms of the droop characteristic (R), time constant of the steam flow (τ_q) given as:

$$G_T(s) = 1/(1 + s\tau_T) \tag{1}$$

Similarly, the fundamental transfer function of the generator is written in terms governor time constant (τ_G) as:

$$G_G(s) = 1/(1 + s\tau_G) \tag{2}$$

The change of maximum power (ΔP_m), reference set point of the steam input, (ΔP_c), angular rotational speed($\Delta\omega$) deviation owing to load change and using the equation (1) & (2) combining written as:

$$\Delta P_m = \left(\Delta P_c - \frac{\Delta\omega}{R} \right) \left\{ 1/(1 + s\tau_T) \times 1/(1 + s\tau_G) \right\} \tag{3}$$

At the steady response of the system reduces using equation (3) yields as:

$$\Delta P_m = \lim_{S \to 0} \left(\Delta P_c - \frac{\Delta\omega}{R} \right) \left\{ 1/(1 + s\tau_T) \times 1/(1 + s\tau_G) \right\} \tag{4}$$
$$= \left(\Delta P_c - \frac{\Delta\omega}{R} \right)$$

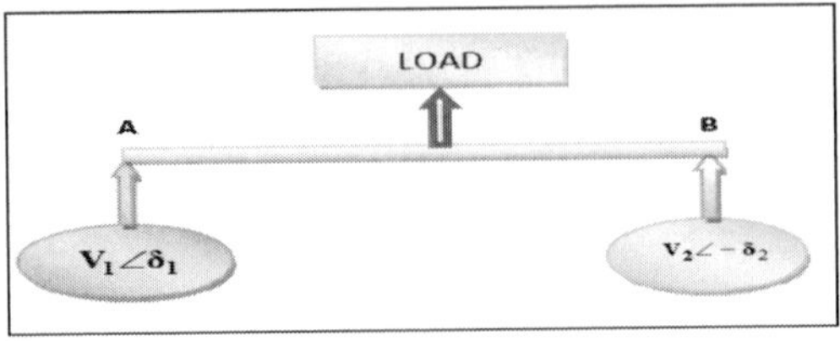

Figure 2. Cantilever modeling of the power flow.

The equation represents the second order system which is required to present on the reheat characteristics of the steam.

3 A MODELING OF THE SYSTEM LOAD

The major load is tremendously reliant of the power frequency of the system so the overall load characteristics may be represented by the load power (ΔP_L), load damping constant (D) as:

$$\Delta P_e = (\Delta P_L + D\,\Delta\omega) \tag{5}$$

4 CANTILEVER MODELING OF THE TIE LINE POWER TO LOAD

The power flows of the two areas are taken into consideration to control two areas as the simplified model as shown in figure.2.

In this modeling, the voltage generation of the generator is considered at generator bus, $V_1\angle\delta_1$ and the load bus voltage is considered $V_2\angle-\delta_2$ through the tie line (AB) of the impedance (jx). Hence, the power flow equation may be established between A and B as:

$$S_{AB} = \frac{V_1\angle\delta_1 \times V_2\angle-\delta_2}{jX} \tag{6}$$

After simplifications, the equation (6) yields as:

$$S_{AB} = \frac{V_1V_2}{X}\left\{\sin(\delta_1 - \delta_2) - j\cos(\delta_1 - \delta_2)\right\} \tag{7}$$

The synchronizing power of the power system of the synchronous generator may be derived as:

$$P_{sp} = \frac{\partial S_{AB}}{\partial \delta_1} = \frac{V_1V_2}{X}\left\{\cos(\delta_1 - \delta_2) - j\sin(\delta_1 - \delta_2)\right\} \tag{8}$$

Hence, the tie line power flow change may write inviting the two equations

5 MODEL OF THE CONTROL CENTER OF THE AGC

The fundamental objective of the auxiliary control is restored balance between load of each area and power generation. If there is change of load of the one area, there should be control action only in one area but not in two areas. Since, the controlled variable of the power system may be involved by the power frequency (f) and direction of line power flows.

Hence, the signal may be expressed in terms of the bias factor (B_1) of the area (A) at which the control area error is written as:

$$A_{E(1)}=\left(\Delta S_{AB}+B_1\Delta f\right) \tag{9}$$

Similarly, the signal may be expressed in terms of the bias factor (B_2) of the area (B) at which the control area error is written as:

$$A_{E(2)}=\left(\Delta S_{AB}+B_2\Delta f\right) \tag{10}$$

6 SELECTION OF THE BIAS FACTOR OF THE SYSTEM

The change of the power frequency variation (Δf) may be same the two areas due to load power charges (ΔP_L) may be written in terms of the drooping characteristics (D_1 & D_2) and equivalent regulations (R_1 & R_2) for the both areas respectively as:

$$\Delta f = \frac{-\Delta P_L}{\left(D_1+D_2\right)+\left(1/R_1+1/R_2\right)} \tag{11}$$

The tie line power flow change (ΔS_{AB}) may be written inviting the equations (10) & (11) while $A_{E(2)} = 0$as:

$$\Delta S_{AB} = \frac{B_2 \times \Delta P_L}{\left(D_1+D_2\right)+\left(1/R_1+1/R_2\right)} \tag{12}$$

The power production reserves reachable for control of power frequency (f) are characteristically only a portion of the capacity of the generation. Moreover, there are reasonable restrictions like boundaries on the rate of rise of the mechanical power of the prime mover power received by the steam turbine to avoid abnormal heat power loss of the pooling coordination. The impulsive changes of 8% may be treated permissible limits and the successive rate of rise is restricted to 1.5% of the plant capacity (MW) per minute. The boiler in a steam prime mover is reasonably sluggish in preserving the steam pressure by escalating fuel input. Thus, the control valves open and restoration of steam pressure is very much unhurried. There are some prohibited operating regions due to effects of the cavitations in hydro turbines.

The AGC may be treated as a control scheme which ensures power frequency digression from the small value to zero and the load flows between the different areas in an interrelated system are regulated. While the governors work comparatively speedy to capture frequency rejection, the AGC comes into play in the power circuit in slow motion.

7 A SMART SCHEME OF THE AGC

A robust but smart system of the AGC is proposed for the multi-area controlled in terms of the different types of the storage and restructured of the power system for the un-interrupted good quality of the power as shown in the figure. 3.

The tertiary power control strategy includes regulating of generator production in terms of the economic load dispatch, unit commitment and power exchange through tie lines to make perfect balance between generation and load demand at any instant of time to control the power frequency in respect of the improved strategy.

The power grids ask for the competent real time electricity markets in terms of the robust power grid services, transient frequency control, excellent regulations, contingency reserves, clogging administration, maximum capacity utilization of the energy storage systems.

When the load demand is lower than the generation, the excess power will be storage in the series of the battery storage, electric vehicle storage, and super capacitor storage using the bi-directional power converters. On the other hand, when the load demand is higher than the power generation, the shortage power will be hurled on grid to meet up demand as perfect balance of load frequency and power quality control. If the storage capacity is filled by the backup systems, the excess power will dispatch to national grids, global grids and pump storage systems to achieve the stability and

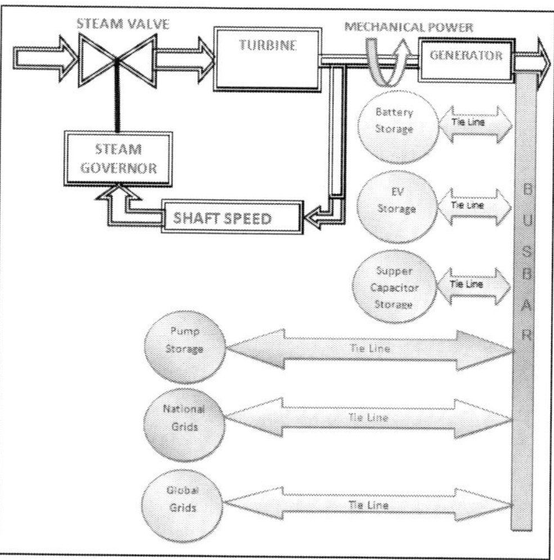

Figure 3. A smart scheme for the AGC.

reliability through the tie line of the power systems as shown in the figure. 3.

The rotational kinetic energy of the generator is written in terms of the moment of inertia (I) angular speed (ω) of the shaft given as:

$$E = 1/2 I\omega^2 \tag{13}$$

If kinetic energy decreases, the angular speed of the shaft decreases vice-versa so the power frequency will reduce with the reduction in terminal speed of the rotor. The speed governor senses the frequency deviation (Δf). Then, Δf is feed back to change the input of the prime mover coupled with generator. The signal (Δf) is fed to prime mover to stay put balance between input real power of the power systems which includes different control components. These segments are coupled to each other through tie lines connecting multi area power system showing by using circles in the figure.3.

8 SMART MATLAB MODEL OF THE AGC

The smart Matlab model is shown to predict on the improved performance of the systems in terms of the performance parameters of the power systems.

This model highly ensures on the AGC using power electronic based storage system as cost effective solutions of the power generation, demand side managements and review of the suitable storage system skills.

9 BATTERY STORAGE SYSTEM FOR AGC

The surplus energy is stored to use for different time when it was generated adequately. The efficient process of power conversions are to be adopted form resources that some energy is vanished due to inefficiency and heat loss. storage is avoided to have a more efficient process of Time-of-Day (TOD) metering expected in the substantial forthcoming as online measurements adopting the huge digital meters for the energy assessments from peak hour to low power utilization times for the benefit of the societies so that power utility and consumers be rewarded by minor tariffs of electricity in terms of the open market policy of the energy market.

10 SUPPER CAPACITOR STORAGE SYSTEM FOR THE AGC

The supper capacitors are the very high value of capacitance banks (C) which are the best competent to store electricity within shortest interval of the time ranging from periods of 0.1 second to 10 seconds upto voltage level (V) as:

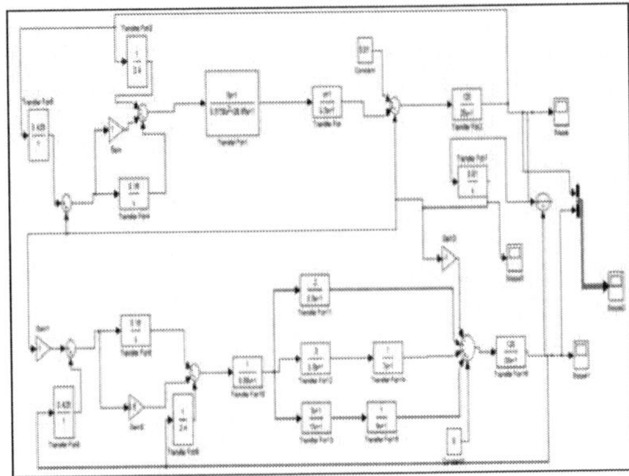

Figure 4. A smart model for the AGC.

$$E_C = \tfrac{1}{2} CV^2 \tag{14}$$

The typical value of the super capacitor is now reached at about 3.1 kF for the carbon electrode surface area ranging from1000 m^2 to 2000 m^2 per gram for the high value of capacitance.

11 PUMPED STORAGE POWER PLANT FOR THE AGC

The extraordinary turbines are use to run either to spin an alternator or to act as a pump. This reversibility agrees to excess electrical energy to be used to pump water to a higher reservoir to be utilized as an energy source later during the time of energy crisis. Since 2.5 foots of the altitude has a bed pressure of one pound per square inch (psi), a head height of 200 ft is considered to an equivalent to 86 psi.

As for example, Japan has built a power plant of 30 MW at seawater pumped hydro system at Yanbaru in 1999 to maintain the generation and demand side managements.

The pumped hydro worldwide scenario is envisaged is about 90 GW storage. The majority pervasive of the high-energy storage technique are not only invited to control AGC but also meet up energy crisis for the green power revolution.

12 RESULTS

The power frequencies change of the generator as well as load are constant due to fluctuation of load and generation inviting the huge energy storage

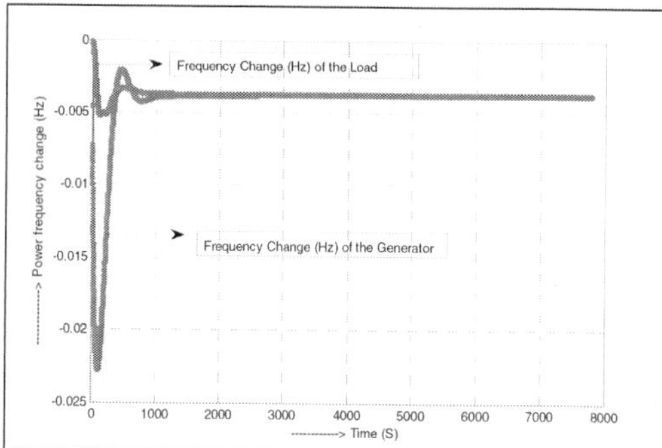

Figure 5. Generator & load frequency change of the power system.

systems to make perfect power balance between load and generator side as any duration of the load-demand managements. The Matlab simulation results are stodgily carried out to forecast the above performance as shown in the figure. 5.

The energy storage is to be kept away from the power losses but it may be a cost effective solutions when load time-shifting is feasible, Energy must be stored in the electric vehicles with smart grids since they can store sufficient power from the standard power systems to advancement the demand response curves inviting the power quality in terms of the performance parameters evaluations.

The energy storage power devices are charged up when they soak up energy either directly from the power generators or indirectly from the power grids maintaining bi-directional power flow. The storage systems discharge when they hurl the stored energy back into the power grids to achieve constant load frequency and AGC control for the multi-areas. Charging and discharging usually need power electronic converters to convert electrical energy (ac or dc) into a different form of the electrical, thermal, mechanical or chemical energy for the unique power storages as closed agreement of the negligible power frequency derivations as also shown in the figure.5.

13 CONCLUSIONS

The smart concepts of load frequency control of the AGC are demonstrated to strengthen the reliability and stability inviting huge backup storage systems. The economic load dispatch of the power system is enriched newer energy storages and managements systems in the modern power generation to save power loss and energy conservations. The newer technologies may swell boost ratio of the energy storage density at a lower cost. Both are needed for a viable product in near and far future for the robust Automatic Generation Control (AGC) and Load Frequency Control (LFC).

REFERENCES

Halder T., "A case file on a smart grid" IEEE International Symposium on Industrial Electronics (ISIE), 2013, pp. 1–6.

Halder T., "A Strategy of Smart and Hybrid Power Distribution Systems" IEEE Proceedings of 2014 1st International Conference on Non Conventional Energy (ICONCE 2014) pp. 210–215.

Lazarewicz M. L. and Ryan T. M., "Integration of flywheel-basedenergy storage for frequency regulation in deregulated markets", Proc. IEEE Power Energy Soc. Gen. Meet., 2010.

Miller N., Manz D., Roedel J., Marken P. and Kronbeck E., "Utility scale battery energy storage systems", Proc. IEEE Power Energy Soc. Gen. Meet., pp. 25–29, 2010.

Suvire G. O. and Mercado P. E., "Combined control of a distributionstatic synchronous compensator/flywheel energy storage system for wind energyapplications", IET Gener., Transm., Distrib., vol. 6, no. 6, pp. 483–492, 2011.

Computational Science and Engineering – Deyasi et al. (Eds)
© 2017 Taylor & Francis Group, ISBN 978-1-138-02983-5

Porter's stemmer—a review

Sourav Kumar Mitra, Payel Saha, Sayan Mukhopadhayay & Debabrata Datta
Department of Computer Science, St. Xavier's College (Autonomous), Kolkata, India

ABSTRACT: Data mining is an application domain used to find any relevant information from a large collection of data. Data mining techniques are used to implement and solve different types of research problems related to knowledge discovery from data. Text mining is one of the research areas related to data mining. Text mining is the process of mining useful information from text documents. Text documents are unstructured. So, for mining useful information from text documents, preprocessing techniques are applied to the text documents to convert the raw textual data into a structured document. Stemming is a preprocessing technique that maps various morphological forms of a word to the stem of the word. In 1980, M. F. Porter proposed an algorithm for stemming words from an English text. This paper discusses about the stemming process and the first algorithm proposed by Porter.

1 INTRODUCTION

Most of the present day systems, which use indexing and searching a text, consider word stemming to be an important feature. This feature is closely attached to an Information Retrieval (IR) environment. In such an environment, there has to be a collection of documents, each described by the words in the document title and possibly the words in the document abstract. Even if the issue of precisely where the words originate is ignored, it can be said that a document is represented by a vector of words or terms (Mohammed, 2002). These terms are used to index a document and find that document. The stem or root of a word is the part of the word that is common to all its inflected variants. Terms with a common stem will usually have similar meaning. For example, the following four words have common part 'CONNECT':

CONNECT
CONNECTED
CONNECTING
CONNECTION

The performance of an IR system can be improved by conflating term groups such as this into a single term. Stemming is the process in which various morphological forms of a word are conflated into its stem. Stemming is usually done by removing any attached suffixes and prefixes (affixes) from index terms before the actual assignment of the term to the index. Since the stem of a term represents a broader concept than the original term, the stemming process eventually increases the number of retrieved documents in an IR system. In the process of stemming, it is assumed that the morphological form of a word and the stem of the word are semantically related and hence the former is converted to the latter. It is not necessary that the stem has to be an existing word in the dictionary but all its variants should map to this form after the stemming has been completed. There are two points to be considered while using a stemmer:

- Morphological forms of a word are assumed to have the same base meaning and hence should be mapped to the same stem.
- Words that do not have the same meaning should be kept separate.

These two rules are good enough as long as the resultant stems are useful for text mining or language processing applications. Stemming is generally considered as a recall-enhancing device. For languages with relatively simple morphology, the influence of stemming is less than for those with a more complex morphology. Most of the stemming experiments that have been conducted so far are for English language as well as some other western European languages (Ayres, 2002).

2 BACKGROUND STUDY

There are different methodologies in the field of text mining. A comparative study of some of the important methodologies has been done in (InjeBhushan, 2013). It has been observed that in the general information retrieval process, the parameters that are required are the models, documents and queries in the form of TF-IDF vectors (Joachims, 1997). There are other ways like sequential pattern mining (Huang, 2003), close pattern mining (Yan, 2003), frequent item-set mining (Borgelt, 2005), maximal pattern mining (Mohammed) etc. Each methodology demands different ways of taking parameters. A comparatively newer approach is based on association rule mining as described in (Agrawal, 1994) and is dependent on parameters like support value and confidence value. Each methodology has its own advantages and limitations. But each method needs to identify the frequent as well as the most important items depending on the requirement of applications for which the text mining is to be used.

3 ERRORS IN STEMMING

There are mainly two errors in stemming—over-stemming and under-stemming. The error related to over-stemming happens when two words with different stems are stemmed to the same root. This error is categorized as a false positive method. On the other hand, the error related to under-stemming happens when two words that should be stemmed to the same root are stemmed to the different roots. This type of error can also be categorized as a false negative one (Ayres, 2002).

4 PORTER'S ALGORITHM

Porters algorithm (Mohammed, 2002) used for stemming was proposed by M. F. Porter in 1980 and since then it has been one of the most popular stemming algorithms. But after its initial proposal, many modifications and enhancements have been made on the basic algorithm. It is based on the idea that the suffixes in the English language (approximately 1200) are mostly made up of grouping of smaller and simpler suffixes. It has five steps, and within each step, rules are applied until one of them passes the conditions. If a rule is accepted, the suffix is removed consequently, and the next step is performed. The resultant stem at the end of the fifth step is returned (Zaki, 2001).

Porter's algorithm is a suffix stripping algorithm. It uses a list of suffixes which is explicitly provided and a criterion is maintained with each suffix under which it may be removed from a word to leave the stem.

In any suffix stripping program is used for applications related to information retrieval, two aspects are to be followed. The first of them is that the suffixes are removed to improve the performance related to information retrieval and not as a linguistic exercise. So, even if it is possible to determine the suffixes of the words, it would not be evident under what circumstances a suffix should be removed. Conceivably, it would have been the best criterion to remove suffixes from the two words W_1 & W_2 to produce a single stem S, is to say that it may be done if there appears to be no difference between the two statements 'a document is about W_1' and 'a document is about W_2'.

So, if W_1 = 'CONNECTION' and W_2 = 'CONNECTIONS', it seems very reasonable to combine them to a single stem.

On the other hand, if W_1 = 'RELATE' and W_2 = 'RELATIVITY', it would be perhaps unreasonable to combine them, especially if the document collection is concerned with applications related to theoretical physics.

There is another issue related to the use of the suffix list with various rules. In such a case, the success rate for the suffix stripping will be considerably less than 100%, irrespective of how the process is evaluated. For example if SAND and SANDER are the words to be conflated, the probability of doing so for WAND and WANDER will also increase. The error here is that "ER" in WANDER has been treated as a suffix when in fact it is part of the stem. Equally a suffix may completely alter the meaning of a word, in which case its removal is not desirable. For example, if two words PROBE and PROBATE are considered, they have quite distinct meanings in modern English (Mohammed, 2002).

Considering all the above issues, the algorithm has five major steps as stated below:

In the first step the algorithm tries to remove the inflectional suffixes. Inflectional suffixes are those suffixes which do not change the meaning of the original word. To exemplify, the following sentences are considered:

"Every day I walk to school" & "Yesterday I walked to school". The word "walk" & "walked" have the same base meaning.

In the next three steps the algorithm tries to remove the derivational suffixes. Derivational suffixes are those that change the meaning of the original word. But the new meaning is related to the old meaning. The new word is usually a different part of speech. For example, derive (verb) + ation → derivation (noun), derivation (noun) + al → derivational (adjective).

The last step is a recoding step.

5 MERITS AND DEMERITS

Like any existing algorithms, Porters algorithm also has its merits as well as demerits. As far as the merits are concerned, the following points may be considered:

1. This algorithm is simpler than an earlier stemmer, the Lovin's stemmer that contains no less than 294 suffixes and each of which is associated

with one of the 29 context sensitive rules that determine when that suffix can be or cannot be removed from the end of a word. In addition, the algorithm also contains 35 recoding rules (Mohammed). Porter's stemmer uses 60 suffixes, two recoding rules and a simple general form of suffix stripping rule.
2. Porter's algorithm is also easier to understand and implement using programming languages such as C or Java.
3. When the stem is too short, the algorithm does not remove a suffix

On the other hand, the demerits of the algorithm may be discussed with the help of the following points:

1. It only removes suffixes. It cannot remove prefixes which are also present in English language.
2. The problem of under-stemming exists in this stemmer.
3. There is no linguistic basis for this approach. It was just observed that the length of the stemmer could be used in an effective way to choose whether or not it would be worthwhile to remove a suffix (Mohammed, 2002).

6 CONCLUSION

Stemming is used as an estimated approach for grouping together the words with a similar basic meaning. But in some cases, different words with the same morphological stem have vernacular meanings which are not closely related. Stemmers are common elements for query systems such as web search engines. The Google search technique has adopted word stemming in 2003. Further, stemmers have been implemented for many languages as well. But stemmers become harder to design as the morphology, orthography, and character encoding of the target language become more complex.

REFERENCES

Agrawal R. and Srikant R., "Fast Algorithms for Mining Association Rules in Large Databases", In Proceedings of 20th International Conference on Very Large Data Bases, pp. 478–499, 1994.

Ayres J., Gehrke J., Yiu T. and Flannick J., "Sequential pattern mining using a bitmap representation", In Proceedings of the 8th ACM SIGKDD, pp. 429–435, 2002.

Borgelt C., "An Implementation of the FP-growth Algorithm", Workshop on Open Source Data Mining Software, ACM Press, USA, 2005.

Gouda Karam and Mohammed J. Zaki, "GenMax: An Efficient Algorithm for Mining Maximal Frequent Itemsets", In proceedings of International Conference on Data Mining and Knowledge Discovery, pp. 1–20, 2005.

Huang Y. and Lin S., "Mining Sequential Patterns Using Graph Search Techniques", In Proceedings of 27th International Conference on Computer Software and Applications, pp. 4–9, 2003.

Han J., Pei J., and Yin Y., "Mining Frequent Patterns without Candidate Generation", In Proceedings of ACM SIGMOD International Conference on Management of Data, pp. 1–12, 2000.

InjeBhushan V. and Prof. Mrs. Ujwalapatil, "A Comparative Studyon Different Typesof Effective Methods in Text Mining: A Survey", International Journal of Computer Engineering & Technology, Volume 4, Issue 2, pp. 535–542, March–April, 2013.

Joachims T., "A Probabilistic Analysis of the Rocchio Algorithm with tfidf for Text Categorization", In Proceedings of the 14th International Conference on Machine Learning, pp. 143–151, 1997.

Li Y. and Zhong N., "Interpretations of Association Rules by Granular Computing", In Proceedings of IEEE Third InternationalConference on Data Mining, pp. 593–596, 2003.

Mohammed J. Zaki and Ching-Jui Hsiao "CHARM: An Efficient Algorithm for Closed Itemset Mining", In Proceedings of the 2nd SIAM International Conference on Data Mining, USA, 2002.

Mohammed J. Zaki, "Mining Closed & Maximal Frequent Itemsets" NSF CAREER Award IIS-0092978, DOE Early Career Award DE-FG02–02ER25538, NSF grant EIA- 0103708.

Yan X., Han J., and Afshar R., "Clospan: Mining Closed Sequential Patterns in Large Datasets,", In Proceedings of International Conference on Data Mining, pp. 166–177, 2003.

Zaki M., "Spade: An Efficient Algorithm for Mining Frequent Sequences", Machine Learning, vol. 42, pp. 31–60, 2001.

Zhong Ning and Li Yuefeng, "Effective pattern discovery in text mining", IEEE Transactions on Knowledge and Data Engineering, Vol. 1, 2012.

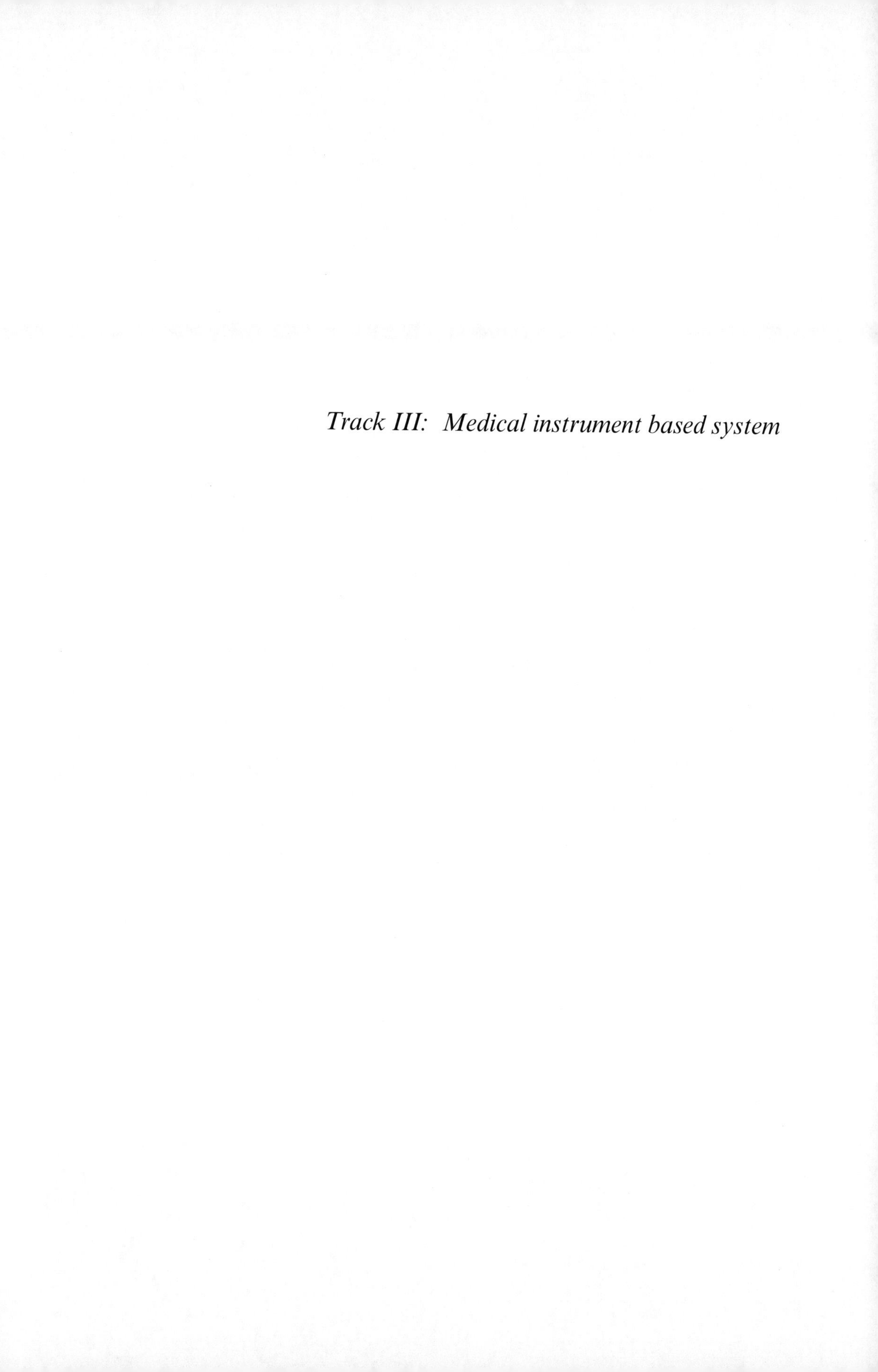

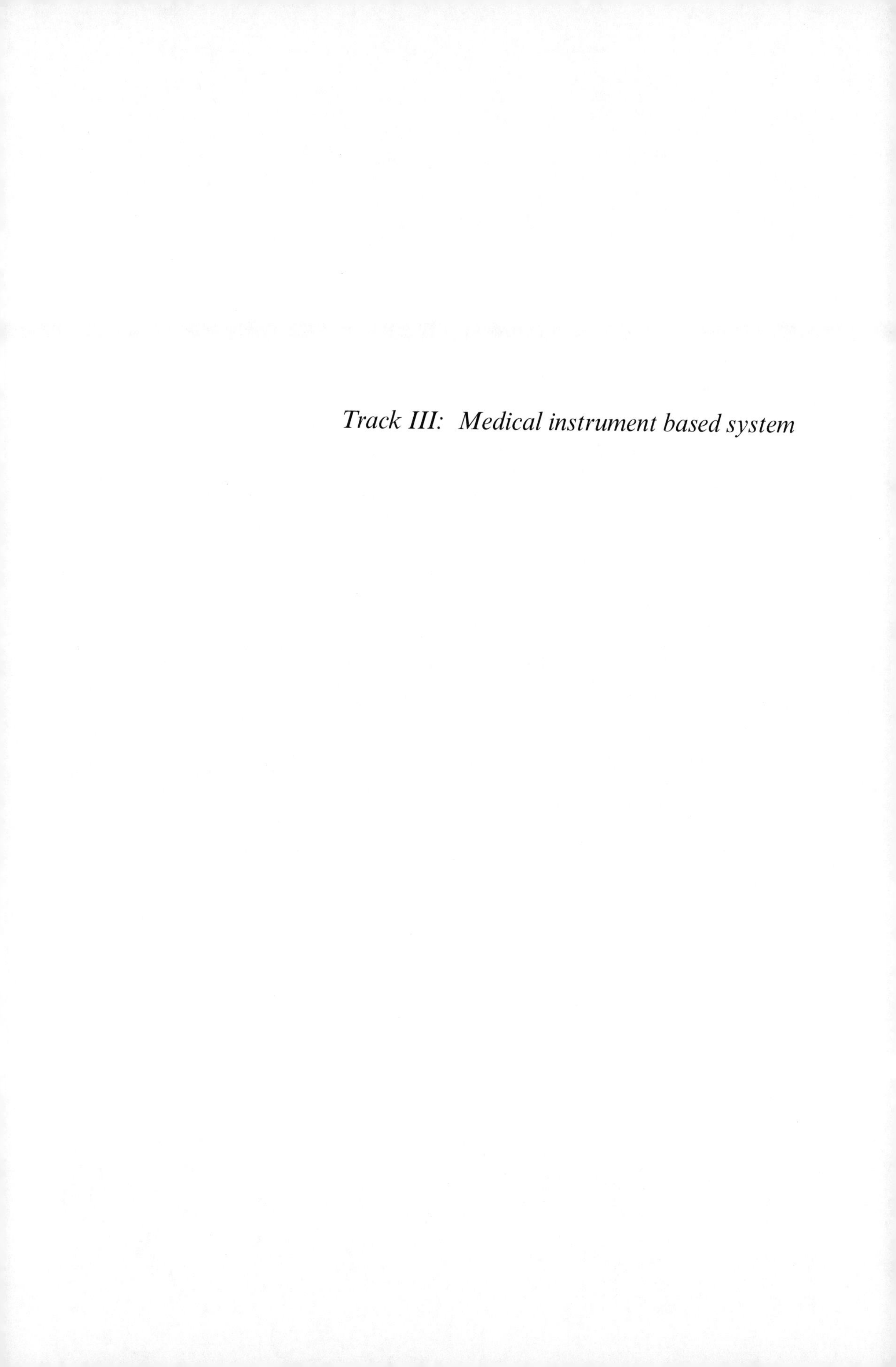

Track III: Medical instrument based system

Computational Science and Engineering – Deyasi et al. (Eds)
© 2017 Taylor & Francis Group, ISBN 978-1-138-02983-5

Electrocardiogram coder using wavelet transform of average beat and residuals

Subhashis Das, Sohel Das & Rajarshi Gupta
Department of Applied Physics, Instrumentation Engineering, University of Calcutta, West Bengal, India

ABSTRACT: Electrocardiography (ECG) is the most popular noninvasive technique for primary stage investigation on cardiac functions. ECG data compression finds wide application in patient monitoring and heart rhythm analysis. In this paper, we present an efficient compression method for single lead Electrocardiogram (ECG) signals based on the Discrete Wavelet Transform (DWT). Since the ECG consecutive beats exhibit sufficient redundant information, DWT was applied on an average beat estimate, computed over 10 beats and the residuals of the corresponding original beats. Selections of the significant coefficients from the average beat as well as the residuals were based on Energy Packing Efficiency (EPE). Finally, the selected coefficients for the average beat were quantized at 8-bit, and the residuals with 5-bit and compressed by a delta coder. The significance maps were encoded by binary coder and run length encoder. The algorithm was validated with one min. mitdb data under Physionet at 360 Hz sampling to obtain an average Compression Ratio (CR) of 8.65, PRD of 2.32, PRDN of 4.30. The reconstructed signals were clinically validated by cardiologists.

1 INTRODUCTION

The Electrocardiogram (ECG) is generated from the Sinoatrial (SA) node of heart, and an important physiological signal for cardiac function diagnosis. Computerized analysis of ECG (Gupta, 2011) can significantly contribute towards assisted diagnosis and in early detection of many cardiac diseases. For monitoring of heart rhythm and presence of abnormal patterns continuous recording of ECG is required. Compression of ECG data is essential in storage memory optimization and real-time health monitoring applications. The primary aim of compression is to reduce the redundancy in ECG data as much as possible while preserving the necessary diagnostic features. Important diagnostic information mainly lie in the amplitudes, durations and curvatures of component waves P, QRS, T and wave segments PQ, ST and TP.

Many ECG data compression methods have been proposed over the last five decades. The algorithms for ECG data compression can be divided into two major categories, viz., direct method and transform methods (Gupta, 2014). In direct methods, the time samples are compressed by eliminating the redundancies between a group of adjacent samples. Some of the direct methods of ECG compression are FAN algorithm (Giallorenzo, 1988), amplitude zone time epoch coding (AZTEC) (Cox, 1968), Coordinate Reduction Time Encoding System (CORTES) (Kortman, 1967), and analysis by synthesis ECG compressor (ASEC) (Zigel, 2000) etc. An excellent review and comparison of ECG compression methods is presented in (Jalaleddine, 1990). In transform methods, time domain ECG signal is mapped into a different domain and the transformed coefficients are compressed. A large number of transform methods have been reported, like Karhunen-Loeve Transform (KLT) (Degani, 1990), wavelet transform (Manikandan, 2014; Rajoub, 2002), Discrete Cosine Transform (DCT) (Allen, 1992) and Fast Fourier Transform (FFT) etc. Parameter extraction based ECG compression techniques extract some clinically significant features from ECG data. These features are used to reconstruct the signal. Methods belonging to this group are: peak picking, linear prediction methods, syntactic methods, neural nets methods and Long Term Prediction (LTP) (Imai, 1985).

In the last decade, wavelet based compression technique became very popular due to its abilities like time-frequency localization, energy compaction in small number of coefficients, cross sub-band similarity, etc. With such attractive properties, wavelet transform is recognized as one of the most powerful tool for modern digital signal processing. Set Partitioning In Hierarchical Trees (SPIHT) algorithm (Lu, 2000) based on wavelet transform achieved better Compression Ratio (CR) along with minimum Percent of Root-mean-square Difference (PRD).

A real time compression of ECG data based on wavelet transform as required for telecardiology applications is presented in (Alesanco, 2006). Most of the signal energy is packed within only few coefficients. The work (Chompusri, 2009) shows the selection of those coefficients based on Energy Packing Efficiency (EPE).

In this paper, we present a compression technique for single lead ECG data based on the Discrete Wavelet Transform (DWT) and residual encoding. An ECG signal comprises of sufficient inter beat redundancy except the occurrence of abnormal beats. Hence, instead of compressing the individual beats, an average beat was generated from a fixed number of beats and the residuals were compressed along with the average beat. The elements of the residual signal, obtained by subtracting an ECG beat from the corresponding average beat, are of very low magnitude. For a fixed quantization error, number of bits required to encode a sample value is directly proportional to its magnitude. Hence, residual encoding technique for ECG data helps to obtain good compression ratio with desired signal quality. Layout of the paper is as follows: Detail methodology of the proposed algorithm is presented in section II. Section III includes results and using benchmark data. The conclusion section highlights salient contribution of the work.

2 METHODOLOGY

The proposed method can be subdivided into a number of steps, as follows:

2.1 *Preprocessing and beat extraction from ECG dataset*

The high frequency noise was removed from the ECG dataset using 4th order Butterworth low pass filter with a cutoff frequency of 0–70 Hz.

R-peak detection was performed using Pan-Tompkins algorithm (Pan, 1985). A beat is defined between two consecutive baseline points, located in the TP segment. Individual beats from the ECG data was extracted to form a beat-matrix, where beats were aligned about their R-peaks. Since the heart rate varies even for a normal patient, the extracted beats were padded by the first and last sample value at the respective ends. The kth ECG beat vector is represented as:

$$beat_k = \left[x_{1k}\ x_{2k} \cdots\cdots x_{pk}\right]^T \tag{1}$$

where, x_{jk} = jth sample of kth ECG beat

After padding and beat alignment with respect to the R-peak, the beat matrix B is represented as,

$$B = \left[beat_1\ beat_2\ \ldots\ldots\ beat_n\right] = \begin{bmatrix} x_{11}\ x_{12} \cdots\cdots x_{1n} \\ x_{21}\ x_{22} \cdots\cdots x_{2n} \\ \cdot \quad \cdot \quad \cdot \\ x_{m1}\ x_{m2} \cdots\cdots x_{mn} \end{bmatrix} \tag{2}$$

where, $beat_k$: modified beat vector, x_{ij} = ith sample of jth beat.

2.2 *Preprocessing and beat extraction from ECG dataset*

An average beat was calculated by taking the running average of ten consecutive beats, as follows:

$$av\_beat_{(i)} = \frac{1}{2}\left[av\_beat_{(i-1)} + beat_i\right] \tag{3}$$

$$x(j)_{av\_beat,\,i} = \frac{1}{2}\left[x(j)_{av\_beat,\,(i-1)} + x(j)_{beat_i}\right] \tag{4}$$

where, $av\_beat_1 = beat_1$ and $i = 2$ to 10. In equation (4) $x(j)_{av\_beat,i}$ is the jth sample value of the average beat at the ith iteration and $x(j)_{beat,i}$ is the jth sample value of ith beat. Residual beat was computed by subtracting the corresponding beat from the final updated average in the group of 10 beats.

$$res\_beat_i = \left[av\_beat_{(final)} - beat_i\right] \tag{5}$$

$$y(j)_i = \left[x(j)_{av\_beat,\,final} - x(j)_{beat_i}\right] \tag{6}$$

where, $i = 1$ to 10. In equation (6), $y(j)_i$ represent the jth sample value of ith *residue* $x(j)_{av\_beat,\,final}$ is the jth sample value of the final updated average beat. Final average beat is the last updated average for a fixed number of beats.

2.3 *Wavelet Transform of the average beat and residuals*

The Discrete Wavelet Transform (DWT) decomposes a signal $f(t)$ into a weighted sum of basis functions which are dilated and translated versions of the analysis function called the "*mother wavelet*". Dilation is accomplished by multiplying t by some scaling factor s^v, where $v \in Z$. Translation is accomplished by varying the parameter k. All these together give the *wavelet decomposition* of a signal.

$$f(t) = \sum_v \sum_k C(v,k)\psi(s^v t - k) \tag{7}$$

where, $k \in z$ and C is the wavelet coefficient. In the current work, "coiflet" was used as the mother wavelet to its high similarity with QRS morphology. At the 4th level of decomposition most of the ECG morphologies were visually prominent. Selection of the wavelet coefficients was done based on the Energy Contribution (EC) of the coefficients towards the signal reconstruction. A predefined threshold was fixed to select the coefficients from the 4th level. The remaining coefficients were discarded (hard thresholding). Energy of a sub-band say A4, containing approximation coefficients, was calculated by the help of the following equation:

$$E(A4) = \sum_{i=1}^{n} a_i^2 \tag{8}$$

Where, a_i is the ith coefficient of A4. Hence, relative energy contribution from a sub-band was estimated as:

$$ECE(A4) = \frac{E(A4)}{[E(A4)+E(D4)+E(D3)+E(D2)+E(D1)]} \tag{9}$$

Figure 1 shows the ECE of mitdb 117 lead II from different sub-bands after performing wavelet transform of the average ECG beat, while the average is estimated for the each set of 10 consecutive ECG beats. It shows that more than 99% energy comes from the sub-band *A4* and remaining 1% energy comes as the added result of other sub-bands. Again, within a sub-band, only few coefficients are dominantly contributing towards total sub-band energy. Hence, these dominant coefficients were selectively chosen using a predefined threshold of 99% by arranging them in descending order of magnitude. The insignificant coefficients were hard thresholded and a Binary Significance Map (BSM) were generated as follows:

$$BSM = [b_1\ b_2 \ldots\ldots\ldots b_n] \tag{10}$$

where, $b_k = 0$ for insignificant coefficients; $= 1$ for significant coefficients.

2.4 *Quantization and encoding of the wavelet coefficients*

The coefficients of average beat and residuals were quantized using 8-bits and 5-bits respectively. The quantization was done as follows:

$$q(k) = \frac{2^b - 1}{c_{\max} - c_{\min}} \times [c(k) - c_{\min}] \tag{11}$$

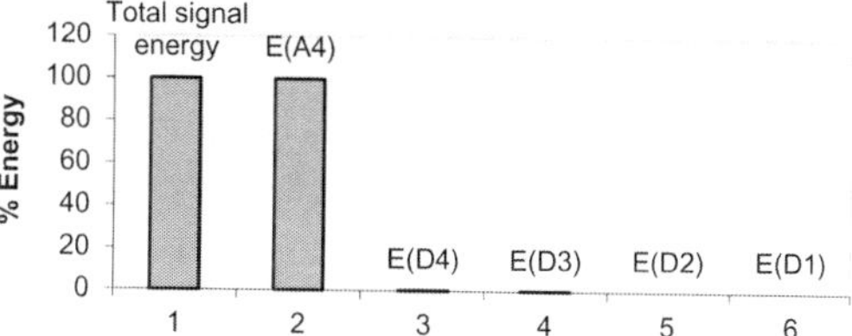

Figure 1. Energy distribution of mitdb 117 in various subbands.

where, $b = 8$ for wavelet coefficients of average beat and $b = 5$ for residuals beats, C_{max} and C_{min} are maximum and minimum of C array.

The quantized value for the average beat was encoded by delta coder as follows:

$$\Delta x_i = x_i - x_{(i-1)} \tag{12}$$

where, i starts from 2. Delta coding reduces the variance (range) of the values as the neighbouring samples are correlated, enabling a lower bit uses for the same data. Δx can take both positive or negative values depending on the operands. In this method we stored the $\Delta x[.]$ by their magnitudes only. The sign map containing the sign information of Δx was generated as follows:

$$\begin{aligned} sign\_map[i] &= 1 \ if\ \Delta x_i < 0 \\ &= 0 \ if\ \Delta x_i \geq 0 \end{aligned} \tag{13}$$

The sign map finally encoded by binary coder. The Binary Significance Map (BSM) regarding the position of the significant wavelet coefficients for both the average and residual beat were encoded by run length encoding technique as follows:

$$\begin{aligned} bsm\_e &= 150 + n\_1; \\ &\quad \text{for consecutive } n\_1 \text{ of 1s} \\ bsm\_e &= 200 + n\_0; \\ &\quad \text{for consecutive } n\_1 \text{ of 0s} \end{aligned} \tag{14}$$

where, *bsm_e*: encoded BSM element.

2.5 *Packet formation of the compressed data*

To embed essential information for decoding, header bytes were constructed and appended with respective compressed bytes to form discrete packets. The header structures are shown in Table I. Decompression was done in exactly inverse way as the compression.

Table I. Wavelet packet structure and header byte allocation.

Primary header	Header for average beat	Compressed data for average beat	Header for residue signals	Compressed data for residues
9-byte	2-byte	Not fixed	7-byte	Not fixed

Table II. Summary of the compression and decompression results for lead ii.

MIT-DB Record no.	CR	PRD	PRDN	MSE × (10^{-3})	MAE (in mV)
100	8.202	1.952	3.293	0.040	0.024
101	8.660	1.810	3.138	0.037	0.023
102	8.950	3.417	6.226	0.174	0.081
111	8.394	3.112	5.857	0.151	0.099
112	7.454	4.041	7.525	0.348	0.057
116	7.937	1.654	2.527	0.028	0.072
117	11.24	1.478	3.187	0.040	0.043
119	7.695	2.722	4.542	0.086	0.048
121	10.02	1.616	2.209	0.020	0.045
234	8.037	1.476	4.503	0.083	0.062

3 RESULTS AND DISCUSSION

The compression and decompression algorithms were tested with one minute records from MIT-BIH arrhythmia (mitdb) data from Physionet (Physionet). The mitdb data contains annotated 30 min two-channel ECG recordings from 48 subjects with arrhythmia. This database has been widely used by researchers working on ECG compression. The quantitative analysis of compression and reconstruction performances was checked by computing the Compression Ratio (CR), Percentage Root mean square Difference (PRD), Percent Root mean square Difference Normalized (PRDN), Mean Square Error (MSE) and Mean Absolute Error (MAE) (Gupta, 2011). Among them, the CR directly presents compressors' efficiency in reducing the volume of original data. PRD, MSE and PRDN shows error in reconstruction computed over whole dataset. MAE shows local sample to sample reconstruction error, in terms of maximum deviation of the original corresponding reconstruction signal.

Table II shows these parameters with mitdb data set. With the proposed method we achieved an average compression ratio (CR) of 8.65 along with average PRD of 2.32, PRDN of 4.30, MSE of 0.10×10^{-3}, and MAE of 0.05 mV. Observation of Table II shows that except 102, 111 and 112, the PRDN is below 5. The highest compression was achieved in record 117 with a value of 11.24.

In the proposed study, the reconstruction was performed with 99% of ECE. The reconstruction quality is represented in Figure 2 for three mitdb data, viz., 100, 121 and 101. The reconstructed signal is superimposed on the original signal to have a direct assessment of reconstruction quality.

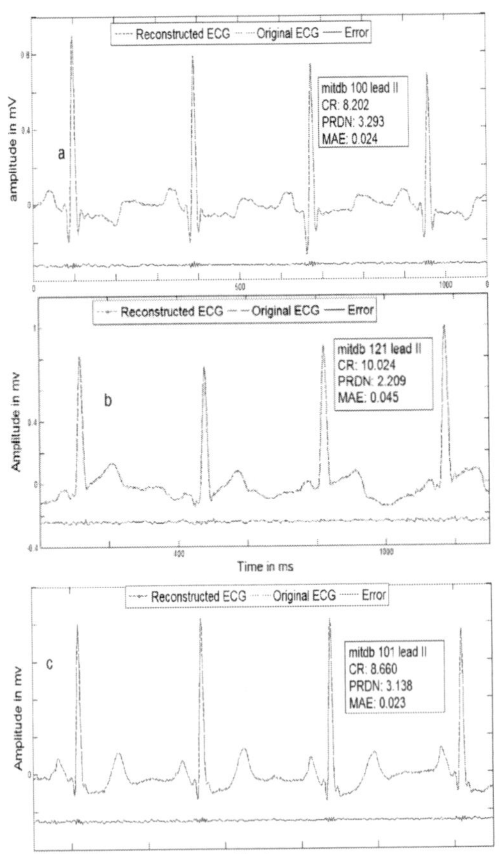

Figure 2. Original and reconstructed data from mitdb database (for clarity in presentation the error signal is plotted with an offset).

Table III. Variation of CR and PRD with different values of energy restoration.

mitdb record no.	ECE	99%	98%	97%	96%	95%
117	CR	11.402	11.704	11.783	11.830	11.837
	PRD	3.674	3.765	4.067	4.417	4.464

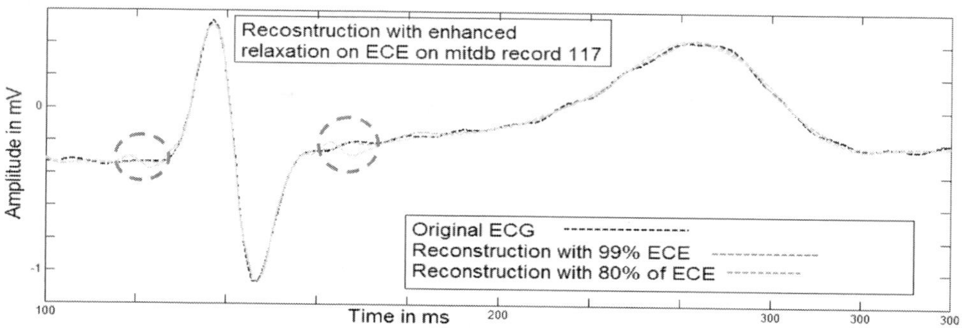

Figure 3. Reconstruction performance with extended ECE up to 80% with mitdb 117 lead II.

Table IV. Comparison with similar works.

Algorithm	CR	PRD (%)
Rajoub (Rajoub, 2002)	9.2:1	0.86
Lu (Lu, 2000)	8.1:1	1.01
Chompusri (Chompusri, 2009)	3.6:1	0.27
Hilton (Hilton, 1997)	8.1:1	2.6
Proposed	8.6:1	2.32

Additionally, a separate plot for sample to sample reconstruction error is shown with a negative offset. The plots show that the reconstruction signal closely follows the original one and the error is negligible.

The CR and PRD values depend on the amount of energy compaction during the compression stage. The less the energy compaction, (i.e., more throwing out of information), the greater the compression ratio and error figures. A study was performed to vary the ECE and record the CR and PRD. Table III shows the results with mitdb data 117. Table III we can conclude that for 4% relaxation in energy compaction, the change in CR is 3.81% but change in PRDN is 21.5% from the respective initial values. Figure 3 shows the reconstruction by reducing the ECE up to 80% total value. It is observed that except very small local distortions in P onset and ST segment (shown in red circles), the reconstruction at 80% also close to original signal.

For clinical validation we showed the unmarked original and reconstructed data to two cardiologists for their comments. They confirmed that the clinical features were retained in the reconstructed signal. Table IV shows the comparison of performance of the proposed algorithm with some other wavelet based works using mitdb data. The proposed work yields equally and in some cases better CR that the sited works in Table IV, and slightly higher PRD value.

4 CONCLUSION

In this paper we proposed a compression algorithm for single lead ECG based on discrete wavelet transform of the average beat and residual signals. With an ECE of 99% for average beat and residuals, an average CR of 8.65, MAE of 0.05 mV and PRDN of 4.30 were obtained with 30 numbers of mitdb records, each tested with one minute duration. For a relaxation of ECE up to 95%, it was found that the CR increases marginally, but PRD increases significantly. For real-time applications the proposed work is efficient in utilizing the channel bandwidth as it uses residual encoding instead of compressing the individual beats. However this algorithm is limited by delay of 10 beats duration. Computational time of the algorithm is around 2 sec with processor speed of 3.07 GHz and 3.18 GB (RAM) memory.

REFERENCES

Alesanco A., Olmos S., Istepanian R.S. H.and Garcia J., "Enhanced Real-Time Coder for Packetized Telecardiology Applications,"*IEEE Trans. Inf. Techno. Biomed*, vol. 10, no. 2, pp. 229–236, Apr 2006.

Allen V.A. and Belina J., "ECG Data Compression Using Discrete Cosine Transform," *Proceedings of Computers in Cardiology*, pp. 687–690, 11–14, Oct. 1992, Durham, NC.

Chompusri Y. and Yimman S., "Energy Packing Efficiency Based Threshold Level Selection for DWT ECG Compression," *International Journal of Applied Biomedical Engineering*, vol. 2, no. 2, pp. 19–28, Dec 2009.

Cox J.R., Nolle F.M., Fozzard H.A., and Oliver G.C., "AZTEC, a preprocessing program for real-time ECG rhythm analysis," *IEEE Trans. Biomed. Eng.*, vol. BME-15, no. 2, pp. 128–129, Apr 1968.

Degani R., Bortolan G. and Murolo R., "Kahunen-Loeve coding of ECG Signal," *Proceedings of Computers in Cardiology*, pp. 395–398, 23–26, Sep 1990, Chicago, IL.

Giallorenzo M., Cohen J., Mora F., Passariello G. and Lara L.O., "Ambulatory monitoring device using the fan method as data-compression algorithm," *Medical & Biomedical Engineering & Computing*, vol. 26, no. 4, pp. 439–443, July 1988.

Gupta R., Mitra M., Mondal K. and Bhowmick S., "A derivative-based approach for QT-segment feature extraction in digitized ECG record", *Proceedings of 2nd International Conference on Emerging Applications of Information Technology (EAIT)*, 2011, Kolkata, India, pp. 63–66.

Gupta R., Mitra M. and Bera J.N., "ECG Acquisition and automated remote processing," Springer, 2014.

Hilton M. L., "Wavelet and Wavelet Packet Compression of Electrocardiograms," *IEEE Trans. Biomed. Eng.*, vol. 44, pp. 394–400, May 1997.

Imai H., Kiraura N. and Yoshlda Y., "An efficient coding method for electrocardiography using spline function," *Systems and Computers in Japan*, vol. 16, no. 3, pp. 85–94, May 1985.

Jalaleddine S., Hutchens C., Strattan R., and Coberly W., "ECG data compression techniques—A unified approach," *IEEE Trans. Biomed. Eng.*, vol. 37, no. 4, pp. 329–343, Apr 1990.

Kortman C. M., "Redundancy reduction-a practical method of data compression," *Proc. IEEE*, vol. 55, no. 3, pp. 253–263, Mar 1967.

Lu Z., Kim D. Y. and Pearlman W. A., "Wavelet Compression of ECG Signals by Set Partioning in Hierarchical Trees (SPIHT) Algorithm," *IEEE Trans. Biomed. Eng.*, vol. 47, no. 7, pp. 849–856, July 2000.

Manikandan M. S. and Dandapat S., "Wavelet-Based electrocardiogram signal compression methods and their performances: A prospective review," *Biomedical Signal Processing and Control*, vol. 14, pp. 73–107, Nov. 2014.

Pan J. and Tompkins W. J., "A Real-Time QRS Detection Algorithm," *IEEE Trans. Biomed. Eng.*, vol. 32, no. 3, pp. 230–236, March 1985.

Physionet: http://www.physionet.org

Rajoub B. A., "An Efficient Coding Algorithm for the Compression of ECG Signals Using the Wavelet Transform," *IEEE Trans. Biomed. Eng.*, vol. 49, no. 4, pp. 355–362, Apr 2002.

Zigel Y., Cohen A., and Katz A., "ECG signal compression using analysis by synthesis coding," *IEEE Trans. Biomed. Eng.*, vol. 47, no. 10, pp. 1308–1316, Oct 2000.

Computational Science and Engineering – Deyasi et al. (Eds)
© 2017 Taylor & Francis Group, ISBN 978-1-138-02983-5

Speech/music discrimination using perceptual feature

Arijit Ghosal
Department of Information Technology, St. Thomas College of Engineering and Technology, West Bengal, India

Suchibrota Dutta
Department of Computer Applications, ECMT, West Bengal, India

ABSTRACT: Multimedia datasets are escalating every day. These dataset includes audio, video, image etc. As the volume of dataset is increasing, a large storehouse is required for storing these dataset. For efficient use these dataset has to be retrieved very easily. To make retrieval easy we require storing them properly. For storing these dataset properly, we need to classify them into respective categories. In this work we have concentrated on classifying audio data into speech and music. We need to do the job of classification automatically for faster and more accurate performance compared to manual process of classification. From the past study of work, we have developed the idea that features play an important role for discriminating different data types. A proper feature generates better accuracy of discrimination. So we have to choose features thoughtfully. Energy of speech signal and music signal (includes instrumental and song both) are not same. So we have used energy based features for our purpose. Zero crossing rates for speech and music is not same. This leads inclusion of Zero Crossing Rates (ZCR) based features in our feature set. We have included perceptual feature also for discriminating speech and music signal as speech and music are perceptually different. RANSAC and Neural-Network have been used for classification purpose.

Keywords: Speech/music discrimination, audio features, Co-occurrence matrix plot of delta energy, periodicity

1 INTRODUCTION

Multimedia data is growing every day. In present time different types of multimedia tools are available for capturing these large multimedia data. Audio is a part of multimedia data. Appropriate management of these audio data is very tough job. To do so we need to store them into their respective category (like speech, music etc) wise. Hence we require classification of audio data into different category e.g. Speech, music etc.

Discrimination of speech and music is considered as the initial job for a speech recognition system. This discrimination filters out speech signal from music signal which will in future be used as input to the speech recognition system. Importance of speech recognition system is increasing in our daily life in biometric security field. So, clearly speech/music discrimination is also gaining its significance.

Due to huge amount of audio data it is not possible to manually distinguish speech and music. Automatic discrimination is very much necessary in this scenario.

For discriminating audio data into different category we have to undergo two successive stages: i) extracting appropriate features from the input audio data and then ii) classify this input audio data into its proper category.

From the previous works we have learnt that huge labors (West, 2004) are employed so far to discriminate audio data correctly into different category.

Different researchers have dealt with different audio features in their respective work (Beigi, 1999, Umbaugh, 2005, Fischler, 1981, Zuliani, 2005, Haralick, 1992, Huang, 1998). We may group these features into different categories like features of low-level type, perceptual features etc. We may categorize this level of features into different sub-category features like features calculated in time-domain and frequency-domain features. Our past study reveals that two features—Zero Crossing Rate, termed as ZCR (West, 2004) as well as Short Time Energy, termed as STE (Saunders, 1996, El-Maleh, 2000) are frequently used by researchers. We know that these two features are time domain features. From the history of previous researches we have seen that a few frequency-domain features like fundamental frequency (Zhang, 1998), signal bandwidth, signal energy (Foote, 1997), spectral centroid, Mel-Frequency Cepstral Co-efficients

(MFCC) (Eronen, 2000, Foote, 1997) etc. are also used as audio features. It is also known from the past study that loudness (Zwicker, 1999), measures for roughness (Zwicker, 1999) are also used by different researchers for discriminating audio data into different category. These features are perceptual features. Saunders (Saunders, 1996) has presented a threshold based two level algorithm. Two-level speech/music classifier has been used by El-Maleh (El-Maleh, 2000). Breebaart and McKinney (Breebaart, 2003) has worked with a large set feature vector which consists of 62 features. Sub-band energy has been used for describing audio data by Liu (Liu, 1997) and Guo (Guo, 2003). Matiyaho and Furst (Matityaho, 1994) have worked with neural network based scheme. We have also seen that Support Vector Machine (SVM) (Ben-Hur, 2001) was also used by different researchers (Guo, 2003, Sadjadi, 2007) in their work for the purpose of audio classification.

Audio signal may be categorized in several types—speech, music, songs etc. This categorization is done based on features that describe the audio data very well.

Ghosal et al. have discriminated speech and music based on Empirical Mode Decomposition (EMD) (Ghosal, 2011). In that scheme they have decomposed audio signal hierarchically using EMD (Huang, 1998). In EMD approach the input signal is broken into many Intrinsic Mode Functions (IMFs) and a residual signal. This EMD approach is iterative by nature and hence time consuming. In their other work (Ghosal, 2009), they have extracted features from the matrix which reflects the nature of co-occurrence of ZCR and STE for same purpose.

In our proposed scheme, we have emphasized our importance on the efficiency of selected features. Instead of directly working with energy, we have considered in this scheme, the change of energy variation. This change of energy variation is termed as delta energy. We have also measured ZCR—a widely used low level time domain feature and correlation-based periodicity. We have always keen to propose a powerful feature vector as we know that powerful feature vector makes the task of classifier very easy.

We have planned this paper as follows—after Introduction, proposed work is explained in the next section. In this section we have discussed the details of feature extraction step as well as our scheme of classification. Then Section III denotes the experimental results and the lastly conclusion is put into section IV.

2 PROPOSED WORK

The major parts of our work are feature computation and classification schemes.

2.1 *Feature computation*

ZCR and STE are mostly used time-domain features for the purpose of speech/music discrimination. If we compare the co-occurrence matrix based approach of Ghosal et al. [19] with their EMD based approach [18], we see that the dimension of feature vector for co-occurrence matrix based approach is large compared to EMD based approach. We have emphasized importance on developing a feature vector which is not so large enough but it will be capable to categorize speech and music splendidly.

We have examined characteristics of speech and music. This observation reveals that major energy is restricted in lower frequency band in case of speech signal. But for music signal it does not happen. So energy distribution of speech signal is quite different from music signal. So it is obvious that the change of energy variation of speech signal between two successive frames will differ from that of music signal. This change of energy variation between two successive frames is denoted by "delta energy". We have seen for speech signal some peaks only when there is no silence, else it comes down to almost zero due to presence of repeated silence in speech signal. But this type of pattern is not seen for music signal because of absence of silence.

Let, energy of ith frame is E_i. For $(i+1)^{th}$ frame, energy is E_{i+1}. Now we have calculated delta energy as –

$$\Delta En = E_{i+1} - E_i \tag{1}$$

Precise study of characteristics of a feature is not possible by mean and standard deviation of that feature as they exhibit only an overall idea. So, for precise study of the characteristics of delta energy, we gave used the idea of co-occurrence matrix (Umbaugh, 2005). In case of an image, the happening of dissimilar intensity values within a locality reflects a nature and this is used to study the appearance/texture of an image. We have implemented this concept here.

By using equation (1), we have calculated delta energy. Now, we have obtained $\{\Delta En_i\}$ which is the chain of delta energy, for the input audio stream. Occurrence of dissimilar values of delta energy within a region reflects a property which characterizes the nature of the signal in better way. So, we have formed a matrix C having dimension L × L. Here $L = max\{\Delta En_i\} + 1$. The number of repeating existence of delta energy p and q in consecutive time instances is represented by an element of the matrix C (p, q). Different statistical measures (Haralick, 1992) which are calculated from the co-occurrence matrix are computed when the co-occurrence matrix is generated. These statistical measures are entropy, energy, contrast, homogeneity and correlation. This represents the pattern of coefficients.

If audio data is considered as discrete signal, we observe that zero crossing occurs when two following samples have opposite signs. Zero crossing rates provide the idea regarding the frequency content. Primarily, the input audio stream is broken into T frames which are overlapped $\{x_d(m): 1 \leq d \leq T\}$, each of specific size. We have considered the frames are overlapped in nature so that no characteristics of a particular frame are missed. Then, we calculate zero crossing rates for d^{th} frame as –

$$z_d = \sum_{m=1}^{n-1} \text{sign}[x_d(m-1) * x_d(m)] \tag{2}$$

Here, the number of samples in the dth frame is represented by n and

$$\text{sign}[v] = \begin{cases} 1, & \text{if } v > 0 \\ 0, & \text{otherwise} \end{cases} \tag{3}$$

We need to calculate ZCR for each of the frames. Then, we calculate standard deviation and mean of ZCR for all the frames.

Music, be it an instrumental or song, exhibits certain type of periodicity in nature. But that type of periodicity we do not observe in speech signal. This examination has motivated us to consider periodicity also as feature. Correlation based periodicity has been used to capture the periodicity of a given signal (Ghosal, 2011, 2016, 2013). Periodicity is a perceptual feature. We first divide the input signal, say S, into frames which are non-overlapping in nature. Each frame contains 100 number of samples. Then we estimate absolute Pearson's correlation coefficients (Walk and Rupp, 2010) between consecutive frames. We have considered only the maximum correlation value for each of the frame. Now, we have considered only the standard deviation and mean for correlation-based feature from the frame level maximum correlation values.

We have calculated 2 ZCR based features, 5 energy based features and 2 periodicity based features. To learn the utility of these features we have generated two set of feature vector F_1 and F_2. F_1 is formed by combining ZCR based features and energy based features. F_2 is formed by combining F_1 with periodicity based features. So F_1 is of 7 dimensions and F_2 is of 9 dimensions. RANdom Sample And Consensus (RANSAC) as well as Neural-Net have been applied for classification purpose.

2.2 *Classification*

Our main aim of this work to propose a feature set that will discriminate speech and music very well.

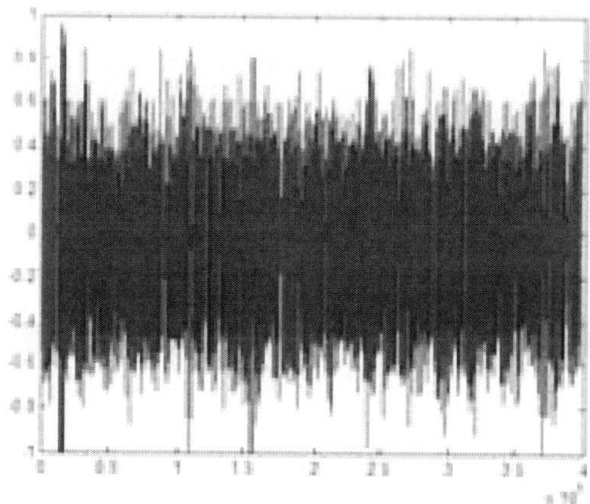

Figure 1. Input speech signal.

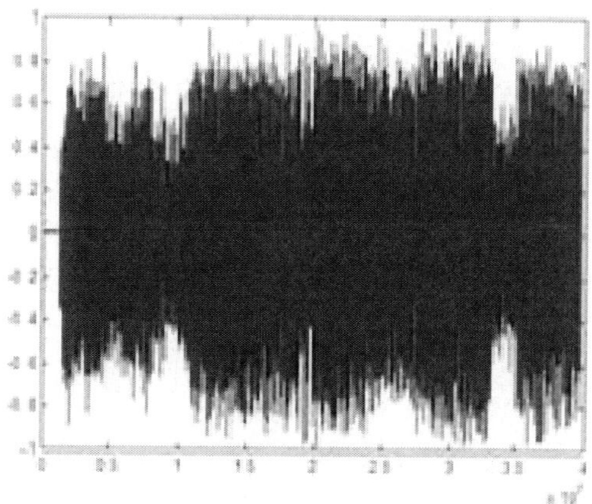

Figure 2. Input music signal.

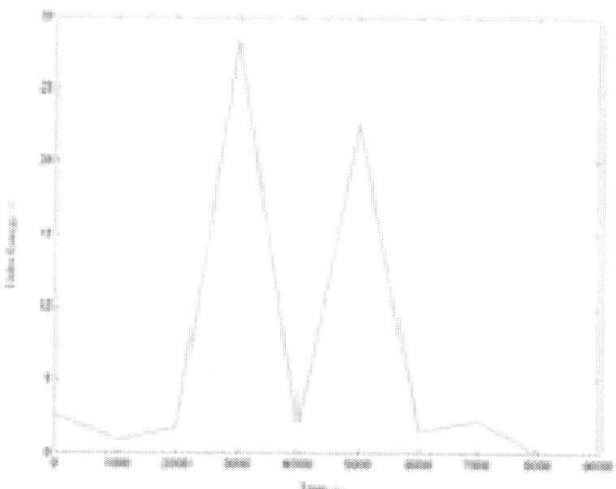

Figure 3. Delta energy plot of speech signal.

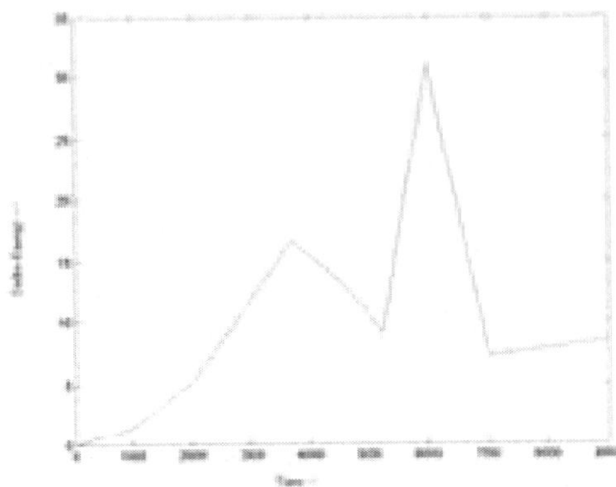

Figure 4. Delta energy plot of music signal.

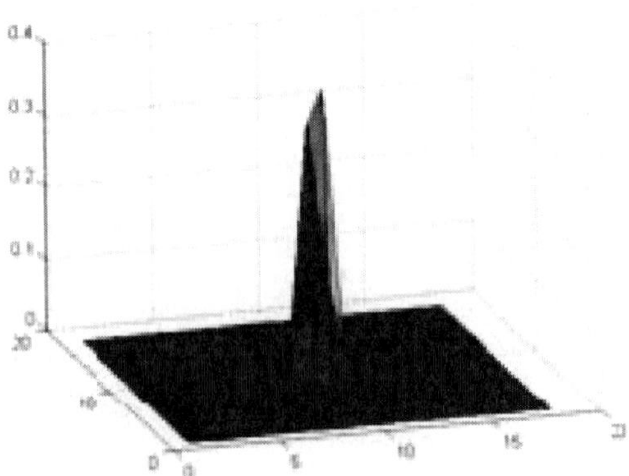

Figure 5. Co-occurrence matrix plot of delta energy for speech signal.

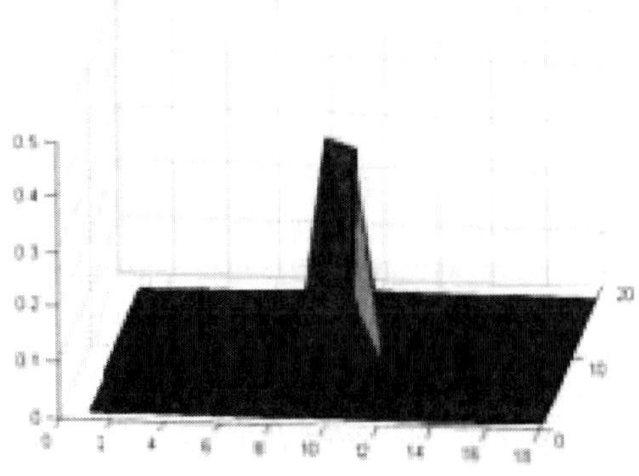

Figure 6. Co-occurrence matrix plot of delta energy for music signal.

From the past study of work by different researchers, we have learned that Support Vector Machine (SVM) is robust. But SVM has a drawback that its parameter tuning is not easy and if they are not properly tuned, SVM fails to generate a good classification result. In this situation, we were searching for a robust estimator which will be able to deal data of wide diversity very well. We have seen that RANdom Sample And Consensus (RANSAC) (Fischler, 1981) is actually a re-sampling method to produce solutions using least number of examinations of data points necessary for estimation of the underlying model parameters. Smallest possible set has been used by RANSAC and proceeds to enlarge this set with steady data points compared to traditional sampling technique that uses large amount of data to get a preliminary solution and then proceed further. The major strength of RANSAC compared to other estimators is that the estimation is based on inlier, not based on outlier. So, RANSAC appears to us as a proper alternative of SVM to attain our goal.

Initially, we arbitrarily select the smallest amount of points which are essential to decide model parameters. Next, parameters of this model are obtained. Now, we decide number of points from the set of all points to fit with a predefined tolerance €. If the number of inliers in the set crosses a predefined threshold value €, we have to re-calculate the model parameters again using all identified inliers and then it is completed. If it is not, we again repeat these steps maximum J times, where J is chosen in such a way such that the probability p (usually set to 0.99) that at least one of the sets of random samples does not include outlier. RANSAC works best with the inliers and it is less affected by noisy data.

Neural-net is widely used standard classifier. Artificial Neural Network (ANN), also called as neural-net (NN), is a mathematical or computational model that is designed based on the structure and/or functional aspects of biological neural networks. An ANN is defined by three types of parameters:

i. The interconnection pattern between different layers of the neurons
ii. The learning process for updating of weights of the interconnections
iii. The activation function that converts a neuron's weighted input to output activation.

The feature vector is input to the network. The dimensions of the feature vector is N (N = no of features), which represents the N parameters of the features extracted from the audio signal. The network has layer of hidden neurons which is not the component of the input or output of the network. These hidden neuron layer permits the network to learn complex jobs by extracting gradually more meaningful features from the input vectors. The number of neurons in each hidden layer is specific in the experiment. The number of the neurons in output layer is designed by the number of output classes what we want to classify. For example, if we want to classify the audio into music and speech, we will have two neurons in output layer corresponding to music and speech respectively. For working with feature vector F_1, we have put 7 nodes in input layer and for F_2 there are 9 nodes in input layer. There is only one hidden layer which will have 5 nodes for F_1 and 6 nodes for F_2 respectively.

3 EXPERIMENTAL RESULTS

For this experiment, we have arranged an audio database that has 200 music files and 200 speech files. Duration of each of these audio files is 90 seconds. These audio files are collected from different sources like CD recordings, live program recordings. Some of these audio files are downloaded from Internet also. The sampling frequency for all the audio files is 22050 Hz. All the sound files are mono type. Both male and female voices of different ages are considered for speech files. These speech files are spoken in different languages also. Some of these audio files are noisy. Music files contain variety of instruments like piano, flute, drum, guitar etc and different types of songs like bhangra (an Indian genre), folk, classical, ghazal, pop, jazz, rock etc.

For feature calculation, we have divided an audio file into several frames. Frames are created in such a way that in each frame there are 150 samples of which 50 samples are overlapped between two successive frames. For implementation of Neural-net we have used MATLAB. For configuration of neural net in MATLAB we have put 9 nodes in input layer and 2 nodes in output layer. For classification work based on feature vector F_1 there are 5 nodes and for feature vector F_2 there are 6 nodes in the one and only hidden layer.

We have done the classification work based on two feature vectors F_1 and F_2. For both the cases, we have used 70% of both speech files and music files for training purpose to determine the model and rest of the data are used for the purpose of testing. The experiment is also done by reversing training and testing data set. We have shown average of these experiments in Table 1 & 2 for the two feature vectors F_1 and F_2 respectively. It may be well-known that using feature vector F_2, significant improvement with respect to F_1 has been achieved for all the classifiers. It represents that the inclusion of features based on periodicity enhances the strength of the feature vector.

Table 1. Accuracy of speech/music differentiation (for feature vector F_1).

	Classification accuracy (in%)		
	Proposed features		
Classifier	Speech	Music	Overall
Neural network	91%	95%	93%
RANSAC	95.5%	92.5%	94%

Table 2. Accuracy of speech/music differentiation (for feature vector F_2).

	Classification accuracy (in%)		
	Proposed features		
Classifier	Speech	Music	Overall
Neural network	93%	97%	95%
RANSAC	97.5%	93.5%	95.5%

Corresponding to a sample speech and music signal, the plot of delta energy has been depicted in Figure 3 and Figure 4 and the plot of co-occurrence matrix of delta energy have been shown in Figure 5 and 6 respectively. From the figures it is very clear that, occurrence patterns of delta energy for speech and music are different.

We have designed our feature set by combining time domain features and perceptual features. The classification result clearly shows the efficiency of our proposed feature set which is based on co-occurrence matrix of delta energy, ZCR and periodicity. From the result it is noticed that RANSAC performs better.

4 CONCLUSION

In speech/music discrimination, a new set of features which is based on features calculated from standard deviation and mean of ZCR, co-occurrence matrix of delta energy and mean and standard deviation of correlation-based feature has been proposed. The result exhibits the potentiality of the proposed feature for discrimination of speech and music. To emphasize the power of the proposed features, we have used simple classification scheme which is based on Neural-Net and RANSAC. We have noticed that RANSAC produces better result compared to neural network. In future, each individual class may further be sub-classified into different sub-categories.

REFERENCES

Ben-Hur A., Horn D., Siegelmann H. T. and Vapnik V., "Support vector clustering," Journal of Machine Learning Research, vol. 2, pp. 125–137, 2001.

Beigi H., Maes S., Sorensen J. and Chaudhari U., "A hierarchical approach to large-scale speaker recognition," in Proceeding of the Int. Computer Music Conference, 1999.

Breebaart J. and McKinney M. "Feature for audio classification," in Int. Conf. on MIR, 2003.

El-Maleh K., Klein M., Petrucci G. and Kabal P. "Speech/music discriminator for multimedia application," in IEEE Int. Conf. on Acoustics, Speech and Signal Processing, 2000.

Eronen A. and Klapuri A. "Musical instrument recognition using ceptral coefficients and temporal features," in IEEE Int. Conf. on Acoustics, Speech and Signal Processing, 2000, pp. 753–756.

Fischler M. A. and Bolles R. C. "Random sample consensus: A paradigm for model for model fitting with applications to image analysis and automated cartography," ACM Communications, vol. 24, pp. 381–395, 1981.

Foote J. T. "Content-based retrieval of music and audio," in SPIE, 1997, pp. 138–147.

Ghosal A., Chakraborty R., Dhara B. C. and Saha S. K. "Song Classification: Classical and Non-classical Discrimination using MFCC Co-occurrence Based Features,"International Conference on Signal Processing, Image Processing and Pattern Recognition, 2011, pp. 179–185.

Ghosal A., Dhara B. C. and Saha S. K. "Speech/Music Classification using Empirical Mode Decomposition," International Conference on Emerging Applications of Information Technology, 2011, pp. 49–52.

Ghosal A., Chakraborty R., Chakraborty R., Haty S., Dhara B. C. and Saha S. K. Speech/music classification using occurrence pattern of zcr and ste. 3rd International Symposium on Intelligent Information Technology Application, pages 435–438, China, 2009. IEEE CS Press.

Ghosal A., Chakraborty R., Dhara B. C. and Saha S. K. "Perceptual Feature Based Song Genre Classification using RANSAC," International Journal of Computational Intelligence Studies, Inderscience Publication, 2015, Vol. 4, No. 1, pp.31–49.

Ghosal A., Chakraborty Rudrasis, Dhara B. C. and Saha S. K. "Genre Based Classification of Song Using Perceptual Features," International Conference on Advanced Computing, Networking, and Informatics, India, June, 2013, pp. 267–276.

Guo G. and Li S. Z. "Content-based audio classification and retrieval by support vector machines," IEEE Transactions on Neural Networks, vol. 14, no. 1, pp. 209–215, 2003.

Haralick R. M. and Shapiro L. G. Computer and Robot Vision (Vol-I). Addision-Wesley, 1992.

Huang N., Shen Z., Long S., Wu M., Shih H., Zheng Q., Yen N., Tung C. and Liu H. The empirical mode decomposition and hilbert spectrum for nonlinear and nonstationary time series analysis. Proc. of the Royal Society of London, 454:903–995, 1998.

Liu Z., Wang J. H. a. and Chen T. "Audio feature extraction and analysis for scene classification," in IEEE Workshop on Multimedia Signal Processing, 1997.

Matityaho B. and Furst M. "Classification of music type by a multilayer neural network," Journal of the Acoustical Society of America, vol. 95, 1994.

Sadjadi S. O., Ahadi S. M. and Hazrati O. "Unsupervised speech/music classification using one-class support vector machines," in In the Proceeding of the ICICS, 2007.

Saunders J. "Real-time discrimination of broadcast speech/ music," in IEEE Int. Conf. on Acoustics, Speech, Signal Processing, 1996, pp. 993–996.

Umbaugh S. E. Computer Imaging: Digital Image Analysis and Processing. CRC Press, 2005.

West C. and Cox S. "Features and classifiers for the automatic classification of musical audio signals," in Int. Conf. on Music Information Retrieval, 2004, pp. 531–537.

Zhang T. and Kuo C. C. J. "Content-based classification and retrieval of audio," in Conf. on Advance Signal Processing, Architectures and Implementations VIII, 1998.

Zuliani M., Kenney C. S. and Manjunath B. S. "The multiransac algorithm and its application to detect planar homographies," in IEEE Conf. on Image Processing, 2005.

Zwicker E. and Fastl H. "Psichoacoustics: Facts and models," Springer Seires on Infromation Science, 1999.

Computational Science and Engineering – Deyasi et al. (Eds)
© 2017 Taylor & Francis Group, ISBN 978-1-138-02983-5

Design of a voice-based system by recognizing speech using MFCC

S. Muhury, G. Neogi, P. Debnath & J. Ghosh Dastidar
St. Xavier's College (Autonomous), Kolkata, India

ABSTRACT: This paper describes the development of automated speech recognition software which will execute various commands to perform basic computer tasks such as opening the task manager or the command prompt, locking the windows, etc. The 'voice-commands' are accepted, processed and Mel Frequency Cepstrum Coefficients (MFCC) are calculated for them. MFCC is used to extract the characteristics of the input voice command and the same is used to recognize it. Successful recognition will trigger the task the command has been associated with. Finally the performance of the proposed algorithm has been analyzed.

1 INTRODUCTION

Of all the man-made sounds which influence our lives, speech and music are arguably the most prolific (Logan, 2000). Speech signals are naturally occurring signals and hence are random signals. Decades of research in the speech department has led to usable system and convergence of the features and models used for speech analysis. Speech recognition is basically a process in which the software recognizes some words which are spoken by a user and matches them with some previously stored voice samples in the database. Pronunciations of different dictionary words are different and they are separated from each other by some special features. The basic job of various speech recognition techniques is to bring out those special features of individual words and compare with the input voice command. The percentage of likeability of both the words (if both words are same) determines the efficiency of the speech recognition technique. Speech based devices have huge application in our daily life especially to those who suffer from various disabilities. Presented in this paper is an approach that uses Mel Frequency Cepstrum Coefficients (MFCC) for feature extraction of the speech samples and then uses the features to recognize simple voice commands.

2 LITERATURE SURVEY

Speech recognition and voice recognition is a very important topic in the field of Pattern Recognition and Audio signal processing. A considerable amount of research has gone into this topic and hence is at a stage where it is now. Isolated speech recognition using MFCC and Dynamic Time Warping has been studied in depth by Dhingra, et al., 2013. Han, et al., 2006 have discussed novel methods for Pre-Emphasizing, pWindowing, etc. for extracting features for the purpose of speech recognition. The same has also been studied by Ittichaichareon, et al., 2012. An integral step during the computation of MFCC is the application of Discrete Fourier Transformation (DFT). The same has been discussed in details by Kido, 2014. Each of the sub-processes required for MFCC and the method to implement them has been described by Klautau & Aldebaro, 2005. The vulnerability of MFCC to accents and noise was pointed out by O'Shaughnessy, 2008. Paliwal & Atal, 2003 have shown the frequency related representation of speech. Sreenivasa, et al., 2015 have discussed the methods for implementing MFCC and LPC in the book edited by them.

3 PROPOSED METHODOLOGY

3.1 *What is MFCC?*

Mel Frequency Cepstral Coefficients (MFCC) are features used in speech recognition, speaker recognition, and also in many other voice based algorithms. They were first introduced by Davis and Mermelstein in 1980. It is less complex in implementation compared to several feature extraction methods. It is designed using the knowledge of human auditory system.

MFCC is a very popular technique for feature extraction. It brings out certain special features depending upon the pronunciation of various dictionary words so that when spoken they can be easily separated from one another. Here we propose a system, which will accept a voice command through a microphone, the MFCC of that incoming signal will be computed and will be matched with a sample, priorly stored in the database. In the case of a successful match, the system will perform the task mapped with that command.

3.2 *Feature extraction using MFCC*

Like any other pattern recognition systems, speech recognition systems also involve two phases namely, training and testing. Training is the process of familiarizing the system with the characteristics of the words that are supposed to serve as voice commands. Testing is the process of matching and recognizing input words with words stored in the database during the training phase. Figure 1 depicts the training phase whereas Figure 2 the testing phase.

First we need to train the system with some voice commands and store them in a database. While during the testing phase a voice command is accepted from a user, then the feature extracted from the input voice and the command stored in the database earlier are matched and finally if the match is found the system will execute the task related to it.

The feature extraction process is a non-invertible (lossy) transformation. Figure 3 shows the computations performed during MFCC.

The processing begins by acquiring a voice command in the form of an audio signal (audio waveform). The signal is then sampled by considering frames in the waveform. The frames in time domain are then transformed into frequency domain by applying the Discrete Fourier Transformation (DFT). The Amplitude Spectrum of the obtained DFT is considered and passed through a Mel filter bank. The output obtained from the filter is known as Mel spectrum. The logarithm operation is then applied on the obtained Mel spectrum. Mel Frequency Cepstral Coefficients (MFCC) is finally obtained by applying the Discrete Cosine Transformation (DCT) to the log of Mel Spectrum.

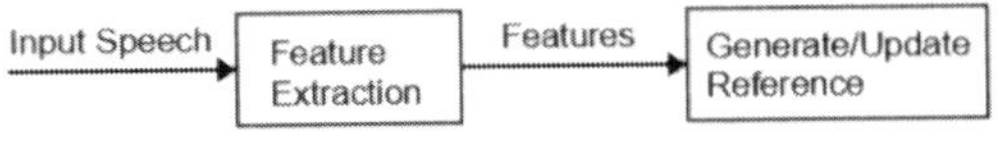

Figure 1. Training phase (Tiwari, 2010).

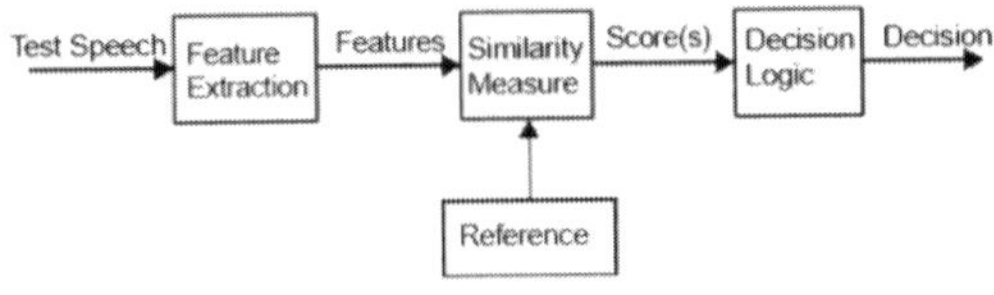

Figure 2. Testing phase (Tiwari, 2010).

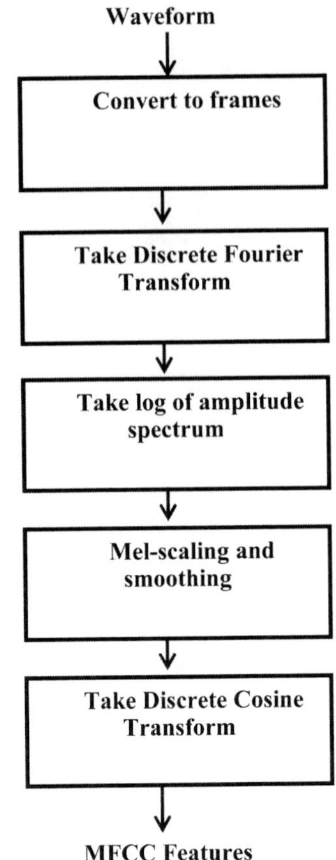

Figure 3. MFCC steps (Logan, 2000).

The steps are explained as follows (Klautau & Aldebaro, 2005):

a. Pre-emphasize the speech signal.
b. Divide the speech signal into frames of 20 ms and a shift of 10 ms. Apply the Hamming window over each of the frames.
c. Compute the Amplitude spectrum for each window by applying DFT.
d. Mel Spectrum is obtained by passing the Amplitude Spectrum through a Mel Filter bank.
e. Logarithm is computed for the Mel Spectrum.
f. DCT is applied to the log Mel frequency coefficients (log Mel spectrum) to derive the MFCC's.

4 IMPLEMENTATION ISSUES

The most important problem related to any sound related application is the presence of noise. It is not always feasible to acquire sound proof environment for testing a software. So, one of the major challenges one needs to face for developing such system is to make the voice input through microphone as the sound to be registered in the database free from surrounding noise through certain changes in its algorithm. At first the voice received needs to be truncated up to the point from where the real command starts and to cut it immediately after it ends. By applying this technique we can somehow diminish the noise and provide the actual command for feature extraction. Though in this kind of system there is always a risk that it may not respond perfectly every time it has been executed. But still such systems are needed to be developed and researched about more so that several ASR (Automated Speech Recognition) systems can be produced. We have implemented and tested the system using MATLAB. Following are the sequence of activities undertaken in order to implement the system:

a. Recording a voice command through a microphone with the help of the software and storing it in the database with the accurate filename which is already mapped with any one of the above mentioned task.
b. The recorded command is first truncated to bring out only the actual command portion from the full recording and then it is stored.
c. MFCC of the recorded sound is finally calculated.
d. Next, the software will receive a command from a random user through a microphone and it will calculate the MFCC for the newly input command.
e. Finally, comes the matching part where the MFCC of all the sounds stored in database will be compared with the newly computed MFCC of the input command. If similarities are found the software will execute the task related to that command.

4.1 *Key issues*

The major issues in implementing the system are the following:

Since matching requires the matching of the input sample with all of the words in the database, the size of the database will affect the performance as the steps required to do the matching are resource intensive. In order to overcome this problem the MFCC's of the database words may be computed in advance and stored. So, whenever the user gives an input command only the MFCC of that particular command is calculated and comparison with the array of stored MFCC values are performed, hence it leads to faster execution of the software.

Another major issue that needs to be handled is noise (Tyagi & Wellekens, 2005). The voice samples should be subjected to pre-processing to remove noise as far as practicable before subjecting it to the MFCC calculation process. Ignoring this will impact the adversely.

Next in the list is diminishing the error percentage, mostly in this kind of speech recognition system error percentage is relatively high until and unless the software is executed in an isolated environment, but in order to control the error rate,the registered commands are categorized according to their length of delivery. Say for example 'Calculator' is a bigger command and definitely will take more time than relatively smaller commands like 'Lock' or 'Start'. So categorizing these list of commands in the database, the final input command by the user will not require to be compared with the whole set of commands in the database but with only with smaller set of commands that are similar to its delivery time. Thus this modification to the software will not increase the speed of execution but will also diminish the error rate.

4.2 *Performance analysis*

The performance of the system has been calculated by computing the False Reject Rate (FRR) and the False Accept Rate (FAR). False Accept (FA) is defined by as the event when a wrong word is accepted in place of a legitimate one. The frequency at which False Accept errors are made is called False Acceptance Rate (FAR). FAR is calculated as a fraction of illegitimate scores over the total number wrong of attempts made.

FAR = NFA/NWA

(NFA = Number of False Acceptances, NIA = Number of Wrong Attempts).

False Reject (FR) is defined as the event when a legitimate word is rejected by the system. The frequency at which False Reject errors are made is called False Reject Rate (FRR).

FRR is calculated as a fraction of genuine attempts being rejected over the total number of legitimate access attempts.

FRR = NFR/NAA

(NFR = Number of failed Rejections, NAA = Number of legitimate Access Attempts).

Table 1 shows the FAR and FRR for a few sampled words.

4.3 *Improvement scope*

MFCC has been used for voice recognition with considerable success in several applications. However MFCC has proved to be not robust enough

Table 1. FAR and FRR for different and same voice.

	Different voice		Same voice	
Words	FAR	FRR	FAR	FRR
Start	0.25	0.20	0.20	0.20
Date	0.35	0.20	0.10	0.20
Keyboard	0.30	0.20	0.20	0.20
Program	0.35	0.30	0.30	0.20
Password	0.30	0.25	0.20	0.20
Shut down	0.30	0.10	0.20	0.10
Calculator	0.25	0.20	0.20	0.10
Task manager	0.25	0.10	0.10	0.0
System status	0.10	0.0	0.0	0.0
File manager	0.20	0.10	0.0	0.0

against noises (Tyagi & Wellekens, 2005). Further, one critical assumption made for the calculation of the cepstral coefficients is that the fundamental frequency is lower than the frequency components of the words being uttered (Kido, 2014). However, many speakers do not fulfill this characteristic. The cepstral coefficients (apart from the first two) cannot be clearly interpreted (O'Shaughnessy, 2008). Thus, the effect of the coefficients on accent and noise is unknown. This may reduce the separability of the different words as is evident from Table 1. These issues can be handled by extending MFCC to include cepstral mean normalization. This extension may reduce external effects such as different microphones or different locations. Alternatively the Mel scale filter bank may be replaced by Nd equidistant band-pass filters on the bark scale to obtain Bark Frequency Cepstral Coefficients (Zheng, 2001). Use of Linear Predictive Coding (LPC) (Deng & O'Shaughnessy, 2003) instead of MFCC may prove to be more rewarding in terms of Speech recognition.

5 CONCLUSION

In this paper here, we have tried to demonstrate an ASR implemented using MFCC. The recognized words may be used as voice-commands given to the computer for performing various tasks. The software may be further fine-tuned and extended to build a computer which will solely run on voice commands.

REFERENCES

Deng, L. & O'Shaughnessy, D. 2003. Speech processing: a dynamic and optimization-oriented approach. Marcel Dekker: New York.

Dhingra, S. D., Nijhawan, G. & Pandit, P. 2013. Isolated Speech Recognition using MFCC and DTW. In International Journal of Advanced Research in Electrical, Electronics and Instrumentation Engineering, Vol. 2, Issue 8: 4085–4092.

Han, W., Chan, C. & Pun, K. 2006. An efficient MFCC extraction method in speech recognition. In IEEE International Symposium on Circuits and Systems. doi: 10.1109/ISCAS.2006.1692543.

Ittichaichareon, C., Suksri, S. & Yingthawornsuk, T. 2012. Speech Recognition using MFCC. In International Conference on Computer Graphics, Simulation and Modeling. Pattaya (Thailand).

Kido, K. 2014. In Digital Fourier Analysis: Advanced Techniques, Berlin, New York, Tokyo: Springer.

Klautau & Aldebaro. 2005. The MFCC. In Technical report, Signal Processing Lab, UFPA, Brasil.

Logan, B. 2000. Mel Frequency Cepstral Coefficients for Music modeling. In International Symposium on Music Information Retrieval.

O'Shaughnessy, D. 2008. In Pattern Recognition, Volume 41 Issue 10: 2965–2979.

Paliwal, K. K. & Atal, B. S. 2003. Frequency related representation of speech. In Proc. EUROSPEECH: 65–68.

Sreenivasa Rao, K., Ramu Reddy, V. & Maity, S. 2015. In Language Identification using Spectral and Prosodic Features. London: Springer.

Tiwari, V. 2010. MFCC and its Applications in Speaker Recognition. In International Journal on Emerging Technologies. 19–22.

Tyagi, V. & Wellekens, C. 2005. On desensitizing the Mel-Cepstrum to spurious spectral components for Robust Speech Recognition, in Acoustics. In IEEE International Conference on Speech, and Signal Processing. Proceedings. (ICASSP '05): 529–532.

Zheng, F., Zhang, G. & Song, Z. 2001. Comparison of Different Implementations of MFCC. In Journal of Computer Science & Technology, vol. 16: 582–589.

Computational Science and Engineering – Deyasi et al. (Eds)
© 2017 Taylor & Francis Group, ISBN 978-1-138-02983-5

Study on systemic modeling for cardiac and respiratory system towards support on prosthesis facets

Susmita Das
Narula Institute of Technology, Kolkata, India

Soumendu Bhattacharjee
KIEM, West Bengal, India

Swapan Bhattacharyya & Biswarup Neogi
JIS College of Engineering, India

ABSTRACT: This paper emphasizes the current status of research on simulation action towards respiratory and cardiovascular system. Mainly, effort is being carried out to present the computational modeling of human lung and heart on consideration of processes flow systemic approaches. Implementation with mimic of internal body organism, primarily heart and lung, are presented here with effective review. In addition, the control model analogies with corrective performance representation for existing generalized models are presented to enrich systems efficacy.

1 INTRODUCTION

Heart and lungs prosthesis on behalf of cardiovascular and respiratory system is a traditional work in biomedical research domain. Heart, a significant muscular part, thrusts oxygen rich pure blood throughout the body parts and simultaneously collects impure deoxygenated blood from there, comprises with four chambers in two pairs: ventricles and arteries (Bhattacharjee *et al.*, 2013). Lungs, an asymmetrical part in shape, helps for purifying the contaminated blood coming from whole body parts and produces the oxygen rich blood for human body blood circulation activities. Control model analogy of human cardiovascular (Bhattacharjee *et al.*, 2014) and respiratory system has already conveyed a new path in biomedical field. This paper portrays a system module with analogical process flow of bloods through human heart and lungs. This one is an interdisciplinary work co-related with medical science and computational domain along with process control analogy in concise.

2 COMPUTATIONAL APPROACH OF HEART PROSTHESIS RELATED TO BLOOD CIRCULATION

An artificial blood circulation system through artificial heart and lungs are represented in Fig. 1. The memory processor insists blood circulation is started with motor driven circuit then it goes to control valves. The pulse generator generates the frequency of heart beat rate that is intended with a memory processor.

An artificial heart consists of four chambers [real-life: right and left atrium (1, 3), right and left ventricles (2, 4)], unidirectional valves, motor device, pulse generator, memory processor shown in Fig. 2.

An artificial lung comprises with controller, control valve, filter, pump, mixed chamber, blood chamber and air chamber. The deoxygenated blood comes at artificial lung for purification shown in Fig. 3 (Saini *et al.*, 2009).

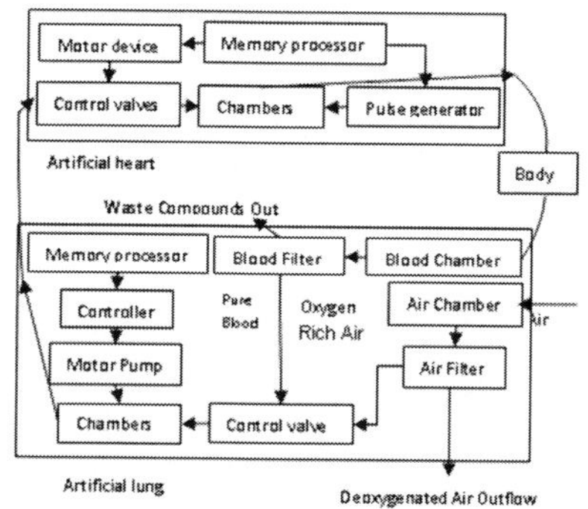

Figure 1. Process flow analysis of human blood circulation.

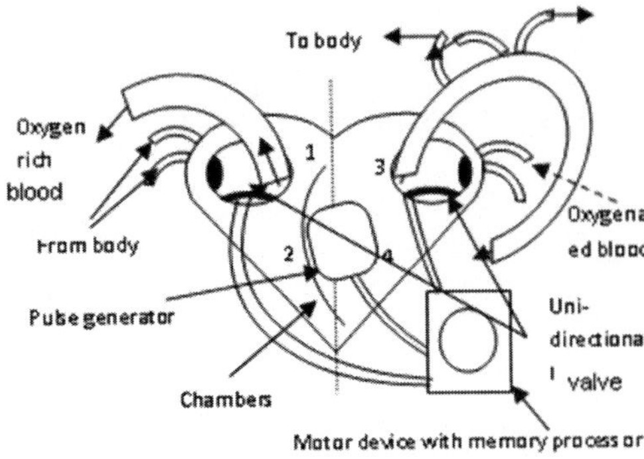

Figure 2. Pictorial representation of prosthetic heart.

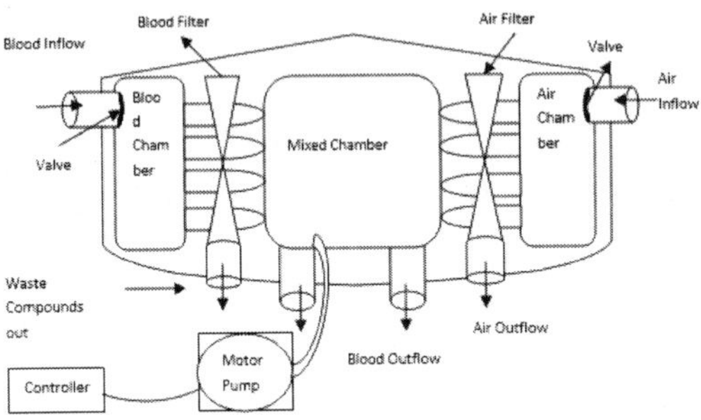

Figure 3. Prosthesis lungs system module.

3 CONTROL MODEL ANALOGY ON EXISTING CARDIOVASCULAR AND RESPIRATORY SYSTEM OF TEXT

The computational analysis of the artificial organs of human body is incorporated utilizing one systemic function of respiratory system and cardiovascular system as well. These two existing systems of control model of lungs and heart has been preferred for the control system analogy. Existing control model proposed by S. Saha et al. on the paper entitled "Investigation of Severity of Lung Fibrosis by Dynamic Modeling of Human Respiratory System" is carried out towards performance study (Neogi *et al.*, 2012). Apart from that, the research extension is being presented for the paper entitled "Study of cardiovascular dynamics with recursive Simulator generation approach", depicted by B. Neogi et al. in the year 2011 (Ghosal *et al.*, 2012). The performance analysis and control system analogy are evaluated through these existing mathematical modeling. In Hybrid modeling numerical and physical models exchange the data in real time.

3.1 *Simulation action on respiratory system*

The existing systemic function of the specified respiratory system is presented [3] here equation no (i).

$$T_p(\sigma) = \frac{0.15s + 1}{s(0.213s + 0.285)} \qquad (\iota)$$

The unit step response of the present respiratory function is represented in Figure 4.

From the above step response "Fig. 4" of the control function it is observed that the respiratory system is unbounded. To obtain a bounded system PID controller is used for tuning of the process. For tuning procedure the tuning parameters such as K_p, K_i, K_d are set with computational simulation action introducing MATLAB simulator. Next, the "Fig. 5" is processed towards bounded and stability.

After PID tuning, the obtained transfer function is

$$T_{p(\sigma)\Gamma\Pi\Delta} = \frac{0.00096s^3 + 0.058s^2 + 0.4s + 0.0077}{s^2(0.213s + 0.285)} \qquad (\iota\iota)$$

For the control analogy the required "Fig. 6" is represented. The controller parameters such as K_p, K_i, K_d values are represented in "Table 1" and the other process characteristics parameters such as rise time, settling time, overshoot and peak are also presented in "Table 2".

The controllability and observability performance analysis of the respiratory system are

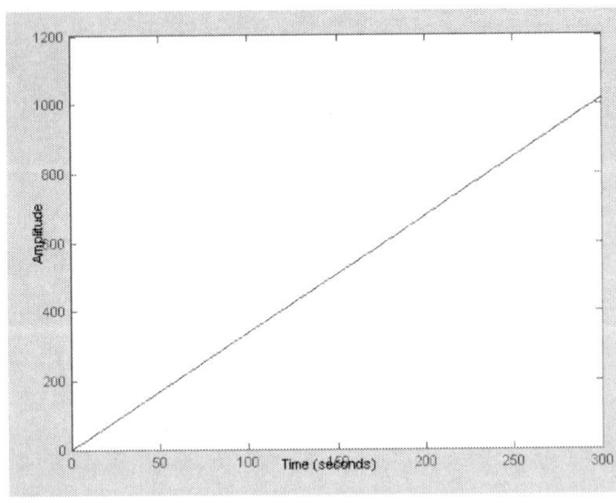

Figure 4. Step response of the mathematical model of the respiratory system.

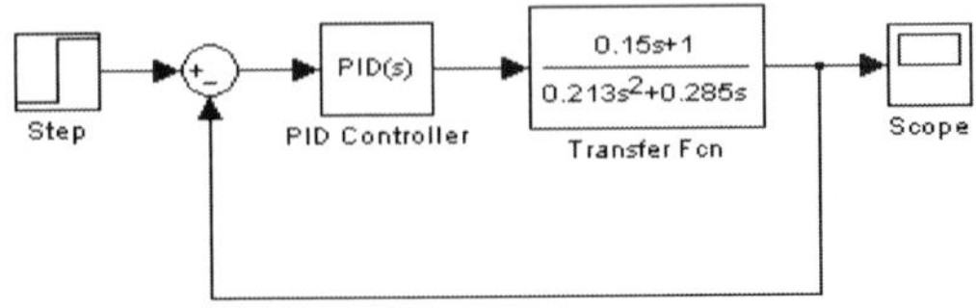

Figure 5. Mathematical model of the respiratory system.

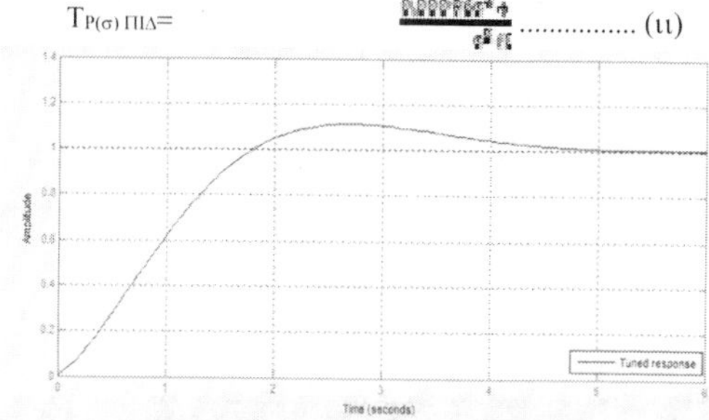

Figure 6. PID controller tuned output curve of the respiratory system.

Table 1. Controller parameters for respiratory system.

Parameters	Values
K_p	0.39885
K_i	0.0076544
K_d	0.0063642

Table 2. Performance and robustness for respiratory system.

Paramters	Values
Rise Time	1.29 Seconds
Settling Time	4.52 Seconds
Overshoot	11.3%
Peak	1.11

observed. State space modeling for the better entities of the system is necessary, as well the controllability and observability determination are carried the information of the system input and output parameter towards better performance (Kumar *et al.*, 2009).

The controllability matrix of the tuned model is

$$XO_{\text{Respiratory}} = \begin{bmatrix} 1.000 & -1.3380 & 1.7903 \\ 0 & 1.000 & -1.3380 \\ 0 & 0 & 1.0000 \end{bmatrix} \quad \text{(III)}$$

The observability matrix of the tuned model is

$$OB_{\text{Respiratory}} = \begin{bmatrix} 0.2663 & 1.8779 & 0.0362 \\ 1.5217 & 0.0362 & 0 \\ -1.9999 & 0 & 0 \end{bmatrix} \quad \text{(IV)}$$

Rank of $CO_{\text{Respiratory}}$ = Rank of $OB_{\text{Respiratory}}$ = System Order = 3.

Since both the rank of the matrices is of same value, so the system "$T_R(s)_{\text{PID}}$" is now controllable and observable as well.

Stability Analysis using Lyapunov Stability Method

Analysis of the Lyapunov stability of closed loop linear system is presented from Fig. 7.

We know, e(s) Gc(s) Gp(s) = Y(s)

$$\text{or}: e(s)\frac{0.15s+1}{s(0.213s+0.285)} = Y(s)$$

or: $(0.213s^2 + 0.285s)Y(s) = (0.15s + 1)e(s)$

Since r(s) – y(s) = e(s) & assuming r(s) = 0, it can be stated that y(s) = –e(s)

or: $(0.213s^2 + 0.285s)e(s) = (0.15s + 1)e(s)$

$$\text{or: } (s^2 + 2.04225s + 4.694)e(s) = 0 \tag{1}$$

Using Inverse Laplace transform:

$$\ddot{e} + 2.04225\dot{e} + 4.694e = 0 \tag{2}$$

$$\text{Let: } x_1 = e,\ x_2 = \dot{e}, x_3 = \ddot{e} \ \text{ Thus: } \dot{x}_1 = x_2 \tag{3}$$

$$\dot{x}_2 = x_3 \tag{4}$$

$$\dot{x}_2 = -4.694x_1 - 2.04225x_2 \tag{5}$$

Equation (5) can be rewritten as,

$$\dot{x}_2 = -ax_1 - bx_2 \tag{6}$$

$$\text{where, } a = 4.694,\ b = 2.04225 \tag{6a}$$

Now taking scalar positive definite function is given by,

$$V(x) = \frac{1}{2}S_1x_1^2 + \frac{1}{2}S_2x_2^2 \tag{7}$$

where $S_1 > 0$, $S_2 > 0$

Now we take the derivative with respect to time *t*, yields

$$\begin{aligned} V(x) &= S_1x_1\dot{x}_1 + S_2x_2\dot{x}_2 \\ &= S_1x_1x_2 + S_2x_2x_3 \\ &= S_1x_1x_2 + S_2x_2(-ax_1 - bx_2) \\ &= x_1x_2(S_1 - aS_2) + S_2bx_2^2 \end{aligned} \tag{8}$$

For the positive definite function V we need another positive definite function U such that $\dot{V}(x) = -U(x)$

Now we take the coefficients in such a manner that

$$\dot{V}(x) = -U(x) \tag{9}$$

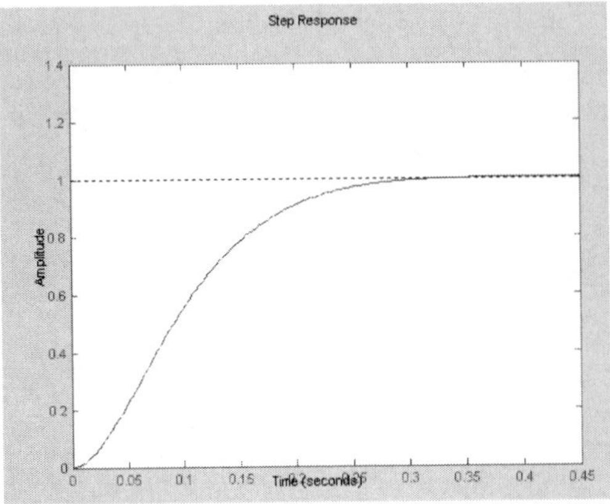

Figure 7. Step response of mathematical modelling of cardiovascular system.

Thus taking $(S_1 - S_2a) = 0$ (10)

$$bS_2 = 0 \tag{11}$$

Therefore, $S_2 = 0$, $S_1 = 0$

$\dot{V}(x) = 0$

Substituting equation (10 & 11)

Also $\dot{V}(0) = 0$, As from equation (8) $V(x) \to \infty$ as $x \to \infty$ & $\dot{V}(x) = 0$

A way of showing that $\dot{V}(x)$ being negative semidefinite is sufficient for asymptotic stability is to show that x_1 axis is not a trajectory of the system.

For $\dot{x}_1 = x_2 = 0$ & $\dot{x}_2 = x_3 = 0$ this shows that x_1 = m (constant).

The equilibrium state at the origin of the system is asymptotically stable.

3.2 *Simulation Action on Cardiovascular System*

The existing system approach of the specified cardiovascular system (Ghosal *et al.*, 2012) is presented equation no (v)

$$T_X(\sigma) = \frac{304.218}{s^2 + 31.39s + 304.218} \tag{ϖ}$$

The existing system step presented in "Fig 7." responded bounded well organized towards stability. To consideration of better performance and system efficacy the tuning method is implemented.

With implementation of controller by MATLAB Simulink (Saini *et al.*, 2009) the proper K_p, K_i, K_d values are construct presented in Table 3.

The tuning cardiovascular function are represented by $T_C(s)_{PID}$

Table 3. Controller parameters for cardiovascular system.

Parameters	Values
K_p	3.427
K_i	25.5966
K_d	0.031092

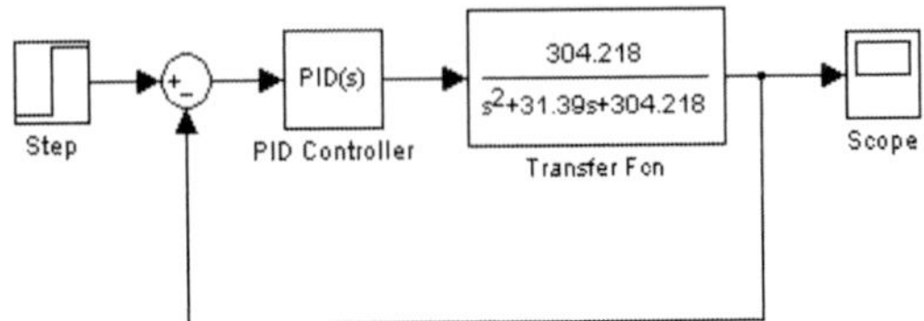

Figure 8. Mathematical model of the cardiovascular system.

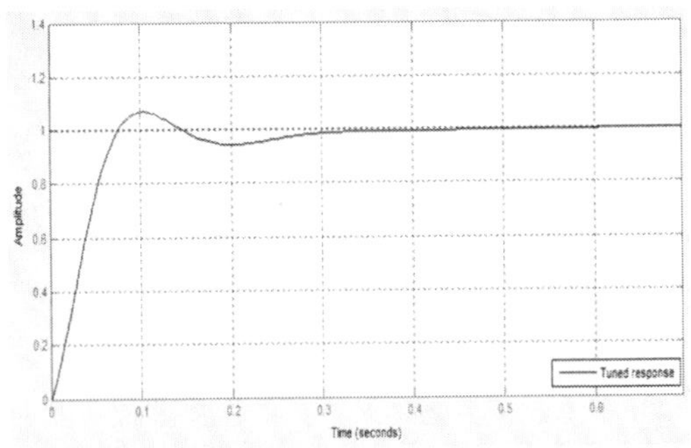

Figure 9. PID controller tuned output curve of the cardiovascular system.

$$T_X(\sigma)_{\Gamma\Pi\Delta} = \frac{9.46s^2 + 1043s + 87.04}{s(s^2 + 31.39s + 304.2)} \tag{ϖl}$$

For the control analogy the required figure is presented in "Fig. 8".

The important transient process characteristics parameters shown in Fig. 9 such as rise time, settling time, overshoot and peak is presented for better robustness of existing system.

The controllability and observability performance analysis of the cardiovascular system is observed. The controllability matrix of the tuned model is:

$$XO_{\chi\alpha\rho\delta\iota o\varpi\alpha\sigma\chi\upsilon\lambda\alpha\rho} = \begin{bmatrix} 1.000 & -31.3900 & 681.1141 \\ 0 & 1.0000 & -31.3900 \\ 0 & 0 & 1.0000 \end{bmatrix}$$

The observability matrix of the tuned transfer function

$$XO_{X\alpha\rho\delta\iota o\varpi\alpha\sigma\chi\upsilon\lambda\alpha\rho} = \begin{bmatrix} 0.0001 & 0.0104 & 0.0009 \\ 0.0075 & -0.0279 & 0 \\ -0.2620 & -2.2683 & 0 \end{bmatrix}$$

On the same way of simulation corresponding respiratory system this cardiovascular system is being analyzed its performance.

Here, Rank of $CO_{Cardiovascular}$ = Rank of $OB_{Cardio\text{-}vascular}$ = System Order = 3.

In conclude of performance analysis, it is stated the tuned cardiovascular system "$T_C(s)_{PID}$" is controllable and observable.

Stability Analysis using Lyapunov Stability Method:

Analysis of the Lyapunov stability of closed loop linear system is presented from Fig. 7.

We know, e(s)Gc(s)Gp(s) = Y(s)

$$\text{or}: e(s)\frac{304.218}{s^2 + 31.39s + 304.218} = Y(s)$$

or: $s^2 + 31.39s + 304.218)Y(s) = (304.218)e(s)$

Since r(s) – y(s) = e(s) & assuming r(s) = 0, it can be stated that y(s) = -e(s)

or: $s^2 + 31.39s + 304.218)e(s) = (304.218)e(s)$

or: $s^2 + 31.39s + 608.436)e(s) = 0$ (12)

Using Inverse Laplace transform:

$$\ddot{e} + 31.39\dot{e} + 608.436e = 0 \tag{13}$$

Let: $x_1 = e, x_2 = \dot{e}, x_3 = \ddot{e}$

Thus: $\dot{X}_1 = X_2$ (14)

$$\dot{X}_2 = X_3 \tag{15}$$

$$\dot{X}_2 = -608.436x_1 - 31.39x_2 \tag{16}$$

Equation (5) can be rewritten as,

$$\dot{x}_2 = -ax_1 - bx_2 \tag{17}$$

where, a = 608.436, b = 31.39 (18)

Table 4. Performance and robustness for cardiovascular system.

Parameters	Values
Rise Time	0.0542 seconds
Settling Time	0.282 seconds
Overshoot	7.02%
Peak	1.07

Now taking scalar positive definite function is given by,

$$V(x) = \frac{1}{2}S_1x_1^2 + \frac{1}{2}S_2x_2^2 \tag{19}$$

where $S_1 > 0,\ S_2 > 0$

Now we take the derivative with respect to time t, yields

$$\begin{aligned} \dot{V}(x) &= S_1x_1\dot{x}_1 + S_2x_2\dot{x}_2 \\ &= S_1x_1x_2 + S_2x_2x_3 \\ &= S_1x_1x_2 + S_2x_2(-ax_1 - bx_2) \\ &= x_1x_2(S_1 - aS_2) + S_2bx_2^2 \end{aligned} \tag{20}$$

For the positive definite function V, we need another positive definite function U such that

$$\dot{V}(x) = -U(x)$$

Now we take the coefficients in such a manner that

$$\dot{V}(x) = -U(x) \tag{21}$$

Thus taking $(S_1 - S_2a) = 0$ (22)

$$bS_2 = 0 \tag{23}$$

Therefore, $S_2 = 0,\ S_1 = 0$

$\dot{V}(x) = 0$ substituting equation (22 & 23)

Also $\dot{V}(0) = 0$, As from equation (20) $V(x) \rightarrow \infty$ as $x \rightarrow \infty$ & $\dot{V}(x) = 0$

A way of showing that $\dot{V}(x)$ being negative semidefinite is sufficient for asymptotic stability is to show that x_1 axis is not a trajectory of the system.

For $\dot{x}_1 = x_2 = 0$ & $\dot{x}_2 = x_3 = 0$ this shows that $x_1 = m$ (constant).

The equilibrium state at the origin of the system is asymptotically stable.

4 CONCLUSIONS

Respiratory and cardiovascular systems are most effective interconnected organic systems of living body. Lots of research activities are carried out by different scientists in this field. Modelling analogy defines a vivid future aspect of entire effective system. This piece of work can bring the light of dawn in the area of prosthetic research and development of the human being. It also deals with the evolution of computational bionic system contributing a new path in the research and development of biomedical domain.

REFERENCES

Anil Kumar, C.L. Varshney, Veer Pal Singh, "Performance Study on Effect of Cognitive States on Heart Rate, Blood Pressure and Respiration Rate", Indian Journal of Biomechanics: Special Issue (NCBM 7–8 March 2009).

Anju Saini, V.K. Katiyar, Pratibha, Mathematical Modeling of Lung Mechanics-A Review, Indian Journal of Biomechanics: Special Issue (NCBM 7–8 March 2009).

Ghosal, S., R. Darbar, B. Neogi, A. Dasand D.N. Tibarewala, "Application of Swarm Intelligence Computation Techniques in PID Controller Tuning: A Review", Book Chapter of Advances in Intelligent and Soft Computing, Springer Series 2012, vol. 132/2012, pp. 195–208.

Neogi, B., Dr. S. Ghosal, S. Ghosh, T.K. Bose, Dr. A. Das, "Dynamic Modeling and Optimizations of Mechanical Prosthetic Arm by Simulation Technique "IEEE Sponsored International Conference on Recent Advances in Information Technology, vol. 01, pp. 883–888, 2012.

Soumyendu Bhattacharjee, Sayanti Roy, Zinkar Das, Abesh Das and Biswarup Neogi, "An Approach towards Error Less ECG Signal Equation Based on Computational Simulation Aspect with Modeling of Cardiovascular Disorder Diagnosis," Int. Conf. on Control, Instrumentation, Energy & Communication, pp. 231–235, 2014.

Soumyendu Bhattacharjee, Zinkar Das, Mainuck Das, Sanjoy Kr. Chakraborty, Sukriti Bhatia, Prapti Kumari, Biswarup Neogi, "Investigation on Study and Modeling Based Analytical Phenomenon of Cardiovascular System towards Heart Disease Detection," International Journal of Advanced Scientific and Technical Research, vol. 6, iss. 3, pp. 599–606, 2013.

Computational Science and Engineering – Deyasi et al. (Eds)
© 2017 Taylor & Francis Group, ISBN 978-1-138-02983-5

Lossless electrocardiogram compression using segmented beats by dynamic bit allocation and Huffman encoders: A study on performances

Priyanka Bera & Rajarshi Gupta
Department of Applied Physics, University of Calcutta, Kolkata, India

ABSTRACT: Lossless compression of Electrocardiogram (ECG) can ensure preservation of clinical features in the dataset. In this work, we applied two such encoders, Huffman coder and Dynamic Bit Allocation (DBA) to compare their performance for single lead ECG compression. Each ECG beat was segmented into three zones P, QRS and T. The encoders were applied to compress biased first order difference for P and T and second order difference for QRS with ptbdb and mitdb data at 8 bit quantization and resampled at 1 kHz. An average compression ratio for of 5.69 and 5.14 were obtained with mitdb and ptbdb data respectively using Huffman coder, where as the same using DBA were 3.99 and 3.97 respectively. The Huffman encoder showed better performance in terms of bits per sample efficiency (1.46 vs. 1.99) and nearly same for data overhead per packet (1.06% vs. 1.04%). However, the DBA encoder showed lower time complexity.

1 INTRODUCTION

Electrocardiography (ECG) is a popular non-invasive evaluation method for preliminary level investigation of cardiac functions. It represents sequential activation of the cardiac chambers, atrium and ventricle through the characteristics waves P, QRS and T, recorded in a pre-calibrated graph paper on 12 traces, each called a 'lead' of ECG.

Recording of high volume of Electrocardiogram data requires digital storage for offline rhythm analysis. Data compression can save memory space for its efficient storage. Telemonitoring application is one other application of compression. Most of the research on ECG compression has been centered on enhancing compressor's efficiency (compression ratio), lowering reconstruction error like PRDN (percentage root mean square difference normalized), and diagnostic acceptability (distortion measures). Two broad approaches for ECG compression are available in the literature, viz., Direct Data Compression (DDC) or time-domain and Transform Domain (TD) methods (Gupta *et al.*, 2014). DDC methods utilize intra-beat and inter-beat redundancy or correlation between a group of samples. They show appreciable distortion in high compression ratios. Popular TD methods employ redundancy reduction (predictors), AZTEC (Cox *et al.*, 1968), Fan (Steward *et al.*, 1973), CORTES (Abenstein *et al.*, 1982), delta (Mitra *et al.*, 2012) and ASCII character encoding (Mukhopadhyay *et al.*, 2012). Transform domain techniques, mostly lossy in nature, prove to be highly efficient for ECG compression due to their ability of energy compaction, and energy preservation in few coefficients and appropriate choice of basis vector. Wavelet Transforms (WT) (Ku *et al.*, 2006, Alesanco *et al.*, 2006, Manikandan *et al.*, 2008), cosine transforms (CT) (Batista *et al.*, 2001), and Karhunen Louve Transform (Degani *et al.*, 1990, Blanchett *et al.*, 1998) has been very popular as lossy ECG compression methods in last two decades. However, unless quality control measures are adopted (Gupta *et al.*, 2016), lossy compression methods may lead to distortions of the ECG at high compression ratio. Furthermore, these methods are computationally complex, which puts a hindrance for their real time deployment in low cost applications. From this angle lossless compression methods, although generating lower compression ratio, are safe to adopt (Koski *et al.*, 1997), since they ensure exact reconstruction of the original data. In our earlier work (Bera *et al.*, 2016) we introduced a simple method, named Dynamic Bi Allocation (DBA) based on zonal complexity of ECG data with the aim to achieve moderate CR and low entropy (or bits per sample) of the encoder. An advantageous feature of a typical ECG waveform is that the most of the energy content and fluctuations are confined in a very small region, the QRS complex. A significant portion of the waveform exhibits near equipotential regions (TP, ST and PQ). If the P and T waves are included, number of distinct 'symbols' (digital values) will be low. Hence, Huffman coder (Chung *et al.*, 1997) can be suitable for lossless ECG compression.

In this work we have done a comparative study performance between two methods, viz., Huffman Encoding and a lossless version of DBA encoder. To enhance the efficiency of the encoders each ECG beat was segmented into three regions, according to heart chamber activities, viz., zone 1 (P-wave), Zone-2 (QRS) and zone-3 (T). Comparison has been done based on entropy and time complexity in addition to the conventional performance metrics. The layout of this paper is as follows. Section II discusses the preprocessing of ECG followed by brief discussion on compression logics of both techniques. Section-III discusses the test results with benchmark ECG data. Section-IV provides conclusions on performance study of both lossless techniques.

2 METHODOLOGY

The proposed algorithm was implemented in 55 second ECG data collected from ptddb and mitdb under Physionet (www.physionet.org), which provide a large repository for different ECG and other signals for testing and validation of signal processing algorithms. The algorithm steps are briefly illustrated in following sub-sections:

2.1 *Preprocessing ECG data*

At first, removal of baseline wander and high frequency noise along with R-peak detection was performed using a Discrete Wavelet Transform (DWT) based approach, as detailed in (Banerjee *et al.*, 2012). In the next stage, baseline points were marked in the TP segments by dividing the R-R interval in 2:1 ratio. The beats were cut and aligned w.r.t. the R-peaks to form an array a_beat. Since heart rate normally varies among beats, shorter beats were padded at the start and tail. The R-peak aligned beat matrix can be represented as:

$$a\_beat = [b_1, b_2, \dots b_n] \tag{1}$$

where n = number of beats and b_k = k[th] beat which can be represented as,

$$b_k = [x_{k1}\ x_{k2}\ x_{k3} \dots x_{lk}] \tag{2}$$

where, l = length of the beat after zero padding. A beat array b_k having length will have R-peak position at r_index and have $(l\text{–}l_k)$ number of total padding elements, x_{kj} (j = 1, 2, … l) are raw ECG samples in beat b_k.

2.2 *Beat Segmentation, Data formatting and presentation*

A regular ECG sequence has some high fluctuation segments (QRS region) and some medium and low fluctuation (P wave, PQ region, T wave, ST region etc). For each beat vector, the QRS region was empirically considered as a 130 ms window around the R-peak index, starting from 60 ms ahead and ending at 70 ms after it.

8-bit quantizer block converts the milli-volt samples of every beat according to the following equation:

$$x_q(k) = \frac{255 \times [x(k) - x_{max}]}{x_{max} - x_{min}} \tag{3}$$

where, x is the millivolt sample, x_q quantized data, x_{max} is the maximum value of x(k) and x_{min} is the minimum value of x(k).

2.3 *First and second order difference computation*

The first delta and biasing in combination converts the x_q sequence into minimum positive elements using the following equations:

$$\begin{aligned} d_1(k) &= x_q(k+1) - x_q(k) \\ bd_1(k) &= d_1(k) - \min(d_1) \\ bd_1(k) &= d_1(k) \end{aligned} \tag{4}$$

Equation [4] was applied for zone1 and zone3. The second delta and biasing in combination is applied for zone 2 using the following equations:

$$\begin{aligned} d_2(k) &= d_1(k+1) - d_2(k) \\ bd_2(k) &= d_2(k) - \min(d_2) \end{aligned} \tag{5}$$

Up to this stage of processing, the logic flow diagram can be represented in Figure 1.

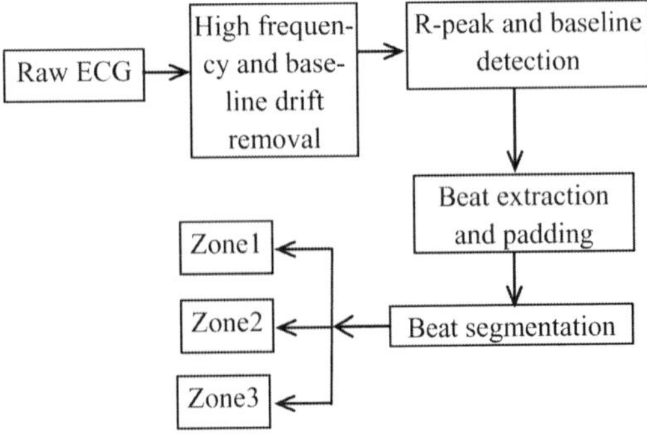

Figure 1. Preprocessing steps.

2.4 *Zonal compression*

In this stage, each of the zones was compressed using two variable length lossless encoding approaches, a. Huffman coding and b. DBA strategy.

Table 1. Huffman table (Typical zone1).

Huffman symbols	No. of occurrences	Probability	Codes
1	258	0.9699	0
2	5	0.0188	10
0	3	0.0133	11

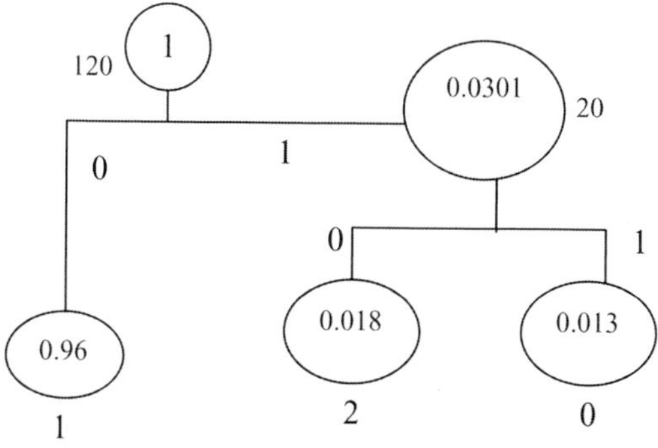

Figure 2. Construction of Huffman table (Physionet data p184/s0551_re lead-II).

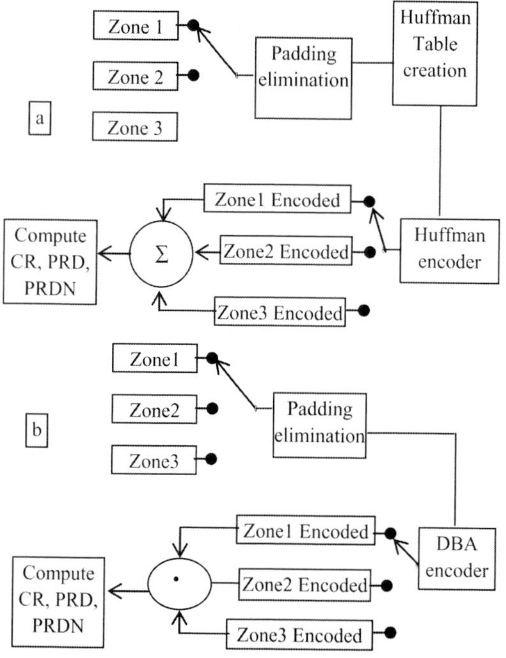

Figure 3. Compression steps for (a) Huffman encoder, (b) DBA encoder.

2.4.1 *Huffman Encoder*

In the proposed work, Huffman encoding was performed bd_1 array for zone 1 and zone 3 and bd_2 for zone 2. Considering a hypothetical example of zone 1, with unique symbol elements 0, 1 and 2, the Huffman table is shown in Table 1 and the corresponding Huffman tree constructed is shown in Figure (Cox *et al.*, 1968). The logic flow diagram for Huffman coder for zonal compression is shown in Figure 3.

The Huffman Encoder appends a header for zone 1, zone 2 and zone 3, the structures of which are shown in Table 2. The detailed methodology of the encoding is available in (Gupta *et al.*, 2015).

2.4.2 *Dynamic Bit Allocation (DBA) Encoder*

Dynamic Bit Allocation (DBA), a variable length encoder proposed by us [15], allocates bits to encode each sample in a group based on its zonal complexity. In a regular ECG sequence, the zone 1(P wave)

Table 2. Header format for Huffman encoder.

Zone	H-byte	Information	No. of bits allocated
Zone1, Zone 3	HB1	Zero pad for Huffman symbol	3
		Zero pad for Huffman bit	3
		Bias flag	1
	HB2	Bias amount	4
		No of Huffman symbol	4
Zone 2	HB1	Zero pad for Huffman symbol	3
		Zero pad for Huffman bit	3
		Bias flag	1
	HB2	First element of del1	4
		No of Huffman symbol	4

Table 3. Header format for DBA encoder.

Zone	H-byte	Information	No. of bits allocated
Zone1, Zone 3	HB1	Zone type	1
		No of bytes	4
		No of zeros for zero padding	3
	HB2	Biasing	1
		Bias amount	3
		No of bit to encode delta samples	4
Zone 2	HB1	Zone type	1
		No of bytes	7
	HB2	Bias amount	5
		No of Huffman symbol	1

and zone 3 (T wave) are having low fluctuations and treated as 'plain' zone, whereas zone 2, having high fluctuations was considered as 'complex' zone. Unlike the work in (Bera *et al.*, 2016), where the zone attribution was based on real-time estimate of complexity, in the current work, the same has been pre-fixed, without down-sampling. The logic flow diagram of DBA encoder is shown in Figure 3 and the header details for DBA for zone 1, 2 and 3 are shown in Table 3.

3 RESULTS AND DISCUSSION

The compression and decompression algorithms were tested with arbitrarily chosen 55 seconds records from Physionet. For comparison of performances, all these datasets were re-sampled at 1 kHz. Table 4 provides comparison study of test results of both techniques with mitdb and ptbdb data respectively. The subjective measures used for evaluation of compression performance were percentage root

Table 4. Compression results with mitdb and ptbdb data.

Source database, Patient ID (record no.), lead	PRD	PRDN	MSE ($\times 10^{-4}$)	RMSE ($\times 10^{-3}$)	Huffman encoder CR	Huffman encoder BPS	DBA encoder CR	DBA encoder BPS
mitdb:(117), v1	7.53	7.53	3.64	19	6.17	1.29	3.89	1.99
mitdb:(124), v1	7.66	7.67	3.67	19	6.12	1.31	4.00	1.99
mitdb:(231), ml2	6.36	6.36	3.48	19	5.03	1.59	3.97	2.00
mitdb:(234), ml2	6.53	6.53	3.61	19	5.49	1.46	3.99	1.99
ptbdb:p222(s0444_re), v5	4.57	4.58	3.41	18	5.49	1.45	3.90	2.04
ptbdb:p232(s0456_re), v1	3.40	3.41	3.54	18	5.18	1.54	4.00	1.98
ptbdb:p232(s0456_re), v2	6.36	6.37	3.56	18	5.03	1.62	4.00	1.99

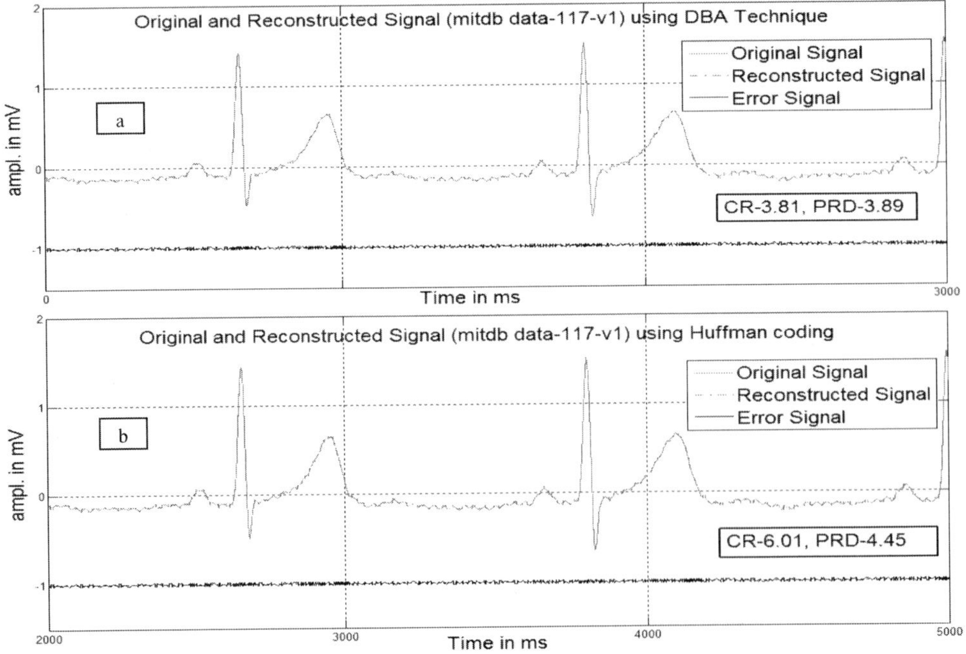

Figure 4. Reconstruction performance of the algorithms: (a) DBA encoder; (a) Huffman encoder with mitdb data (for ease of interpretation, the error signal has been shown with a fixed offset of -1 mV).

mean squared difference (PRD), PRD normalized, Mean Square Error (MSE), and root MSE (RMSE) (Gupta *et al.*, 2014). Figure 4 shows the reconstruction efficiency of both the encoders. It is observed that the reconstruction signal closely follows the original one and in both the cases. For each dataset the digital error, computed between the quantized original and reconstructed digital bit pattern were found as zero. Hence, clinical validation was not necessary with the qualified cardiologists. Furthermore, since both the compression techniques are lossless, the error figures in Table 4 are due to 8 bit quantization. On mitdb, the average CR achieved by Huffman coder and DBA encoder was 5.50 and 3.98 respectively. In terms of Bits Per Sample Efficiency (BPS) Huffman encoder performs superior (1.46) compared to DBA encoder (1.99). This is attributed to the zonal compression where each wave (zone) was considered at a time. Hence, for each wave, the maximum amplitude of the first (or second) order difference determined the span of the zone and hence, the numbers of bits to encode each sample. However, the computational simplicity of the DBA over Huffman encoder is reflected by its time complexity (41.09 μs/sample vs. 48.36 μs/sample for DBA encoder at MATLAB2009b® using Intel core i3–2350M CPU at 2.30 GHz supported by 4 GB RAM). We also computed data overhead in terms of extra bytes that the encoders put in addition to the compressed characters. In this respect both the encoders yields nearly same performance (1.04% by DBA encoder and 1.06% by Huffman encoder).

4 CONCLUSION

In this paper we have compared two lossless compression techniques, namely, DBA encoder and Huffman encoder for single lead ECG compression using segmented ECG beats. The MATLAB® simulation result is provided with ptbdb and mitdb data at 8 bit resolution. It was observed that in terms of compression ratio and bit allocation efficiency, the Huffman coder outperforms the DBA encoder (Cr of 5.50 vs. 3.98 and BPS of 1.46 vs. 1.99). However, the DBA encoder works better in terms of computational complexity. Since an ECG signal contains more than 60% flat or near flat region, and Huffman encoder can be useful for low cost standalone applications.

ACKNOWLEDGMENT

The authors acknowledge the UGC SAP DRS-II project (2015–2020) at Dept of Applied Physics, CU. Priyanka Bera acknowledges the financial support from DST-WB funded project under the Department of Applied Physics, University of Calcutta.

REFERENCES

Abenstein, J. P., Tompkins, W. J., (1982) New data-reduction algorithm for real-time ECG analysis, IEEE Trans. Biomed. Eng, BME-29, 43–48.

Alesanco, A., Olmos, S., Istepanian, S. H., Garcia, J., (2006) Enhanced real-time ECG coder for packetized telecardiology applications, IEEE Trans Inf Tech. Biomed, 10(2), 229–236.

Banerjee, S., Gupta, R., Mitra, M., (2012) Delineation of ECG characteristic features using multiresolution wavelet analysis method, Measurement, 45(3), 474–487.

Batista, L. V., Melcher, E. U. K., Carvalho, L. C., (2001) Compression of ECG signals by optimized quantization of discrete cosine transform Coefficient, Med Eng & Phy, 23, 127–134.

Bera, P., Gupta, R., (2016) Real-time compression of electrocardiogram using dynamic bit allocation strategy, First IEEE International Conference on Control, Instrumentation & Measurement (CMI) 2016.

Blanchett, T., Kember G. C., Fenton, G. A., (1998) KLT-based quality controlled compression of single-lead ECG, IEEE Trans Biomed Eng, 45(7), 942–945.

Chung, K. L., Lin, Y. K., (1997) A novel memory-efficient Huffman decoding algorithm and its implementation, Signal Processing, 62, 207–213.

Cox, J. R., Nolle, F. M., Fozzard, H. A., Oliver, G. C., (1968) AZTEC, a preprocessing program for real-time ECG rhythm analysis, IEEE Trans. Biomed. Eng., BME-15, 128–129.

Degani, R., Bortolan, G., Murolo, R., (1990) Karhuman-Loeve coding of ECG signals, Proc. Computers in Cardiology, 395–398.

Gupta, R., (2015) Lossless compression technique for real time photoplethysmographic measurements, IEEE Trans. Inst. Meas., 64(4), 975–983.

Gupta, R., (2016) Quality aware compression of electrocardiogram using principal component analysis, Jour Med Sys (Springer), 40(5), 112,.1–11.

Gupta, R., Mitra, M., Bera, J. N., (2014) ECG acquisition and automate remote processing, Springer, India.

Koski, A. (1997) Lossless ECG encoding, Computer Methods and Programs in Biomedicine (Elsevier), 52, 23–33.

Ku, C. T., Wang, H. S., Hung, K. C., Hung, Y. S., (2006) A novel ECG data compression method based on non-recursive discrete periodized wavelet transform, IEEE Trans Biomed Eng, 53, 2577–2583.

Manikandan, M. S., Dandapat, S., (2008) Wavelet threshold based TDL and TDR algorithms for real-time ECG signal compression, Biomed. Sig. Proc. Cont., 3, 44–66.

Mitra, M., Bera, J. N., Gupta, R., (2012) Electrocardiogram compression technique for global system of mobile-based offline telecardiology application for rural clinics in India, IET Sc, Meas Tech, 6(6), 412–419.

Mukhopadhyay, S. K., Mitra, S., Mitra, M., (2012) An ECG signal compression technique using ASCII character encoding, Measurement, 45(6), 1651–1660.

Physionet website - www.physionet.org.

Steward, D., Dower, G. E., Suranyi, O., (1973) An ECG compression code, Electrocardiology, 6(2), 175–176.

Track IV: Image processing

Computational Science and Engineering – Deyasi et al. (Eds)
© 2017 Taylor & Francis Group, ISBN 978-1-138-02983-5

Identification of commercially available turmeric powders using color projection features

Dipankar Mandal
Department of Applied Electronics and Instrumentation Engineering, Future Institute of Engineering and Management, Kolkata, India

Arpitam Chatterjee
Department of Printing Engineering, Jadavpur University, Kolkata, India

Bipan Tudu
Department of Instrumentation and Electronics Engineering, Jadavpur University, Kolkata, India

ABSTRACT: Turmeric powder is an essential ingredient in Indian cuisine. It has found several applications in traditional medicinal applications as well. It is widely available in the market supplied by different manufacturers. The prices of different brands vary according to the purity and quality which also arises the chances of mixing low quality with the high quality turmeric powder and selling at the price of high quality one. Detecting such mixing is important to prevent but difficult to achieve manually. This paper presents a machine vision based approach to perform such discrimination between the original and mixed grades of market available turmeric powder. The frequency transform of color features along With Principle Component Analysis (PCA) based identification being performed. The experimental results show that the presented method may be considered to discriminate between original and mixed varieties of commercially available turmeric powders.

1 INTRODUCTION

Turmeric *(curcuma longa L)* is popularly used in various applications due to its medicinal values (Du et al., 2006). It can be used both internally and externally. In ayurvedic as well as Chinese medicinal application, it is widely used to treat inflammatory conditions. It is used as tonic and blood purifier when it is consumed internally and used in prevention and treatment of skin disease even for wound healing when it is applied externally. Apart from medicinal usages, it is very popular rather essential in its grounded form in Indian household and sub-continental cuisine. Turmeric powder extracted from dry turmeric rhizomes is extensively and widely used as food colorant and in food industry it is one of the most vital ingredients in curry powder, mustard paste etc. In household turmeric powder is extensively used as spice which imparts color, aroma and taste to different food preparations and sometimes is used for masking undesirable ordours of food.

The main source of turmeric production is in countries like India, Indonesia, Jamaica and china. In India there are many companies which manufacture different types of branded turmeric powder. Turmeric rhizomes contain curcuminoids which is responsible for its yellow color. The final color of the packed turmeric powders in the market also varies in terms of their color although in general all are yellow in appearance. At the same point of time it may be difficult to trace whether the desired quality of turmeric powder is mixed with low quality or not. Normally the branded turmeric powders should be such that it should not contain artificial colors such that its quality remains high and it should be up to the mark so that it does not pose any health threats. but the loose non branded local turmeric powders may contain extraneous coloring material which may be harmful to health. This paper aims to discriminate between different packed turmeric powders available in Indian market based through machine vision.

Food quality evaluation using machine vision is becoming important research field due to its many advantages such as low cost, faster processing, non-invasive nature, etc. etc. (Dixit et al.,2009). In a very general way the major components of a machine vision based food quality evaluation system are image acquisition, image processing, image feature extractions and identification or classification. The presented method also comprises all of the men-

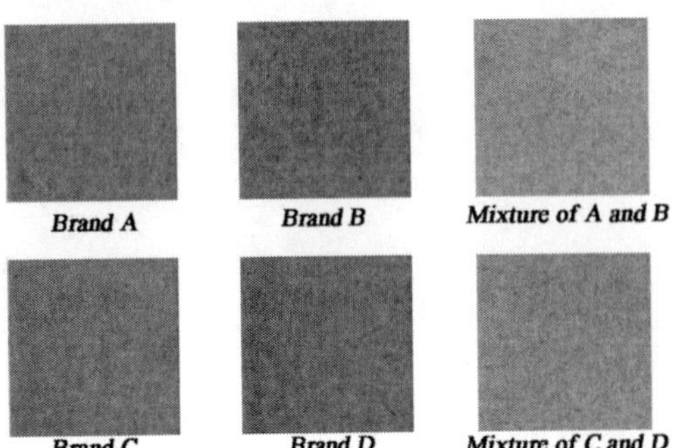

Figure 1. The pictures of turmeric powders used of experiments.

tioned steps as described in following sections. The Fourier transform of horizontal and vertical projections of color channel-wise information is used as feature sets and a principle component analysis (PCA) based identification is realized. The results show that the presented method can be found potential in order of grading turmeric powder which is difficult and tedious to perform manually.

2 PRESENT METHOD

The experiments were carried with different branded as well as non-branded turmeric powders available in market. The different samples were

Horizontal projection sample 1

Horizontal projection sample 2

Vertical projection sample 1

Vertical projection sample 2

FFT of Horizontal projection sample 1

FFT of Horizontal projection sample 2

FFT of Vertical projection sample 1

FFT of Vertical projection sample 2

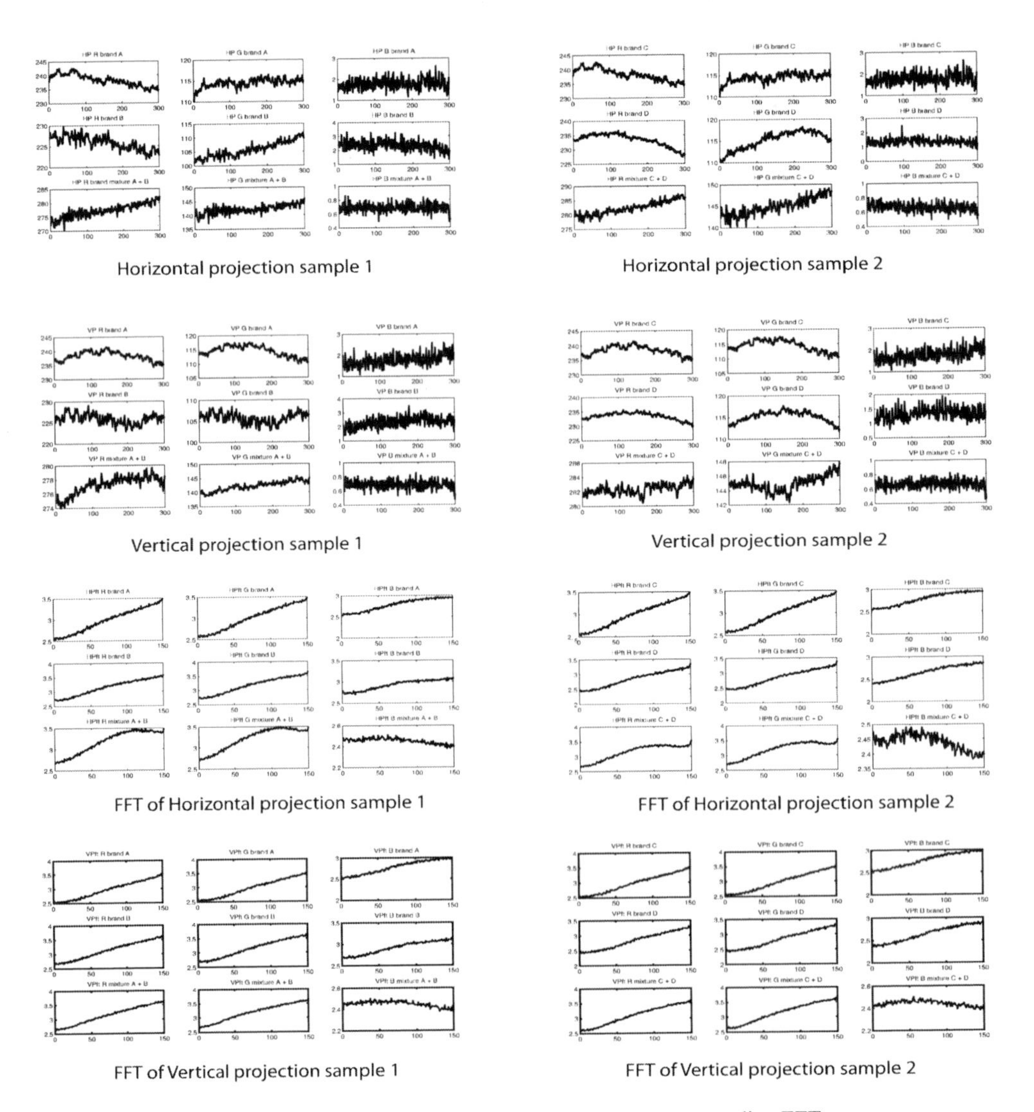

Figure 2. Plots of channel-wise horizontal and vertical projections with corresponding FFTs.

also created by mixing them for experimental purpose. Following the general workflow of a machine vision system the presented method comprises of four major components and each of them are described in following sections.

Image acquisition and processing: Here images of some branded packaged, one non branded loose and uniform mixture of branded and non branded loose turmeric powder of some particular volume and particular weight are taken with the help of SONY DSC-W530 Camera. To nullify the effect of illumination and maintaining constant distance for all the images a simple imaging chamber, developed by the research team, was used. The images captured images were further cropped to a fixed size of 300 × 300 pixels. Some of such cropped samples are shown in Fig. 1.

2.1 *Image feature extractions*

The color features were considered for this work. The different color channels i.e. R-G-B were separated and the horizontal and vertical projections (Chandra et al., 2006) for each channel were calculated. Plots of such projections are shown in Fig. 2. The plots show significant difference in terms of the trend characteristics. Sample 1 consists of two different branded turmeric powders and their mixture. Similarly sample 2 consists of other two different branded turmeric powder and their mixture. The horizontal projections of both the samples reveal that R channel projection shows significant difference among three brands of sample 1 as well as three brands of sample 2. The characteristic of vertical projections of both the samples reveal that the differences are prominent for both R and G channels.

This was further extended by taking the frequency transformation of the projections by means of FFT. The plots of corresponding FFTs are also shown Fig. 2. The horizontal projection of FFT for the both samples reveals that there is visible difference in B channel projection. Similarly it is also clear from the figure that the vertical projection of FFT shows significant trend difference in B channel.

Based on the above analysis the feature set was generated. The set consisting of the basic statistical measures i.e. mean, median, mode, standard deviation, variance, etc. derived from the R channel both horizontal and vertical projections without FFT, G channel horizontal projection vertical projection without FFT and B channel both vertical and horizontal channel with FFT. This feature sets were subjected to PCA for evaluating the potential of this dataset towards identification of different grades.

2.2 *Identification of different grades*

PCA (Ham et al., 2002) plotting is an statistical method which can determine an optimal linear transform matrix such that given input vector which is considered to be from a zero mean wide sense stationary stochastic process, the data can be compressed. PCA transforms large amount of correlated input data to a set of statistically décor related features. The components are usually ordered according to decreased variance. PCA plot using the extracted feature sets were examined to investigate whether the grades are creating separable classes from the mixture of grades.

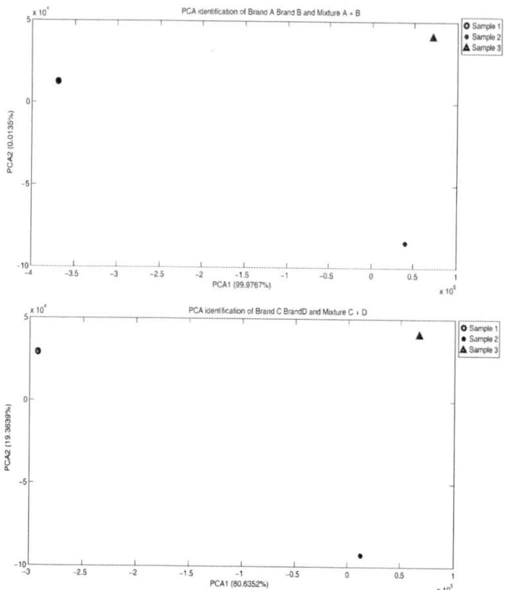

Figure 3. The PCA plots of different samples for identification of different grades.

3 RESULTS AND DISCUSSIONS

The PCA plots for two samples each comprising of two different grades of turmeric powders and their mixtures are shown in Fig. 3. Both cases it is clearly seen that with the extracted feature-set PCA plot shows three different classes which are considerably separate from each other.

4 CONCLUSIONS

This paper presents a color feature based identification process for different grades of turmeric powders. This can be useful not only for grading but also to identify detection of so called high market priced turmeric powder mixed with low priced one. The results show that the presented can be potentially useful for the stated purpose. The work can be further extended to detection of additive colors with the dry rhizomes of turmeric which can significantly affect the quality of turmeric powder.

REFERENCES

Babu, P.R., (2004) Digital Signal Processing, Scitech.

Chandra, B., & Dutta Majumder, D., (2006) Digital image processing and analysis, PHI, New Delhi, India.

Dixit, S., Puroshottm, S.K., Khanna, S.K., & Das, M., (2009). Surveillance of the quality of turmeric powders from city markets of India on the basis of cur cumin content and the presence of extraneous colors, Food Additives and Contaminants,1227–1231.

Du, C.J., Sun, D.W., (2006) Learning techniques used in computer vision for food quality evaluation: a review, Journal of Food Engineering 72, 39–55.

Ham, F., Kostanic, I., (2002) Principles of Neurocomputing for Science & Engineering, TMH.

Li, S., Yuan, W., Deng, G., Wang, P., Yang, P., Aggarwal. B., (2011) Chemical composition and product quality control of turmeric (Curcuma longa L.) Pharmaceutical Crops, 28–54.

Computational Science and Engineering – Deyasi et al. (Eds)
© 2017 Taylor & Francis Group, ISBN 978-1-138-02983-5

Color feature based approach towards maturity discrimination of medicinal plants: Case study with Kalmegh (*Andrographis paniculata*)

G. Mukherjee
Department of MCA, Regent Education and Research Foundation, Kolkata, India

A. Chatterjee
Department of Printing Engineering, Jadavpur University, Kolkata, India

B. Tudu
Department of Instrumentation and Electronics Engineering, Jadavpur University, Kolkata, India

ABSTRACT: Indian medicinal plants are popularly used for medicinal purposes. To get optimum effects from these plants, it is important to pluck leaves in appropriate maturity state. This paper presents a computer vision based approach for maturity detection of Indian medicinal plants where the studies being performed with Kalmegh (*Andrographis paniculata*) leaves. The leaves of different maturity stages were imaged and different statistical features from leaves were extracted with different combinations of R, G and B color channels. The resultant features were converted to frequency domain and followed by their horizontal and vertical projections. The discrimination of different maturity levels were performed using PCA analysis while the cluster centers were calculated by K–means. The class seperability measures are also presented. The results show that the presented color feature based method can be a potential tool for maturity discrimination of medicinal plants.

1 INTRODUCTION

Medicinal plant has made a mark in the domain of medicine due to its rich and varied contribution towards remedy of many critical and chronic diseases worldwide. The legacy of medicinal plant cultivation and research has started from very ancient time and has become in advanced stage at present. This popular medicine system is known as the Ayurveda and has the root in the very ancient time. Wide variety of different plants Neem (Azadirachtaindica), Tulsi (*OcimumSanctum*), Kalmegh (*Andrographis Paniculata*), Basaka (Justicia adhatoda), etc. are popularly used for their medicinal values (Joy et al. 1998). In order to get the good quality of medicinal extracts from the leaves of medicinal plant, we need to know the exact maturity of plant. Detection of maturity and segregation of the mature leaves from the premature and overmature one can be done by applying computer vision. Number of veins, leaf size, leaf blade etc can be used to compute the age and maturity level of the leaves. (Tsuyuzaki 2000) gave a detailed account for maturity detection on the basis of number of veins, but this process is very much temperature dependent.

In this work, one of the popular medicinal plant Kalmegh (*Andropgraphis paniculate*), has been choosen for experimental purpose where three kinds of kalmegh leaves have been collected at their premature, mature and overmature stage. Images of each of these group of leaves are subjected to the Fast Fourier Transform (FFT) followed by the statistical analysis to get horizontally and vertically projected dataset of red, green and blue channels. Each of these datasets of respective channels are analyzed with PCA to form the respective cluster and the cluster centers of each such clusters are formed by applying the K—means algorithm.

Inter cluster separability has been performed to assess the good distinction between the different leaves of Kalmegh belonging to different classes of maturity.

The experiments were carried out in two ways. First, the cluster triads of premature, mature and overmature leaves for red, green and blue channels were formed, respectively. Afterwards the cluster triads were formed by taking different combination of red, green and blue channels each representing leaves of different levels of maturity. Secondly, the unknown leaves were taken and included into respective cluster of each category of leaves. The PCA analysis was carried out for the newly formed group and centre of respective clusters were determined by K-means algorithm once again. Finally, the shifts of centers from the original cluster center positions were measured. The unknown leaves were considered to belong to one of the defined level of maturity class depending on the principle that shift of center for the class, to which the unknown leaf belongs, is minimal.

2 PROPOSED METHODOLOGY

The proposed methodology consists of the following stages 1) image acquisition, 2) denoising of the captured images, 3) Discrete Fourier transform 4) statistical feature extraction 5) Formation of feature vectors 6) Calculation of class separability.

2.1 *Image acquisition*

Forty five different leaves of Kalmegh plant were collected. Group of fifteen leaves for each category were labeled as premature, mature and overmature classes. The images were collected with the help of digital camera of 15 megapixels against white background keeping the fixed distance of 40 cm. Samples of the captured images of different categories of leaves are shown in Fig. 1. The captured images were further cropped for feature extraction.

2.2 *Denoising of images*

Gaussian noise is the very commonly seen noise, whose likely cause is the poor illumination or high temperature. Gaussian filter as expressed in Eq. (1) was applied to denoise the captured images of leaves.

$$G(x,y)=\frac{1}{2\pi\sigma^2}e^{\frac{x^2+y^2}{2\sigma^2}} \tag{1}$$

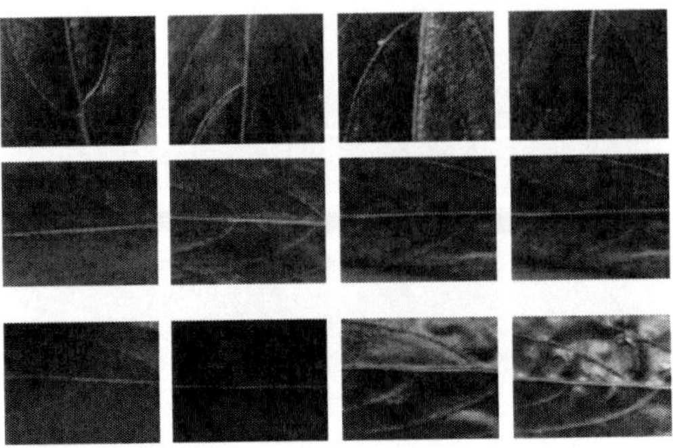

Figure 1. Sample of cropped images for (top) premature, (middle) mature and (bottom) over mature leaves.

2.3 *Discrete Fourier Transform*

Discrete Fourier Transform converts the finite sequence of equally spaced samples of function into the finite list of coefficients of a finite combination of a complex sinusoids ordered by their frequencies containing the same sample values. By this transformation the sampled function converts to frequency domain and it can be realized using FFT (Cooley et al. 1969). FFT was applied to convert

Table 1. Combinations of color channels for different cluster.

Combination No	Premature	Mature	Overmature
1	Red	Red	Red
2	Red	Green	Blue
3	Red	Blue	Green
4	Blue	Blue	Blue
5	Blue	Green	Red
6	Blue	Red	Green
7	Green	Green	Green
8	Green	Red	Blue
9	Green	Blue	Red

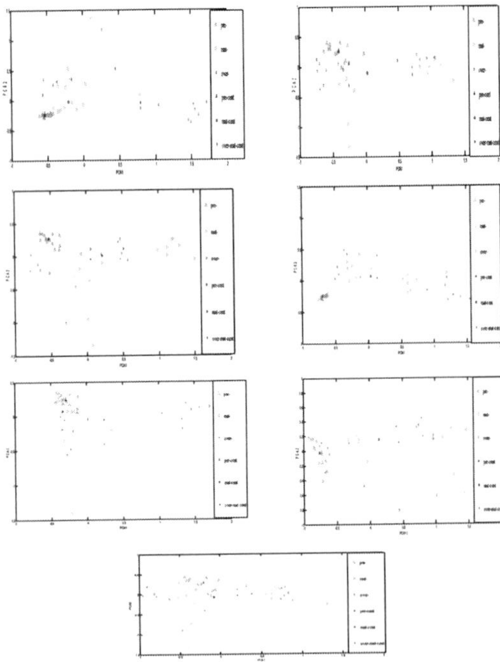

Figure 2. PCA plots for (top row left) blue channel based horizontal projection, (top row right) green channel based horizontal projection (2nd row left) GRB combination based horizontal projection (2nd row right) RBG combination based horizontal projection (3rd row left) BRG combination based horizontal projection (3rd row right) GBR combination based vertical projection. (Bottom row) RGB combination based horizontal projection.

the intensity of captured image into the frequency domain. The frequency components were divided into the horizontally and vertically projected components before being applied with the statistical functions for the purpose of feature extraction.

2.4 *Statistical feature extraction*

Feature extraction implies extracting features like texture or color from the images of leaves. Texture features of the leaves can be extracted by applying other different algorithms like Fractal based algorithm, Gabor filter and Markov Random field. In this paper, the textural features from the images of Kalmegh leaves have been extracted using statistical features for their respective category of maturity. Textures are the quantitative measure of the arrangements of intensity levels in the image region. The statistical properties of intensities used in this work are; mean, median, mode, kurtosis, moment, skewness, standard deviation, variance, inter quartile range, covariance and energy (Singpurwala, 2013). The set of those eleven statistical features were collectively used to generate respective feature vectors for respective color channel corresponding to each of the leaves.

2.5 *Feature vector for combination of different color channels*

In this work, the six sets of feature vectors have been calculated for each individual leaf. Horizontal

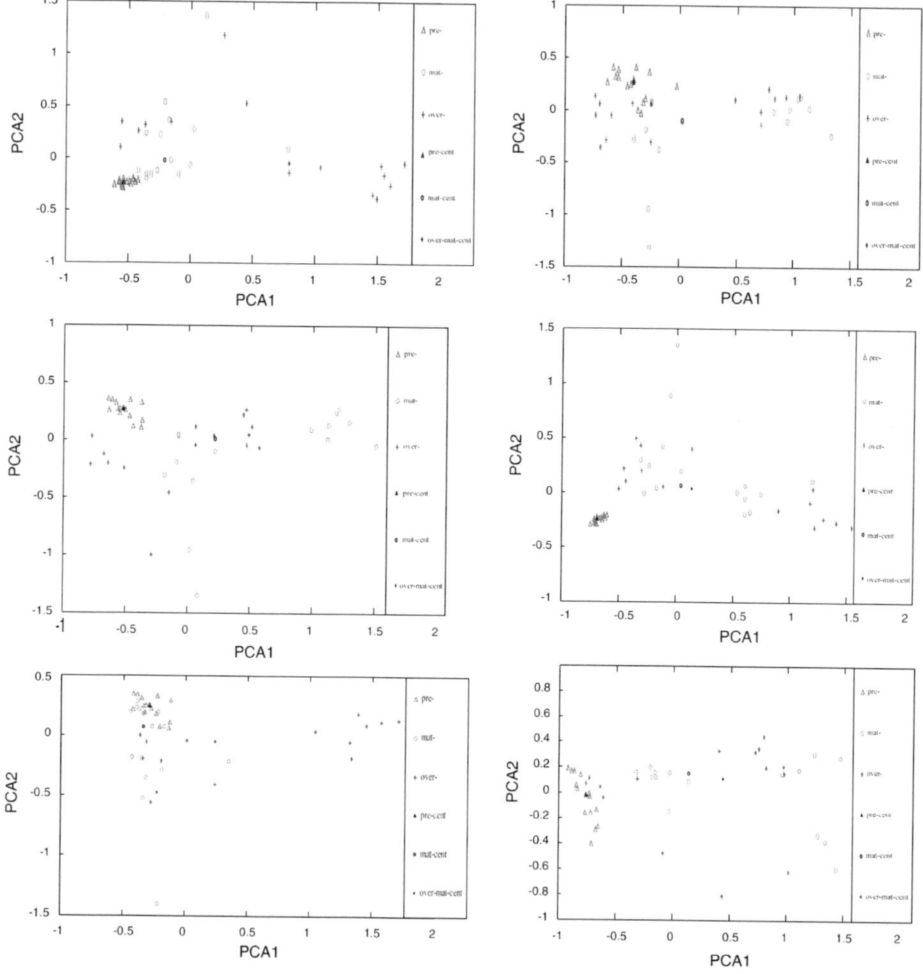

Figure 3. PCA plots for (top to bottom) Green channeled horizontal projection with shifted center, GBR channeled vertical projection with shifted center, BRG channeled horizontal projection with shifted center, PCA plot for GRB channeled horizontal projection with shifted center.

Table 2. Maturity detection performance.

Combination of channels	Projection type	Detection performance for prematurity	Detection performance for maturity	Detection performance for overmaturity
GBR	Vertical	5%	5%	70%
RBG	Vertical	80%	5%	5%
BRG	Horizontal/Vertical	5%	100%	60%
GRB	Horizontal/Vertical	100%	5%	5%
RGB	Horizontal	80%	10%	70%(vertical components)
B	Horizontal/Vertical	5%	100%	5%
G	Horizontal/vertical	5%	50%	60%

Table 3. Class separability measure for respective combination.

	Class Seperability Measure	
Combination of Channels	Horizontal Projection	Vertical Projection
GBR	0.061519222	0.04875268
RBG	0.047391064	0.035652699
BRG	0.034592829	0.031906528
GRB	0.043137445	0.030521373
RGB	0.010181547	0.007457853
B	0.043364	0.038054
G	0.022306	0.017382

projection of the feature vectors of R, G and B channels separately and vertical projection of the feature vectors of R, G and B channels for respective categories. Nine different combinations of RGB color channels as shown in Table 1, each representing combination of three classes premature, mature and overmature leaves, have been considered.

2.6 *Feature vector for combination of different color channels*

In this work, the dataset of three different classes of leaves were subjected to the PCA (Nurhayati et al. 2013; Kadir et al. 2012) analysis to form the cluster and the centre of each of the clusters have been found by means of K-means (Li et al. 2005) algorithm. In experiment the centers of the three different categories of leaves have been found. The Eq. (2) shows the objective function for K-means algorithm(Likas et al. 2003).

$$j(v)=\sum_{i=1}^{c}\sum_{j=1}^{c_i}\left(\|x_i-\mu_i\|\right)^2 \tag{2}$$

where, $\| x_i - \mu_j \|$ is the Euclidean distance between x_i and μ_j. c_i is the number of data points in the ith cluster. c is the number of cluster center.

2.7 *Class separability*

Class separability is the measure that gives an insight of classification of different related classes.

The class separability measure (Mthembu et al. 2004; Wolfel 2007) has been applied to different cluster combinations of premature, mature and overmature datasets.

3 RESULTS AND DISCUSSIONS

From the captured images of 45 leaves comprising of 15 each for the premature, mature and overmature categories, the FFT value were calculated and then the statistical features were extracted from the results of FFT analysis. The eleven statistical values for each channel and for respective type of leaves thus calculated and were grouped to form the respective feature vectors. Different combinations of red, green and blue channel based clusters are formed and graphically tested along with the single channel based cluster system. The following plots in the Fig. 2 shows horizontal and vertical projections of PCA plotting for different combinations of color channels, which gives good performance towards the detection of level of maturity. These plots also show the distinct cluster center. The plots given in the Fig. 3 shows the respective clusters along with their shifted centers for some of the selected combinations due to inclusion of unknown leaf within the cluster. For each of the combinations, red marked parts indicate the cluster of premature leaves, green marked part indicate the cluster of matured leaves and blue marked part shows the cluster of overmature leaves.

Table 2 shows the performance of the classification i.e. classifying an unknown Kalmegh leaf as premature, mature or overmature. The result is shown for different combination of channels. It is evident that the result of classification of leaf according to their maturity is good for some of the combination of channels and for the specific projection like horizontal or vertical. The result with very poor percentages have not been considered.

Table 3 shows the class seperability measure for respective cluster triads for combinations of different color channels.

4 CONCLUSION

The paper presented a computer vision based approach for identifying the maturity level of medicinal plants. For experiment, the leaves of Kalmegh have been considered and the scope of classifying them as premature, mature or overmature have been explored. The color channel based statistical features are considered and the identification is performed on the basis of magnitude of shift of the cluster center after inclusion of the feature vector of the unidentified Kalmegh plant leaf within the cluster formed by the feature vectors of known premature, mature and overmature leaves. Further the experiment has been carried out for different combination of color channels and for different types of projection of the statistically extracted feature vectors of different Kal-megh leaves. The work can be extended to other varieties of medicinal plants leaves with more robust type of features, feature optimization towards improved classification, etc. The results show that the presented technique can be a possible alternative for maturity detection of medicinal plants.

REFERENCES

Cooley, W. James & Lewis, Peter A. W. & Welch, Peter D. 1969 The Fast Fourier Transform and Its Applications IEEE TRANSACTIONS N EDUCATION, VOL.12, NO. 1.

Joy, P. P & Thomas, J & Baby S. Mathew & Skaria, P. 1998. Medicinal plants, *Kerala Agricultural University, Aromatic and Medical Plants Research Station.*

Kadir, A & Nugroho, L. E & Susanto, A & Santosa, P. I. 2012. Performance Improvement of Leaf Identification System Using Principal Component Analysis. *International Journal of Advanced Science and Technology* Vol. 44.

Li, J & Tao, D. & Hu, W, & Li. X. 2005. "Kernel principle component analysis in pixels clustering," in *Proc. Int. Conf. on Web Intelligence*, pp.786–789.

Likas, A &, Vlassis, N & Verbeek, J. J. Feb.2003. "The global k-means clustering algorithm," Pattern Recognition, vol.36 no. 2, pp. 451–461.

Mthembu, L. S & Greene J. 2004. A comparison of three class separability measures. In Proc of the *15thAnnual symposium of the Pattern Recognition Association of South Africa (PRASA)*, Grabouw, South Africa.

Nurhayati, O. D & Susanto, A & Widodo, T. S & Tjokronagoro M. 2013. Principal Component Analysis combined with First Order Statistical Method for Breast Thermal Images Classification, Oky Dwi Nurhayati et./*International Engineering and Technology Research Journal*, Vol. 1(2), 72–78.

Singpurwala, Darius. 2013. A Handbook of Statistics An Overview of Statistical Methods.

Tsuyuzaki, Shiro. 2000. Characteristics of "Number of Veins" to Estimate Leaf Maturity in Pteris mutilafa (Pteri daceae), *Journal of Plant Research by The Botanical Society of Japan.*

Wolfel, Matthias. 2007. Channel Selection by Class Separability Measures for Automatic Transcriptions on Distant Microphones, INTERSPEECH.

Computational Science and Engineering – Deyasi et al. (Eds)
© 2017 Taylor & Francis Group, ISBN 978-1-138-02983-5

Computer vision based identification of nitrogen and potassium deficiency in FCV tobacco

R.V.V. Krishna
Department of ECE, Aditya College of Engineering and Technology, Kakinada, India

S. Srinivas Kumar
Department of ECE, JNT University, Kakinada, Andhra Pradesh, India

ABSTRACT: Commercial crops often face nutritional deficiency. Two major nutrient deficiencies, Nitrogen and Potassium, which hamper plant growth, are identified in Flue Cured Virginia (FCV) tobacco. In the proposed research work, a computer vision based identification of nutritional deficiency in tobacco is presented. First, a dynamic thresholding technique is used to haul out tobacco leaves from the background. Then the color features of leaves are extracted by colour percentiles, and texture features are extracted by a novel texture descriptor, named Power Law Descriptor (PLD). A Naïve Bayes classifier is used to classify the segmented leaf images into three classes as Nitrogen (N) deficient, Potassium (K) deficient or Normal, using extracted colour and texture features. This experiment is conducted on a data set of 33 images per class, obtained from Central Tobacco Research Institute (CTRI), Rajahmundry, India. The results obtained in this experiment are compared with Support Vector Machine (SVM), and k-Nearest Neighbor Classifier at the classification stage. Experiments showed that the classification accuracy of this model, using Power Law Descriptor is nearly equal to 90% and it can also diagnose the nutrition deficiency with higher speed and accuracy compared to the other methods.

1 INTRODUCTION

Tobacco is a major crop of high commercial value, cultivated in different parts of the world. Tobacco plants often appear to be nutrition deficient in resource poor soils of India, where tobacco is being cultivated. It is observed that the symptoms of the nutrition deficiency are embodied in leaves colour and texture. Therefore the diagnosing system of nutrition deficiency based on computer vision is proposed in this paper. By extracting colour and texture features, the plant nutrition deficiency could be diagnosed in early stage, and then some effective measures could be taken well in time to cut down the losses. This research also provides theory and technology support for future online supervision of nutrients supply in Tobacco.

Till date, no instance of a model to detect Nutrient deficiencies in Tobacco leaves, using machine vision has been identified. However, works on other areas such as automated seedling disease diagnosis, automated grading, and automated harvesting of tobacco leaves could be traced in the literature (Zhang *et. al.*, 2008, Zhang *et. al.*, 2011). Researcher (Xu *et al.*,2011) presented a model on nutritional deficiency identification in their work on "Use of leaf color images to identify nitrogen and potassium deficient tomatoes". They extracted color and texture features of leaves by methods such as percent intensity histogram, percent differential histogram, fourier transform, and wavelet packets. Then Genetic Algorithm (GA) has been used to select features to get the best information for diagnosing the plant disease.

There are many feature extraction techniques, nevertheless,the modus operandi in this paper yielded better segmentation and classification results. The classifier used in our experimentation is the Naïve Bayes classifier. The Naive Bayes algorithm affords fast, highly scalable model building and scoring. It scales linearly with the number of predictors and rows.The build process for Naive Bayes is parallelized.

The major research contributions in this paper are as follows:

i. A novel attempt is made to automate and address the nutrient deficiency identification problem in Tobacco Leaves.
ii. A new texture descriptor named as Power Law Descriptor (PLD) is proposed by the authors. This Descriptor is a modification of the Weber Local Descriptor proposed by Chen (Chen *et al.*, 2008).

The remaining part of the paper is organized as follows. In Section 2 the characteristic features of Nitrogen and Potassium Deficient Tobacco Leaves is discussed. The proposed classification model, including colour and texture feature extraction, for nutrient deficiency identification in Tobacco is discussed in Section 3. The Naïve Bayes Classifier is discussed in Section 4. The results of the proposed model are presented in Section 5.The Concluding remarks are given in Section 6, followed by References in Section 7.

2 VISUAL CHARACTERISTIC FEATURES OF NITROGEN AND POTASSIUM DEFICIENT TOBACCO LEAVES

Nitrogen has a major role in the content of nicotine, chlorophyll, and other important substances in the leaf that are essential for growth. Some symptoms of nitrogen deficiency are when the leaves are more of a pale yellow versus a bright green colour. From the growing process of nitrogen-deficient leaves, it is found that the leaves grew yellow with the development of disorder. Hence the amount of yellow-pixels in leaves images could reflect the feature of nitrogen deficiency.

When potassium levels are not up to par the plant will develop cankers of dead tissue forming on the plant body, and the tips of the leaves will turn a brownish colour and severely affect the quality of FCV Tobacco [Figure 1].

3 PROPOSED CLASSIFICATION MODEL

The proposed model is divided into three stages. They are segmentation followed by feature extraction and then classification. A data base of (100) tobacco leaf images which are nutrient deficient and normal were acquired by the authors from CTRI, India. The leaves are initially segmented from the background using a dynamic thresholding technique. Then the colour features are extracted from the segmented image by colour percentiles model proposed by Niskanen (Niskanen *et al.*, 2001). Texture features are extracted using a novel descriptor called Power Law Descriptor, which is a modification of Weber Local Descriptor (Chen *et al.*, 2008). The extracted features of different sample images are labelled into three groups as Nitrogen Deficient (ND), Potassium Deficient (PD) and Normal Leaves (NL). 30% of these samples are grouped as training samples and the remaining 60% of the samples are grouped as testing samples. A Naïve Bayes Classifier is trained using the training samples. More the size of training data, more accurate will be the throughput. Post training, when a new leaf image from test samples (test features) is presented as input to the trained classifier, it classifies as to which class the presented leaf belongs. Naive Bayes is a simple probabilistic classifier which uses Bayes' theorem (or Bayes rule) with strong independence (naive) assumptions (Laursen *et. al.*, 2014).

Figure 1. Nitrogen (a) and Potassium (b) Deficient Leaves.

3.1 *Segmentation*

The leaf image is converted to gray level. Segmentation is used to separate leaf from its background. It can be accomplished by using dynamic threshold technique. Firstly, an intensity histogram of image is built with 20 bins. Secondly, two major peaks in the histogram that represent the leaf and its background respectively are obtained. Lastly, a bin with the smallest value that lies between the two major peaks is found. Then, the median of the bin is used as a threshold to separate leaf and its background [Figure 2]. After that, several morphological operations are performed to remove holes in the leaf caused by previous thresholding. The leaf is obtained by using AND operation between RGB image and binary image.

3.2 *Feature extraction*

The colour and texture features vary from leaf to leaf. 33 typical samples from each class of leaves were chosen. The typical features of the leaves (training samples) are extracted. The extracted features are needed to train the Naïve Bayes Classifier.

A. Colour feature extraction

The colour percentiles are used to extract colour features. A percentile is the value that cuts the distribution of a random variable into two parts so that a given percent of observations fall below that value.

Figure 2. Segmentation of Tobacco Leaves.

For Ex. If a certain data is given by X = [2, 4, 6, 8, 10, 12, 14]. Then the first, second and third quartile are given by [4.5000 8.0000 11.5000]. Colour percentiles are generally computed from each colour channel. Niskanen et al.,[5] used colour percentiles in combination with either co-occurrence features or Local Binary Patterns for identification of knots in wood inspection. In the experiments presented herein, quartiles and quintiles of each colour channel are used. Chromatic features are the first, second and third quartile of each R, G and B channel (each channel is normalized between 0 and 1). The method produces nine chromatic features, three per each chromatic channel.

3.1.1 *Texture feature extraction Power Law Descriptor (PLD)*

Ernst Weber observed that the ratio of incremental threshold to the background intensity is a constant. This relation known since as Weber's law can be expressed as:

$$\frac{\Delta I}{I} = k \tag{1}$$

where ΔI represents the increment threshold (just noticeable difference for discrimination); I represents the initial stimulus intensity and k signifies that the proportion on left side of the equation remains constant despite variations in the 'I' term. The fraction $\Delta I/I$ is known as the Weber fraction. Weber's law, says that the size of a just noticeable difference is a constant proportion of the original stimulus value.

Chen (Chen *et al.*, 2008) proposed Weber Local Descriptor as a texture descriptor, by considering the concepts of Weber's law. But Augustein (Augusterin 2009) observed that empirical data such as an image does not always fit well into Weber's law. He suggested a modification to Weber's law as follows, and hence called as Guilford power law.

$$\frac{\Delta I}{I^{\alpha}} = k \tag{2}$$

where α is an exponent slightly less than one. The perceived brightness of the human eye is proportional to the logarithm of actual pixel value, rather than the pixel value itself. The power law is also scale invariant. Hence the proposed power law descriptor models the perception of human beings better than Weber local descriptor. The Power law descriptor consists of two components differential excitation (ξ) and orientation (θ).

Differential excitation finds the salient variations within an image to simulate the pattern perception of human beings.

$$\xi(x_c) = \arctan\left[\frac{v_s^{00}}{[v_s^{01}]^{\alpha}}\right] \tag{3}$$

It is defined as the ratio between two terms V_s^{00} and $[v_s^{01}]^{\alpha}$. Here V_s^{00} at any pixel is the sum of the differences between the neighbors and the current pixel, whereas V_s^{01} is the value of the current pixel to a power of α.

$$V_s^{00} = \sum_{i=0}^{p-1}(\Delta x_i) = \sum_{i=0}^{p-1}(x_i - x_c) \tag{4}$$

$$\begin{bmatrix} 1 & 1 & 1 \\ 1 & -8 & 1 \\ 1 & 1 & 1 \end{bmatrix} \begin{bmatrix} 0 & 0 & 0 \\ 0 & 1 & 0 \\ 0 & 0 & 0 \end{bmatrix} \begin{bmatrix} x_0 & x_1 & x_2 \\ x_7 & x_c & x_3 \\ x_6 & x_5 & x_4 \end{bmatrix}$$

Filters used to realize V_s^{00} and V_s^{01} *Template*

These values are obtained by convolving the image with the following filters.

The orientation component is the gradient orientation which is computed as

$$\theta(x_c) = \gamma_s^1 = \arctan\left[\frac{v_s^{11}}{v_s^{10}}\right] \tag{5}$$

$$V_s^{10} = x_5 - x_1 \text{ and } V_s^{11} = x_7 - x_3 \tag{6}$$

$$\begin{bmatrix} 0 & 0 & 0 \\ 1 & 0 & -1 \\ 0 & 0 & 0 \end{bmatrix} \begin{bmatrix} 0 & -1 & 0 \\ 0 & 0 & 0 \\ 0 & 1 & 0 \end{bmatrix}$$

Filter used to realize V_s^{10} Filter used to realize V_s^{11}

where V_s^{11} and V_s^{10} are obtained using the following filters.

Both the orientation and excitation values range in the interval $[-\pi/2, \pi/2]$. Finally the two dimensional histogram of the differential excitation and orientation component is the texture descriptor used in the segmentation process. The Texture feature is expressed as

$$TF_{ij}^{k} = 2D\,Histogram\ [\xi(x_c), \theta(x_c)]_{i\,j} \quad k = L, a, b \tag{7}$$

4 CLASSIFICATION USING NAIVE BAYES CLASSIFIER

Bayes theorem provides a way of calculating the posterior probability p(c|x), from p(c), p(x), and p(x|c). Naive Baye's classifier assumes that the effect of the value of a predictor (x) on a given class (c) is independent of the values of other predictors. This assumption is called class conditional independence and is the basis for naming the classifier as Naive (Laursen *et. al.*, 2014)

$$p(c\mid x)=\frac{p(x\mid c)p(c)}{p(x)} \quad (8)$$

p(c|x) is the posterior probability of class (target) given predictor (attribute).
p(c) is the prior probability of class.
p(x|c) is the likelihood which is the probability of predictor given class.
p(x) is the prior probability of predictor.

Training phase

The first step in the classification procedure is the training phase. The combined colour and texture features of all the leaves in the database are obtained. The class prior probabilities for all the three classes, Nitrogen Deficient, Potassium Deficient and Healthy Leaves are assumed to be equal and hence

$$p(ND)=p(PD)=p(NL)=0.33$$

The prior conditional probabilities $p\left(\frac{x}{ND}\right), p\left(\frac{x}{PD}\right), p\left(\frac{x}{NL}\right)$ are found by the process of kernel density estimation.

A probability density function is generated for each feature from the feature set. The function is given as

$$p(F_k=x\mid C)=\frac{1}{L.h}\sum_{i=1}^{L}K\left(\frac{x-x_i}{h}\right)$$
$$k(x)=\frac{1}{\sqrt{2\Pi}}e^{\frac{-x^2}{2}} \quad (9)$$

4.1 *Testing phase*

When a new leaf feature is given as an input to the classifier, the posterior probabilities $p\left(\frac{ND}{x}\right), p\left(\frac{PD}{x}\right), p\left(\frac{NL}{x}\right)$ are found using the relevant formulae. A feature vector x is labelled as belonging to a particular class, if the posterior probability for that particular class is maximum.

	Classification accuracy		
Models	Exp-1	Exp-2	Average
Naive Bayes Classifier	90.419	85.4354	87.926
SVM	78.932	83.467	81.199
k-NN Classifier	76.321	74.145	75.233

5 EXPERIMENTAL RESULTS

The Tobacco leaf images were obtained with the aid of Central Tobacco Research Institute (CTRI), Rajahmundry, India. A total of 33 images per class were captured using Sony digital camera, under invariable illumination conditions. The obtained images are divided into training images and testing images. The obtained images are resized to 640*480 and used for the approach.

Tobacco leaves are classified as Nitrogen Deficient (ND), Potassium Deficient (PD) and Normal Leaves (NL) using Naïve Baye's classifier. The feature extraction is done using (Colour Quartiles + PLD). Out of 100 sample images collected, 11 images from each class are given for training, and the rest are used for testing.

Extensive experimentation, i.e. two experiments each, with 30 trails was conducted on the leaf images and the average classification accuracy was noted and presented in Table 1. The experimentation shows a clear dominance of the novel colour texture feature (Colour Quartiles+Power Law Descriptor) feature proposed by the authors in classifying the tobacco leaves as ND, PD, and NL.

6 CONCLUSION

In this paper a classification model based on computer vision is proposed and implemented to identify the primary nutritional deficiencies in commercial Tobacco. The colour and texture features were extracted using colour quartiles and a novel texture descriptor named Power Law Descriptor. These features were used to train a Naive Bayes Classifier. The trained classifier is used to classify tobacco leaves as Nitrogen Deficient, Potassium Deficient or Normal Leaves. The efficacy of the proposed model is compared with other benchmark classifiers, SVM and k-NN Classifiers. The experimental results show that Power Law Descriptor and Bayes Classifier exhibits better classification accuracy compared to other classifiers.

REFERENCES

Augustin, T. (2009) The problem of meaning fulness: Weber's law, Guilford's power law, and the near-miss-to-Weber's law. Mathematical Social Sciences, 109, 271–277.

Chen, Jie, *et al.* (2008) A robust descriptor based on Weber's law." Computer Vision and Pattern Recognition, CVPR 2008. IEEE Conference on. IEEE.

Guru, D. S., Mallikarjuna, P. B., Manjunath, S., Shenoi,M. M., (2012) Machine Vision Based Classification of Tobacco Leaves For Automatic Harvesting." Intelligent Automation & Soft Computing, 18(5).

Laursen, M. S., Midtiby, H. S., Krüger, N., Jørgensen, R. N., (2014) Statistics-based segmentation using a continuous-scale naive Bayes Approach, Computer and Electronics in agriculture, 109, 271–277.

Niskanen, M., Silven, O., & Kauppinen, H. (2001). Color and texture based wood inspection with non-supervised clustering. In proceedings of the 12th scandivanian conference on image analysis (SCIA 2001). 336–342, Bergen, Norway.

Xu, G. *et al.* (2011) Use of leaf colour images to identify nitrogen and potassium deficient tomatoes." Pattern Recognition Letters. 32.11, 1584–1590.

Zhang, F., Zhang., X. (2011) Classification and quality evaluation of tobacco leaves based on image processing and fuzzy comprehensive evaluation, Sensors 11.3, 2369–2384.

Zhang, X. J., Zhang. F., (2008) Images Features Extraction of Tobacco leaves", Congress on Image and Signal Processing, 773–776.

Computational Science and Engineering – Deyasi et al. (Eds)
© 2017 Taylor & Francis Group, ISBN 978-1-138-02983-5

Recognition of plant leaves with major fragmentation

J. Chaki & R. Parekh
School of Education Technology, Jadavpur University, Kolkata, India

S. Bhattacharya
Department of Electrical Engineering, Jadavpur University, Kolkata, India

ABSTRACT: The current work proposes a methodology for the recognition of plant species using features obtained from digital leaf images which are fragmented. Due to various environmental and biological factors leaves are fragile and prone to fragmentation. The paper studies how recognition of leaves can be effectively done when only a portion of the leaf can be obtained. The situation is considered: major fragmentation, where less than 25% of the leaf area is present. Since part of the leaf contour would be missing, shape based techniques will not be effective for recognition, due to which a combination of color and texture features using Normalized Cross Correlation have been used. Experimentations involving classification of 1500 fragmented leaf slices varying in shape, size and fragmentation, to 30 pre-defined classes, demonstrate effectiveness of the approach.

1 INTRODUCTION

Plants play a crucial role in Earth's ecology as they help to provide sustenance, shelter, medicines, fuel and maintain a healthy breathable atmosphere. More and more plants are however at the brink of extinction due to incessant de-forestation to pave the path for modernization. Building a plant database for quick and efficient classification and recognition is an important step towards their conservation. In recent years, computer vision and pattern recognition techniques have been utilized to catalogue a large variety of plant species efficiently, which provides ways to search for known and unknown flora types. Most of the techniques extract visual properties like color, texture and shape from different parts of a plant like leaf, flower, fruit, bud and root. Among them leaf recognition is the simplest and most effective way for the recognition of plant species.

A number of visual features, data modeling techniques and classifiers have been proposed for plant leaf classification. Visual parameters include leaf contour shape (Wang et al, 2000), color and shape (Perez et al, 2000), shape and texture (Beghin et al, 2010). Different data modeling techniques used include fuzzy logic (Wang et al, 2002), fractal dimensions (Du et al, 2013), Fourier descriptors (Yang et al, 2012), wavelets (Wang et al, 2010), linear discriminant analysis (Zhang et al, 2011), Local Binary Descriptors (LBD) (Wang et al, 2014), combination of local and global descriptors (Shabanzade et al, 2011), radial basis functions (Arunpriya et al, 2014), histogram of oriented gradients (HoG) (Xia et al, 2014), pyramid histogram of oriented gradients (PHoG) (Zhao et al, 2015). A variety of classifiers have been used viz. neuro-fuzzy (Chaki et al, 2015; Chaki et al, 2015; Chaki et al, 2015), multi-scale distance matrix (Hu et al, 2012), k-means clustering (Valliammal et al, 2012).

The current work proposes schemes for plant leaf recognition where the test image is fragmented i.e. parts of it might be missing. Recognizing the fact that leaves are fragile and prone to various fragmentation due to environmental and biological factors, we consider the situation to simulate this viz. major fragmentation: when the amount of fragmentation is large and about 75 to 80% of the leaf area might be missing. It needs to be mentioned here that all the previous works surveyed by us, pertains to recognition of whole leaf images and we have not come across any other work aimed specifically to recognize partial leaf portions. The organization of the paper is as follows: the proposed approach with discussions on feature computation and classification schemes is in section 2, the details of the dataset and experimental results obtained is in section 3, the comparison of the proposed approach vis-à-vis other contemporary approaches is in section 4, while the overall conclusions and scopes for future research is in section 5.

2 PROPOSED APPROACH

The current work proposes schemes for plant recognition based on features automatically extracted from images of partial or fragmented leaves. It has been assumed here that the whole (entire) leaf image samples of all classes have been previously obtained during a training phase and features extracted from these images are available in a plant database. The unknown test images are however fragmented (partial) and features extracted from a fragmented portion of the test image is subsequently compared with the stored whole leaf image for class recognition. The features used are sensitive to the size and orientation of the leaf image, and hence a preprocessing step is first performed to normalize images before feature extraction.

2.1 *Preprocessing*

The objective of the pre-processing step is to standardize the rotation, translation and scaling factors of the image before feature computation. The raw image is typically a color image oriented at a random angle and having a random size, Fig. 1(a). The angle, by which the major axis of the leaf is oriented, is extracted from the image and used to rotate it to align the leaf along the horizontal direction, Fig. 1(b). The major and minor axes of a leaf are defined as being equal to the corresponding axes of an ellipse that has the same normalized second central moments as the leaf shape. To standardize the translation factors with respect to the origin, the background is shrunk until the leaf just fits within its bounding rectangle, Fig. 1(d).

2.2 *Major fragmentation*

As an extension to our previous work, in this paper we focus on the changes needed in the system design, when the amount of fragmentation exceed those mentioned above and only a small portion of the leaf, typically about 25% of the total area, is available. To simulate such situations, we split an entire leaf image into four quadrants and a fifth equal sized portion from the center. See Fig. 2. The partitions are referred to as "slices" and are named S–1 to S–5.

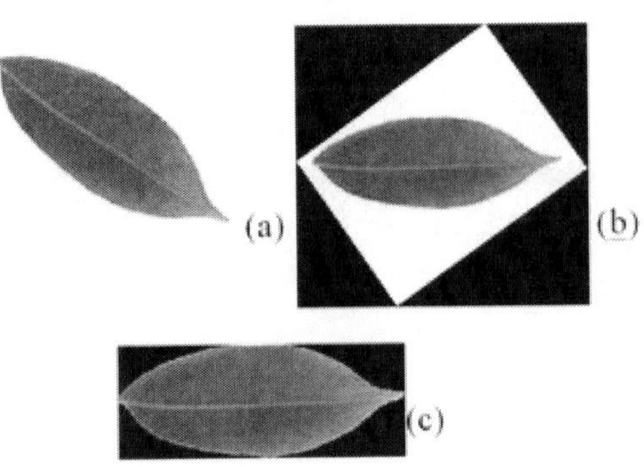

Figure 1. Pre-processing steps of a leaf image.

2.2.1 *Feature calculations*

Due to the large extent of the fragmentation, shape based features are no longer reliable and effective for classification, since only a small portion of the shape information is available from each of the peripheral slices. The central slice does not provide any shape information whatsoever.

We therefore resort to the use color and texture based information for comparisons. To localize color information, the R, G, B color channels of both the train and test samples are separated out. This generates three 2D matrices from each image which can be visually represented as grayscale versions of the original image. See Fig. 3.

The variation of gray shades over each grayscale version of a sample image embodies the texture patterns of the leaf surface which can be compared with the corresponding patterns of another image for possible matches. We use each test slice as a template, and perform a Normalized Cross Correlation (NCC) operation with each of the train images to search for maximized correlation coefficients. The correlation coefficient $C(u, v)$ which ranges in value from −1 to +1 is given by (1),

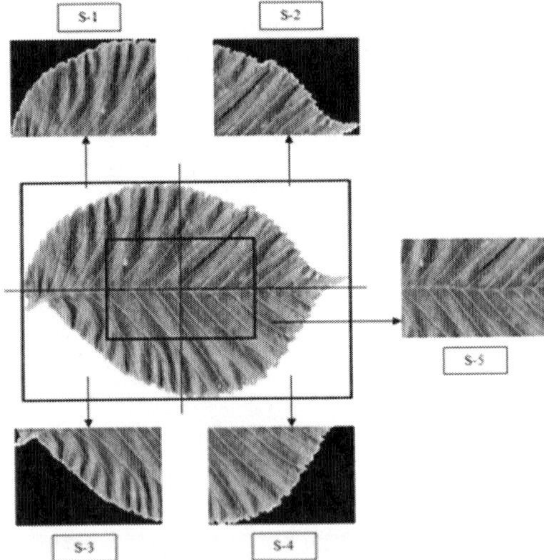

Figure 2. Test image split into 5 slices S-1 to S-5.

Figure 3. R, G, B channels of a sample image.

where $I(x, y)$ is the 2-D matrix of a specific color channel of the original color image in the database, $T(x, y)$ is the template corresponding to the 2-D matrix of the corresponding color channel of the test slice, I_m is the mean of $I(x, y)$ in the region under T, T_m is mean of T (Lewis, 1995).

$$C(u,v) = a/b, \text{ where} \quad (1)$$

$$a = \sum_{x,y}\left[\{I(x,y) - I_m\}\{T(x-u, y-v) - T_m\}\right]$$

$$b = \sqrt{\sum_{x,y}\left[\{I(x,y) - I_m\}^2 \sum_{x,y}\left[\{T(x-u,y-v) - T_m\}^2\right]\right]}$$

2.2.2 *Classification scheme*

The database of trained images would typically contain leaf samples of varying height and width. This fact is utilized to reduce computational loads during the matching stage. Prior to initiating NCC operations between the whole train image and fragmented test image, a check is performed to determine whether the height or width of the fragment is larger than the corresponding dimensions of a whole image in the database. If this condition is found to be true for a specific class, that class is discarded from being considered for a possible match, as the fragment cannot belong to that class. Thus typically correlation coefficients are computed only over a subset of the actual classes. The maximum correlation coefficient between a test fragment and a train image is obtained by combining together the coefficients for each color channel. For each test sample, the process is repeated over all training samples and the correlation values summed up for each class. The test slice is considered a member of that class for which the sum is maximum.

3 EXPERIMENTATIONS AND RESULTS

To test system performance, experimentations are done using 600 leaf images divided into 30 classes, collected from Flavia and Plants can dataset. For the 20 images per class, 10 are used for training and the remaining 10 for testing. Each image is 300 × 225 pixels in size and stored in JPEG format. Fig. 4 shows first training samples of all the classes.

Each of the test samples is split into 5 slices as shown in Fig. 5 generating a total of 1500slices. Sample slices (S-1 to S-5) of test samples of the first 15 classes (C-1 to C-15) are shown in Fig. 5.

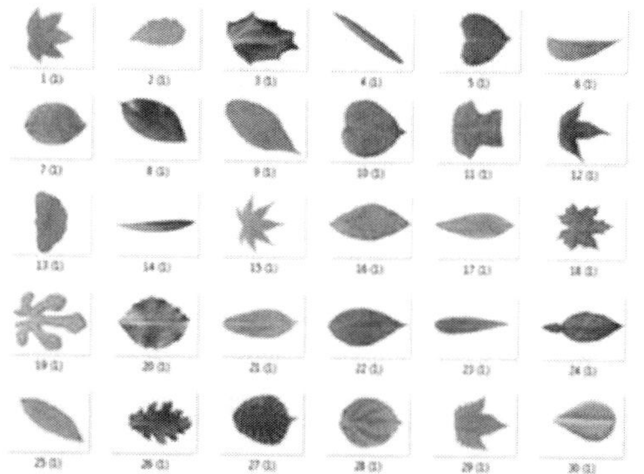

Figure 4. First samples of each of the 30 classes of the training set.

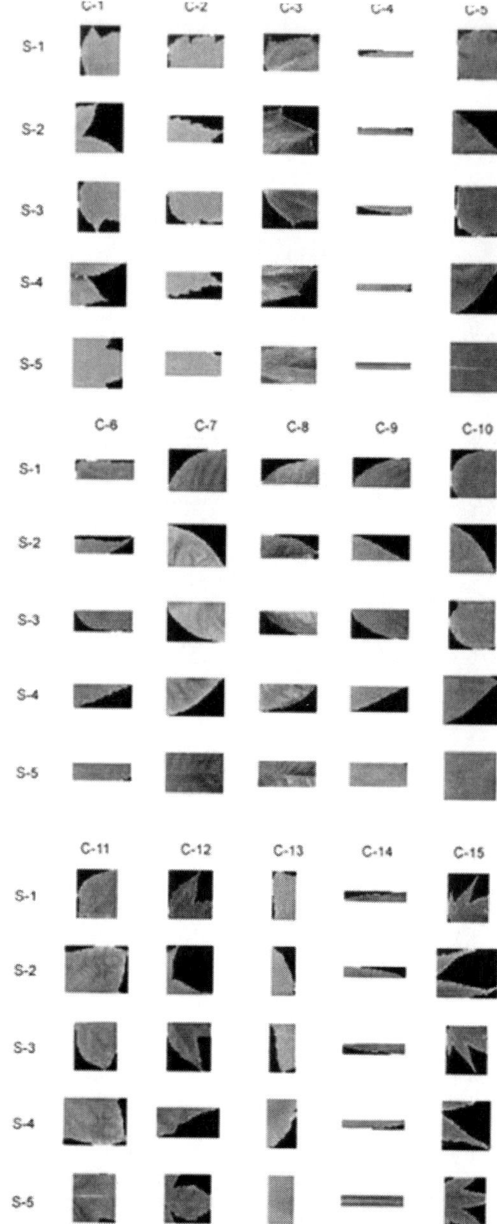

Figure 5. Sample slices for the first 15 classes of the test set.

Each test slice is compared with a maximum number of 300 train images before a classification decision can be made. Table 1 shows the percentage of correct recognition rates averaged for all 300 S-1 test slices. The overall accuracy is 81.66%.

Table 1. Percent recognition rates for slice S-1.

Class	Accuracy	Class	Accuracy	Class	Accuracy
1	90	11	50	21	100
2	90	12	100	22	50
3	90	13	100	23	0
4	70	14	90	24	100
5	80	15	100	25	0
6	90	16	100	26	80
7	100	17	100	27	90
8	0	18	100	28	90
9	90	19	100	29	100
10	100	20	100	30	100

Table 2 shows percentage of correct recognition rates of S-2 test slices. The overall accuracy obtained is 80%.

Table 2. Percent recognition rates for slice S-2.

Class	Accuracy	Class	Accuracy	Class	Accuracy
1	100	11	100	21	100
2	100	12	100	22	100
3	100	13	100	23	100
4	100	14	70	24	0
5	100	15	100	25	100
6	100	16	0	26	80
7	0	17	100	27	0
8	0	18	100	28	100
9	100	19	100	29	100
10	100	20	50	30	100

Table 3 shows percentage of correct recognition rates of S-3 test slices. The overall accuracy obtained is 81%.

Table 3. Percent recognition rates for slice S-3.

Class	Accuracy	Class	Accuracy	Class	Accuracy
1	100	11	70	21	100
2	100	12	100	22	70
3	100	13	100	23	0
4	50	14	0	24	100
5	100	15	100	25	0
6	60	16	100	26	100
7	80	17	100	27	100
8	0	18	100	28	100
9	100	19	100	29	100
10	100	20	100	30	100

Table 4 shows percentage of correct recognition rates of S-4 test slices. The overall accuracy obtained is 80%.

Table 4. Percent recognition rates for slice S-4.

Class	Accuracy	Class	Accuracy	Class	Accuracy
1	100	11	100	21	100
2	70	12	60	22	100
3	60	13	100	23	100
4	100	14	80	24	0
5	100	15	100	25	100
6	100	16	100	26	0
7	100	17	0	27	70
8	0	18	100	28	100
9	100	19	100	29	100
10	100	20	60	30	100

Table 5 shows percentage of correct recognition rates of S-5 test slices. The overall accuracy obtained is 85%.

Table 5. Percent recognition rates for slice S-5.

Class	Accuracy	Class	Accuracy	Class	Accuracy
1	100	11	100	21	100
2	100	12	100	22	100
3	100	13	20	23	100
4	100	14	100	24	0
5	100	15	100	25	10
6	100	16	70	26	100
7	30	17	100	27	20
8	100	18	100	28	100
9	100	19	100	29	100
10	100	20	100	30	100

The highest recognition rates for the S-5 slice is possibly due to the larger amount of texture information contained within the central slice which leads to a better matching performance. Fig. 6 illustrates the variations of the correlation values of slice S-1 / C-1. The upper plot indicates the aver-

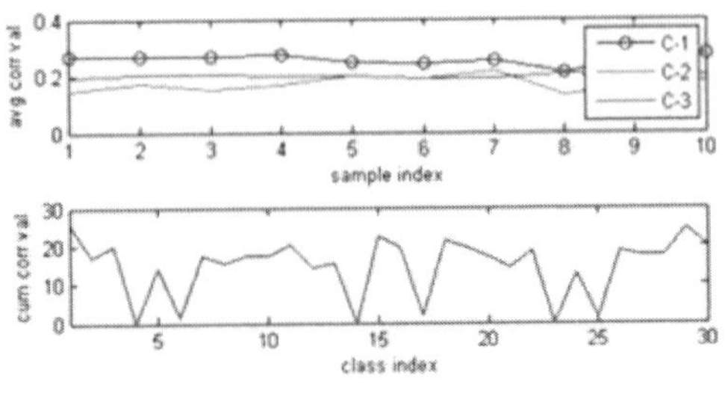

Figure 6. Variation of correlation values of S-1/C-1 over samples and classes.

age correlation values of the test slice with each of the 10 training samples of classes 1, 2, 3. The lower plot shows the variation of the cumulative correlation values for all training samples with each of the 30 classes. The top four matches are 25.68 (C-1), 24.85 (C-29), 22.55 (C-15), 21.51 (C-18). The values do indicate visual similarities between the concerned classes.

4 ANALYSIS

To compare the current work vis-à-vis other contemporary works, some of the recent approaches found in extant literature are applied on to the current dataset to observe their recognition performances. It should be mentioned here that these approaches have been designed for whole leaf images and none of them seem to have been specifically designed for handling fragmented or partial leaves. This would account for their low accuracies. In [15] Fourier descriptors have been used for data modeling. Fourier basis functions are however sinusoidal in nature with infinite length making them unsuitable to model transient signals with sharp changes that occur along fragmented leaf contours. Experimentations with this approach produce overall accuracies of 4.33%, 3.33%, 3.33%, 3.33% and 0% for slices S-1 to S-5. In [20] authors have used LBD which are however very sensitive to noise and different binary patterns are seen to be classified into the same class when fragmented leaves are used. Accuracy values produced are 7%, 17%, 6.67%, 8.33% and 3.66%. HoG used in [22] depends largely on the block size, cell size and number of orientation bins, which tend to be different for whole and fragmented leaf images and also vary across various fragment shapes. Accuracy values produced are 3.33%, 3.33%, 3.33%, 2.67% and 3.33%. In [8] a set of local and global features have been used, which are however seen to vary widely for whole and partial leaves even for the same class. Accuracy values produced are 6%, 5.67%, 5.67%, 6.33% and 7.33%. Table 6 summarizes the recognition values of all these approaches.

Table 6. Comparison chart.

Approach	S-1	S-2	S-3	S-4	S-5
Approach [15]	4.33	3.33	3.33	3.33	0
Approach [20]	7	17	6.67	8.33	3.66
Approach [22]	3.33	3.33	3.33	2.67	3.33
Approach [8]	6	5.67	5.67	6.33	7.33
Current approach	81.66	80	81	80	85

5 CONCLUSIONS AND FUTURE SCOPE

This article discusses a method of automatically characterizing and recognizing fragmented plant leaves where the entire leaf surface is not available. The case is studied: leaf with major fragmentation, where shape based approaches would not be reliable and color and texture based techniques are used. A normalized cross correlation operation between the color channels of the whole and fragmented leaves indicates a possible direction for solution.

One limitation of the current approach is that for leaves with major fragmentation, there is no way to automatically normalize the orientation and scaling factors of the fragments since the major/minor axis information of the leaf surface cannot be retrieved from the slices. Hence manual pre-processing needs to be resorted to. For the sake of experimentations, the slices of the leaf fragments have been obtained from previously pre-processed whole test samples. These provide directions for future research and are currently being investigated into for a more robust plant recognition system.

REFERENCES

Arunpriya, C. & Thanamani, A.S. 2014. An effective tea leaf recognition algorithm for plant classification using radial basis function machine. Int. Journal of Modern Engineering Research 4: 35–44.

Beghin, T., Cope, J. S., Remagnino, P. & Barman, S. 2010. Shape and texture based plant leaf classification. International Conference on Advanced Concepts for Intelligent Vision Systems (ACVIS) 345–353.

Chaki, J., Parekh, R. & Bhattacharya, S. 2015. Plant Leaf Recognition using Ridge Filter and Curvelet Transform with Neuro Fuzzy Classifier. In *Proc. of 3rd Int. Conf. on Advanced Computing, Networking, and Informatics (ICACNI '15)* 37–44.

Chaki, J., Parekh, R. & Bhattacharya, S. 2015. Plant leaf recognition using texture and shape features with neural classifiers. *Pattern recognition letters* 58: 61–68.

Chaki, J., Parekh, R. & Bhattacharya, S. 2015. Recognition of whole and deformed plant leaves using ststistical shape features and neuro-fuzzy classifier. *IEEE ReTIS* 189–194.

Du, J. X., Zhai, C. M. & Wang, Q. P. 2013. Recognition of plant leaf image based on fractal dimension feature. Neurocomputing 116: 150–156.

Flavia Plant Leaf Recognition System (http://sourceforge.net/projects/flavia/files/Leaf%20Image%20Dataset/).

Hu, R., Jia, W., Ling, H. & Huang, D. 2012. Multiscale distance matrix for fast plant leaf recognition. IEEE transactions on Image Processing 21: 4667–4672.

Lewis, J.P. 1995. Fast Template matching. *Canadian Image Processing and Pattern Recognition Society* 120–123.

Perez, A. J., Lopez, F., Benlloch, J. V. & Christensen, S. 2000. Color and shape analysis techniques for weed detection in cereal fields. Computers and Electronics in Agriculture 25: 197–212.

Plantscan Dataset (http://imedia-ftp.inria.fr:50012/Pl@ntNet/plantscan_v2/).

Shabanzade, M., Zahedi, M. & Aghvami, S.A. 2011. Combination of local descriptors and global features for leaf recognition. Int. Journal of Signal & image Processing, 2: 23–31.

Valliammal, N. & Geethalakshmi, S. N. 2012. Leaf image segmentation based on combination of wavelet transform and K-means clustering. J. Advanced Research in Artificial Intelligence 1: 37–43.

Wang, Q. P., Du, J. X. & Zhai, C. M. 2010. Recognition of leaf image based on ring projection wavelet fractal feature. LNCS 6216: 240–246.

Wang, X., Liang, J. & Guo, F. 2014. Feature extraction algorithm based on dual scale decomposition and local binary descriptors for plant leaf recognition. Digital Signal Processing 34: 101–107.

Wang, Z., Chi, Z. & Feng, D. 2002. Fuzzy integral for leaf image retrieval. IEEE Int. Conf. on Fuzzy Systems 372–377.

Wang, Z., Chi, Z., Feng, D. & Wang, Q. 2000. Leaf image retrieval with shape feature. LNCS 1929: 477–487.

Xia, Q., Zhu, H. D. & Shang, L. 2014. Plant leaf recognition using histograms of oriented gradients. Springer LNAI ICIC 8589: 369–374.

Yang, L W. & Wang, X F. 2012. Leaf Image Recognition Using Fourier Transform Based on Ordered Sequence. Springer LNCS 7389: 393–400.

Zhang, S. & Lei, Y. K. 2011. Modified locally linear discriminant embedding for plant leaf recognition. Elsevier Neurocomputing 74: 2284–2290.

Zhao, Z. Q., Ma, L. H., Cheung, Y. M., Wu, X., Tang, Y. & Chen, C. L. P. 2015. ApLeaf: An efficient android based plant leaf identification system. *Elsevier Neurocomputing* 151: 1112–1119.

Computational Science and Engineering – Deyasi et al. (Eds)
© 2017 Taylor & Francis Group, ISBN 978-1-138-02983-5

Parallel thinning of a digital binary image

Debi Prasad Bhattacharya, Pabani Das, Soham Manna & Swagata Sadhukhan
Department of Computer Science, Barrackpore Rastraguru Surendranath College, West Bengal, India

ABSTRACT: This paper describes a novel approach to thin digital binary images. This approach uses the contour of the image to create the skeleton of the image. In this approach the connectivity of the input images have been maintained. Depending on the perspectives of thinning and additional erosion the experimental results so produced are better in terms of visual performance.

1 INTRODUCTION

In digital image processing, binary digital image can be represented by a matrix, where each element in matrix is either zero (white) or one (black) and the points are called pixels. Image thinning is a fundamental pre-processing step. Thinning is a process that deletes the unwanted pixels and transforms it into a one pixel thick image. It reduces the requirement of large amount of memory for storing the structural information. The thinning approach has a history of more than sixty years.

Thinning is a type of a basic morphological technique which is very important in the field of pattern recognition, data compression and data storage, and also used for extracting skeleton of any region. And skeleton can be used for recognition of any character. So thinning indirectly plays an important role in character recognition also.

Although there are many types of thinning approach available, many of them have problems of producing poor result because of excessive erosion, noise and disconnections in the resultant thinned image. In this paper we have introduced a novel thinning approach manipulating the contour or boundary of the image. The basic idea of thinning in our approach involves the generation of a skeleton which has a single-pixel width. Our approach can solve the problem of excessive erosion, disconnections to a greater extent producing visually better results. Our proposed technique is scalable and style independent of the input character image.

The rest of the paper will be structured as, the Design and Implementation of the heuristics will be described in section 2. After that in section 3 we will discuss about the heuristics with the results shown in section 4. Finally the paper will be concluded in section 5.

2 DESIGN AND IMPLEMENTATION

Before discussing about the thinning heuristics we have to know about the properties we will be using to determine whether a pixel belongs to the contour and can be removed to get the thinned image. A point P will be in the boundary, only if P = 1 and either P_0 = 0 or P_2 = 0 or P_4 = 0 or P_6 = 0 which represents north, east, south or west boundary pixels of P respectively. N(P) is a property that determines the number of black neighbors of a point P where black is represented by value 1 and T(P) is another property that determines the number of 0–1 clockwise transitions among the neighbors of P.

So, knowing about N(P) and T(P) properties of boundary point P of any region, we can now keep it or remove the point P according to our need in heuristics.

With reference to the 8-way neighborhood notation shown in Figure. 1(a), first we will check whether the point P is on the north boundary i.e. if the value of P = 1 and $P_0 = 0$. Now, let us assume a set C_n that contains all the north boundary points. Now, for each pixel P∈ C_n, we will check whether

i (N(P) > 2 & T(P) = 1) OR
ii (T(P) = 2 & 2 < N(P)< 3 & (P_2 * P_4 = 1 | P_4 * P_6 = 1)) is satisfied by P. If one of the two proposed conditions are being satisfied by P, then P remains in the set C_n otherwise P will be subtracted from C_n. Now C_n will be subtracted from the original image.

For the condition (i) the significance of N(P) is that, each point must have more than one neighbors i.e. N(P)> 2, which prevents the deletion of the end points. This condition will be violated when the contour point P only has one neighbor. Having only one neighbor i.e. N(P) = 1 implies that P is the end point of a skeleton stroke and

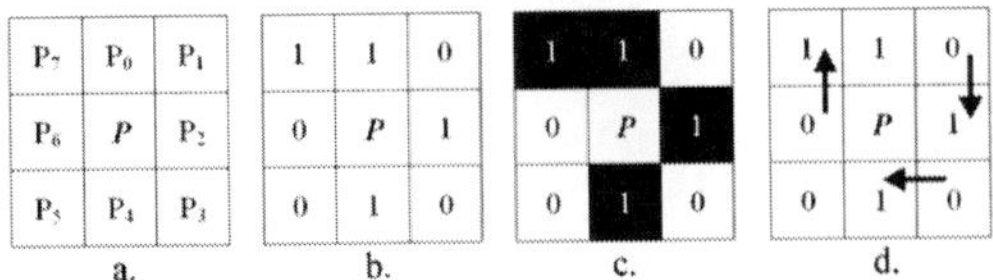

Figure 1. (a) 8 way neighborhood of P. (b) An example of (a). (c) N(P) = 4. (d) T(P) = 3.

obviously should not be deleted. The significance of T(P) is that, neighbors of each points must be adjacent, which prevents deletion of critical points. Because critical points are those, which makes the connection or junction between two regions and the deletion of them can cause disconnections and so to prevent unnecessary erosion, that critical point must be preserved. This condition is violated when it is applied to points on a stroke one pixel thick. Hence this condition prevents disconnection of segments of a skeleton during the thinning operation.

The condition (ii) is being used to remove unnecessary points from the skeleton and which also helps to prevent deletion of diagonal lines.

Now, similarly checking (P = 1 & $P_2 = 0$); (P = 1 & $P_4 = 0$) and (P = 1 & $P_6 = 0$) we will get the set C_e, C_s, C_w, containing east, south and west contour points respectively. Here also we will check the above two conditions. The first condition will remain same for construction of each set, but the second condition will vary. The second condition will be ((T(P) = 2 & 2 < N(P)< 3) & ($P_0 * P_6 = 1$ | $P_4 * P_6 = 1$)) for the set C_e, ((T(P) = 2 & 2< N(P) < 3) & ($P_0 * P_2 = 1$ | $P_0 * P_6 = 1$)) for the set C_s and ((T(P) = 2 & 2 < N(P) < 3) & ($P_0 * P_2 = 1$ | $P_2 * P_4 = 1$)) for the set C_w. After constructing each set it is subtracted from the original image before constructing the next set. Hence after the construction of north boundary pixels C_n it is removed from the original image, then similar process is carried on with east, south, west boundaries respectively.

The above steps will be repeated until there are no more pixels to be deleted from the given image to obtain the thinned image as output. That means $C_n = C_e = C_s = C_w = \phi$.

3 HEURISTICS

Now the pseudo code for the above mentioned process is described below.

Input: Image to be thinned

Output: The thinned image of the given image

Step 1: Repeat the following steps

1.1 Construct a set C_n of north boundary points (i.e. P = 1 & $P_0 = 0$) from the image, those satisfy also one of the two following condition:-

N(P) ≥ 2 & T(P) = 1

(T(P) = 2 & 2 ≤ N(P) ≤ 3) & ($P_2 * P_4 = 1$ | $P_4 * P_6 = 1$)

1.2 Subtract the set C_n from the image

1.3 Construct a set C_e of east boundary points (i.e. P = 1 & $P_2 = 0$) from the image, those satisfy also one of the two following condition:-

N(P) ≥ 2 & T(P) = 1

(T(P) = 2 & 2 ≤ N(P) ≤ 3) & ($P_0 * P_6 = 1$ | $P_4 * P_6 = 1$)

1.4 Subtract the set C_e from the image

1.5 Construct a set C_s of south boundary points (i.e. P = 1 & $P_4 = 0$) from the image, those satisfy also one of the two following condition:-

N(P) ≥ 2 & T(P) = 1

(T(P) = 2 & 2 ≤ N(P) ≤ 3) & ($P_0 * P_2 = 1$ | $P_0 * P_6 = 1$)

1.6 Subtract the set C_s from the image

1.7 Construct a set C_w of west boundary points (i.e. P = 1 & $P_6 = 0$) from the image, those satisfy also one of the two following condition:-

N(P) ≥ 2 & T(P) = 1

((T(P) = 2 & 2 ≤ N(P) ≤ 3) & ($P_0 * P_2 = 1$ | $P_2 * P_4 = 1$))

1.8 Subtract the set C_w from the image

1.9 If $C_n = \phi$ & $C_e = \phi$ & $C_s = \phi$ & $C_w = \phi$ then goto Step 2

Step 2: return

4 EXPERIMENTAL RESULTS

Table 1. Here the type of input character is Arial Black and size is 200 X 200.

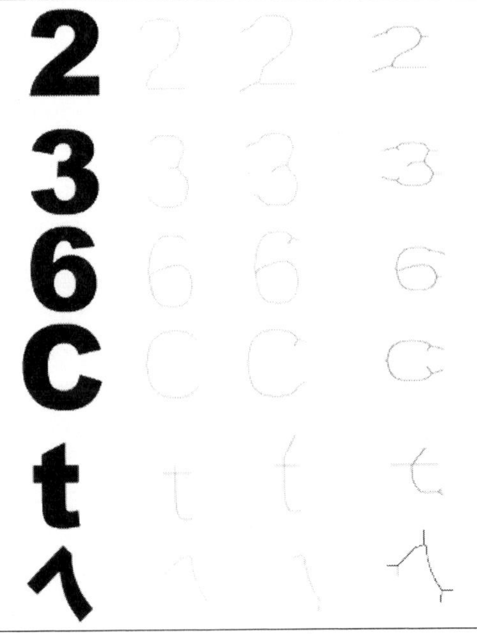

5 CONCLUSION

The proposed heuristics has advantages in terms of some important characteristics: thinning quality and the clarity of processing stages. We have also removed the problem of disconnection to a greater extent.

There are some branches that extended out from the skeleton at the resulting image of ZS algorithm and Stentiford thinning algorithm. But in our case there are no branches at the end points.

The ZS algorithm has problems preserving connectivity with diagonal lines and identifying line endings but we have removed this problem in our heuristics.

REFERENCES

Gonzalez, Woods. (2002). Digital Image Processing. 2nd Edition, 672–675.

Jagna, Kamakshiprasad. (2010). New Parallel Binary Image Thinning Algorithm, ARPN Journal of Engineering and Applied Sciences, 3(4), 64–67

Zhang, Suen (1984) A Fast Parallel Algorithm For Thinning Digital Patterns. Communications of the ACM, 27(3), 236–239.

http://Www.Mathworks.In/Matlabcentral/Fileexchange/34712-Stentiford-Thinning-Algorithm/Content/Stentiford-Thinning/Stentifordthining.M (22.10.2015)

http://Www.Mathworks.In/Matlabcentral/Fileexchange/34712-Stentiford-Thinning-Algorithm (22.10.2015)

Http://Www.Mathworks.In/Matlabcentral/Fileexchange/34712-Stentiford-Thinning-Algorithm/Content/Stentiford-Thinning/Connectivityfun.M (22.10.2015)

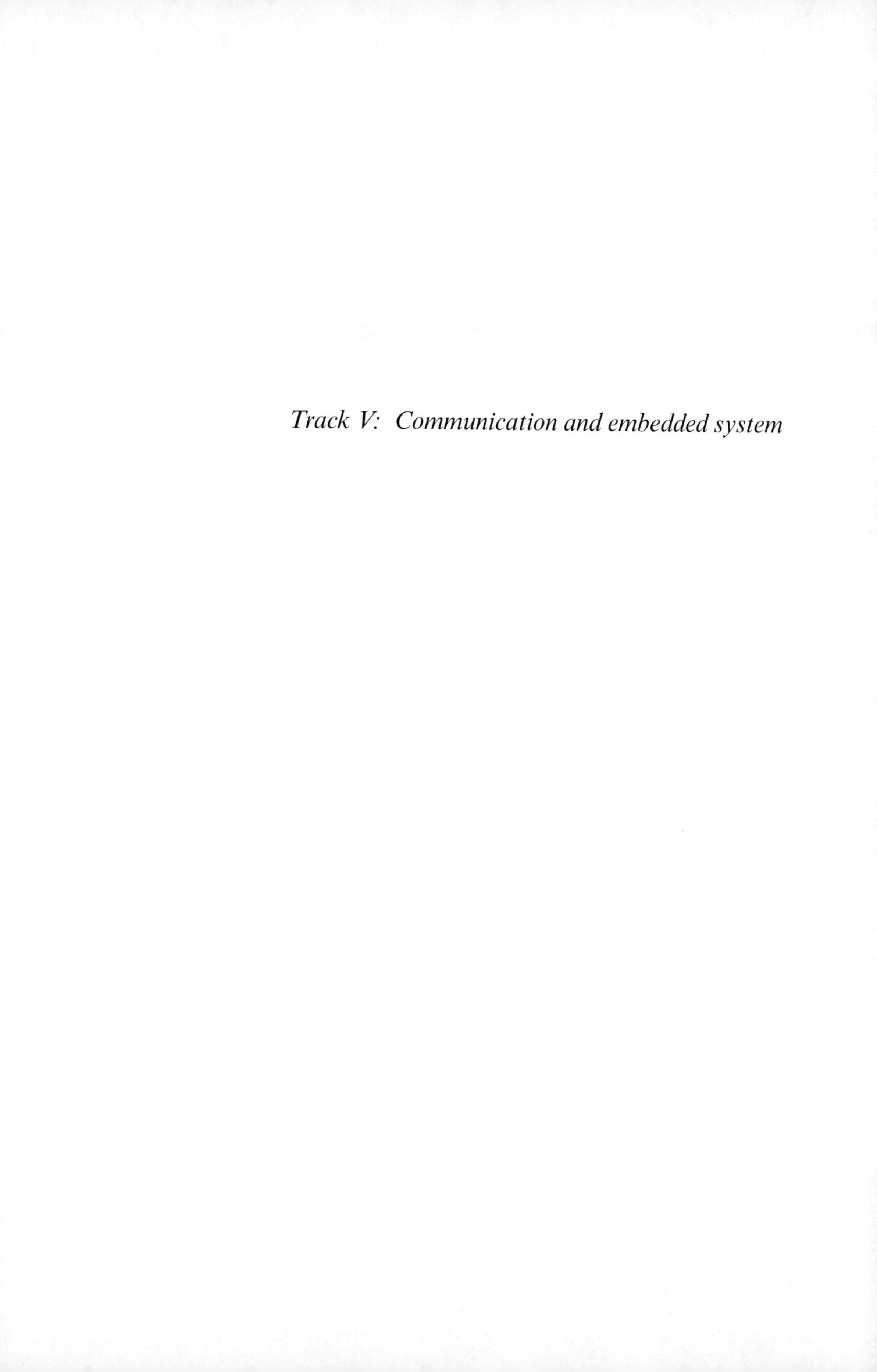

Track V: Communication and embedded system

Computational Science and Engineering – Deyasi et al. (Eds)

A method of arbitrary waveform generation with microcontrollers

Sourish Haldar, Sreyansh Bhupal, Sreeparna Sen, Priyankar Sandel,
Urmi Bose & Swapan Bhattacharyya
ECE Department, JIS College of Engineering, Kalyani, West Bengal, India

ABSTRACT: Low cost microcontrollers with integrated flash memory have enabled novel solutions for implementation of abstract mathematical functions. Arbitrary waveforms are a class of signal where the end-user defines the shape and time-period of required waveform. These waveforms are of high demand in signal processing and communication applications. The present work aims at synthesis of such waveforms by defining the wave-shape in terms of waypoints. Waypoints are specific points in the time-period of synthesized wave through which the synthesized wave should pass. Presently, the work is incorporating eight way-points in a time-period. This is being done by presetting time vs. voltage points at a particular value by potentiometers. The ADC integrated in the micro-controller IC would read these voltages and store them in 8-bit internal registers. Time-period is user programmable. The micro-controller would then repetitively generate the required waveform. External DAC would then read the output from a digital port of the micro-controller and generate continuous analog arbitrary wave.

Keywords: micro-controller, direct-digital synthesis, arbitrary waveform, waypoint, microcontroller, mixed-signal, communication

1 INTRODUCTION

Arbitrary waveforms have their use in Communication, Control Systems, Digital Signal Processing, and Signal Analysis. Direct Digital Synthesis (DDS) of wave-shapes enables the user to have precision control over amplitude and frequency together with absolute repeatability (Xiaodong et al. 2007, Liu et al. 2007, Quintáns 2006, Cordness 2004). The bench-top arbitrary wave generators are expensive, space inefficient, and not system integrable. This work introduces real-time variation of synthesized waveform. Most of the Arbitrary Waveform Generation methods reported till now have static programmability (Kulkarni et al. 2011), whereas some have dynamic waveform manipulation mechanism but use preprogrammed wave-shape characteristics (Liu et al. 2007, Quintáns 2006).

The input method that defines the shape of waveform is discussed in section 3, user input acquisition by the system is discussed in section 4. The time-period, frequency variation and clock control is discussed in section 5 of this article along with the details of the micro-controller used. The Section 6 & 7 deals with the timing parameters and generation of output consisting of Digital-to-Analog Conversion. Filtering/Band-limiting of signal is mentioned in section 8. The summarized concept of working of the proposed system is described in section 9. Scopes of improvement and possible applications of this work are mentioned in section 10 with a comparative study in section 11.

2 LITERATURE REVIEW

Most of the Arbitrary Waveform Generation methods reported till now have static programmability (Kulkarni et al. 2011), whereas some have dynamic waveform manipulation mechanism but use preprogrammed wave-shape characteristics (Liu et al. 2007, Quintáns 2006).

3 METHOD OF INPUT

3.1 *Resistor array implementation*

As shown in Fig. 1, the method of providing input to the system involves a set of potentiometer (pots). The two extreme terminals of the pots are connected with supply voltage and ground. The third terminal of the pots with variable output voltage, which define waypoints of the waveform, are connected to the ADC terminals of the MCU. In this work, presently eight such pots are used.

3.2 *Preprogramming*

The values of the waypoints can also be provided during the programming of microcontroller chip

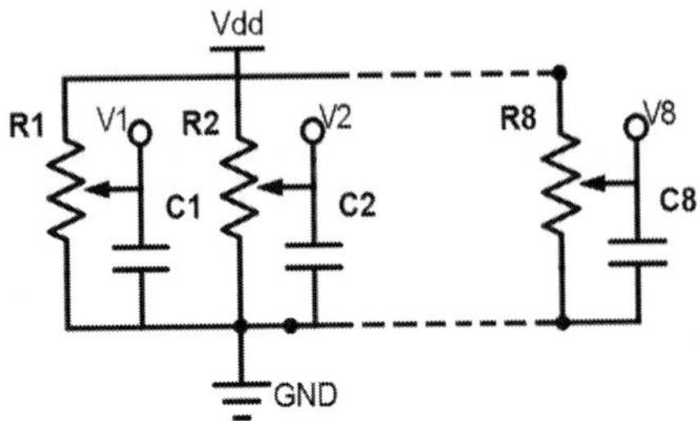

Figure 1. Resistor bank.

by the user but that would be static and would not allow real time variation in wave-shape (Liu et al. 2007, Yeary et al. 2004).

3.3 *Real time computer communication and secure data communication*

The other method for more detailed shaping of arbitrary wave is to define more way-points by communicating the time-vs.-voltage information from file in a computer. This is possible for ATMEGA series of microcontrollers as they have a USART bus. This increases the cost of system realization but, certain applications become possible which would be discussed later.

To analyze the number of arbitrary shapes possible through this method of direct digital synthesis, ATMEGA series has 8-bit register system; each way-point in the waveform can be of any value between 0 and 255. Assuming 8-way points in a time period the total number of possible arbitrary shapes that could be realized is $(256)^8$ or $(2)^{64}$. A low-cost 8 bit DAC and only a single port of the microcontroller of 8 bit output has been used for generating the wave in digital form. More precise waveform with higher number of way-point definition and higher number of quantization levels is a possibility at the cost of more registers per way-point and higher number of Microcontroller machine cycle per time-period. Alternately an expensive microcontroller with 16-bit register and higher clock frequency could be used as per the requirement of a specific application. The current work aims at proof-of-concept. This method opens up several possible applications including secure data transmission through arbitrary line coding. One can now transmit digital data stored in the computer through arbitrary line coding. This has immense potential for secure data communication.

4 DIGITIZATION OF USER INPUT

The inputs provided by user from the resistor bank are in analog form which has to be digitized in order to be processed by the microcontroller. This is done by the built-in 8 channel, 10 bit, Analog-to-Digital converter of the microcontroller which in this case is an ATMEGA32 A of ATMEL.

5 MICROCONTROLLER DESCRIPTION

In this work an ATMEL's ATmega32 microcontroller has been used which follows Harvard architecture,40-pin DIP, with a built in 8-channel, 10-bit Analog-to-Digital Converter, Byte-Oriented Two-wire Serial Interface (I2C), Master/Slave SPI Serial Interface, Programmable Watchdog Timer, USART Interface and 32 Programmable I/O Lines. The block diagram of the ADC is shown in Fig. 2.

At the input of the ADC is an analog multiplexer, which is used to select between eight analog inputs. After conversion the corresponding value is transferred to the registers ADCH and ADCL. The reference voltage is an external voltage which must be supplied at the A_{ref} pin of the chip. The value the voltage at the input is converted to can be calculated with the formula (1).

$$ADC\,conversion\,value = round\left(\left(-\frac{V_{in}}{V_{ref}}\right)\times 1023\right) \quad (1)$$

6 CLOCK DESCRIPTION OF MICROCONTROLLER, FREQUENCY AND TIME PERIOD MANIPULATION

The new generations of microcontrollers are capable of working on internal as well as external clock option. These options are selected with the microcontroller's fuse bits. The advantage of fuse bits is that once they are programmed, the configuration would remain stored even after the MCU is turned off. The fuse bits are configured to use the built in oscillator of 8 MHz frequency by default which can be changed to use any external crystal

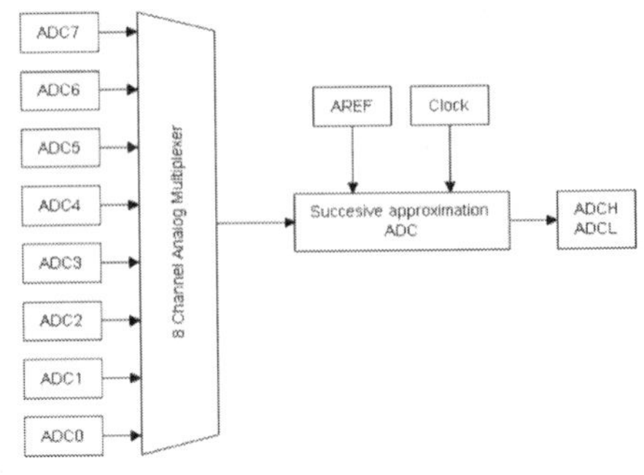

Figure 2. Block diagram of ADC in ATMEGA32.

oscillator as resonator. After the clock has been set up and configured, the time period of the output signal can be provided during the programming of microcontroller using delay functions or appropriate assembly code.

7 GENERATING OUTPUT

The microcontroller is programmed to take input from each ADC bus or directly from a computer, store it in the memory and then furnish it digitally to another output port of 8-bits, continuously with a constant time delay, hence enabling the control the frequency of the generated arbitrary wave. These 8 bits are now provided as input to an 8-bit Digital-to-Analog converter. A typical DAC converts the abstract numbers into a concrete sequence of impulses that are then processed by a reconstruction filter using some form of interpolation to fill in data between the impulses. The outputs generated from the microcontroller in Digital form defining the 8 points of each time cycle is to be converted to an analog signal which can be done using a Digital to analog converter (In this case a DAC8EP 8 bit DAC chip has been used as shown below) as the AVR microcontroller used does not contain a built-in DAC. The Digital waveform obtained from the output of the DAC is shown in Figure 3. A sample unipolar RZ signal wave and its FFT is shown in Figures 3 and 4 respectively. A sample arbitrary wave is shown in Figure 5; the frequency of this wave is 20 KHz (Approx.).

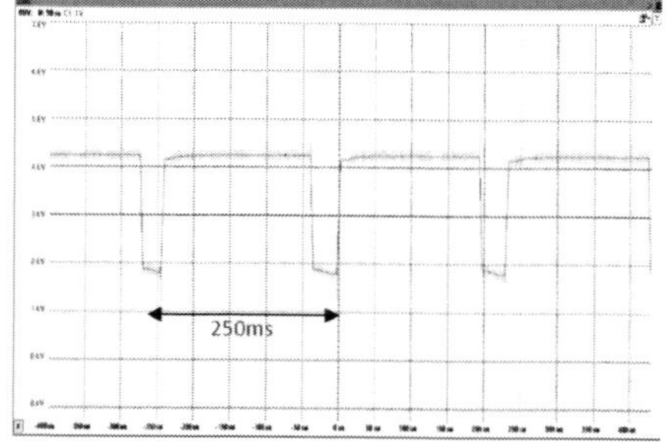

Figure 3. Unipolar return to zero signal.

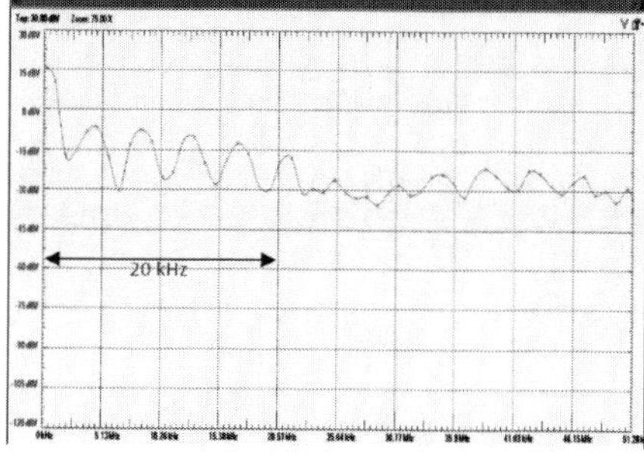

Figure 4. FFT of unipolar RZ signal.

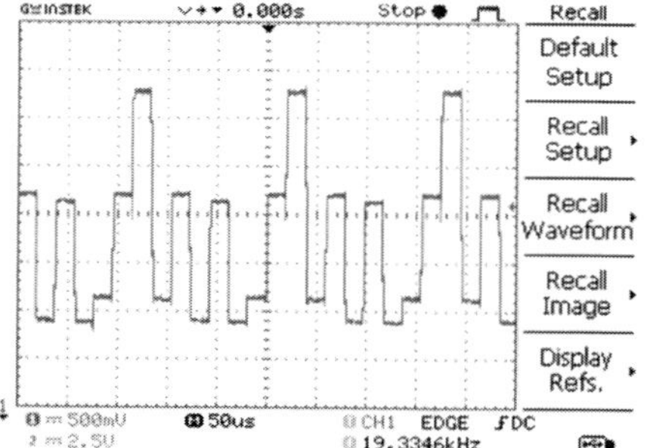

Figure 5. DAC output (before filtering).

8 FILTERING AND BANDLIMITING

The waveform obtained from the DAC has step edges; the frequency signature of waveform with sharp edges is wide causing adjacent channel interference. To eliminate this undesired response filtering of the waveform was necessary. A third order low pass filter was designed as shown in Fig. 6. Standard approach was followed to synthesize the filter and calculate the component values [4]. The response of the filter is shown in Figure 7. The filter forced a suppression of 15 dB at 96 kHz (Fig. 8). The suppression increases which frequency and thus helps in band-limiting the arbitrary waveform. Due to the L-C nature of the filter ringing is observable. This could be controlled by decreasing the quality factor of the inductor by adding a small amount of resistance in series.

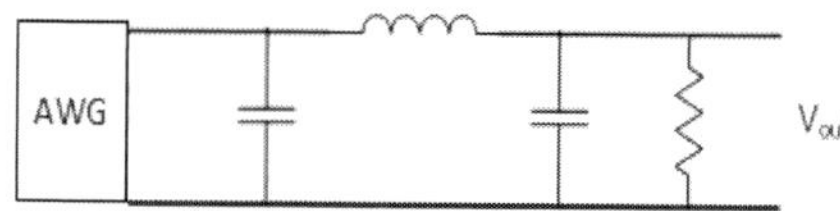

Figure 6. Third order low pass filter.

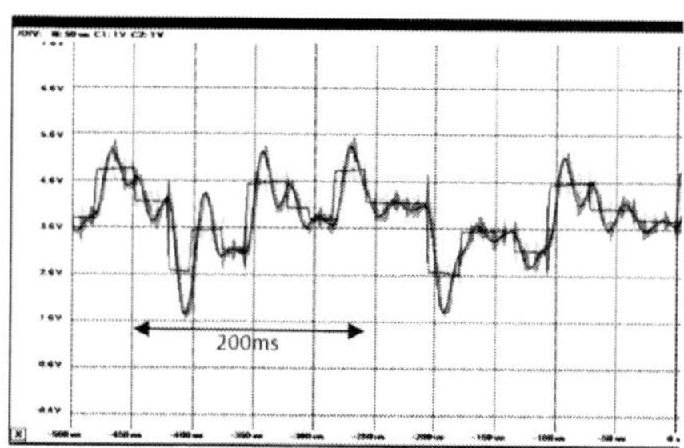

Figure 7. Filtered and unfiltered output (time domain).

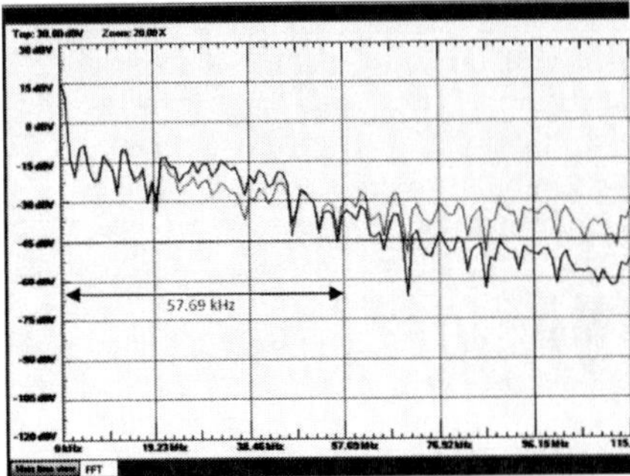

Figure 8. FFT of filtered and unfiltered waveform.

9 DIGTAL COMMUNICATION WITH ARBITRARY LINE CODING

That the synthesis of arbitrary wave-shapes with low cost hard-ware is shown to be feasible, the USART port of the Micro-controller could be linked with Serial Port or Universal Serial Bus of a Computer opening a custom channel for data communication. If used in Line Communication, the output of the filter could be amplified and fed into a transmission line with adequate bandwidth. After proper hand-shaking, the receiver would know the wave-shape for "ONE" and "ZERO", and would extract them from the received time vs. signal data through correlation.

10 FURTHER DEVELOPMENT

A software application customized for data communication through the developed hardware would be designed and implemented enabling seamless secure data-transfer. The developed hardware is mostly composed of discrete elements. The system would be integrated into a hand-held appliance with or without modulation option. The correlation receiver section would be implemented. This could be complete solution for secure communication of high priority data.

11 COMPARATIVE STUDY

The nearest previous example of this work is of Liu et al. 2007, which has used much costlier hardware but has achieved DDS up to 10 KHz only. Whereas, present work uses off-the-shelf components to achieve real-time arbitrary waveform synthesis up to 20 KHz.

12 CONCLUSION

The use of potentiometers for input of waypoints is not efficient as the precision level of voltage supplied is not fine. Conversely this method makes a system comparatively cheaper and simpler to build. This also enables the user to establish a very secure form of communication. The short comings can be solved by using Integrated Circuits to attain reliability and precision.

REFERENCES

Ashkan Ashrafi, Reza Adhami, Laurie Joiner, Parisa Kaveh. "Arbitrary Waveform DDFS Utilizing Chebyshev Polynomials Interpolation," IEEE Trans. Circuits and Systems. Vol. 51. August 2004.pp1468–1475.

Chakraborty, Dr. A. an Introduction to Network Filters and Transmission Lines, Dhanpat Rai & Co. Tech. Publ. 2002.

Cordness, L. "Direct Digital Synthesis: A Tool for Periodic Wave Generation (Part 1)", IEEE Sig. Proc. Mag., vol. 21, pp. 50–54, 2004.

Jen-Wei Hsieh, Guo-Ruey Tsai and Min-Chum Lin, "Using FPGA to Implement a N-channel Arbitrary Waveform Generator with Various Add-on Functions", Department of Electronic Engineering, Kun-Shan University of Technology, pp. 296–298.

Jouko Vankka, Mikko Waltari, Marko Kosunen, Kari A.I.Halonen. "A Direct Digital Synthesizer with an On-chip D/A-Converter," IEEE Journal. Solid-state Circuits. VOL.33. FEBUARY 1998. pp 218–227.

Karmakar, N.C., R. Koswatta, P. Kalansuriya, Rubayet-E-Azim, "Chipless RFID Reader Architecture", 5.2 99.

Kulkarni, V., et al. "A Multilevel NRZ Line Coding Technique", International Conference on Technology Systems and Management (ICTSM) 2011, Proceedings published by International Journal of Computer Applications® (IJCA).

Liu X.; S. Li; Y. Shi "An Arbitrary Waveform Generator for Switch-Linear Hybrid Flexible Waveform Power Amplifier", Industrial Electronics and Applications, 2007. ICIEA 2007. 2nd IEEE Conference on, on page(s): 2732–2736.

Liu Xiaodong; Shi Yanyan; Li Shubo "A MCU-Based Arbitrary Waveform Generator for SLH Power Amplifier Using DDS Technique", Electronic Measurement and Instruments, 2007. ICEMI '07. 8th International Conference on, on page(s): 4-895-4-899.

Quintáns, C. "Frequency accuracy improvement in the Direct Digital Synthesis for the wave form generators", Proc. 2006 Instrumentation and Measurement Technology Conference, pp. 1380–1385, Apr. 2006.

Tie-liang L. and Q. Yu-lin, "An approach to the single-chip arbitrary waveform generator (AWG)", Proc. 4th Int. Conf. ASIC, pp. 506–509, 2001.

US 4514761 A, "Data encryption technique for subscription television system"

Valls, J.; Sansaloni, T.; Perez-Pascual, A.; Torres, V.; Almenar, V. "The use of CORDIC in software defined radios: a tutorial", Communications Magazine, IEEE, on page(s): 46–50 Volume: 44, Issue: 9, Sept. 2006.

Yeary, M.B., J.Fink, and David Beck, "A DSP-Based Mixed-Signal Waveform Generator", IEEE Trans. Instrumentation and Measurement. Vol.53, June 2004. pp. 665–671.

Computational Science and Engineering – Deyasi et al. (Eds)
© 2017 Taylor & Francis Group, ISBN 978-1-138-02983-5

Intrusion detection using SVM in a video sequence and signaling an alarm through Arduino UNO

J. Ghosh Dastidar, P. Guha, S. Pal & N. Ahmed
St. Xavier's College (Autonomous), Kolkata, India

ABSTRACT: This paper presents an overview of a system which will detect human intrusion and signal an alarm using a GSM (Global System for Mobile Communications) module through an Arduino UNO. A simple case of intrusion is detected using correlation analysis between two images—a static background image and an image with the intruder. Background subtraction method is then applied to obtain the object of intrusion. Skeletonization procedure is used to extract features of the object, in order to train a Support Vector Machine (SVM). Thereafter, an unknown pattern is classified into one of five classes using the Minimum Distance Classifier method. Finally, an alarm in the form of an SMS (Short Message Service) is signaled to an authorized entity.

1 INTRODUCTION

This paper mainly focuses on the realm of automated security. Humans do not fare well in carrying out mundane tasks which are repetitive in nature and require constant attention to details. On the other hand, computers outperform human beings in such domains.

Achieving the automation would work wonders for the current security procedures in place. Not only does the cost come down, but also the efficiency shoots up a few notches. Once automaticity is achieved, the question of "human error" no longer arises, thereby improving the overall quality of the system.

Preventing any loss whatsoever and also nabbing the attacker before it is too late forms the core of automated surveillance systems. The signaling of alarm in the form of an SMS makes this possible.

Figure 1 illustrates how the system has been implemented. A closed circuit television monitoring the area under surveillance feeds live footage to a software at the back-end, which carries out the detection of intrusion and signaling of an alarm. If a human intruder steps into the monitoring zone of the CCTV, the software (S/W) detects it and through an Arduino UNO sends an alarm in the form of GSM signals to an appropriate authority. The software in question has been implemented using MATLAB.

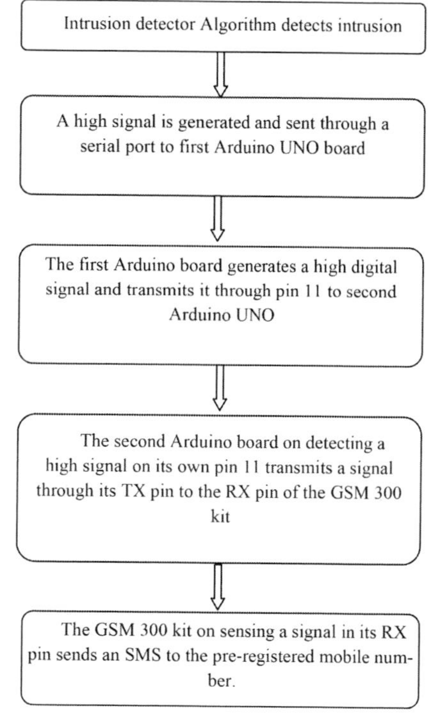

Figure 1. Overview of proposed system.

2 LITERATURE SURVEY

There are direct as well as indirect methods for analyzing human presence in images and videos.

Direct methods work on the source image itself without performing any preprocessing, thereby classifying objects on the basis of colour, motion, shape, contour, etc.

Indirect methods also attempt to classify objects based on their colour, shape (Belongie, et al., 2002), motion or contour (Shotton, et al., 2008) but only after performing some amount of preprocessing on the input image. The simplest preprocessing technique employed is background subtraction. After eliminating the background, the foreground object can be processed for human presence.

Another popular human detection scheme is the facial recognition system, which employs selected facial features to draw a conclusion regarding human presence.

In order to extract the foreground intruding object from the overall scenario, the use of Canny Edge Detection algorithm has been made in some work (Adhikari, 2014). It involves computing the intersection of the row-wise and column-wise representations of the image in question.

3 PROPOSED METHODOLOGY

The task of detecting human intrusion and signaling an alarm is as complex as it sounds. Due to the cumbersome nature of the process, it is conveniently divided into a number of sub-tasks and each of those sub-tasks is implemented individually.

3.1 *Detecting intrusion*

The method adopted in this case does not require storing of frames, except a fixed background. Successive frames are then extracted from real-time video one at a time after fixed intervals of time (say 2 seconds) and compared with the static background. To identify intrusion, correlation analysis is employed (Sengupta, et al., 2014). A high correlation coefficient indicates more similarity whereas a low value indicates little or no resemblance. Thus, two images having a significant value of correlation coefficient states the fact that they are very similar or almost the same (Gonzalez & Woods, 2009). So, if we set a threshold for intrusion and compare two images for their correlation value and find that the value is below the benchmark set, we can safely conclude that intrusion has occurred. To determine this limiting value for correlation, we employ a trial-and-error heuristic technique.

$$r = \frac{\sum_m \sum_n (A_{mn} - \bar{A})(B_{mn} - \bar{B})}{\sqrt{\left(\sum_m \sum_n (A_{mn} - \bar{A})^2\right)\left(\sum_m \sum_n (B_{mn} - \bar{B})^2\right)}}$$

In the given formula, r is the correlation coefficient whose value lies in the range [-1, 1]; A is an image matrix with m rows and n columns; B is also an image matrix with m rows and n columns; $\bar{A}$ is the average value of the pixels in image matrix A; $\bar{B}$ is the average value of the pixels in image matrix B. After extensive experimentation we set the benchmark for r as 0.9023. Correlation is calculated for the static background and a sampled frame. If the value is found to be less than the previously set threshold, it is indicative of potential intrusion or significant difference in the background and foreground. But if the coefficient exceeds the threshold, it is understood that no drastic changes have taken place (Kundu, et al., 2014). The background which is kept fixed needs to be periodically refreshed because it is subject to changes resulting from varying light conditions, interference and noise. Thus, it must be time-invariant.

3.2 *Extracting object of intrusion*

A pixel-by-pixel subtraction of the background image from the foreground image gives the image of the intruding object. The pixels constituting the foreground are set to white, while the remaining pixels are all black. The result is the intruding object in white with its surrounding all black.

3.3 *Skeleton extraction and processing*

In-order to classify the intruder we compute the skeleton. Skeleton is the set of points that are equidistant from the nearest edges of the image (Bai, et al., 2007, Bai & Latecki, 2008). Obtaining the skeleton of the object of intrusion and computing certain measures like endpoints and fork points has proved to be effective.

A skeleton point having only one adjacent point is called an endpoint; a skeleton point having more than two adjacent points is called a fork point; a skeleton segment between two skeleton points is called a skeleton branch; every point which is not an endpoint or fork point is called a branch point (Ghosh Dastidar & Biswas, 2015).

The methodology adopted requires computation of endpoints as well as fork points followed by calculating certain distance measures which are known to vary for skeletons of different objects. However, skeletons of two similar objects are also known to vary widely. To overcome this drawback, the height to width ratio (aspect ratio) of the skeleton is found and the fork points calculated. Human skeletons are known to have two forks—one at the neck and the other at the waist (Blum, 1973). These points give a fair idea of where the skeleton is broken into branches, thereby indicating the positions of limbs, neck and head. One skeleton is distinguished from another through the positions of these points.

3.4 *Use of Support Vector Machine (SVM)*

Using a supervised learning model like Support Vector Machine, aids in the classification of an unknown pattern into an appropriate pattern class. In the proposed system, five pattern classes have been defined to identify human intruders (Ghosh Dastidar 2016). Each such class has the following attributes:

a. possibility—indicates the possibility of an intruder being a human
b. aspect ratio—height to width ratio obtained from the skeleton
c. shapeneck—position of neck
d. shapewaist—position of waist
e. shape_pos—possibility of the shape resembling a human form
f. final_score—score given to an intruder

After obtaining a number of such patterns for each class, they are stored in a file. This forms the training set of the SVM. Based on the values of the defined attributed five pattern classes have been created as shown in Table 1.

Classifying an object of intrusion, which in this case can be treated as an unknown pattern with particular values for possibility, aspect_ratio, shapeneck, shapewaist, shape_pos, final_score, consists of the following steps:

a. All pattern vectors of each class are extracted from the spreadsheet and stored in respective variables.
b. The mean vector for each class is calculated.
c. The Euclidean distance is found for each pair: mean vector of ith class, m_i and unknown pattern vector, x. It is given by $||x - m_i||$.
d. The unknown pattern belongs to the class for which the distance is the least.

This method of classification is known as the Minimum Distance Classifier (Gonzalez & Woods, 2009).

Table 2 shows a sample of two pattern vectors each for the five pattern classes. The differences in the values for the attributes are worth noting.

3.5 *Signaling an alarm*

In order to signal an alarm, there is need to interface an Arduino UNO with the system containing the procedures for human intrusion detection. Thereafter, establishing communication merely requires setting a pin of the Arduino microcontroller in output mode and writing a bit indicating human intrusion. This bit, received in the form of a signal, is to be relayed to the second Arduino microcontroller which is used to interface the GSM module.

Once the microcontroller on breadboard receives a positive signal from its UNO counterpart regarding the occurrence of human intrusion, it relays this information to the GSM module which uses attention commands to generate an alarm in the form of a text message. Shown below in Figure 2 is the interfacing circuit with the two Arduino UNO chips along with the GSM module and the intrusion detection software.

Table 1. Class descriptions.

Class number	Description
1	Not human
2	Probably not human
3	Most probably human
4	Human
5	Definitely human

Table 2. Sample pattern vectors for different classes.

Class	Possibility	Aspect ratio	Shape neck	Shape waist	Shape pos	Final score
1	0	1.89	1	0	0.4	0.4
	0	0.98	0	0	0	0
2	0	1.48	1	1	0.8	0.8
	0	1.73	1	1	0.8	0.8
3	1	2.42	0	0	0	1
	1	2.63	0	0	0	1
4	1	2.45	0	1	0.4	1.4
	1	2.98	1	0	0.4	1.4
5	1	2.32	1	1	0.8	1.8
	1	2.39	1	1	0.8	1.8

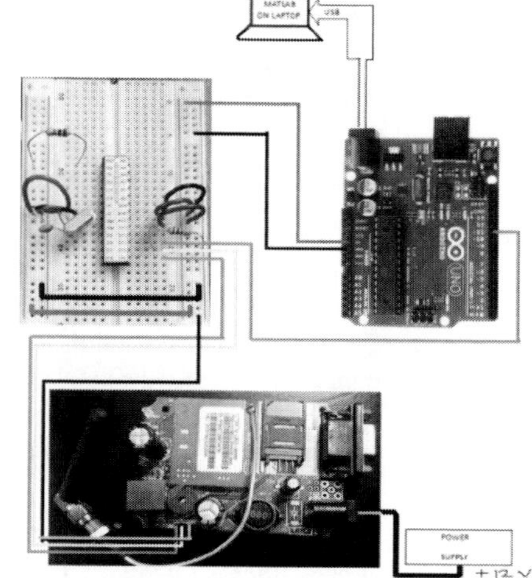

Figure 2. Interfacing circuit.

4 RESULT AND DISCUSSION

The algorithm for the detection of intrusion and then classifying the intruder has been implemented using the MATLAB software. Two ATMEGA

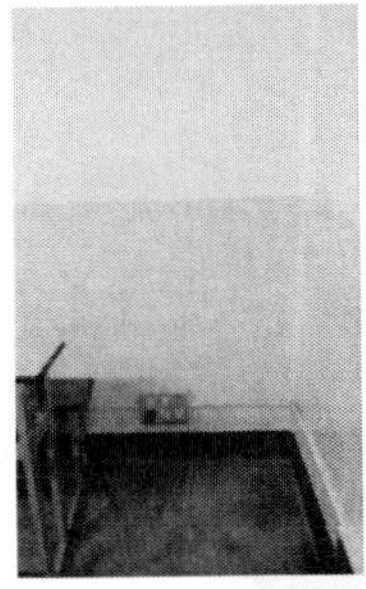
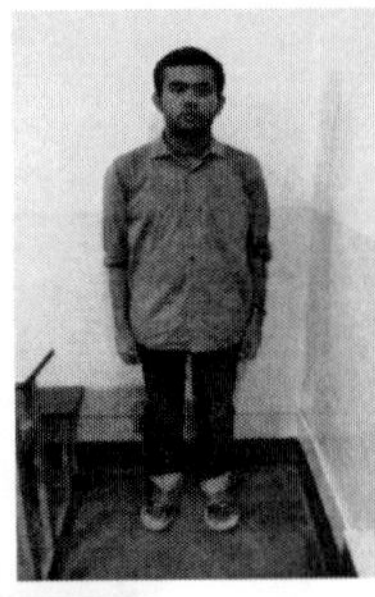
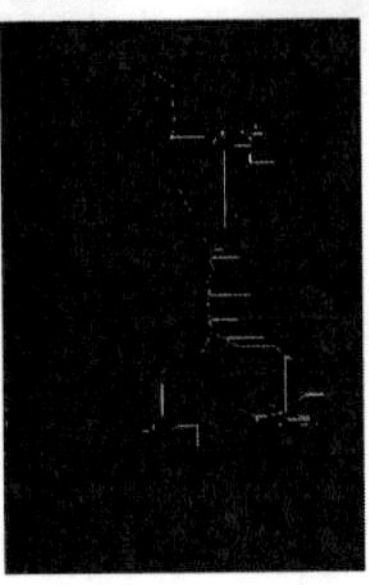

Figure 3. Background frame, Intruder, Skeleton of intruder (clockwise from top-left).

microcontrollers have been used—one on a bread board and another on a Arduino UNO kit. MATLAB connects to Arduino, which stores the procedures necessary to communicate with modules written in MATLAB, through a USB cable.

These procedures are written using the IDE (Integrated Development Environment) of Arduino. ATMEGA microcontroller on bread board and the associated circuitry is built by having a separate crystal clock on the bread board. The code for interfacing a GSM module is kept in this microcontroller. Once human intrusion is detected, a signal is sent from MATLAB to UNO which is relayed to the AT MEGA on the bread board. Due to the interfacing, the GSM SIM300 is also informed of the same. On receiving the signal, an alarm in the form of a text message is sent to a particular mobile number of the authority enabling rapid action and prevention of harm. The flow of data is from MATLAB on a computer to Arduino UNO to ATMEGA on breadboard and finally to GSM module which initiates an SMS. The Figure 3 shows a scenario of an area being intruded and the skeleton being computed.

5 CONCLUSION

The system proposed in this paper detects all forms of intrusion—human as well as non-human. It is able to detect multiple intrusions, i.e., if consecutive intruders pose a threat over a given area, they are identified and an alarm is generated for each intruder separately. This also gives an indication of the level of trespassing that is taking place in the zone being monitored. The availability is on the higher side because the system never stops monitoring. Also, the classification of the intruder into one of five classes with the help of SVM increases the precision level of intrusion detection. It is not merely restricted to human intrusion or no human intrusion but to finer details indicating the probability of human intrusion.

REFERENCES

Adhikari, S., Kar, J. & Ghosh Dastidar J. Mar-Apr 2014. An automatic and efficient foreground object extraction scheme. In International Journal of Science and Applied Information Technology Volume 3, No. 2: 40–43.

Bai, X., Latecki, L. J. & Liu, W. Y. 2007. Skeleton pruning by contour partitioning with discrete curve evolution. In IEEE Trans Pattern Analysis and Machine Intelligence (PAMI), 29(3): 449–462.

Bai, X. & Latecki, L. 2008. Path similarity skeleton graph matching. In IEEE Transactions on Pattern Analysis and Machine Intelligence (PAMI), Vol 30, No 7: 1282–1292.

Belongie, S., Malik, J. & Puzicha, J. 2002. Shape matching and object recognition using shape contexts. In PAMI, 24(4): 509–522.

Blum, H. Biological shape and visual science. 1973. In J. Theoretical Biology. Vol 38: 205–287.

Ghosh Dastidar, J. & Biswas, R. Dec 2015. Tracking human intrusion through a CCTV. In IEEE conference CICN2015, doi - 10.1109/CICN.2015.95, ISBN: 978-1-5090-0076-0/15: 461–465.

Ghosh Dastidar, J. Jan 2016. Human form identification from a video stream using support vector machine. In 3rd. IEEE conference ICBIM2016, ISBN: 978-1-5090-1228-2/16.

Gonzalez, R. C. & Woods, R. E. 2009. Digital Image Processing, (3rd. ed): Pearson Education.

Kundu, M., Sengupta D. & Ghosh Dastidar, J. Jun 2014. Tracking direction of human movement – an efficient implementation using skeleton. In International Journal of Computer Applications, Vol 96 – No 13: 27–33.

Sengupta, D., Kundu, M. & Ghosh Dastidar, J. Mar 2014. Human shape variation—an efficient implementation using skeleton. In International Journal of Advanced Computer Research, Vol 4 Number 1, Issue 14:145–150.

Shotton, J., Blake, A. & Cipolla, R. 2008. Multi-scale categorical object recognition using contour fragments. In PAMI, 30(7): 1270–1281.

Computational Science and Engineering – Deyasi et al. (Eds)
© 2017 Taylor & Francis Group, ISBN 978-1-138-02983-5

Design and implementation of wireless sensor network using ARDUINO

Eshita Sarkar
VLSI Design, ECE Department, MCKVIE, Howrah, India

Swarup Kumar Mitra & Sagar Mukherjee
ECE Department, MCKVIE, Howrah, India

ABSTRACT: The popularity of Wireless Sensor Networks (WSN) has increased excessively due to the vast potential of the sensor networks to connect the physical world with the virtual world. As these devices lean on battery power so very less power required for this devices. The circuit consist of sensor parts build using various sensors like LDR. The ARDUINO UNO board is used to control the whole system by monitoring the sensors. Development of these interfacing of sensors needed hardware as well as software. We have designed a wireless sensor network with the support of ARDUINO board. The different types of wireless communication mainly include IR wireless communication, Satellite communication, Bluetooth, Zig-bee etc. For communication we have installed HC05 Bluetooth module which can detect and monitor a system of industrial and commercial application. The main focus of our work is on the hardware implementation of WSN.

Keywords: WSN, ARDUINO, LDR, HC05 BT Module

1 INTRODUCTION

A **wireless Sensor Network** (WSN) (Akyildiz, 2002, Chong, 2003, Gowrishanka, 2008, Korkalainen, 2009, Leelavati, 2013) is a **wireless network** consisting of spatially distributed autonomous devices using **sensors** to monitor physical or environmental conditions such as **temperature, sound, pressure**. A **WSN** system is a gateway that provides **wireless** connectivity back to the wired world and distributed nodes. In the most cases, the sensor forming these networks and deployed randomly and left unattended to and are expected to perform their mission properly and efficiently. The communication devices on these sensors are small and have limited power and range.

The design issues and constraints on WSNs along with the software based design rules and simulations has been described in (Akyildiz, 2002, Chong, 2003, Gowrishanka, 2008, Korkalainen, 2009, Leelavati, 2013). The focus of the work is to pay special attention on the hardware based design and implementation with the help of BT module and microcontroller for WSN mote design. The WSN consist of three major components i.e. Sensor Nodes, Base Station, power supply unit.

2 DESIGN ISSUES AND CONSTRAINTS

Hardware devices that enable wireless sensor network are Microcontroller, Sensors and/or actuators, Wireless development tool.

2.1 *Microcontroller*

The microcontroller board are used in this design is ARDUINO UNO Board (www.arduino.cc). ARDUINO is an open-source prototyping platform based on easy-to-use hardware and software. It is using the ARDUINO programming language (Software IDE version 1.6.3).

2.2 *Sensor*

In this work for sensor nodes are using LDR sensor (en.wikipedia.org) because it is easily available and cost effective. It is Light-controlled variable resistor. The resistance of a photoresistor decreases with increasing incident light intensity. It is also known as photocell.

2.3 *Wireless development tool*

In this work wireless environment are established using HC05 Bluetooth module (www.bluetooth.

com, en.wikipedia.org). HC-05 module is an easy to use Bluetooth SPP (Serial Port Protocol) module, designed for transparent wireless serial connection setup. Permit pairing device to connect as default. Auto-pairing PINCODE:"1234" as default.

3 PROPOSED MODULE

This design module comprises of wirelessly send LDR sensor data (Zolkapli, 2013, Noor, 2012, Awale, 2016) from one point to another point using HC05 BT module and ARDUINO. For example in (Santoshkumar, 2015, Darp, 2015) it has been described the interfacing of Raspberry Pi with HC05 BT module and how it will interface with Android phones. So a basic idea of interfacing of HC05 BT module has been obtained. This working module mainly comprises with two parts first wirelessly transmit the sensor data from PC to BT module and second wirelessly transmit the sensor data from one BT module to another using Master-Slave connection (Cotta, 2016).

3.1 *Wirelessly transfer LDR sensor data from PC to BT module using ARDUINO*

In this working module one can find that the LDR sensor data are sending and receive wirelessly using Bluetooth module and ARDUINO board. For example in (m.instructables.com) the detail of connection between ARDUINO and HC05 BT module. The receiving or sending LDR sensor data are shown in the serial monitor of the IDE software and at the part of Bluetooth module the output is shown in TERA-TERM software.

- *TERA-TERM Software*
  TeraTerm (rarely TeraTerm) is an open-source, free, software implemented, terminal emulator (communications) program. It emulates different types of computer terminals, from DEC VT100 to DEC VT382. It supports telnet, SSH 1 & 2 and serial port connections (en.wikipedia.org).
- *Block Diagram*

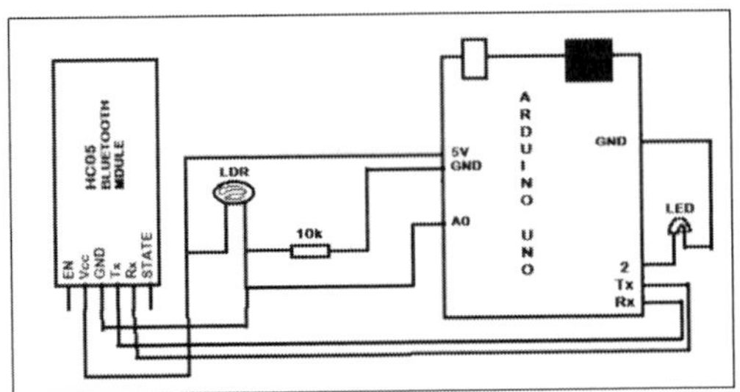

Figure 1. Wirelessly transfer sensor data to BT module.

- *Circuit Description*
  The top of the Potential Divider is 5V; the bottom is at 0V and the middle (connected to A0) is some value between 5V and 0V that varies as the LDR resistance varies. Now for BT module connection is:

Table 1. Circuit connection.

ARDUINO	HC05 bluetooth module
5V–3.3V	Vcc
GND	GND
Tx	Rx
Rx	Tx
Vcc(AT Mode Operation)	EN

Now in the serial monitor of the IDE software shows the LDR sensor reading and the voltage reading for that particular sensor and it will shown into the TERA-TERM software as BT module output.

- *Circuit Implementation*

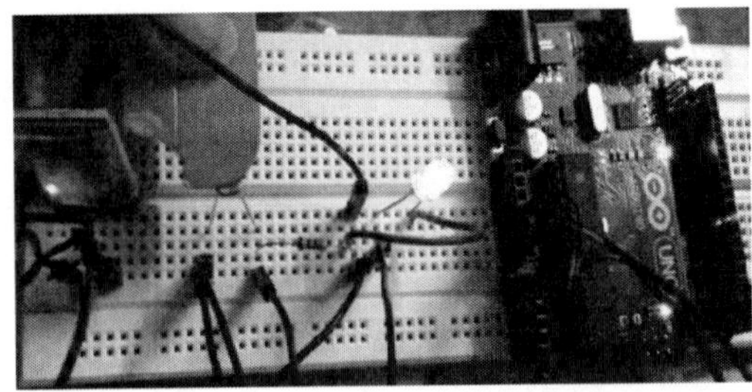

Figure 2. Hardware implementation.

- *Output*

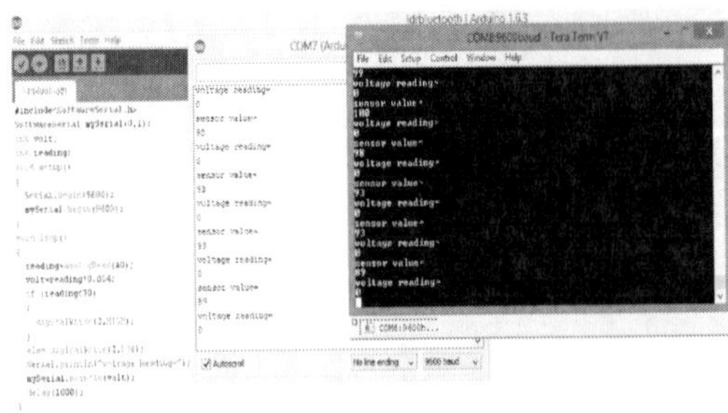

Figure 3. Overall output.

3.2 *Point to point wireless network using BT module as Master-Slave*

In this work by taking the help of the AT Commands the HC05 module worked as Master and Slave with the modification of the AT Commands. In (Cotta, 2015) the Bluetooth network topology of Master-Slave connections is described. Interfacing and circuit description with ARDUINO with HC05 BT module is also discuss in detail.

- *Block Diagram*

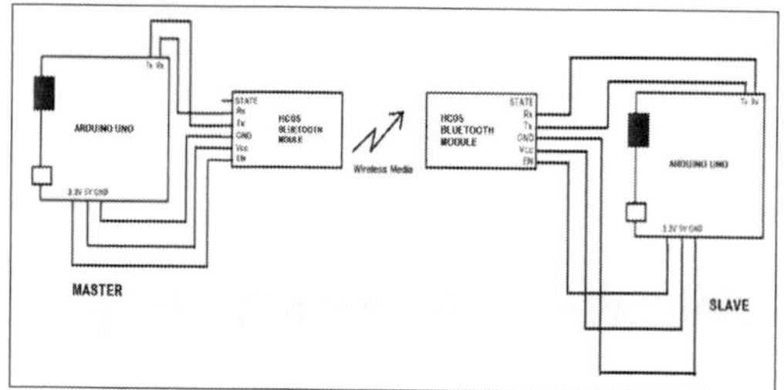

Figure 4. Wireless network.

- *HC05 BT module with AT commands*

How to get to AT COMMAND mode:

1. Connect EN pin to VCC.
2. Supply power to module. Then the module will enter into AT MODE. In this mode you have to use baud rate at 38400. In this way user should change the baud rate for SLAVE AND MASTER mode (www.ram.com.eg, www.techbitar.com).

- *Procedure for SLAVE-Device role*

It is a bit tricky to pair one BT module with another. In this work it will neatly describe the method of pairing two BT modules. One of the modules is assigned ROLE as MASTER and the other left as SLAVE.

Here IDE software serial monitor are used to modify one HC05 BT module to serve it in Slave mode by just write AT+MODE = 0.

By default all HC05 modules are SLAVEs. To configure the SLAVE one can make use of an ARDUINO UNO board. Not much of configuration needed for slave but to know the ADDRESS of the slave has to follow this procedure.

Step 1: Before connecting the HC05 module, upload an empty sketch to ARDUINO with IDE software. This bypasses the Boot loader of UNO & the ARDUINO is used as USB-UART converter.

1. void setup() {}
2. void loop() {}

Step 2: After uploading this empty sketch, remove USB power from ARDUINO & do the following connections of BT module with ARDUINO board.

Table 2. Circuit connection.

ARDUINO UNO	HC05 BT MODULE
5V	Vcc
GND	GND
Rx(Pin 0)	Rx
Tx(Pin 1)	Tx
3.3V	EN

Step 3: Now provide the USB cable power to ARDUINO. The HC05 module enters the Command mode with Baud Rate 38400. Then open the Serial Monitor of ARDUINO.

Ensure to select "BOTH NL & CR" and Baud Rate as 38400. This is very important as the Bluetooth module HC05 expects both Carriage Return and Line Feed after every AT command.

If you type in AT you should get an OK response.

AT+PSWD? To know the default password that is "1234".

AT+ROLE? Default role is "0" leave it "0" for Slave

AT+ADDR? Find out the address of Slave BT module so that Master BT module can communicate.

Find response as +ADDR: 98d3:31:502d96 that is the address of Slave.

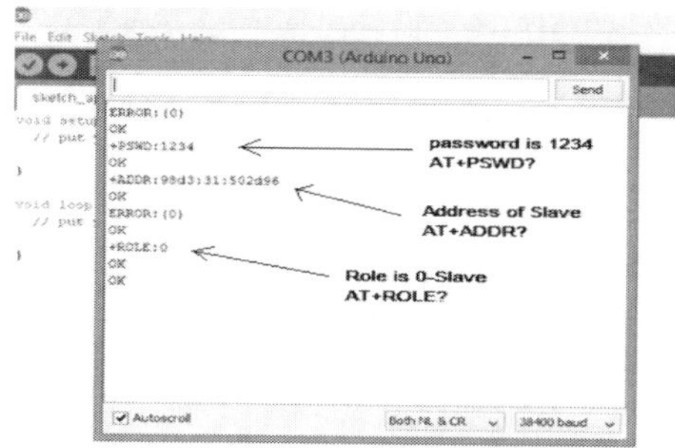

Figure 5. Slave device configuration.

- *Procedure for MASTER-Device role*

Now the emulator software interface are used to modify the HC05 BT module to serve it in Master mode by just write AT+MODE = 1, after that only it can serve as Master and try to find the nearest Slave connected to it and after the connection is established then it will send the data to the Slave which it has receive. For Master device connection also has some steps to follow those are:

Step 1 & Step 2 are the same as used in Slave device role. Differences are identified in the interfacing software implementation which has to follow for this purpose.

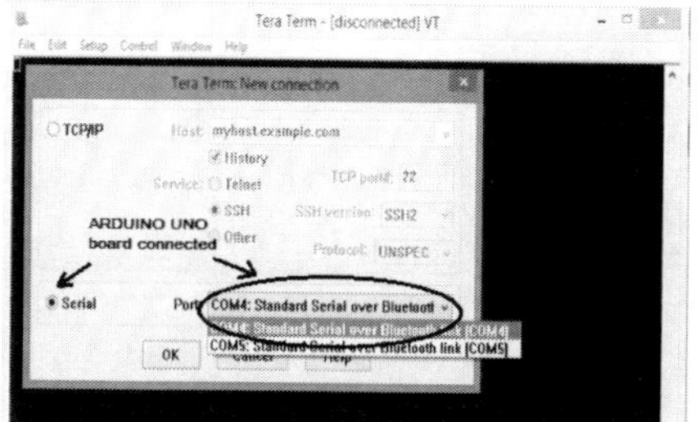

Step 3: Open the TERATERM terminal & select SERIAL & the port number where ARDUINO UNO board is connected.

Step 4: Now for communication purpose go through the setup of terminal and serial port setup.

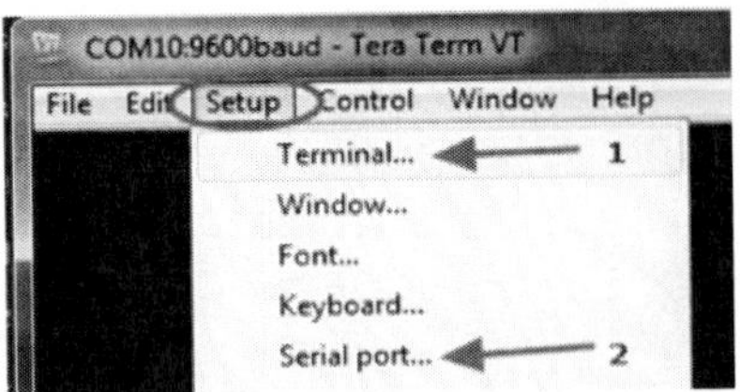

Step 5: Under SETUP ->Terminal select CR+LF for TRANSMIT

Step 6: Under SETUP -> SERIAL PORT select the Baud Rate as 38400, 8 N1.

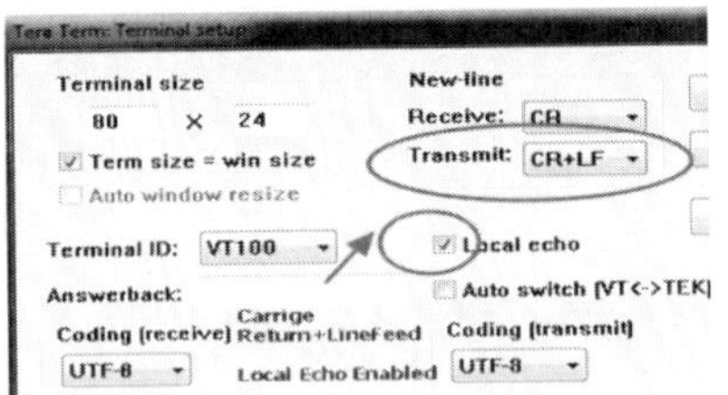

Step 7: Now it is time to put the AT Commands for Master mode setup of HC05 BT module

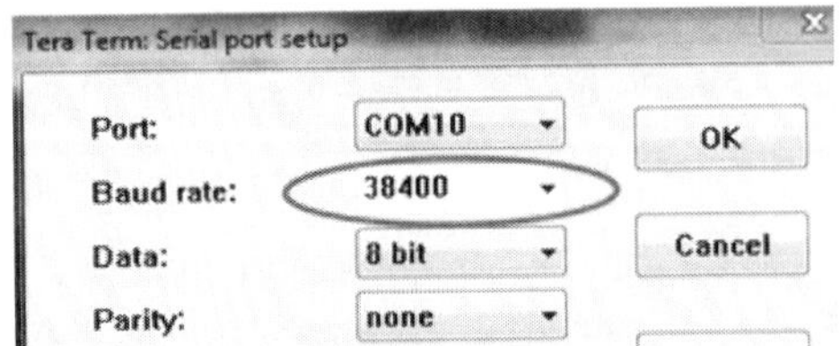

If you type in AT you should get an OK response.

Reset the module by issuing AT+ORGL which restores the module to original state.

You can change the name to user friendly one by typing AT+NAME = usergivenname

AT+RMAAD will release the module from any previous PAIR.

AT+PSWD = 1234 to set the password as 1234

AT+ROLE = 1 changes the ROLE of the module to MASTER

AT+CMODE = 1 Allows connecting to any address.

AT+INIT that is search surrounding BT devices.

AT+LINK = 98d3, 31,502d96 to link or to communicate with Slave BT module.

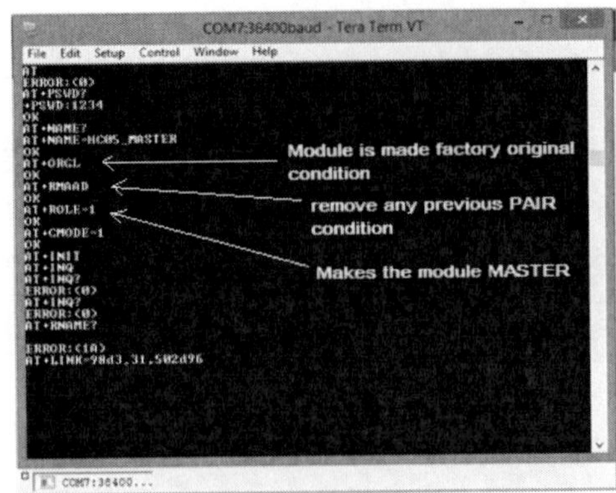

Figure 6. Master Device Configuration.

- Circuit Implementation

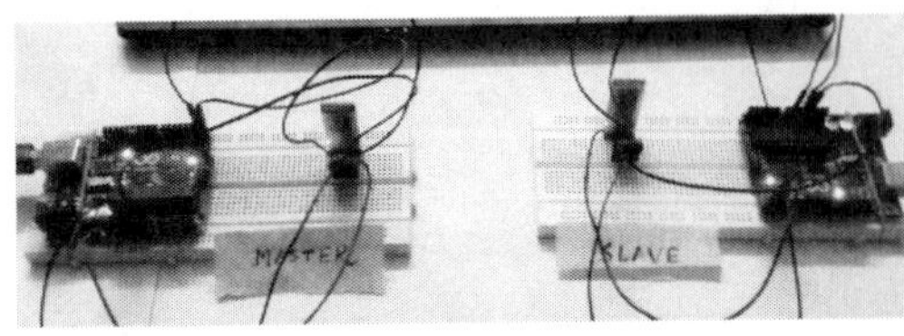

Figure 7.

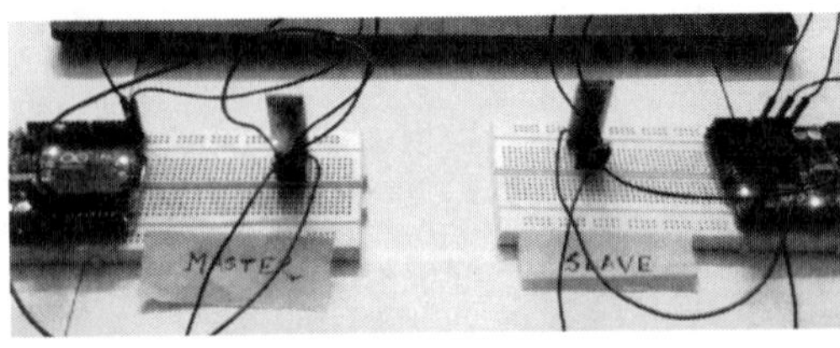

Figure 8.

In Figs. 7 and 8, shows that when two HC05 BT modules are being connected to each other the indicating LED of Slave BT module blinking once in every two second and the indicating LED of Master BT module remains ideal throughout the communication.

4 CONCLUSION

Design and implementation of Wireless Sensor Network using ARDUINO was successfully established and tested. As at last the Master-Slave configuration was successfully done and the wireless connections in between two BT modules are made. So that by using that process one can easily transmit the sensor data wirelessly.

REFERENCES

Akyildiz, I.; Su, W.; Sankarasubramaniam, Y.; Cayirci, E. 2002. *"A Survey on Sensor Networks"*. IEEE Commun. Mag vol. 40, pp. 102–114.

Awale, Sudeshna., Sarode, Ashwini., Patil, Divya., Naik, Samridhi., 2016. *" Solar Street light using microcontroller"* IJETEMR, Vol. 2, Issue 1, January.

Chong, C Y and Kumar, S P. 2003. *"Sensor Networks: Evolution, Opportunities and Challenges,"* Proceedings of the IEEE, vol. 91, pp. 1247–1256.

Cotta, Anisha., Devidas, Naik Tripati, Kalides, Varda., Esoskar, Naik.,2016 *" Wireless Communication using HC05 Bluetooth Module Interfacing with ARDUINO"* IJSETR, Vol. 5, Issue 4, April.

Darp, Dinesh D. & Hirekhan, Sunil R., 2015 *"e-MPHA-SYS Monitoring Healt Parameters using Android Operating System"*, International Journal of trend in research and development, Vol. 2(3), May.

Gowrishanka, S, Basavaraju, T G, Manjaiah, D H. & Sarkar, Subir Kumar. 2008. *"Issues in Wireless Sensor Networks"*, In Proceedings of the World Congress on Engineering WCE 2008, vol I, July, London, U.K.

https://www.arduino.cc/en/main/arduinoBoardUno

https://en.wikipedia.org/wiki/Photoresistor

http://www.bluetooth.com/what-is-bluetooth-technology/bluetooth-technology-besics

http://en.wikipedia.org/wiki/Bluetooth

http://m.instructables.com/id/Arduino-AND-Bluetooth-HC-05-Connecting-easily

https://en.wikipedia.org/wiki/Tera_Term

http://www.techbitar.com/modify-the-HC-05-Bluetooth-module-defaults-using-at-commands-html.

http://www.ram.com.eg *"Bluetooth Transceiver RF Module Wireless Serial TTL V1.05 Manual."* A product from RAM electronics Integrated solution at one place,

Korkalainen, Marko, Sallinen, Mikko, Kärkkäinen, Niilo & Tukeva, Pirkka. 2009. *"Survey of Wireless Sensor Networks Simulation Tools for Demanding Applications"* Fifth International Conference on Networking and Services Survey of Wireless Sensor Networks Simulation Tools for Demanding Applications.

Leelavati G, Shaila K, Venugopal K R, and L M Patnaik. 2013. " Design Issues on Software Aspects and Simulation Tools for Wireless Sensor Network" IJNSA, Vol. 5, No. 2.

Noor, Hakim bin, M.Z. Fac. MARA, Shah Alam, Mohd Zainuddin, Malaysia bin, M.S.; Saaid, M.F.; Megat Ali, Amin bin, M.S. 2012 *" Design and conceptual development of a sunbathe laundry robot"* Signal Processing and its Applications (CSPA), IEEE 8th International Colloquium. May.

Santoshkumar, G., & MNVLM Krishna., 2015. *"Low Cost Speech Recognition System Running on Raspberry Pi to support Automation Applications*" IJETT, Vol. 21, No. 5, March.

Zolkapli, M. Fac. MARA, Shah Alam, Malaysia,Al-Junid, S.A.M.; Othman, Z.; Manut, A.; Mohd Zulkifli, M.A. 2013. *" High-efficiency dual-axis solar tracking development using Arduino"* Technology, Informatics, Management, Engineering, and Environment (TIME-E), International Conference. March.

Computational Science and Engineering – Deyasi et al. (Eds)
© 2017 Taylor & Francis Group, ISBN 978-1-138-02983-5

ARDUINO based water level detector for visually impaired people

Somshuddha Datta, Tamalika Chakraborty, Tamanna Roy & Subhayan Haque
RCC Institute of Information Technology, Kolkata, India

Arpita Ghosh
Department of ECE, RCC Institute of Information Technology, Kolkata, India

ABSTRACT: A device that detects different water levels in a container is quite an available scenario but we have designed and implemented a circuit which generates distinct audible frequencies in different levels. It helps visually impaired people to understand the level of water in a tank. The ARDUINO UNO board is used for the hardware and software implementation with four input pins for the detection of four distinct water levels. As the water reaches each level, tones of distinct frequencies are generated at the 8Ω speaker. Apart from small scale implementation of the system in households, there can be large scale implementations too like in water reservoirs, large containers and docks, where visual determination of water level is not possible.

Keywords: Water level detector, ARDUINO UNO, ATmega328 microcontroller

1 INTRODUCTION

Designing a water level detector is not a new phenomenon. There are previous reported works on design and implementation of water level detector using microcontrollers and pressure sensors. In the work of (Clifford 2006) a pressure sensor has been developed to replace the mechanical switch for applications like water level sensing, water flow detection and leak detection. The system was used to monitor water level and water flow using the temperature compensated MPXM2010GS pressure sensor in the low cost MPAK package, a dual op-amp, and the MC68HC908QT4, eight-pin microcontroller. In another work, a PIC16F84A microcontroller was used to automate the process of water pumping (Reza et al. 2010) in an over-head tank storage system and detect the level of water in the tank and switch on/off the pump accordingly as well as display the status on an LCD screen. A calibrated circuit was used for water level indication and it was implemented using the DC instead of AC power supply to eliminate the risk of electrocution. Similarly in all previous works related to the water level monitoring was mainly designed for the level detection and indication with either LED or LCD display or with buzzer. Previously, AT mega32A microcontroller (Mani et al. 2014) was used to indicate water level and display the output on LCD display. A transistor circuit was used to detect the present level of water and feed it to the microcontroller, which generated output text (displayed on LCD). If water level was full, the circuit beeped through the buzzer, notifying that the water level was full. In another research, PIC16F84A microcontroller (Ebere et al. 2013) was used to automate water pumping in an overhead tank and to detect the water level in the tank, switch on/off the pump accordingly and display the status on LCD screen. This work used calibrated circuit and DC instead of AC power which eliminated the risk of electrocution. In another work, PIC 18F452 microcontroller (Abdullah et al. 2015) was used to design a water level indicator. An LCD was used to show the level of water and a buzzer was used to create a siren to stop the water channel.

But so far no work has been done for identifying the water levels for the visually impaired people. We have taken a slightly different approach and implemented a circuit which generates distinct audible frequencies to detect different water levels to help the visually impaired people to understand the level of water. We have used ARDUINO UNO 1.6.1 cc to compile and run a program in the ARDUINO UNO circuit board. Apart from small scale implementation of the system in households, this proposed circuit can be also used in large scale implementations too such as in the water reservoirs, large containers and docks, where visual determination of water level is not possible.

The complete work in this paper is organized in the following manner. The first part or the introduction part is based on the previous works done in this regard and also the approach of our implementation and how it is different from the other available approaches. The second and the third section deal with the detailed hardware and the software implementation of the proposed system respectively. In the fourth section of the paper the results of the work is discussed. The complete work is concluded in the last section.

2 HARDWARE IMPLEMENTATION OF THE SYSTEM

Here in this work we have used ARDUINO Uno board for the complete system implementation. The microcontroller used in this is ATmega328. The power supply requirement for the board is +5 V. The ARDUINO Uno is powered with an external power supply. The clock speed of the microcontroller is 16 MHz. The size of flash memory, SRAM and EEPROM is 32 KB, 2 KB and 1 KB respectively.

The DC current in input/output pin is 40 mA and for the 3.3 V pin is 50 mA. The 3V3 pin where a 3.3 volt supply is generated by the on-board regulator, is connected to the tank water. Four digital pins-2, 3, 4 and 5 are used as input pins and they are connected with four different levels of water of the tank. The output is taken from pin 8. The output of the ARDUINO is applied to a speaker (8Ω) from which we can hear distinct sounds. The block diagram of the designed system is shown in Figure 2.

Figure 2 shows the overall circuit implementation using the ARDUINO UNO board along with the output line connected to the speaker and the input lines are connected for determining the water levels. The four sensing probes are connected to the container with water at four different levels. Mainly four water levels are determined from this circuit- level 1 (lowest), level 2, level 3 and level 4 (highest). The analog signal based on the water level is sent to the on board microcontroller and accordingly the program determines the different conditions or states of the water level. According to the decision generated from the program flow chart the audible tones are generated using the speaker connected to the microcontroller output port.

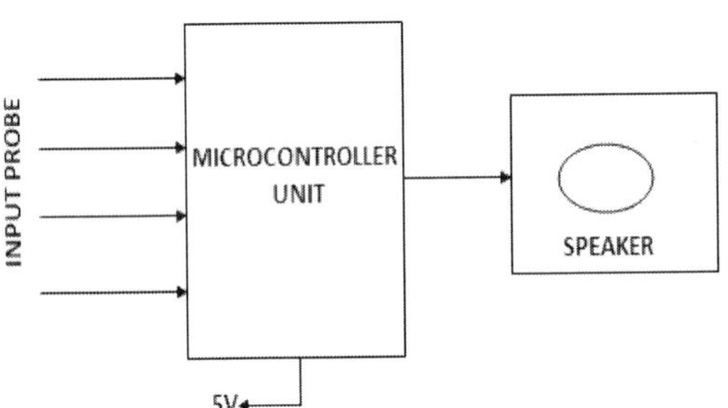

Figure 1. Block diagram of the water level detector with audible tunes.

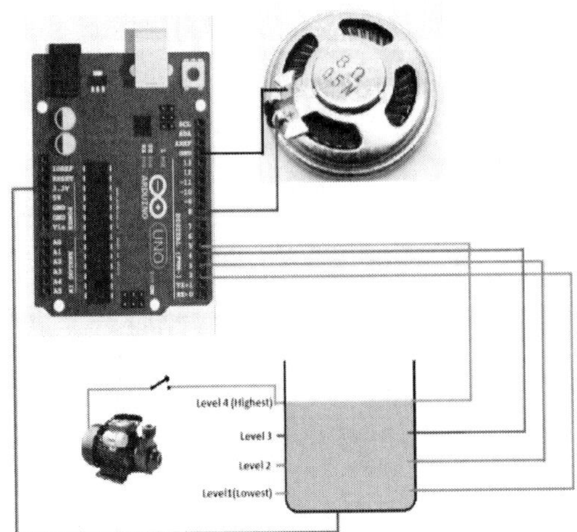

Figure 2. Overall circuit implementation.

3 SOFTWARE IMPLEMENTATION

The complete software part is written in Embedded C and run in ARDUINO 1.6.1 cc software. In ARDUINO UNO, the digital pins Pin2, Pin3,

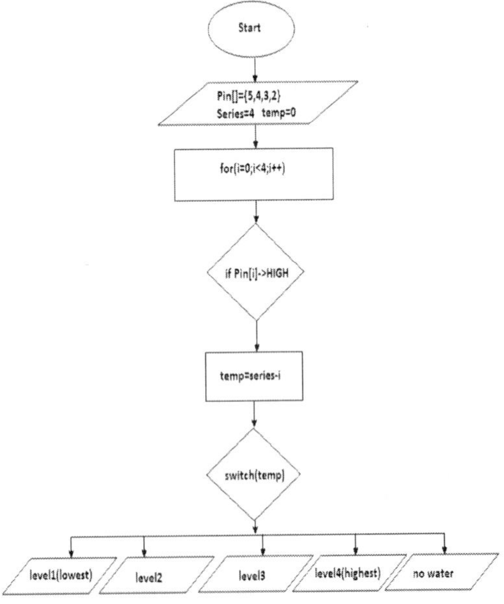

Figure 3. Flow chart of the software.

Pin4, Pin5 are considered as level 1 (lowest), level 2, level 3, level 4 (Highest) respectively.

The complete flow chart of the software part is shown in Figure 3. The initialization is done with declaring an array Pin[] = {5,4,3,2}, two variables temp = 0 and series = 4. At first we consider that i = 0. The range of i is considered from 0 to 3 as we need to determine four levels. For including several more water levels we can increase the range of i and also change the values of the variables.

For i = 0, if the water is at the highest level, the condition Pin[0] = High is satisfied. The value of temp becomes 4(4–0). switch(4): Case 4 (Level 4_ Highest) is considered.

If the water is at Level 2, then the condition Pin[0] = = HIGH is not satisfied, The value of i is incremented by 1 and if in the next iteration also the condition Pin[1] = = HIGH is also not satisfied the value of i is further incremented by 1.Now if the condition is satisfied, temp value becomes 2(4–2). switch(2): case 2(Level 2) is considered.

For all values of i (i = 0, i = 1, i = 2, i = 3) if the condition fails, then default case, that is there is no water in the container, appears.

4 RESULTS

The results of the proposed systems are shown in the Figure 4(a), (b), (c) and (d). Figure 4 shows

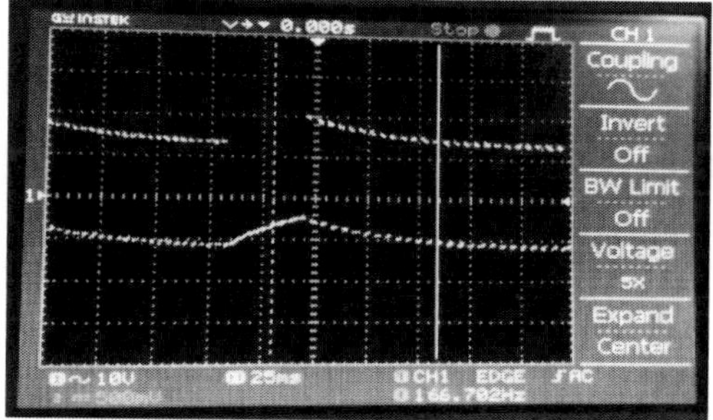

(a)

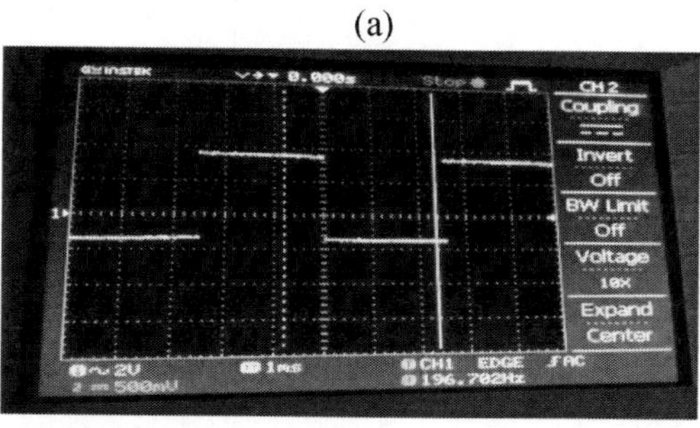

(b)

Figure 4. (*Continued*)

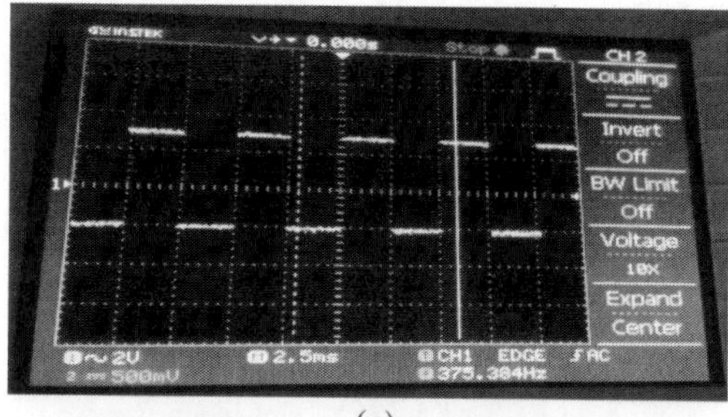

(c)

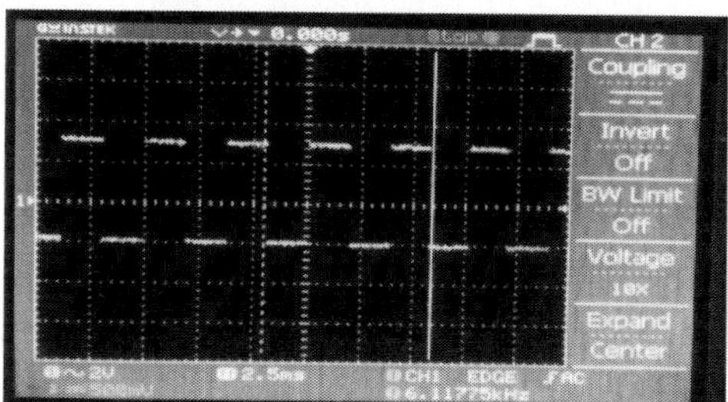

(d)

Figure 4. Output for (a) level 1, (b) level 2, (c) level 3 and (d) level 4.

Table 1. Generated waveforms and their amplitude frequency.

Waveform	Frequency	Amplitude	Ton	Toff
Level 1	166.7 Hz	4 V	2.99 mS	2.99 mS
Level 2	196.7 Hz	4.4 V	2.55 mS	2.55 mS
Level 3	375.3 Hz	4.8 V	1.33 mS	1.33 mS
Level 4	6.12 KHz	5.2 V	0.817 mS	0.817 Hz

the generated output waveforms in the DSO with the different frequency and amplitude values for the four different water levels. Figure 4(a) is the output waveform for the detection of the level 1, with 4 V amplitude and frequency is 166.7 Hz and 50% duty cycle with on and off time of 2.99 m sec. Similarly Figure 4(b), (c) and (d) are for level 2, level 3 and level 4 respectively. The different amplitude and frequency, on time, off times are shown in a tabular manner for the different waveforms generated in Table 1. The highest level is detected using a waveform with 5.2 V amplitude, 6.12 KHZ frequency and on—off time of 0.0817 m sec.

5 CONCLUSION

We have successfully implemented and tested the proposed water level detector circuit for the visually impaired people so that they can easily identify the level of the water in a container or tank by only hearing the different audible tones generated for different water levels. The system that we have designed and implemented is quite economical and easily configurable which can solve the problem of water wastage. The use of very few components reduces the overall system cost to a large extent. An ARDUINO UNO1.6.1 cc is used to run the program in the ARDUINO UNO board with a 5 V power supply. Four input pins are used to detect four separate water levels. As the tones of distinct frequencies are generated at the 8 Ω speaker, when water level reaches each level the prediction of water level can be done by the visually impaired people easily.

REFERENCES

Abdullah. A., Anwar. M.G., Rahman. T. & Aznabi. S. 2015 Water Level Indicator with Alarms Using PIC Microcontroller *American Journal of Engineering Research (AJER), Vol. -4, Issue-7, pp-88–92.*

Clifford. M. 2006 Water Level Monitoring, *Freescale Semi-conductor Application Note, AN1950 Rev Nov.*

Ebere. E. V. & Francisca. O.O. 2013 Microcontroller based Automatic Water level Control System *International Journal of Innovative Research in Computer and Communication Engineering Vol. 1, Issue 6, Aug.*

Mani. N, Sudheesh T.P, Joseph. V., Titto V.D & Shamnas P.S 2014 Design and Implementation of a Fully Automated Water Level Indicator *International Journal of Advanced Research in Electrical, Electronics and Instrumentation Engineering Vol. 3, Issue 2, Feb.*

Reza. S. M. K., Ahsanuzzaman. S., Md. Tariq & Reza. S.M. M. 2010 Microcontroller Based Automated Water Level Sensing and Controlling: Design and Implementation Issue *Proc. of the World Congress on Engineering & Computer Science 2010, Vol I (WCECS) Oct.* San Francisco, USA

Computational Science and Engineering – Deyasi et al. (Eds)
© 2017 Taylor & Francis Group, ISBN 978-1-138-02983-5

Active noise cancellation using adaptive filter algorithms and their comparative performance analysis

Mayuri Borah
R&D Engineer, Toshniwal Enterprises Controls Limited, Kolkata, India

Arpita Banerjee
RCC Institute of Information Technology, Kolkata, India

ABSTRACT: In present decade in the field of communication Adaptive filtering is a wide area of research. Noise is the most important environmental factor which determines the reliability of the system operation in practice. Noise Minimization with Adaptive Filtering is an approach where received signal is continuously corrupted by noise and both received and noise signal changes continuously. Its advantage lies in that, with no prior estimates of signal or noise, the levels of noise minimization or rejection are attainable. Different adaptation algorithms have been adopted considering the noise pattern and other factors. This paper includes a study of Adaptive Noise Cancellation and comparative analysis of adaptation algorithms.

1 INTRODUCITON

In Signal Processing the signal is susceptible to the noise interference that can arise from a wide variety of sources. With the high level of technology development now days, the real-time processes became more and more necessary and popular. Those types of processes are the most vulnerable to the action of noise interference. The usual method of estimating a signal corrupted by noise is to pass it through a filter that tends to suppress the noise while leaving the signal relatively unchanged i.e. direct filtering. The design of such filters is the domain of optimal filtering, which originated with the pioneering work of Wiener [1] and was extended and enhanced by Kalman, Bucy and others [2]. Filters used for direct filtering can be either Fixed or Adaptive. The design of fixed filters requires a priori knowledge of both the signal and the noise, on the other hand Adaptive filters require no prior knowledge of both the signal and noise. The main advantage lies in the working capability in an unknown environment with a very good tracking property, being capable of detecting time variations of the system variables. Very less computational complexity and hardware cost also influences to selecting adaptive filtering concept rather than normal or fixed filtering for Noise Minimization.

2 ADAPTIVE NOISE CANCELLATION

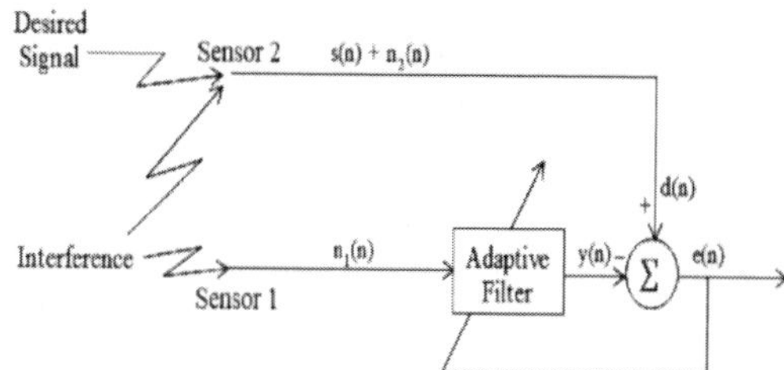

Figure 1. Block diagram—adaptive noise canceller.

Analyzing the Figure 1, it can be seen that the noise $n_2(n)$ which corrupts the desired signal s(n) is represented by the same initial characters as the input signal $n_1(n)$.This noise is conveniently represented as shown, to make easier to understand that the noise $n_1(n)$ and $n_2(n)$ have some similarities. Indeed, the noise $n_1(n)$ is the noise source and the noise $n_2(n)$ is a secondary noise which is correlated with n1(n). Both of noises are also uncorrelated with the signal s(n), which implies in the following conditions:

$$d(n) = s(n) + n_2(n) \quad (1)$$

$$E[s(n)\, n_1(n)] = 0, \text{ for all } n \quad (2)$$

$E[s(n)\, n_2(n)] = 0$, for all n (3)

$E[n_1(n)\, n_2(n)] = p(n)$, for all n (4)

where d(n) is the desired signal, E[*] is the expectation and p(n) is an unknown correlation between $n_1(n)$ and $n_2(n)$. Those necessary conditions can be obtained, for example, if it is installed a sensor sensor1 in a place where only the noise source $n_1(n)$ is detected by the sensor. The sensor sensor2 will detect the desired signal d(n), and the noise $n_2(n)$ could be correlated with n1(n)because of the delay between them, for exam-ple, or by applying a filter.

The error signal e(n), which is also the system output should contain the original signal s(n) in an optimum sense. The aim of this algorithm is to make the output y(n), which is equal to the filter's tap-weight transposed wT(n) times x'(n) (vector formed by the reference noise signal $n_1(n)$ having the same size as w(n)),or y (n) = wT (n) x' (n), be equal to the noise $n_2(n)$ (y(n) = (n)). Having this equi-valence, it is easy to deduce that the error is equal to the desired signal s (n).

$$e(n) = d(n) - y(n) = s(n) + n_2(n) - n_2(n) = s(n)$$

To make it sure that $n_1(n)$ and $n_2(n)$ are truly correlated, it was created a noise v(n) uncorrelated with the desired signal having variance 0.8. This signal was filtered in order to create the both $n_1(n)$ and $n_2(n)$ noise signals.

Adaptive Noise Cancellation is done by two processes—Filtering process and the Adaptation process. For adaptation process the adaptation algorithm is required. LMS, NLMS and RLS are some Adaptation Algorithms. Here in this paper we are considering three algorithms (LMS, NLMS, RLS) and analyzing their performance in Noise Minimization/Cancellation using Matlab Simulink.

3 LMS SOLUTION

Here is presented a LMS solution for noise cancellation. Doing necessary alterations to reach the interference cancelling problem, we found that:

$$x(n) = s(n) + n_2(n) \quad (5)$$

$$d(n) = x(n) \quad (6)$$

$$y(n) = w^T(n) + x'(n) \quad (7)$$

$$e(n) = d(n) - y(n) \quad (8)$$

$$w(n+1) = w(n) + 2\mu e(n)\, x'(n) \quad (9)$$

where x(n) is a signal composed by the original signal s(n) added to the noise signal n1(n); d(n)is the desired system output; y(n) is the ANC's output; w(n)is the tap-weight vector; x'(n) is the vector formed by the reference noise signal n1(n) having the same size like w(n); e(n) is the error signal and μ is the step-size parameter. In Fig 2 and 3 we can see the result of LMS algorithm for a given problem. In the figure the input signal s(n) is represented in blue, signal after noise corruption is shown in green and the error signal e(n) is represented in red colour.

L=5, μ=.0002

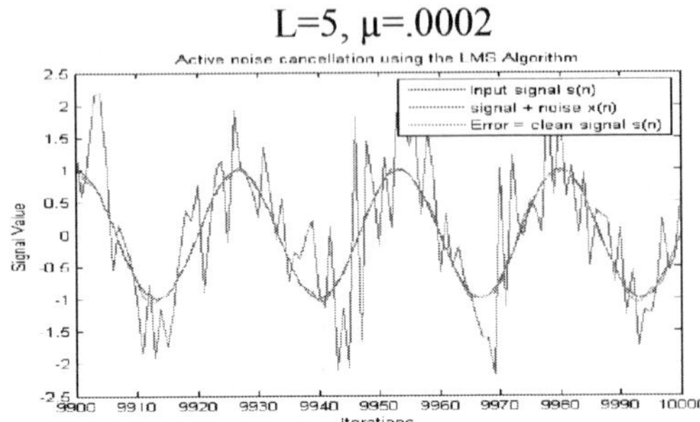

Figure 2. LMS solution for ANC.

Analyzing the Fig. 2 it is seen that the LMS algorithm has not a very good performance, having the error signal e(n) tending to the original signal s(n), free of the noise interference $n_1(n)$.

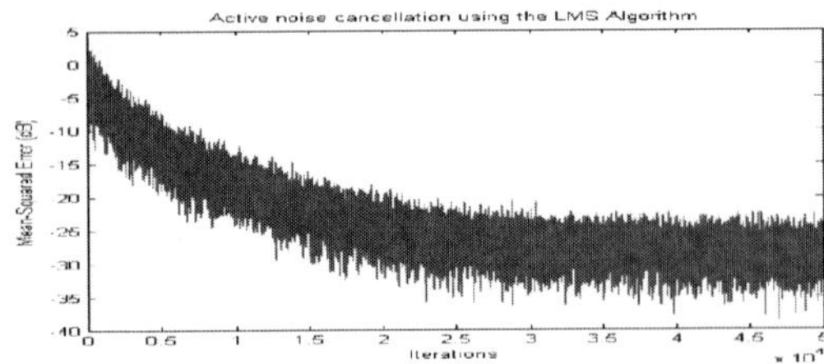

Figure 3. Mean-squared error of the ANC using the LMS algorithm.

3.1 *For NLMS algorithm*

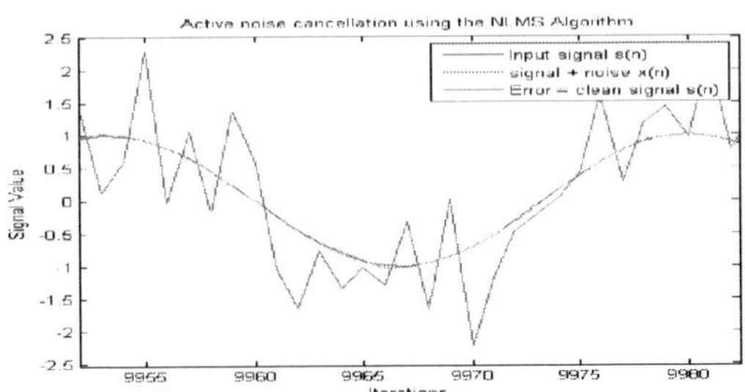

Figure 4. NLMS solution for ANC.

By analyzing the above figure the NLMS has good response in removing additional white noise which was corrupting the original signal.

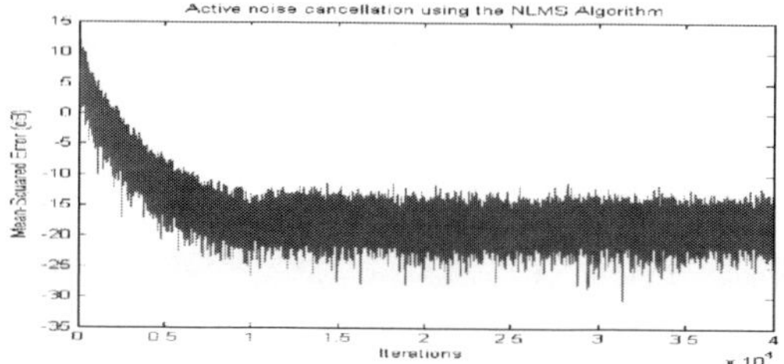

Figure 5. Mean-squared error of the ANC using the NLMS algorithm.

Figure 5 depicts the MSE of the difference between the error signal e(n) and the input signal s(n).It can be observed that the algorithm presents a conversion after about 10000 iterations. After this time, the algorithm presents some variance but the conversion is not compromised.

3.2 *For RLS algorithm*

$$L = 5, \lambda = 1$$

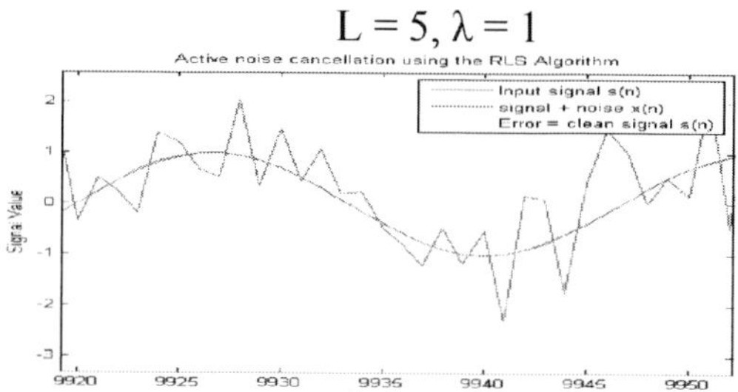

Figure 6. RLS solution for ANC.

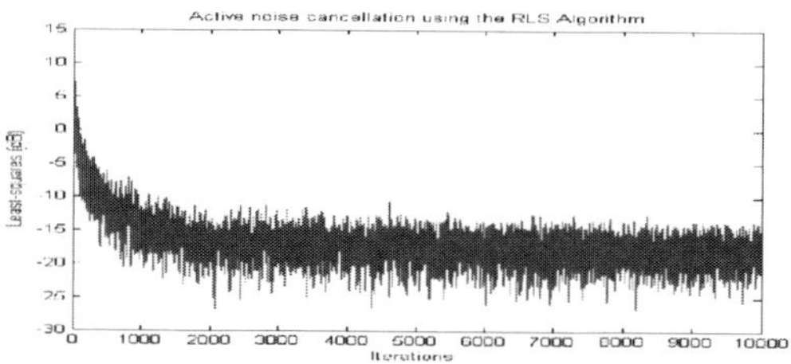

Figure 7. Mean-squared error of the ANC using the RLS algorithm.

From Figure 6 it is clear that RLS has very good performance in noise cancellation. From figure it is analyzed that the algorithm has its convergence after the first few thousand iterations.

3.3 *Result comparison*

$$L = 5,\ \mu = 0.0002(\text{LMS}),\ \delta = 0.9,$$
$$\mu = 0.0083(\text{NLMS}),\ \lambda = 1$$

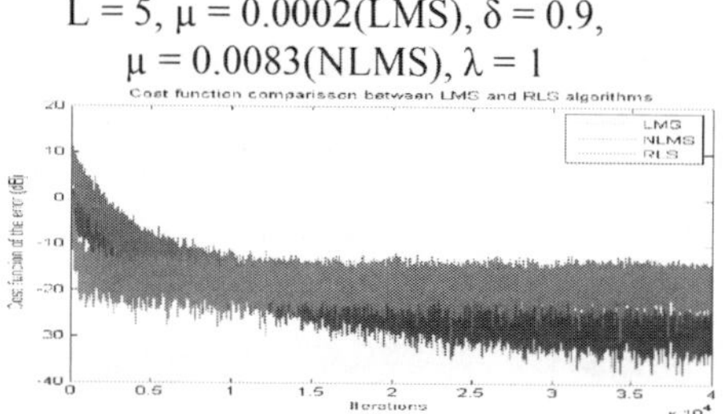

Figure 8. Cost function comparison.

The RLS algorithm has a fixed computational cost, derived from its way of calculation, higher than the two other algorithms. Figure 8 depicts the comparison between the three algorithms "cost function". Analyzing this figure, it can be noticed that the LMS algorithm has a very slow convergence, compared to the convergence of the NLMS and RLS algorithms, the LMS also converge to a higher error value, approximately 30 dB against 20 dB for the other two. In this case, if it is needed an algorithm which the convergence speed is important, the LMS is not a good choice. The RLS algorithm presents a convergence three times faster than the NLMS algorithm, being the fastest algorithm for the ANC problem and having a good efficiency.

From the above analysis RLS become the best and LMS become the worst in Noise Cancellation. So now we have again compared the both in Simulink Platform considering the system is corrupted by White Gaussian Noise and then by Colored Noise.

After testing many times we have selected filter length as 32, noise variance as 0.1 and step size parameter μ as 0.0008, keeping the fact in mind that input signal and the error signal should be similar.

3.4 *Simulation block for ANC (LMS/RLS)*

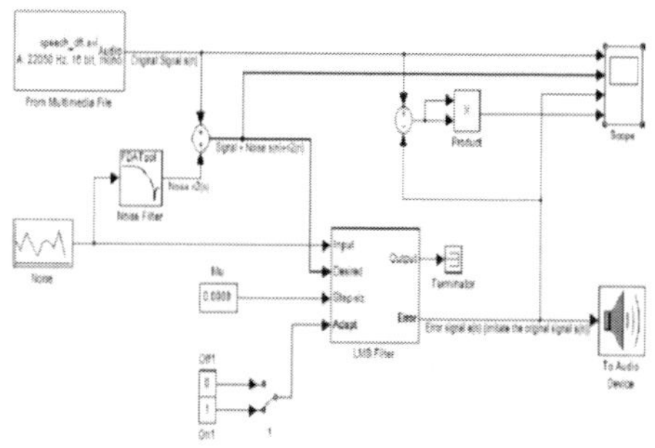

Figure 9. Active Noise Canceller using White Gaussian Noise.

3.5 *LMS results*

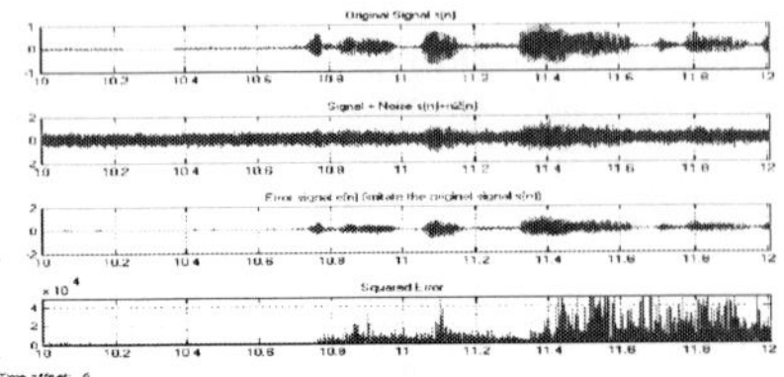

Figure 9a. Relevant signals and squared error after 10 seconds.

The figure illustrates the behavior of the algorithm after 10 seconds of work. After this time, the algorithm presents an error near to its optimum value, with average error from the order of 2×10^{-4}, without having big variations.

3.6 *RLS results*

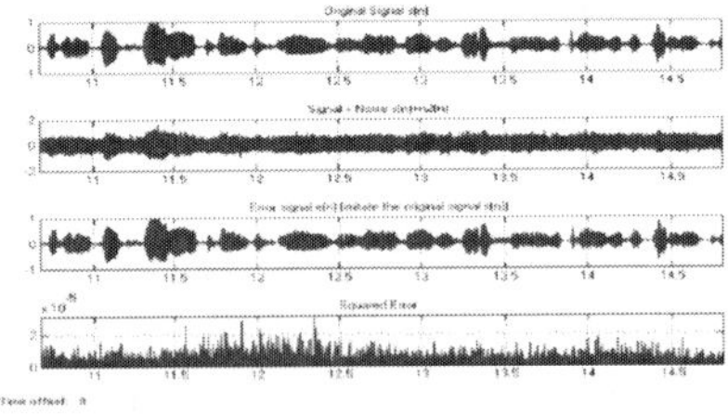

Figure 9b. Relevant signals and squared error a few seconds after convergence.

Fig. 9b presents behavior of the algorithm a few seconds after its convergence. It can be noticed that the algorithm presents a very good performance, with the squared error having values near to 1×10^5 and small variation in the squared error.

3.7 *Input signal corrupted by a colored noise*

The filter length was chosen to be 32. The input signal is now corrupted by a colored noise. This colored noise is random and detected by the microphone input. The figure below illustrates the LMS program using a colored noise to corrupt the input signal.

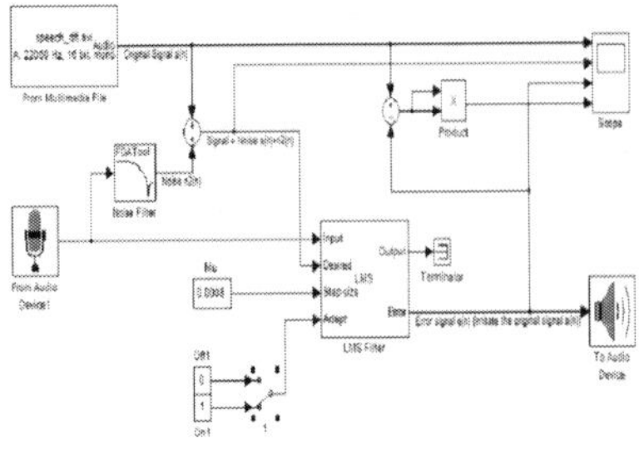

Figure 10. Active noise canceller using colored noise.

Analyzing this figure, it can be noticed that the algorithm takes about 0.3 second to present an acceptable conversion characteristic, having the squared error tending to zero. In the case of the colored noise, the convergence time is not a fix value and it will vary depending on the power of the interference at that time.

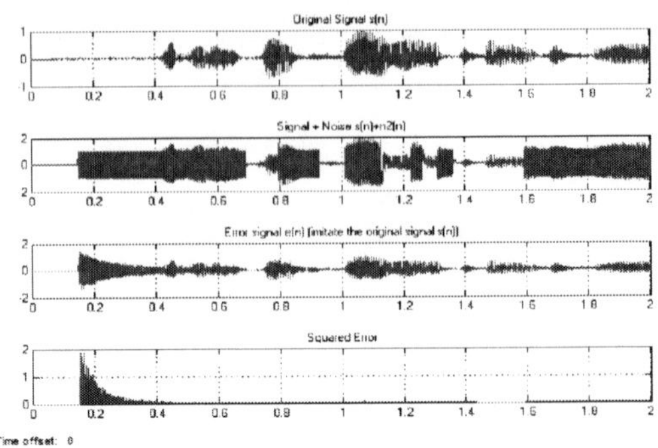

Figure 11. Relevent signal and algorithm convergence.

In case of RLS the Simulink block is same as the previous one. Just the algorithm we have to select is RLS algorithm.

Result of RLS algorithm in case of colored noise is shown in the below figure

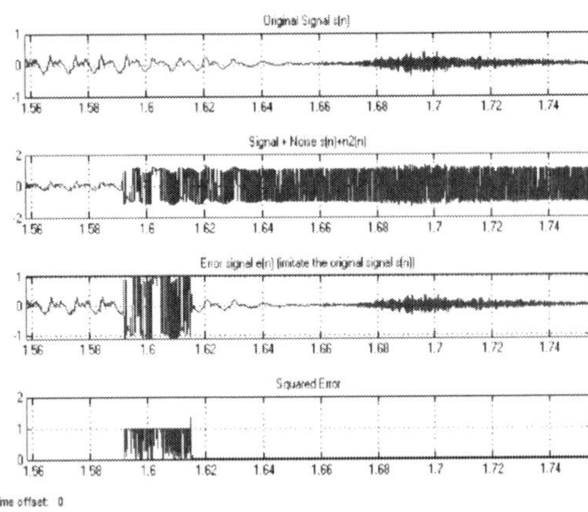

Figure 12. Relevant signals and algorithm convergence.

4 CONCLUSION

Number The study proves that the RLS algorithm is more robust than the LMS algorithm, having a smaller error and a faster convergence for the case of the White Gaussian Noise Interference.. For the colored noise interference problem, the RLS has presented a part from the previous advantages, a powerful stability, being capable of keeping its cancellation quality even with non-white variation in the noise source. It is the opposite to the LMS algorithm, which has proved its inefficiency in such environment, having big variations in the noise cancellation error when the colored noise presented a strong signal. Those error variations are big enough to be listened in the error output signal. The RLS algorithm has a bigger complexity and computational cost, but depending on the quality required for the ANC device, it is the best solution to be adopted.

The proposed continuation of the work is to proceed towards hardware implementation with a DSP processor.

ACKNOWLEDGEMENT

We convey our heartiest thanks to Mr. Anindya Sankar Roy, R & D Engineer, Toshniwal Enterprises and Controls Limited for his continuous support in this research. His contribution for this paper is invaluable.

REFERENCES

Diniz, P. S. R., Adaptive Filtering—Algorithams and Practical Implementation, Third Edition, Kluwer Academic Publishers, 2008.

Farhang-Boroujeny, B. Adaptive Filters: Theory and Applications", Wiley, 1998.

Gale, Z., GMC Terrain Crossover uses Noise Cancelling Tech to Quiet Cabin, Help Efficiency, Truck Trend Magazine, February 25, 2011.

Górriz, J.M.; Ramírez, J.; Cruces-Alvarez, S.; Puntonet, C.G.; Lang, E.W.; Erdogmus, D., A Novel LMS Algorithm Applied to Adaptive Noise Cancellation, IEEE Signal Processing Letters, VOL. 16, NO. 1, January 2009.

Haykin, S., Adaptive filter theory, 3rd Edition, Prentice Hall, 1996.

Pota, H.R.; Kelkar, A.G., "Analysis of Perfect Noise Cancelling Controllers", International Conference on Control Applications, Anchorage, Alaska, USA 9 September 25–27, 2000.

Sayed, A.H., Fundamentals of Adaptive Filtering, John Wiley & Sons, New Jersey, 2003.

Shin, D.C.; Nikias, C.L., Adaptive noise canceller for narrowband/wide band interferences using higher-order statistics, IEEE International conference on Acoustics, Speech, and Signal Processing, 1993.

Tienan, L.; Limo, W.; Baochang, X.; Aihua, X.; Hang, Z., Adaptive Noise Canceller and Its Applications for Systems with Time-variant Correlative Noise, IEEE Proceedings of the 4° World Congress on Intelligent Control and Automation, June 10–14, 2002.

Widrow, B.; Glover Jr., J.R.; McCool, J.M.; Kaunitz, J.; Williams, C.S.; Hearn, R.H.; Zeidler, J.R.; Dong Jr., E.; Goodlin, R.C., Adaptive Noise Cancelling: Principles and Applications, Proceedings of the IEEE, Vol. 63, no. 12, December 1975.

Computational Science and Engineering – Deyasi et al. (Eds)
© 2017 Taylor & Francis Group, ISBN 978-1-138-02983-5

IRNSS capabilities: An initial study using IGS receiver

Sujoy Mandal, Koushik Samanta, Debipriya Dutta & Anindya Bose
Department of Physics, The University of Burdwan, Burdwan, West Bengal, India

ABSTRACT: The Indian Regional Navigation Satellite System (IRNSS) is an autonomous and independent regional navigational system under development by Indian Space Research Organization (ISRO). IRNSS will provide position, velocity and timing services for India and surrounding regions. All the planned 07 satellites of IRNSS constellation are now in space and within short period of time the system would be declared to be operational. IRNSS GPS SBAS (IGS) receiver developed by ISRO can use both S and L band of the IRNSS signals along with GPS signal in L1 band, and also GAGAN (SBAS) signals. This paper presents the first experimental results using the IRNSS in stand-alone and hybrid multi-GNSS operation modes with the GPS and SBAS from the University of Burdwan, India for providing the position solution and the available signal geometry. These primary results are encouraging and would be helpful in understanding the capabilities of IRNSS.

1 INTRODUCTION

Currently the usefulness and popularity of Global Navigation Satellite System (GNSS) is increasing FOR its manifold advantages. As of now, two global GNSS systems are fully operational- one is Global Positioning System (GPS) developed by USA and the other is Global Navigation Satellite System (GLONASS) developed by Russia. Other upcoming global navigation satellite systems- GALILEO of European Union, BEIDOU of China are in active development phases and two regional systems- QZSS of Japan and the Indian Radio Navigation Satellite System (IRNSS) are in active deployment phases. Importance and advantages of GNSS may be understood from the interest shown by so many countries in developing their own system.

All the proposed 07 satellites of the IRNSS constellation have been successfully placed in the orbit and by 2017 the system would be declared fully operational GPS World. (2016), ISRO News (2016), Thombre et al. (2014). IRNSS is also known as Navigation Indian Constellation (NAVIC) GPS World. (2016). So, as a new regional system, scope for studies emerges to understand and explore the potential and positioning capabilities of IRNSS. Researchers from different parts of the globe have reported their experiences with IRNSS position solutions and the advantages of presence of IRNSS satellites Thombre et al. (2014), Ganeshan et al. (2015), Zaminpardaz et al. (2016), Rao et al. (2016), Mandal et al. (2016) - though many of these efforts are based on theoretical studies. Aim of this paper is to study the initial solution accuracies provided by IRNSS in standalone operation mode and in integrated mode of operation with GPS and SBAS (GAGAN) by using the IRNSS IGS receiver from the University of Burdwan, West Bengal, India using 06 operational satellites in constellation. The results may be helpful for understanding the potentials of fully operational IRNSS, which is expected to occur soon. The paper is organised in the following way. As a relatively new topic, the IRNSS system structure is briefly described in section 2. The experimental setup for this study is described in the next section. Observations, analysis and the experimental results are described in section 4; the overall summary of the contribution is described in section 5.

2 BRIEF DESCRIPTION OF IRNSS

IRNSS is a regional satellite based navigation system being developed by ISRO to provide Position, Velocity and Time (PVT) information for the Indian and the surrounding region. It consists of 07 satellites in constellation; by April 2016, ISRO has successfully launched all the 07 IRNSS satellites using the Indian launcher PSLV. Among these, 03 satellites are located in Geostationary Earth Orbit (GEO) and the rest 04 satellites are located in Geosynchronous Orbit (GSO) with an inclination of 29° relative to the equatorial plane. Such an arrangement would ensure continuous radio visibility of all the 07 satellites from the operational zones.

All the satellites in the constellation will be continuously visible in the Indian region for the

Table 1. IRNSS satellites brief description with their launching history.

Satellite	Launch date	Placed in	Location (Longitude)	Expected lifetime (Mission life), years
IRNSS 1A	01 July, 2013	GSO	55° E	>10.0
IRNSS 1B	04 April, 2014	GSO	55° E	
IRNSS 1C	16 October, 2014	GEO	83° E	
IRNSS 1D	28 March, 2015	GSO	111.75° E	
IRNSS 1E	20 January, 2016	GSO	111.75° E	
IRNSS 1F	10 March, 2016	GEO	32.5° E	
IRNSS 1G	28 April, 2016	GEO	129.5° E	

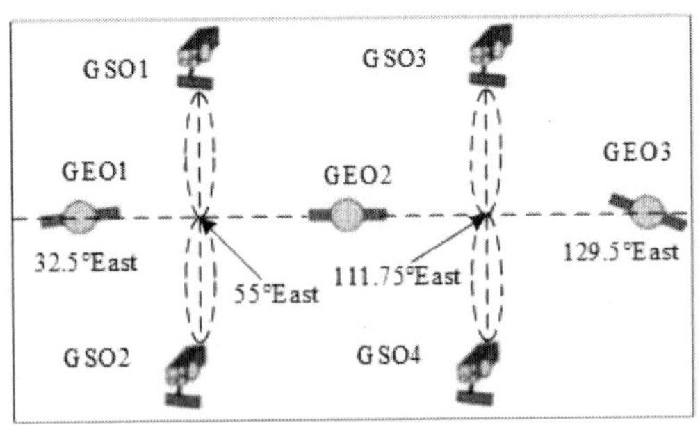

Figure 1. IRNSS constellation plan.

24 hours in a day. The IRNSS is expected to provide position accuracy of better than 20 meters over India and a region extending outside the land mass up to about 1,500 kilometers. The system will provide two types of services, a Standard Positioning Services (SPS) and the restricted/authorized service or RS. Both of these two services are provided using two frequencies, one in the L5 band and the other in the S1 band. ISRO has filed for 24 MHz bandwidth of spectrum in the L5 band (1164–1189 MHz) for IRNSS and for the second signal in the S band (2483.5–2500 MHz).

The launching history and brief description of all the IRNSS satellites are shown in the Table 1 and the schematic diagram of location geometry of the IRNSS satellites is shown in the Figure 1.

3 EXPERIMENTAL SETUP

GNSS Activity Group, The University of Burdwan, India is engaged in exploring the benefits of multi-GNSS signal for the Indian region. The group monitored IRNSS signal for obtaining position solution using an IRNSS/GPS/SBAS (IGS) receiver jointly developed by Space Application Centre (SAC), ISRO, Ahmedabad and M/S Accord Software Systems, Bangaluru, India and is installed at Burdwan in late April, 2016. This IGS receiver is a high performance Global Navigation Satellite System (GNSS) receiver capable of tracking and acquiring IRNSS (dual frequency L5 and S band), GPS (L1) and SBAS (GAGAN) signals. IGS receiver can estimate user PVT solution using standalone IRNSS or GPS signals and using signals from more than one systems together in hybrid operation mode (e.g. IRNSS+GPS) Accord Software user manual (2016). The schematic for the experimental set-up used for the studies is shown in the Figure 2. This figure illustrates how GNSS data is recorded from the satellites using the IGS antenna and receiver; the complete setup consists of a roof mounted antenna having a clear view of the sky, the IGS receiver installed inside the laboratory and a data recording computer with necessary communication facility with the receiver and vendor supplied GUI software for data logging.

3 OBSERVATIONS, ANALYSIS AND RESULTS

Firstly, IRNSS solution data is provided by the IGS receiver is recorded in 1 Hz frequency for both L5 and S1 bands independently and also using SBAS (GAGAN) signals together with IRNSS signals in the month of May, 2016. Though the constellation of IRNSS is completed but the seventh satellite of its constellation is yet to be used for solution and therefore, 06 IRNSS satellites are used for solution in this initial results presented in this paper. Results obtained by analyzing the recorded data is shown in the Table 2 where the peak to peak variation (P-P) and standard deviation (σ) of different position solution parameters (viz. latitude, Longitude and Altitude) converted to meters are presented

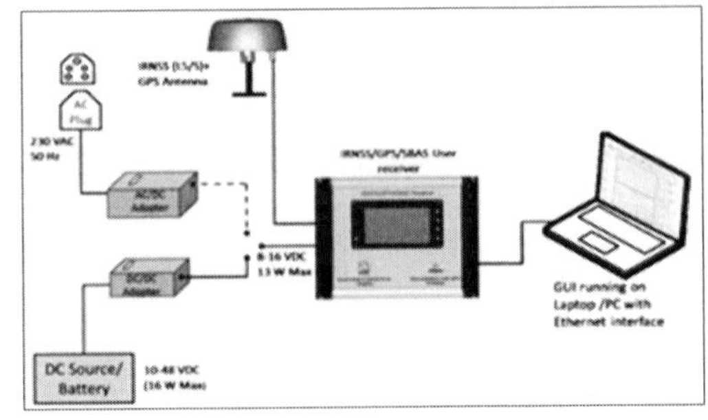

Figure 2. Experimental setup.

for different modes of observations. In this table IR-S1 implies use of IRNSS S1 frequency only, IR-L1 implies use of IRNSS L1 frequency only while SB suggests use of IRNSS corresponding signal with SBAS (GAGAN) augmentation information for position solution.

Inspection of results shown in Table 2 reveals that the IRNSS provides modestly accurate 3d position solution in standalone mode using the available 06 signals. Using the augmentation signal (GAGAN), position solution accuracy improves slightly.

In the Tables P-P indicates peak to peak (maximum) variation and σ indicates standard deviation of observations.

While calculating the accuracy of the position solution provided by the satellites, geometrical configuration of the satellites used for navigation solution plays a vital role in accuracy of position solution provided by the satnav systems. Satellite geometry is quantitavely expressed in terms of Dilution of Precision (DOP); for 3-dimensioanl solution the associated term is Position DOP (PDOP) [13]. From the initial results presented in Table 2, slightly degraded Position Dilution of Precision (PDOP) values are observed for IRNSS-only or for IRNSS+GAGAN solutions.

Further, IRNSS data is recorded at 1 Hz frequency while operating in hybrid mode with GPS with and without support from GAGAN augmentation signal. Results obtained by analyzing such data obtained is shown in Table 3. In this table peak to peak (maximum) variation of solution coordinates and the standard deviation of observations are shown.

Results obtained from the observations as presented in Table 3 point towards a successful integration of IRNSS (both L5 and S5 signal) with GPS for position solution.

Results shown in Table 3 suggests better position solution obtained in IRNSS-GPS hybrid mode of operation compared to that of the initial results using IRNSS-only signals in L5 or S1 frequency bands. Use of GAGAN signals further improves the situation. IRNSS operation in tandem with other available GNSS signals (e.g., GPS, as is the case here) definitely would attract attention of the users for future successful integration of all available systems towards a robust and efficient multi-GNSS environment.

Improved satellite geometry using IRNSS-GPS hybrid mode of operation is observed in comparison to both individual IRNSS and GPS operation that suggests the other possible advantages of IRNSS-GPS hybrid operation.

For a better comprehension, available results are pictorially represented in Figure 3 wherein the latitude and longitude variations (converted in meter) are shown. Figure 3 presents a comparison of two dimensional (2d) position solutions using IRNSS L5 signals, IRNSS L5+GPS L1 signals, and IRNSS S1+GPS L1+SBAS signals in hybrid modes of operation. A detailed inspection of the graph reveals that the position solution obtained using only IRNSS L5 is less accurate as compared to the solution obtained from the GPS L1+ IRNSS

Table 2. Position solution results obtained using standalone IRNSS and IRNSS with SBAS support.

Constellation used	No. of samples	Latitude (m)		Longitude (m)		Altitude (m)		PDOP
		σ	P-P	Σ	P-P	σ	P-P	
IR-S1	4604	2.5	10.5	0.55	3.49	1.9	9.88	4.3
IR-L5	7580	1.3	8.77	0.81	4.29	1.8	7.65	3.8
IR-L5+SB	4173	0.49	2.87	0.51	2.59	1.4	6.49	3.4
IR-S1+SB	5050	1.0	6.68	0.45	2.86	1.4	9.09	4.2

Table 3. Position solution accuracy analysis using IRNSS, GPS and SBAS.

Constellation used	No of Samples	Latitude (m)		Longitude (m)		Altitude (m)		PDOP
		Σ	P-P	σ	P-P	σ	P-P	
GPL1	2887	0.52	2.3	0.49	2.4	1.55	7.7	1.9
GPL1+IRL5	7648	0.68	3.4	0.34	2.0	1.77	12.0	1.3
GPL1+IRS1	9049	0.62	6.7	0.82	3.7	1.65	13.2	1.4
GPL1+SB	6942	0.47	2.8	0.72	4.0	1.03	7.2	1.6
GPL1+IRL5+SB	6272	0.49	2.8	0.59	2.9	0.738	4.7	1.6
GPL1+IRS1+SB	6484	0.31	2.0	0.36	2.0	0.577	4.8	1.2

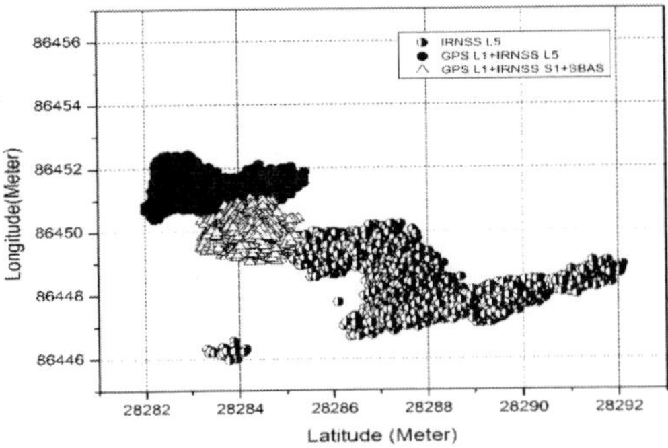

Figure 3. Comparison of the position solution obtained using three different combinations of IRNSS, GPS and SBAS (GAGAN) signals.

L5. The experimental data shown in Table 2 and Table 3 also suggested this finding.

Data presented in Table 2 and 3, and inspection of of Figure 3 also indicates the support of SBAS (GAGAN) signal to augment the position solution provide by the GPS L1 and IRNSS S1 mode of operation.

4 CONCLUSION

IRNSS, in the current developing stage provides position solution in standalone mode and also in integration mode of operation together with GPS. This would be helpful for all the GNSS users in the region towards system independence and redundancy. On the other hand, potential of hybrid IRNSS-GPS operation mode would boost the popularity and advantages of multi-GNSS operation in the Indian region.

IRNSS alone provides relatively worse satellite geometry but with GPS the satellite geometry improves. The results present here are based on preliminary and short-term observations based on signals from 06 IRNSS satellites currently in operation. Within short period of time with all of the 07 operational satellites of IRNSS, the system would be declared to be operational. A full and operational IRNSS is expected to enhance the benefits.

This study is based on small amount of data collected and calls for more exploration and real-time data analysis using the operational IRNSS in near future for a longer period to explore the benefits of IRNSS. So, the researchers and other stakeholders are waiting to explore and to utilize the full constellation of IRNSS when it would be declared to be operational soon.

ACKNOWLEDGEMENT

The authors would like to acknowledge Space Application Centre (SAC), ISRO Ahmedabad for providing the IGS receiver used for the studies.

REFERENCES

Accord Software and Systems Pvt Ltd. 2016. IRNSS/GPS/SBAS receiver: User manual 1.2: 1–92

Banerjee, P. Bose, Anindya. Mathur, B S. 1997. A study on GPS PDOP and its impact on position error. *Indian J. Radio and Space Physics* 26(2): 107–111

Ganeshan, A S. Ratnakara, S C. Srinivasan, Nirmala. Rajaram, Babu. Tirmal, Neeth. Anbalagan, Kartik. 2015. First Position Fix with IRNSS. Inside GNSS 10(4): 48–52.

GPS World. April 29, 2016. With IRNSS-1G launch, India completes and renames it's navigation constellation. *http://gpsworld.com/with-irnss-1g-launch-india-completes-and-renames-its-navigation-constellation/*

IRNSS - Indian Regional Navigation Satellite System. March 08, 2016. *http://www.isac.gov.in/navigation/irnss.jsp*.

ISRO Satellite Centre. ISRO. Bangaluru. June 2014, Indian Regional Navigation Satellite System Signal in Space ICD for Standard Positioning Service.

List of Navigation Satellites, March 08, 2016. *http://www.isro.gov.in/spacecraft/list-of-navigation-satellites*

Mandal, Sujoy. Samanta, Kousik. BOSE, Anindya. 2016. IRNSS and Possible Benefits in Hybrid Operation with GPS. Proc. National Conference on Materials, Devices and Circuits for Communication Technology (MDCCT 2016): 72–75.

Rao, Vyasaraj Guru. Lachapelle, Gerard. Kumar, Vijay. 2016. Analysis of IRNSS over Indian Subcontinent. *http://plan.geomatics.ucalgary.ca*

Sarma, Achanta D. Sultana, Qudussa, Srinivas, Vemuri Satya. 2010. Augmentation of Indian Regional Navigation Satellite System to improve Dilution of Precision. The Journal of Navigation 63:313–321.

Thombre, Sarang. Bhuiyan, Mohammad zahidul H. Soderholm, Stefan. Jaakkola, Martti Kirkko. Ruotsalainen, Laura. Kuusniemi, Heidi. 2014. Tracking IRNSS Satellites for Multi- GNSS Positioning in Finland. Inside GNSS 9(6): 52–56.

Zaminpardaz, Safoora. Teunissen, Peter J G. Nadarajah, Nandakumaran. 2016. IRNSS stand-alone positioning: first results in Australia. Journal of Spatial Science 61(1): 5–27.

Track VI: Soft computing

Computational Science and Engineering – Deyasi et al. (Eds)
© 2017 Taylor & Francis Group, ISBN 978-1-138-02983-5

Large scale real parameter optimization using genome grouping

Abhishek Basu
Department of Electronics and Communication Engineering, RCCIIT, Kolkata, India

Ankur Mondal, Sharbari Basu & Niladri Mondal
Department of Computer Science and Engineering, GNIT, Kolkata, India

ABSTRACT: This paper is intended to examine large-scale fully separable continuous optimization problems of real-valued functions, using genome grouping. Genome or chromosome or namely the vector is the smallest unit of an entire population in the world of Differential evolution. The main focus of this paper is the non-interactive and separable genes, which can form fittest groups with fittest chromosomes/genomes and can give the most optimized result of a large-scale problem with minimum space and time complexity and have compared the result with an existing adaptive algorithm called MLCC, and shown that Genome Grouping per-form better on a large-scale fully separable problems. This paper illustrates a differential evolution algorithm without using crossover.

1 INTRODUCTION

In large scale optimizations problems, finding the solutions are quiet a difficult task while handling with large number of decision variables. Also the numbers of steps to get the most optimized values becomes a very important point. So the decision variables are very important factors are shown by T. Weise, R. Chiong, and K. Tang (2012), and so the number of solving steps. There are various approaches to handle the large scale optimization problems dealing with large number of decision variables.

One of these approaches is to disintegrate the given large-scale problem into a set of smaller sub problems, which are more manageable and easier to solve. This method is known to be "divide-and-conquer" strategy and can be found in famous book 'A Discourse on Method' by R. Descartes (1956). In "Local convergence analysis for partitioned quasi-newton updates" by A. Griewank and P. L. Toint, "Decomposition principle for linear programs" by G. B. Dantzig and P. Wolfe, and Nonlinear programming by D. Bertsekas, this decomposition strategy has been established. With reference to this idea, the focus of this paper is to examine large-scale global optimization of real-valued functions, using grouping of genomes. T. B¨ack, D. B. Fogel, and Z. Michalewicz, in 1997, R. Sarker, M. Mohammadian, and X. Yao in 2003, and Y. Liu, X. Yao, Q. Zhao, and T. Higuchi in 2001shown that though Evolutionary Algorithms (EAs), are effective optimization methods, and have been widely used in many optimization problems but the performance graph deteriorates when the dimension of the problem increases. This fact is called as "the curse of dimensionality" shown by R. E. Bellman.

Cooperative Co-evolution (CC) has been proposed by Potter and De Jong as an explicit means of decomposition problem in the arena of evolutionary algorithms, also described by Hsin-Chuan Kuo, Ching-Hai Lin, and Jeun-Len Wu. The algorithm of CC is more or less equivalent to the ordering of genes as was in early days of Genetic Algorithm research. D. E. Goldberg. Y. Ping Chen, T. li Yu, K. Sastry, stated that ordering of genes on a chromosome is a very important contributing factor as per the performance of EAs is concerned. In an experiment conducted by Goldberg et al., it was shown "ordering of genes" is an important factor of the success of a simple genetic algorithm.

In the reality, gene interaction problem is the decision key between ordering of genes and the performance of EAs. Based on this science, in this paper we propose a decomposition method called **Genome Grouping** that allows an automatic, near-optimal decomposition of decision variables. Here, after the process of initialization, the genes are being plotted on a graph and focuses on those genes that share the same coordinate value, thus forming groups. The coordinate values of each group are unique in nature. In this paper, the decomposition strategies are being followed after considering the non-interactive and separable criterion.

This process not only diminishes the huge population size but also provide highly optimized values with minimum time complexity.

2 STRATEGIES OF DECOMPOSITION

2.1 *Gene interaction and cooperative Co-evolution*

M. Ptashne (1989) has described that the interaction between two genes are said to happen in natural genetics, when (i) as a whole, they represent a feature at phenotype level, (ii) when value of one gene activates or deactivates an effect of other genes. **Epistasis** is the term used to refer any means of gene interaction that is described by W. S. Klug, M. R. Cummings and Y. Davidor. Genetic algorithm also refers this as linkage. A. Auger, N. Hansen (2007) developed that non-separability also refers the same idea, but it is more widely applied in the continuous optimization literature. CC is an effective algorithm as per large scale optimization is being considered. The secret of effectiveness lies in decomposition of large scale problems into smaller sets of sub-problems.

2.2 *Classifications*

Random Methods: These algorithms rely more on randomly permuted decision variables rather than systematic approach to find out the interdependencies. These algorithms try to increase the probability to bring the interacting variables close to each other for a few evolutionary cycles.

Perturbation: These methods unsettle the path of decision variables by using various heuristics, and closely monitor the deflection of the objective functions, thus attempting to detect the interaction between decision variables.

Interaction Adaptation: In these methods, the interaction is being monitored at the chromosome level and also the order of genes and the decision variables of the original optimization problem are being evolved.

Model Building: In these methods, the probabilistic model is being sketched based on the desired solution of the optimization problem.

3 GENOME GROUPING

3.1 *Genome initialization*

The goal of Differential Evolution is to search a globally optimized value in a D-dimensional real parameter space D. The steps start with a randomly initiated population of NP D dimensional real-valued parameter vectors. Each vector also known as genome or chromosome. These genomes form the candidate solutions as a result of large-scale optimization problems. Mohammad Nabi Omidvar has described that the chances of putting two interacting variables in the same subcomponents increase when the decision variables are being grouped randomly at the beginning of each cycle. Equation (1) explains the same. Here, the probability of v interacting variables lying in the same subcomponent for at least k cycles is being calculated.

$$P(X \geq k) = \sum_{r=k}^{N} \binom{N}{r} \left(\frac{1}{m^{v-1}}\right)^{r} \left(1 - \frac{1}{m^{v-1}}\right)^{N-r} \quad (1)$$

Where P is the probability of assigning v interacting variables x1, x2,........., xv into one subcomponent for at least k cycles. N is the number of cycles, v is the total number of variables, m is the number of subcomponents, and the random variable X is the number of times that v interacting variables are grouped in one subcomponent. In this paper we have focused on calculating the probability of v interacting decision variables those form the group in at least k cycles and X holds the value greater than or equal to k. From this equation, we can get the probable values of interacting genes and so the rest are non-interacting ones. The next step is, grouping the non-interacting genes, and from there we can probe to find the coinciding genes. This very step is being carried out by plotting the genomes on a 2-D parametric space after initialization is being completed. So now, we are able to decompose the components into sub-components and hence into sub problems carrying the non-interactive variables or genes. In case of dimension having values 1000 and above, it creates a large population size. There is high probability, that many of the genes coincide on the same coordinate of a X-Y graph. Those falling on the same coordinates, repeats the coincidence over the population in every (almost) iteration, thus increasing the space complexity. If we treat them as a single then ambiguity will decrease. Now we can further proceed based on only these separable and non-interactive coinciding genomes. We will be considering the number of coinciding genomes as to be fall on the same group and the number of counts of such genomes. We have tried to optimize the real-valued continuous functions only considering these fully separable and non-interactive genes.

3.2 *Genome mutation*

As previously mentioned, each genome is considered to be plotted on space of 2D parameter of X-Y axis. So each genome must have a value. The number of same valued genomes can be considered to be fall on the same group if and only if the genes can't interact with each other at the phenotype level or one gene can't activates or deactivates the effect of other genes. Also, each gene, within a single population,(for example, when NP = 1) must be separable if and only if it is possible to find the globally optimized value of a function by considering one dimension at time and optimizing it, without considering the values taken by other dimensions, otherwise it is non-separable that has been described by A. Auger, N. Hansen (2007). If these conditions met, we can proceed then and can consider these group values are unique in nature. The unique and fittest values are used for the process of mutation by pivoting one group and selecting randomly two other groups. The result is the mutant vector.

$$\mathbf{G[m]_l = G_i + F.\,(G_a - G_b)} \tag{2}$$

where G_i is the pivot-group coordinate value, and G_a and G_b are the two other groups randomly selected with fittest values and F is the scale factor.

The donor vector groups are combined and recombined in the combination of

$$\mathbf{(nC_1\ (nC_1 {*}\ {}^{n-1}C_2))/{}^{n}c_3} \tag{3}$$

As we are selecting each genome with the fittest value, crossover is not required.

3.3 *Genome selection*

Selection is done between the pivoted group value and the mutant genome. The fittest one is selected. When the population develops, one of the above mentioned decomposition strategies is being used to decompose the initial population into number of subpopulations, which is followed by integration of search results obtained from each subpopulation. From here, we can identify the promising space of each variable of each subpopulation and then searching separately in each promising space. For example, if the population contains 1000 individuals, among which 500 are better ones, the promising space is the interval of each gene in these 500 better ones. So, after every evaluated generation number, the parent population is being decomposed into three subpopulations, where each subpopulation is within the limit of search space.

1. The first child population: *The search space U1is the initial search space, Ui0 = [i0, i0], i = 1,2,3,…n, where i0, i0 are the lower and upper bounds, respectively.*
2. The second child population: *U2 is the promising search space, where Uip = [ip, ip] is then broadened.*
3. The third child population: *U3 is the promising search space, which is being reduced by applying roulette wheel selection model. The resultant value is Uipr = [ipr, ipr] and subsequently broadened.*

For example, each variable is being divided into five segments. As per probability model of a roulette wheel, the better function-valued individual from each segment are being assigned with high probabilities, and those individuals are being selected. After completion of the process, one of the five segments is taken resulting fast convergence. By applying the genome search strategy, in this paper, another new version of DE, which is called the Crossover-Less Differential Evolution algorithm (CLDE) is proposed. When we relate the human behavioral model of decision making with computer science and engineering, four parameters seek importance, namely (i) the evaluated generation number (ii) the number of better individuals(iii) the broaden ratio (iv) the number of partitioned segments. In this paper, the number of partitioned segments is set to 5.

4 ALGORITHM: GENOME GROUPING

Step 1: Read the control parameters of DE; scale factor F, crossover rate Cr;
Step 2: Get the values of population NP, dimension D dynamically.
Step 3: Set generation Gen = 0;
Step 4: Initialize fitness();
Calculate fitness of each genome over the entire population;
Step 5: form Group()
Check separable or non-separable;
Call Random();
Call fitness();
Initialize Input_Group[size_NP][size_D] with the fittest value;
Step 6: perform compare()
Step 7: Select the pivot group G_i where i = {1,……, size_group};
{
7.1 selec the genome from Gi;
Step 8: Select the other genomes randomly from groups G{1,……….i-1};
Step 9: Perform mutation;
Step 10: Perform Crossover;
Step 11: Perform Selection;
}
Step 12: increase Gen = Gen+1 until stopping criterion reached;

5 EXPERIMENTAL RESULTS

In this section we compare the performance of MLCC against Genome Grouping. Z. Yang, K. Tang, and X. Yao (2008) explained that in Multilevel Cooperative Co-evolution (MLCC) as d-dimensional problem is being uniformly decom-

Table 1. Experimental results F = 0.6; Cr = 0.4; Dynamic NP = 40; Dynamic D = 1000.

Group	Group size	f1	f2	f3	f4	f5	f6
G1	55	2.93E–02	5.64E + 01	1.09E + 02	5.19E + 00	2.09E–03	1.02E + 00
G2	127	1.01E–02	2.64E + 00	8.19E + 01	4.93E + 00	4.19E–03	3.02E + 00
G3	199	1.14E–02	5.22E + 01	2.29E + 01	4.1712 1E + 00	5.04E–03	6.03E + 00
G4	271	3.06E–02	7.80E + 00	1.39E + 01	9.07E–01	5.56E–03	9.05E + 00
G5	43	2.36E–02	4.04E + 01	2.49E + 01	3.976 8E–01	1.07E–03	5.21E + 00
G6	115	2.12E–03	2.30E + 00	1.58E + 01	8.879 6E–01	5.92E–03	8.51E + 00
G7	187	2.29E–03	1.56E + 00	6.83E + 01	3.782 4E–01	6.11E–03	2.81E + 00
G8	59	5.11E–03	3.02E + 01	2.18E + 01	8.685 2E–01	1.63E–03	1.11E + 00
G9	31	.79E–03	2.01E + 01	7.92E + 00	3.588E–01	3.15E–03	6.41E + 00
G10	3	8.27E–03	6.32E + 01	8.81E + 01	8.490 77E–01	5.67E–03	8.71E + 00

Table 2. Experimental results F = 0.6; Cr = 0.4; Dynamic NP = 60; Dynamic D = 1200.

Group	Group size	f1	f2	f3	f4	f5	f6
G1	121	2.32E–02	2.56E + 01	3.21E + 02	3.2611 E + 00	1.45E–03	2.11E + 00
G2	136	1.80E–02	1.30E + 01	2.56E + 02	3.2517 E + 00	4.19E–03	3.31E + 00
G3	10	2.13E–02	1.22E + 02	4.58E + 02	8.2584 E + 00	9.21E–03	6.13E + 00
G4	185	1.02E–02	4.10E + 01	1.05E + 02	3.2455 E + 00	4.62E–03	2.73E +00
G5	122	3.52E–01	2.54E + 02	2.87E + 02	1.5214 E + 00	4.74E–03	3.84E + 00
G6	230	2.14E–02	2.29E + 02	1.58E + 02	1.2587 E + 00	1.14E–03	1.41E + 00
G7	289	1.26E–02	–2.33E + 02	1.91E–02	1.8415 E + 00	1.25E–03	1.47E + 00
G8	347	1.02E–02	1.02E + 02	1.80E + 01	1.4258 E + 00	1.32E–03	1.14E + 00

Table 3. comparisons between MLCC and Genome Grouping results when D = 1000.

Test functions	MLCC	Genome grouping
f1	1.76E–2	2.93E–02
f2	9.33E + 2	5.64E + 01
f3	1.66E + 03	1.09E + 02
f4	7.46E–01	5.19E + 00
f5	5.51E–03	2.09E–03
f6	1.05E + 00	1.02E + 00

posed into sub-problems and maintain the list of those having the same sizes (named decomposer). Next step includes assigning performance score based on their performance in each co-evolutionary cycle. At the end, it selects a decomposer by applying a probabilistic method, thus subdividing a large-scale problem in each cycle. MLCC, at beginning of each cycle, rearranges every decision variables so as to increase the probability to group the interacting variables in same subcomponents. It was initially being designed to arrange the non-separable problems. However, it has replaced differential grouping due to its high accuracy level of handling interacting variables. But it is not very good with dealing non-separable variables.

6 CONCLUSION

In this paper we have proposed Genome Grouping. This algorithm decomposes a given optimization problem into subproblems having no interdependencies between decision variables. The very algorithm is being evaluated using fully-separable benchmark functions and the results have shown that it is capable of grouping non-interacting variables with great accuracy for the majority of the benchmark functions. This work can be used for voice detection, image detection as well as fingerprint detection. The results can be accurate as we are optimizing the coordinates of vector's path. So further experiments can be done on these detections using co-ordinate optimization.

REFERENCES

Auger, A., N. Hansen, N. Mauny, R. Ros and M. Schoenauer, "Bioinspired continuous optimization: The coming of age," Invited talk at IEEE CEC, Piscataway, NJ, USA, September 2007.

Bellman, R. E. *Dynamic Programming*, ser. ser. Dover Books on Mathematics. Princeton University Press, 1957.

Bertsekas, D. *Nonlinear programming*, ser. Optimization and neural computation series. Athena Scientific, 1995.

Bäck, T., D. B. Fogel, and Z. Michalewicz, Eds., *Handbook of Evolutionary Computation*. Institute of Physics Publishing, Bristol, and Oxford University Press, New York, 1997.

Dantzig G. B. and P. Wolfe, "Decomposition principle for linear programs," *Operations Research*, vol. 8, no. 1, pp. 101–111, 1960.

Davidor, Y. "Epistasis Variance: Suitability of a Representation to Genetic Algorithms," *Complex Systems*, vol. 4, no. 4, pp. 369–383, 1990.

Descartes, R. *Discourse On Method*, 1st ed. Prentice Hall, January 1956, translated by: Laurence J. Lafleur.

Goldberg, D. E. *Genetic Algorithms in Search, Optimization, and Machine Learning*. Addison-Wesley, 1989.

Goldberg, D., B. Korb, and K. Deb, "Messy genetic algorithms: Motivation, analysis, and first results," *Complex Systems*, vol. 3, no. 5, pp. 493–530, 1989.

Griewank A. and P. L. Toint, "Local convergence analysis for partitioned quasi-newton updates," *Numerische Mathematik*, vol. 39, pp. 429–448, 1982, 10.1007/BF01407874.

Hsin-Chuan Kuo, Ching-Hai Lin, and Jeun-Len Wu, A Creative Differential Evolution Algorithm For Global Optimization Problems, Journal Of Marine Science And Technology, Vol. 21, No. 5, Pp. 551–561 (2013).

Ke Tang 1, Xiaodong Li 2, P. N. Suganthan 3, Zhenyu Yang 1, and Thomas Weise, Benchmark Functions for the CEC'2010 Special Session and Competition on Large-Scale Global Optimization, November 21, 2009

Klug, W. S., M. R. Cummings, C. Spencer, C. A. Spencer, and M. A. Palladino, *Concepts of Genetics*, 9th ed. Pearson, 2008.

Liu, Y., X. Yao, Q. Zhao, and T. Higuchi, "Scaling up fast evolutionary programming with cooperative coevolution," in *Proc. of IEEE Congress on Evolutionary Computation*, 2001, pp. 1101–1108.

Mohammad Nabi Omidvar, Member, IEEE and Xiaodong Li, Senior Member, IEEE and Zhenyu Yang and Xin Yao, Fellow, IEEE, Cooperative Co-evolution for Large Scale Optimization Through More frequent Random Grouping

Ping Chen, Y., T. li Yu, K. Sastry, and D. E. Goldberg, "A survey of linkage learning techniques in genetic and evolutionary algorithms," Illinois Genetic Algorithms Library, Tech. Rep., April 2007.

Potter M. A. and K. A. De Jong, "A cooperative coevolutionary approach to function optimization," in *Proc. of International Conference on Parallel Problem Solving from Nature*, vol. 2, 1994, pp. 249–257.

Ptashne, M. "How Gene Activators Work," *Scientific American*, pp. 40–47, January 1989.

Sarker, R., M. Mohammadian, and X. Yao, *Evolutionary Optimization*, ser. International Series in Operations Research & Management Science. Boston: Kluwer Academic Publishers, 2003, vol. 48.

Weise, T., R. Chiong, and K. Tang, "Evolutionary Optimization: Pitfalls and Booby Traps," *Journal of Computer Science and Technology (JCST)*, vol. 27, no. 5, pp. 907–936, 2012, special Issue on Evolutionary Computation.

Yang, Z., K. Tang, and X. Yao, "Multilevel cooperative coevolution for large scale optimization," in Proc. of IEEE Congress on Evolutionary Computation, June 2008, pp. 1663–1670.

Computational Science and Engineering – Deyasi et al. (Eds)
© 2017 Taylor & Francis Group, ISBN 978-1-138-02983-5

A framework for personal selection process using trapezoidal intuitionistic fuzzy sets

S. Pahari & D. Ghosh
Department of Computer Science and Engineering, Om Dayal Group of Institutions, Howrah, India

A. Pal
Department of Mathematics, National Institute of Technology, Durgapur, India

ABSTRACT: In today's world one the most important part of human resource management is recruitment or personal selection. HR team have to determine those who possess required quality among the candidates applying for the job. Selecting the best candidate is a multi criteria decision making problem. In this paper, a novel personnel selection framework is proposed to find the best possible personnel for a specific job. The proposed work starts with a predefined job announcement which defines required skill set in linguistic form converted into Trapezoidal Intuitionistic Fuzzy Numbers (TrIFNs). After decision makers give their opinion the framework will rank them using Intuitionistic Fuzzy Ordered Weighted Geometric (ITFOWG) aggregation operator. Finally using hamming distance method the best candidate is found.

1 INTRODUCTION

The recruitment and selection of competent personnel is essential for the success of any organization. Various strategies were developed to help organization to take right decision to select best person for any post. The recruitment process is to recruit the applicants who do meet the essential requirements which are a multi criteria decision making problem with uncertainty. After the introduction of fuzzy sets by Zadeh (Zadeh 1965), a number of applications of fuzzy set theory came into existence which deals with uncertainty in different real world problem. The concept of Intuitionistic Fuzzy Sets (IFSs) was introduced by Attansov (Atanassov 1986, Atanassov 1994). The benefit of IFSs is their property to deal with the hesitancy that must be present due to information impression. This can be achieved by including a second function, along with the membership function of conventional fuzzy sets, called non-membership function. In this way, not only the degrees of the belongingness, the IFSs also combine the concept of the non-belongingness in order to better explain the real status of the information. IFSs as a generalization of fuzzy sets can be helpful in situations when a problem description by a linguistic variable, given in term of a membership function only, appears to be insufficient. In this paper, a novel personnel selection framework is proposed to find the best possible personnel for a specific job. The proposed work starts with a predefined job announcement which defines required skill set in linguistic form converted into Trapezoidal Intuitionistic Fuzzy Numbers (TrIFNs). After decision makers give their opinion the model will rank them using Intuitionistic Fuzzy Ordered Weighted Geometric (ITFOWG) aggregation operator. Finally using hamming distance method the system can find the best candidate.

2 LITERATURE SURVEY

In any recruitment or selection process generally experts in the interview committee should state their opinion about the candidates linguistically. A relationship is formed between the basic requirements for the job (objectives) and the experts opinion about the candidates (data elements) using fuzzy subset representation (Siler and Buckley, 2005).The consequence due to wrong decision become problematic and difficult to correct (Liao and Chang, 2009). In any selection process qualification, communication skill, age etc are criteria that can be considered as linguistic variable whose linguistic values may be low, medium, good etc. In 2005 Chen and Cheng showed personnel selection problem as a combination of Group Decision Support System (GDSS) and Multi Criteria Decision Making (MCDM) in fuzzy domain. They used this approach for information system project manager selection. The evaluation of candidate and the weights of criteria are given in linguistic terms and converted into Triangular Fuzzy Numbers (TFNs). In 2007 Golec and Kahya developed a hierarchical structure and use a fuzzy model for personnel selection. The model selects the personnel using the fuzzy rule base approach.

3 PRELIMINARIES

3.1 *Intuitionistic Fuzzy Sets (IFSs)*

Atanassov first introduce the concept of intuitionistic fuzzy set as follows:

Let X be the universe of discourse then an intuitionistic fuzzy set I on U is defined as $\bar{x}_t=(x,y,z)^T, i=1,\dots n$ where functions $\mu_t(x): X\to[0,1]$ and $\upsilon_t(x): X\to[0,1]$ are the degree of membership and non membership of the element $x\in X$ and $\forall x\in X, 0\le\mu_t(x)+\upsilon_t(x)\le 1$.

3.2 *Intuitionistic Fuzzy Number (IFN)*

An intuitionistic fuzzy number is an intuitionistic fuzzy set $\bar{i}$ on $\Re$ with membership and non-membership function are represented is given by:

$$\mu_{\bar{i}}(x)=\begin{cases}\mu^L_{\bar{i}}(x) & ;a_1\le x\le b_1\\ w_{\bar{i}} & ;b_1\le x\le c_1\\ \mu^R_{\bar{i}}(x) & ;c_1\le x\le d_1\\ 0 & ;otherwise\end{cases}$$

and $$\upsilon_{\bar{i}}(x)=\begin{cases}\upsilon^L_{\bar{i}}(x) & ;a_1\le x\le b_1\\ u_{\bar{i}} & ;b_1\le x\le c_1\\ \upsilon^R_{\bar{i}}(x) & ;c_1\le x\le d_1\\ 0 & ;otherwise\end{cases}$$

such that $\forall x\in\Re, 0\le\mu_{\bar{i}}(x)+\upsilon_{\bar{i}}(x)\le 1$, where $\mu^L_{\bar{i}}(x):[a_1,b_1]\to[0,w_{\bar{i}}]$ is continuous and strictly increasing, $\mu^R_{\bar{i}}(x):[c_1,d_1]\to[0,w_{\bar{i}}]$ is continuous and strictly decreasing. $\upsilon^L_{\bar{i}}(x):[a_1,b_1]\to[u_{\bar{i}},1]$ is continuous and strictly decreasing, $\upsilon^R_{\bar{i}}(x):[c_1,d_1]\to[u_{\bar{i}},1]$ is continuous and strictly increasing.

3.3 *Trapezodial Intuitionistic Fuzzy Number (TrIFN)*

A Trapezoidal Intuitionistic Fuzzy Number $\tilde{s}=\langle(s_1,s_2,s_3,s_4);w_{\tilde{s}},u_{\tilde{s}}\rangle$ is a special case of intuitionistic fuzzy number defined on the set of real number $\Re$, whose membership and non-membership functions are defined as:

$$u_{\tilde{s}}(x)=\begin{cases}\dfrac{(x-s_1)}{(s_2-s_1)}w_{\tilde{s}};s_1\le x\le s_2\\ w_{\tilde{s}};s_2\le x\le s_3\\ \dfrac{(s_4-x)}{(s_4-s_3)}w_{\tilde{s}};s_3\le x\le s_4\\ 0;x<s_1 \text{ and } x>s_4\end{cases}$$

and

$$\upsilon_{\tilde{s}}(x)=\begin{cases}\dfrac{(s_2-x)+u_{\tilde{s}}(x-s_1)}{(s_2-s_1)};s_1\le x\le s_2\\ u_{\tilde{s}};s_2\le x\le s_3\\ \dfrac{(x-s_3)+u_{\tilde{s}}(s_4-x)}{(s_4-s_3)};s_3\le x\le s_4\\ 1;x<s_1 \text{ and } x>s_4\end{cases}$$

The graphical representation of which is shown in Fig. 1. The values $w_{\tilde{s}}$ and $u_{\tilde{s}}$ represents the maximum membership degree and minimum non-membership degree respectively such that $0\le w_{\tilde{s}}\le 1, 0\le u_{\tilde{s}}\le 1$ and $0\le w_{\tilde{s}}+u_{\tilde{s}}\le 1$ holds. The parameters $w_{\tilde{s}}$ and $u_{\tilde{s}}$ define the confidence level and the non-confidence level of the TrIFN $\tilde{s}=\langle(s_1,s_2,s_3,s_4)w_{\tilde{s}},u_{\tilde{s}}\rangle$ respectively.

Let $\rho_{\tilde{s}}(x)=1-w_{\tilde{s}}(x)-u_{\tilde{s}}(x)$ which is called hesitation degree or intuitionistic fuzzy index of whether x belongs to $\tilde{s}$. If the hesitation degree is small then knowledge whether x belongs to $\tilde{s}$ is more certain, while if the hesitation degree is large then the knowledge about x belonging to $\tilde{s}$ is more certain.

Wang and Zang (Wu 2011) developed certain arithmetic operators on TRIFNs.

The accuracy functions of $\tilde{A}$ and $\tilde{B}$ are $H(\tilde{A}),H(\tilde{B})$ As proposed by Jianqiang and Zhong, the scoring and accuracy function of aTrIFN, $\tilde{A}$ and $\tilde{B}$ are calculated as:

$$S(\tilde{A})=\left[\left(\frac{1}{8}\times(a_1+a_2+a_3+a_4)\times(1-w_1-u_1)\right)\times(w_1-u_1)\right] \tag{1}$$

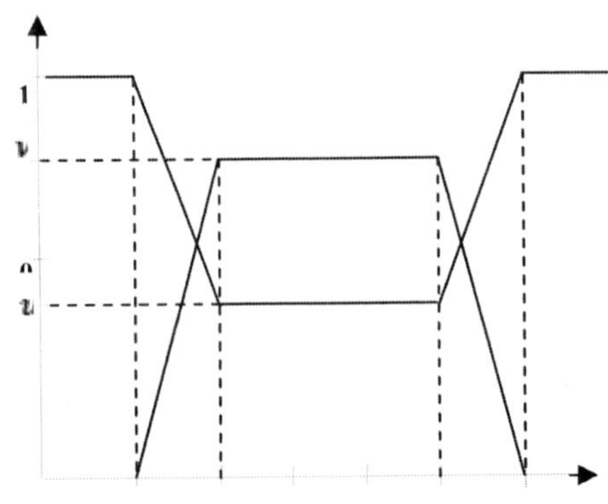

Figure 1. Membership degree and non membership degree of TrIFN.

$$S(\tilde{B}) = \left[\begin{pmatrix} \frac{1}{8} \times (b_1 + b_2 + b_3 + b_4) \times \\ (1 - w_2 - u_2) \end{pmatrix} \times \\ (w_2 - u_2) \right] \tag{2}$$

$$H(\tilde{A}) = \left[\begin{pmatrix} \frac{1}{8} \times (a_1 + a_2 + a_3 + a_4) \times \\ (1 - w_1 - u_1) \end{pmatrix} \times \\ (w_1 + u_1) \right] \tag{3}$$

$$H(\tilde{B}) = \left[\begin{pmatrix} \frac{1}{8} \times (b_1 + b_2 + b_3 + b_4) \times \\ (1 - w_2 - u_2) \end{pmatrix} \times \\ (w_2 + u_2) \right] \tag{4}$$

Based on these scoring and accuracy function the ranking between two TrIFN $\tilde{A}$ and $\tilde{B}$ are done as per the following algorithm:

if $S(\tilde{A}) > S(\tilde{B})$,
then $\tilde{A} > \tilde{B}$
if $S(\tilde{A}) < S(\tilde{B})$,
then $\tilde{A} < \tilde{B}$
if $S(\tilde{A}) = S(\tilde{B})$,
then
if $H(\tilde{A}) > H(\tilde{B})$,
then $\tilde{A} > \tilde{B}$
if $H(\tilde{A}) < H(\tilde{B})$,
then $\tilde{A} < \tilde{B}$
else
$\tilde{A} = \tilde{B}$

3.4 *Ordered Weighted Averaging (OWA) operator*

An Ordered Weighted Averaging (OWA) operator (Mohammadi *et. al.,* 2014) of dimension m is a function τ, $\tau : \mathfrak{R}^{+m} \rightarrow \mathfrak{R}^{+}$ which has associated a set of weights or weighting vector $\omega = [\omega_1, \omega_2, \ldots, \omega_m]^T \forall \omega_i \in [0,1]$ and $\sum_i \omega_i = 1$. Furthermore, $\tau(y_1, y_2, \ldots, y_m) = \omega^T.B = \sum_{j=1}^{m} \omega_j b_j$ where B is the associated ordered value vector and each element $b_j \in B$, is the *jth* largest value among $y_j (j = 1,2,3,\ldots,m)$.

3.5 *Weights of OWA operator*

One of the most important factor in the OWA operator is to determine its associated weight vector, ω; where $\omega = [\omega_1, \omega_2, \ldots, \omega_m]^T \forall \omega_i \in [0,1]$ and $\sum_i \omega_i = 1$. Xu (Xu *et. al.,* 2006) develop a normal distribution based method which is defined as

$$\omega_i = \frac{e^{-\frac{(i-\mu_m)}{2\sigma_m^2}}}{\sum_{k=1}^{m} e^{-\frac{(k-\mu_m)}{2\sigma_m^2}}}, k = 1,2,\ldots,m \tag{5}$$

where μ_m is the mean of the collection of $1,2,\ldots,m$ and $\sigma_m (> 0)$ is the standard deviation of the collection of $1,2,\ldots,m$. So mathematically we define

$$\mu_m = \frac{1+n}{2} \tag{11}$$

$$\sigma_m = \sqrt{\frac{1}{m} \sum_{l=1}^{m} (1 - \mu_m)^2}. \tag{6}$$

3.6 *Geometric mean operator*

A geometric mean operator (Mohammadi *et al.,* 2014) of dimension m is function ρ, $\mathfrak{R}^{+m} \rightarrow \mathfrak{R}^{+m}$ such that

$$\rho(y_1, y_2, \ldots, y_m) = \prod_{k=1}^{m} (y_k)^{\frac{1}{m}} \tag{7}$$

3.7 *Ordered Weighted Geometric (OWG) operator*

OWG operator of dimension m [3] is a function ρ, $\mathfrak{R}^{+m} \rightarrow \mathfrak{R}^{+m}$ that has associated a set of weights or exponential weighting vector $\omega = [\omega_1, \omega_2, \ldots, \omega_m]^T \forall \omega_i \in [0,1]$ and $\sum_i \omega_i = 1$ and is defined for aggregating a list of values $\{a_1, a_2, \ldots, a_m\}$ according to the following expression $\delta(y_1, y_2, \ldots, y_m) = \prod_{i=1}^{m} c_i^{\omega_i}$ where C is the associated ordered value vector, and each element $c_i \in C$ is the *ith* largest value in the collection $a_1, a_2, \ldots, a_m$.

3.8 *Intuitionistic Trapezoidal Fuzzy Ordered Weighted Geometric (ITFOWG) operator*

Let $\tilde{\alpha}_j$ (j = 1, 2, …, n) be a collection of intuitionistic trapezoidal fuzzy numbers. An Intuitionistic Trapezoidal Fuzzy Weighted Geometric (ITFOWG) operator of dimension n is a mapping ITFOWG: $\Omega^n \rightarrow \Omega$, that has an associated vector $w = (w_1, w_2, \ldots, w_n)^T$ such that $w_j \in [0, 1]$ and $\sum_{j=1}^{n} w_j = 1$. Furthermore $ITFOWG_W (\tilde{\alpha}_1, \tilde{\alpha}_2, \ldots, \tilde{\alpha}_n) = \tilde{\alpha}_{\sigma(1)}^{w_1} \otimes \tilde{\alpha}_{\sigma(2)}^{w_2} \ldots \otimes \tilde{\alpha}_{\sigma(n)}^{w_n}$

where ($\sigma(1)$, $\sigma(2)$, …, $\sigma(n)$) is a permutation of (1, 2, …, n) such that $\tilde{\alpha}_{\sigma(j-1)} \geq \tilde{\alpha}_{\sigma(j)} \forall j$.

Let $\tilde{\alpha}_j$ (j = 1, 2, …, n) be a collection of intuitionistic trapezoidal fuzzy numbers, then their aggregated value by using the ITFOWG operator is also an intuitionistic trapezoidal fuzzy number and

$$ITFOWG_W(\tilde{\alpha}_1,\tilde{\alpha}_2,\ldots,\tilde{\alpha}_n) = \begin{pmatrix} \begin{bmatrix} \prod_{j=1}^{n} a_{\sigma(j)}^{w_j}, \prod_{j=1}^{n} b_{\sigma(j)}^{w_j}, \\ \prod_{j=1}^{n} c_{\sigma(j)}^{w_j}, \prod_{j=1}^{n} d_{\sigma(j)}^{w_j} \end{bmatrix}; \\ \prod_{j=1}^{n} \mu_{\tilde{\alpha}_{\sigma(j)}}^{w_j}, \\ 1-\prod_{j=1}^{n}\left(1-v_{\tilde{\alpha}_{\sigma(j)}}\right)^{w_j} \end{pmatrix} \quad (8)$$

where $w = (w_1, w_2, \ldots, w_n)^T$ is the weight vector of the ITFOWG operator, with $w_j \in [0, 1]$ and $\sum_{j=1}^{n} w_j = 1$.

4 NUMERICAL ILLUSTRATION

We have taken three candidates c_1,c_2,c_2 and three experts e_1,e_2,e_2 as decision maker who will give their opinion in linguistic term. The linguistic values are Very Low (VL), Low (L), Medium (M), High (H) and Very High (VH). The trapezoidal intuitonistic fuzzy value is given in table1. The skill set over which experts will give their opinion are communication skill, educational qualification and age represented as p_1, p_2, p_3.

The trapezoidal fuzzy set for each criteria is represented by three numbers: membership μ, non-membership ϑ, hesition margin π.

Using equation (5), (6) and (7) we calculate the weight vector $\omega = [0.2429, 0.5142, 0.2429]$.

Now we apply the ITFOWG (8) operator on those linguistic questionnaires associated with each candidate C_i to get the aggregated weight.

Table 2 represents the desired skill set in terms of intuitionistic trapezoidal fuzzy set and Table 3 represents the normalised Hamming distance (9) of each candidate from the desired skill set.

In our proposed methodology we explained that all three parameters of a trapezoidal fuzzy set the membership function, the non membership function and the hesitation margin play a vital role for calculating the distances between intuitionistic fuzzy sets. To be more specific, the normalized Hamming distance for all the symptoms of the i-th desired criteria of the candidates from the k-th weight of the criteria of the candidate is equal to

$$l(s(p_i),d_k) = \tfrac{1}{6}\sum_{j=1}^{16}(|\mu_j(p_i)-(\mu_j(d_k)| + |(v_j(p_i)-(v_j(d_k)| + |(\pi_j(p_i)-(\pi_j(d_k)) \quad (9)$$

Table 1. Aggregated linguistic weight A_{ggci} of the given set.

Criteria	Candidate	Expert	Linguistic variable	Intuitionistic trapezoidal fuzzy set	Aggregated weight $Agg_{\tilde{c}_i}$
P_1	C_1	e_1	VL	$\langle[0,0.05,0.15,0.25];0.55,0.21\rangle$	$\langle[0,0.195,0.317,0.4265]; 0.6169,0.0001\rangle$
		e_2	L	$\langle[0.15,0.25,0.35,0.45];0.56,0.2\rangle$	
		e_3	M	$\langle[0.35,0.45,0.55,0.65];0.85,0.1\rangle$	
	C_2	e_1	M	$\langle[0.35,0.45,0.55,0.65];0.72,0.11\rangle$	$\langle[0.5313,0.6344,0.7366, 0.8284]; 0.8177,0.000001\rangle$
		e_2	H	$\langle[0.55,0.65,0.75,0.85];0.83,0.12\rangle$	
		e_3	VH	$\langle[0.75,0.85,0.95,0.1];0.9,0.03\rangle$	
	C_3	e_1	H	$\langle[0.55,0.65,0.75,0.85];0.85,0.1\rangle$	$\langle[0.3179,0.4265,0.5313, 0.6344]; 0.7574,0.00006\rangle$
		e_2	M	$\langle[0.35,0.45,0.55,0.65];0.76,0.15\rangle$	
		e_3	L	$\langle[0.15,0.25,0.35,0.45];0.67,0.23\rangle$	

(Continued)

Table 1. (*Continued*)

Criteria	Candidate	Expert	Linguistic Variable	Intuitionistic trapezoidal fuzzy set	Aggregated weight $Agg_{\tilde{c}_i}$
P_2	C_1	e_1	H	$\langle[0.55,0.65,0.75,0.85];0.7,0.25\rangle$	$\langle[0.5315,0.6345,0.7366,0.8284];0.6986,0.0829\rangle$
		e_2	VH	$\langle[0.75,0.85,0.95,0.1];0.81,0.09\rangle$	
		e_3	M	$\langle[0.35,0.45,0.55,0.65];0.6,0.03\rangle$	
	C_2	e_1	H	$\langle[0.55,0.65,0.75,0.85];0.7,0.25\rangle$	$\langle[0.3179,0.4265,0.5313,0.6344];0.7574,0.00006\rangle$
		e_2	M	$\langle[0.35,0.45,0.55,0.65];0.6,0.03\rangle$	
		e_3	L	$\langle[0.15,0.25,0.35,0.45];0.5,0.4\rangle$	
	C_3	e_1	M	$\langle[0.35,0.45,0.55,0.65];0.6,0.03\rangle$	$\langle[0.5313,0.6344,0.7366,0.8284];0.6986,0.0829\rangle$
		e_2	H	$\langle[0.55,0.65,0.75,0.85];0.7,0.25\rangle$	
		e_3	VH	$\langle[0.75,0.85,0.95,0.1];0.81,0.09\rangle$	
P_3	C_1	e_1	O	$\langle[0.55,0.65,0.75,0.85];0.7,0.25\rangle$	$\langle[0.3179,0.4265,0.5313,0.6344];0.5959,0.1166\rangle$
		e_2	M	$\langle[0.35,0.45,0.55,0.65];0.6,0.03\rangle$	
		e_3	Y	$\langle[0.15,0.25,0.35,0.45];0.5,0.4\rangle$	
	C_2	e_1	M	$\langle[0.35,0.45,0.55,0.65];0.6,0.03\rangle$	$\langle[0.5313,0.6344,0.7366,0.8284];0.6986,0.0829\rangle$
		e_2	O	$\langle[0.55,0.65,0.75,0.85];0.7,0.25\rangle$	
		e_3	VO	$\langle[0.75,0.85,0.95,0.1];0.81,0.09\rangle$	
	C_3	e_1	O	$\langle[0.55,0.65,0.75,0.85];0.7,0.25\rangle$	$\langle[0.5313,0.6344,0.7366,0.8284];0.6986,0.0829\rangle$
		e_2	VO	$\langle[0.75,0.85,0.95,0.1];0.81,0.09\rangle$	
		e_3	M	$\langle[0.35,0.45,0.55,0.65];0.6,0.03\rangle$	

Table 2. The desired criteria set given by the experts in the form of TrIFN.

Skill	Communication skill	Educational qualification	Age
Desired Fuzzy value	$\langle[0.55,0.65,0.75,0.85];0.8,0.1\rangle$	$\langle[0.75,0.85,0.95,0.1];0.75,0.12\rangle$	$\langle[0.15,0.25,0.35,0.45];0.55,0.3\rangle$

Results are given in Table 3. According to fuzzy rule the lowest measured distance 0.25817 which is the score of candidate 3 which is the best candidate in personal selection.

5 CONCLUSION AND FUTURE WORK

In our work we have gathered data from the experts in form of TrIFN and representing in matrix from of those personal selection criteria. With the help

Table 3. Category wise distance between aggregated weight and desired fuzzy set of each candidate.

Candidate			
Criteria	C_1	C_2	C_3
Communication Skill	0.4321	0.38973	0.25817
Educational Qualification	0.92925	0.64613	0.59715
Age	0.5774	0.82237	0.83444

of this database we have ranked the opinions of experts using ITFOWG aggregation operator and using resemblance measure of trapezoidal intuitionistic fuzzy sets we have find out the best candidate. As future work, we plan to consider if standard measures and distance on intuitionistic fuzzy sets are applicable in the context of hesitant fuzzy set or fuzzy multiset and optimize the result by more simulation.

REFERENCES

Aggarwal, A., Thakur, G. S. M., (2014) Design and Implementation of Fuzzy Rule Based Expert System for Employees Performance Appraisal in IT Organisations, I.J. Intelligent Systems and Application, 6(8), 77–86.

Atanassov, K. T., (1986) Intuitionistic fuzzy sets, Fuzzy sets and systems, 20, 87–96.

Atanassov, K. T., (1994) New operations defined over the intuitionistic fuzzy sets, Fuzzy sets and Systems, 61, 137–142.

Ghosh, D., Majumder, S., Pal, A., (2014) A Novel approach to find the Shortest Path of a Network using A* Search Algorithm, Proceeding of the IEEE CALCON 2014, November, National Conference on Electrical, Electronics and Computer Engineering, 7–8.

Gui-Wu, W. (2011) Some generalized aggregating operators with linguistic information and their application to multiple attribute group decision making, Computers & Industrial Engineering, 61(1), 32–38.

Mohammadi, F., Bazmara, M., Pouryekta, H., (2014) A New Hybrid Method for Risk Management in Expert Systems, I.J. Intelligent Systems and Application, 6(7) 75–80.

Szmidt. E., Kacprzyk, J., (1997) On measuring distances between intuitionistic fuzzy sets, Notes on IFS, 3(4), 1–13.

Szmidt. E., Kacprzyk, J., (2000) Distances between intuitionistic fuzzy sets, Fuzzy Sets and Systems, 114(3), 505–518.

Wu., J., Cao, Q. W. (2013) Same families of geometric aggregation operator with intuitionistic Trapezoidal fuzzy numbers, Applied Mathematical Modelling, 37, 318–327.

Xu. Z., Yager, R. R. (2006) Some geometric aggregation operator based on intuitionistic fuzzy Sets", International Journal of General Systems, 35(4) 417–433.

Zadeh, L. A., (1965) Fuzzy Sets, Inform. and Control, 8, 338–353.

Computational Science and Engineering – Deyasi et al. (Eds)
 ISBN 978-1-138-02983-5

Camera calibration from known correspondences using PSO

A. Mukhopadhyay
Gargi Memorial Institute of Technology, Kolkata, India

S. Hati
Budge Budge Institute of Technology, Kolkata, India

ABSTRACT: Most pose estimation methods use weighted least square techniques. For the detection of outliers robust statistics have been used. In this work, the pose of 3D objects is estimated using state-of-the-art Particle Swarm Optimization (PSO) technique. For the detection of outliers, PSO has built in robustness. The method described in this paper, demonstrates the advantages of PSO by thorough testing on both synthetic and real data.

1 INTRODUCTION

Photogrammetry is the art, science and technology of obtaining reliable information about physical objects and the environment through the process of recording, measuring, and interpreting photographic image (Wolf *et al.*, 2000). Camera calibration is a necessary step in 3D computer vision and is a part of the *metric* photogrammetry. It extracts metric information from 2D images of 3D world. Formally camera calibration is the process of computing the 3D transformation from the known world coordinates of a set of points in object space to the given pixel co-ordinates of the corresponding set of points in image space. In this paper resultant camera parameters identify uniquely the transformation between the camera co-ordinates reference frame and the known world co-ordinates reference frame. Extrinsic camera parameters are the parameters that define the location and orientation of the camera with respect to a known world co-ordinates reference frame. Intrinsic camera parameters describe the geometry of the camera. Each intrinsic camera parameter is necessary to link the pixel coordinates of an image point with the corresponding coordinates in the camera coordinates reference frame.

2 PREVIOUS WORK

Fischler and Bolles (Fischler *et al.*, 1981) determine the pose of the camera using a method of geometry. Yuan (Joseph *et al.*, 1989) derives the exterior orientation parameter solution from the knowledge of four coplanar points. He shows that camera calibration from a non-coplanar feature point set outperform that from coplanar point set in both accuracy and robustness in the presence of noise. The four different problems on pose estimation is given by Researchers (Haralick *et al.*, 1985) using weighted least square method. They have obtained a linear solution, and a robust solution using robust statistics. In robust statistics they have considered Tukey's bi-weight function to detect outliers. In this work it is being proved that the robustness is built in PSO. Xinhua and Huang (Xinhua *et al.*, 1994) solve many pose estimation problems in 3D space using weighted least square technique. Abidi and Chandra (Abidi *et al.*, 1966) apply a geometrical closed form technique to determine both extrinsic and intrinsic camera parameters. Kang and Ikeuchi (Kang *et al.*, 1993) describe a method which does not require correspondences between object and its image in determining pose of the object. Hati and Sengupta (Hati *et al.*, 2001) first derive the solution for the calibration of exterior orientation parameters of a camera using Genetic Algorithm (GA). Later Ji and Zhang (Li *et al.*, 2001) find both the internal and external orientation parameters of the camera using GA.

Like GA, the population of PSO are randomly initialized and searched for optimal solution in every generation. PSO has a lot of advantages over the other stochastic optimization techniques. PSO has no operators like crossover or mutation that is available in GA and it assumes simple algorithms. Only few parameters of PSO are updated in each generation. It is based solely on the movement of flying birds. Each flying bird, called particle, flies over the bounded search space. Worker (Elbeltagi *et al.* 2005) compares in detail five evolutionary-based algorithms: GA, Memetic Algorithms, Particle Swarm Optimization, Ant Colony

Optimization, Shuffled Frog Leaping. Based on this comparative analysis, they show that PSO method performs better than other algorithms in terms of success rate, solution quality and the processing time. PSO is robust in the sense of finding optimal in continuous optimization problems with uncertainty or fluctuations in the input variables or finding optima in dynamic environments or finding optima given noisy or uncertain objective functions. In view of these advantages mentioned above, PSO based methods are applicable to finding optimum of multi-modal functions. In (Ji *et al.*, 2001), it is shown that the error surface of the solution of the camera calibration problem is multi-modal in nature. Researcher (Lepetit *et al.* 2008) find a linear time solution of *n*-points camera calibration problem. Each 3D point is expressed as linear combination of four virtual control points. Then the method determines the coordinate of four virtual points in the camera coordinate systems. In this work it is shown that robustness is built in PSO and there is no need to use another method to detect *rejection points* like the method by Researcher (Haralick *et al.* 1985) that uses the methods of robust statistics (Hoaglin *et al.*, 1983). It is also found that the methods found in (Haralick *et al.* 1985, Joseph *et al.*, 1989, Fischler *et al.*, 1981) fail miserably when there is a mismatch in the correspondence problem of image points and object points. But PSO can bear with and gives reasonable good results when there are mismatches between object points and image points. However, PSO is *robust* up-to a certain number of mismatched points and the error skyrockets after a certain number of mismatched pairs of points.

There are two papers (Song *et al.*, 2009, Kumar *et al.*, 2009) on camera calibration using particle swarm optimization technique. In (Song *et al.*, 2009), they have computed intrinsic parameters of the camera only. In this work, both internal and external parameters of the camera are estimated. They have not considered the effect of noise i.e. the presence of outliers in the image data. In this paper it is shown the effect of noisy data on the camera parameters. The effect of noise is particularly useful for applications such as in remote sensing and war. In another work (Kumar *et al.*, 2009), they have not calculated directly camera parameters using PSO. They have trained Artificial Neural Network (ANN) using PSO and the ANN system has calibrated the camera.

The paper is organized as follows: Camera calibration problem is discussed in Section 3. Section 4 presents the formulation of camera calibration problem using PSO followed by Results and Discussions are described in Section 5. Conclusion is discussed in Section 5.

3 CAMERA CALIBRATION PROBLEM

Let $\bar{x}_i = (x, y, z)^T, i = 1, \ldots n$ be the n points in object space. Let $\left(P_{i1}, P_{i2}\right), i = 1, \ldots, n$ be the corresponding image points in 2D image space after perspective projection of the points in object space. The correspondence problem of image points and object points is given by (Haralick *et al.* 1993),

$$p_{i1} = f\,\frac{r_1 x_i + t_1}{r_3 x_i + t_3} \tag{1}$$

$$p_{i2} = f\,\frac{r_2 x_i + t_2}{r_3 x_i + t_3} \tag{2}$$

$$t = \left(t_1, t_2, t_3\right) \tag{3}$$

$$\boldsymbol{R} = \begin{pmatrix} r_1 \\ r_2 \\ r_3 \end{pmatrix} = \begin{pmatrix} r_{11} & r_{12} & r_{13} \\ r_{21} & r_{22} & r_{23} \\ r_{31} & r_{32} & r_{33} \end{pmatrix} \tag{4}$$

where $r11 = \cos\varphi\cos\psi$, $r12 = \cos\varphi\sin\psi$, $r13 = -\sin\varphi$, $r21 = -\cos\theta\sin\psi + \sin\theta\sin\varphi\cos\psi$, $r22 = \cos\theta\cos\psi + \mathrm{Sin}\theta\sin\varphi\sin\psi$, $r23 = \sin\theta\cos\varphi$, $r31 = \sin\theta\sin\psi + \cos\theta\sin\varphi\cos\psi$, $r32 = -\sin\theta\cos\psi + \cos\theta\sin\varphi\sin\psi$, $r33 = \cos\theta\cos\varphi$.

$\boldsymbol{R}$ is the rotation matrix, $\boldsymbol{t}$ is the translation vector, and f denotes camera focal length. θ, ϕ, and ψ represent tilt, pan, swing angles respectively (Haralick *et al.* 1993). Let $\left(\mathrm{u}, \mathrm{v}\right)^T$ denote the coordinates of (P_{i1}, P_{i2}). Hence

$$u = fs_x \frac{r_{11}x + r_{12}y + r_{13}z + t_x}{r_{31}x + r_{32}y + r_{33}z + t_z} + u_0 \tag{5}$$

$$v = fs_y \frac{r_{21}x + r_{22}y + r_{23}z + t_y}{r_{31}x + r_{32}y + r_{33}z + t_z} + v_0 \tag{6}$$

where s_x and s_y are indicate scale factors (pixels/mm) due to spatial quantization, u_0 and v_0 are indicate coordinates of the principle point in pixels relative to image frame.

Let us now define the camera calibration problem with respect to two coordinate frames, viz., object space coordinate system and pixel space coordinate system. The exterior orientation parameters of the camera are $[t_x, t_y, t_z, \theta, \phi, \psi]^T = \boldsymbol{a}_{ex}$ (say) and interior orientation parameters of the camera are $[f, s_x, s_y, u_0, v_0]^T = \boldsymbol{b}_{in}$ (say). Thus, camera calibration problem is to determine the two vectors $\boldsymbol{a}_{ex}, \boldsymbol{b}_{in}$ from the knowledge of the control points in object space and the corresponding points in image space.

4 PROBLEM FORMULATION OF CAMERA CALIBRATION PROBLEM USING PSO

In regression analysis, the basic principle is to determine a function for a given set of noisy data points. In this paper, the problem is of the nature of regression. Hence, in this work the parameters of the model are estimated.

PSO mimics the process of migration of birds from one place to another. The "birds" in PSO are called particles and these particles move in search space iteratively. The particles are characterized by their velocities and positions.

$$v_{id}(t+1) = \omega v_{id}(t) c_1 R_1 (P_{id}(t) - x_{id}(t)) + c_2 R_2 (P_{gd}(t) x_{id}(t)) \quad (7)$$

$$x_{id}(t+1) = x_{id}(t) + v_{id}(t+1) \quad (8)$$

where v_{id} represents d-th element of the velocity vector of the i-th particle; t denotes iteration counter; R_1 and R_2 are random variables uniformly distributed within [0,1]; c_1 and c_2 are weighting factor and the value of each is taken as 0.2, also called cognitive and social parameter respectively. $p_{id}(t)$ is the best position along the d-th dimension of particle in iteration t. $p_{gd}(t)$ is the best position among all particles along d-th dimension in iteration t. $\omega \geq 0$ is defined as inertia factor whose value is taken as 0.9. Each particle flies through the search space with an adaptive velocity that depends on flying experience of the particle under consideration and also that of the neighboring particles. The search process uses two methods to enhance the search (i) deterministic and probabilistic rules (ii) information sharing among the population members. There is an information sharing mechanism among the particles to communicate their experiences. The algorithm finds the global optimizer with the best position ever visited by all the particles. In this work, two vectors $\mathbf{a}_{ex}$ and $\mathbf{b}_{in}$ are merged to get a decision vector,

$$s = \begin{bmatrix} \boldsymbol{a}_{ex} \\ \boldsymbol{b}_{in} \end{bmatrix} = \left[t_x t_y t_z \theta \phi \psi f s_x s_y u_0 v_0 \right]^T$$

This decision vector s denotes the position of a particle flying in search space.

Let $(p_{i1}^e, p_{i2}^e), i = 1, \ldots, n$ be the corresponding image points of the model points after transformation. Let d be the distance function defined as:

$$d = \sum_{i=1}^{n} [Abs(p_{i1}^e - p_{i1}) + (p_{i2}^e - p_{i2}) \quad (9)$$

$Abs(x)$ denotes absolute value of its argument x. It is aimed to optimize the values of $\boldsymbol{a}_{ex}$ and $\boldsymbol{b}_{in}$ such that $\boldsymbol{d}$ is minimum. In PSO, parameters are used as column vector of the swarm matrix, such that d tends towards minimum.

4.1 *Results and discussions*

In this section the results of the computer experiments is experienced using both synthetic image data and real image data. The robustness and the performance of the mentioned algorithm are tested using 20 dB and 40 dB noise are added to the image data.

The particle swarm optimization algorithm (Eq. (7) and Eq. (8) are used for optimization of Eq. (9). Experiments are run for defined number of generations, i.e., 700 times with swarm size of 300. In this problem eleven camera parameters are taken as no. of inputs to the objective function and hence eleven is the dimension of the optimization problem using PSO.

4.2 *Camera calibration using synthetics and real data*

The described problem in this paper is applied on SOFA synthetic image sequences which can be obtained from Computer Vision Group at Heriot—Watt University, Scotland. The sequences have been developed for optical flow analysis. The results obtained from the algorithm mentioned in Eq. (9) are compared with provided ground truth with the image sequences. Thus it is being shown pictorially the comparison of the results with the ground truth given for SOFA sequences at various degrees of noise.

The wire-frame models of the cubes are constructed using the knowledge of the world co-ordinates of the vertices of the cubes of SOFA sequences. The wire-frame models are constructed based on Winged-Edge data structure given in (Weiler et al., 1985). The camera parameters are computed from the knowledge of the co-ordinates of the vertices and then align the wire-frame models of the cubes on the images. It is being observed that alignment is more or less perfect at 40 dB (Fig. 1(a)) noise labels. But there is a marked deviation of the alignment at 20 dB (Fig. 1(b)) noises. Thus it is being noted that PSO can calibrate the camera parameters from the knowledge of *small* number of points when the points are perturbed by good amount of noise.

In this work QUICAM FAST 1394 digital CCD camera is used. Its specification is compatible

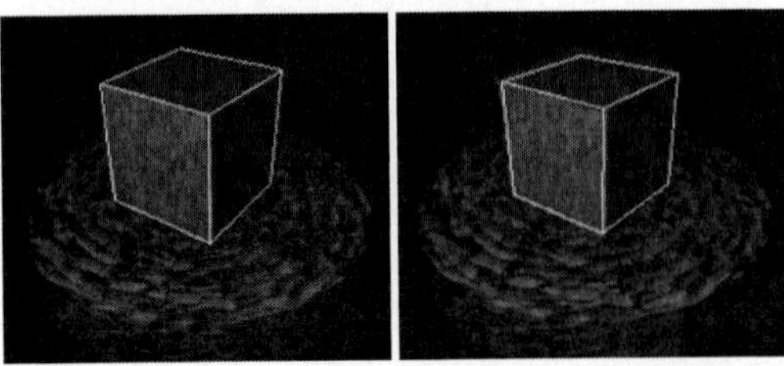

Figure 1. Pose Estimation of a cube placed on a cylinder at 10 degree angle at different noise level. Fig. (a) 40 dB (b) 20 dB noise level. **(Synthetic data).**

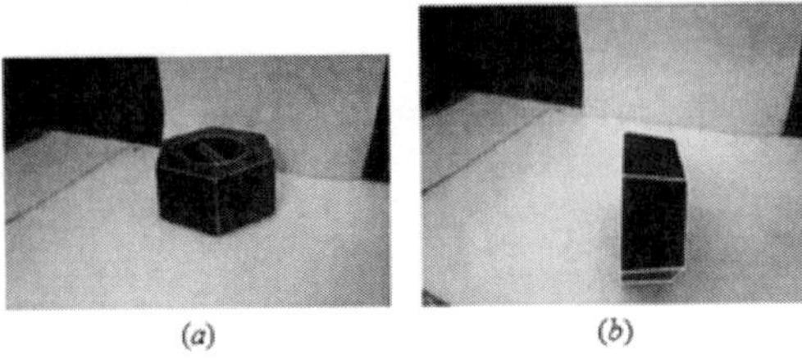

Figure 2. Alignment of the wireframe models with the objects in real image.

with IEEE 1394 firewall digital CCD camera. The images of the same object are taken from different viewpoints. Fig. 2(a) and Fig. 2(b) show two such images from different viewpoints. In these two images the excellent alignment of the original images of the models of the objects are achieved. The results are shown in Fig. 2(a) and Fig. 2(b). These results show PSO based camera calibration are well fitted for scientific and industrial applications.

5 CONCLUSIONS

The goal of this paper is to find solution of camera calibration problem. Both intrinsic and extrinsic parameters of the camera are estimated with reasonable accuracy applicable for industries using Particle Swarm Optimization (PSO) technique. Classical techniques using gradient descent found in the literature such as in (Tsai, 1987) perform miserably due to the effect of noise and mismatch of points. In this paper the reasonable accuracy of computing the camera parameters is achieved even in presence of noise as well as the mismatch of points in image space with those of in object space. The method mentioned in this paper shows the robustness of PSO to find the solution of one of the fundamental problems in Robotics and Computer Vision, viz., camera calibration problem.

REFERENCES

Abidi, M.A., Chandra, T., 1995. A new efficient and direct solution for pose estimation using quadrangular targets: Algorithm and evaluation. IEEETrans. on Pattern Analysis and Machine Intelligence 17:129–141.

Elbeltagi. E., Hegazy. T., Grierson. D., 2005. Comparison among five evolutionary based optimization algorithms. Advanced Engineering Informatics 19, 43–53

Fischler. M.A., Bolles. R.C., 1981. Random sample consensus: A paradigm for model fitting with application to image analysis and automated cartography. Communication ACM 24: 381–395.

Haralick R.M., Joo. H., Lee. C., Zhuang. X., 1985. Pose estimation from corresponding data. IEEE Trans. on SMC 19: 1426–1446.

Haralick. R.M., Shapiro. L., 1993. Computer and Robot Vision. volume2. Addison-Wesley Publishing Company.

Hati S., Sengupta. S., 2001. Robust camera parameter estimation using genetic algorithm. Pattern Recognition Letters 22:289–298.

Hoaglin, Mosteller. C., Tukey. J., 1983. Understanding Ro bust and Exploratory Data Analysis. Newyork: JohnWiley.

Ji. Q., Zhang,Y., 2001. Camera calibration with genetic algorithms. IEEE Trans. Syst., Man, Cybern.-PartA: Syst., Humans31:120–130.

Joseph S.C., Yuan, 1989. A general photogrammetric method for determining position and orientation. IEEE Trans. on Robotics and Automation 5,129–141.

Kang. S.B., Ikeuchi. K., 1993. The complex egi: A new representation for 3-D pose determination. IEEETrans. On Pattern Analysis and Machine Intelligence15:707–721

Kumar. S., Raman. B., Wu. J., 2009. Neuro- calibration of a camera using particle swarm optimization, in:ICETET, pp.273–278.

Lepetit. V., Moreno-Noguer. F., Fua. P., 2008. Epnp: an accurate o(n) solution to the pnp problem. International Journal of Computer Vision 81: 151–166.

Song X., Yang B., Feng Z., Xu T., Zhu D., Jiang Y., 2009, Camera Calibration based on Particle Swarm Optimization, in: CISP: 1–5.

Weiler K, 1985, Edge based Data Structure for Solid Modelling in Curved Surface Environment, IEEE Computer.

Wolf, P.R., Dewitt, B.A., 2000, Elements of Photogrammetry, McGraw Hill Inc.

Xinhua Z., Huang Y., 1994, Robust 3D-3D Pose Estimation, IEE Trans. On Pattern Analysis and Machine Intelligence, 16: 813–824

Computational Science and Engineering – Deyasi et al. (Eds)
 ISBN 978-1-138-02983-5

Solution of insect population model by using Laplace Adomian decomposition method

P. Bhattacharya & S. Pal
Mathematics Department, NIT Agartala, Tripura, India

ABSTRACT: In this paper, the Laplace Adomian Decomposition Method is implemented to give an approximate solution of nonlinear ordinary differential equation systems, such as a model for Insect population. The technique is described and illustrated by numerical examples. For reliability and simplicity of this method some plots are shown. We modify an insect population model by introducing Holling type III functional response and intraspecific competition term and hence we solve it by this numerical technique. We try to compare this method with Runge-Kutta Method of order 4 (RK4) and with the exact solution.

1 INTRODUCTION

In the last decade, many mathematical models have been proposed for describing the prey-predator relationship in population dynamics. The growth of a population (of men) was developed in (Malthus, 1798). Then Verhust (Verhulst, 1838) first proposed a three-parameter model for the growth of single-species populations that presented a logistic sigmoid growth curve over time. Though in the area of theoretical ecology the research was started by Lotka (Lotka 1925) and Volterra (Blumberg 1968). The Lotka-Volterra model is an intervention, competition model: two species are assumed to diminish each other's per capita growth rate by direct intervention. These kinds of models are also discussed by Erbe (Erbe et al. 1986), Chaudhur (Chaudhuri 1988), D. Jana et al. (Jana 2014). There are different kinds of functional response defined in predator prey models. This paper presents an over view of the mathematical models of population growth for insect populations where we pioneer Holling type III functional response.

In the recent period, a great deal of interest has been focused on the application of Runge-Kutta Method of order 4 (RK4) to solve an extensive variety of linear and nonlinear problems this is the most popular method for ordinary differential equation.

Nowadays a great deal of interest is focused on the application of Laplace Adomian Decomposition Method to solve an extensive variety of linear and nonlinear problems. Unlike in numerical methods, Laplace Adomian decomposition method is free from rounding off errors. So we emphasized on this method. The Laplace Adomian Decomposition Method (LADM) (Khuri 2001, Kiymaz 2009) was firstly introduced by Suheil A. Khuri and has been successfully used to find the solution of linear and nonlinear differential equations. This method picks out a solution in the form of a series whose terms are determined by a recursive relevance using the Laplace transform and Adomian Polynomials (Adomian 1992a, 1993b). The important characteristic of this method is that it can obtain the exact solution for a nonlinear equation by combining the two powerful methods and it is a significant advantage of this method.

The present articles consider the one species insect population model. The principal aim of this paper is to perform systematic analysis of the comparison among the exact solution, Runge-Kutta Method of order 4 (RK4)and Laplace Adomian Decomposition Method (LADM) on the dynamics of the nonautonomous insect population which shall be made to determine the performance of Laplace Adomian Decomposition Method (LADM).

2 LAPLACE ADOMIAN DECOMPOSITION METHOD

Consider the following nonlinear differential equation.

$$Lu(t)+Ru(t)+Nu(t)=g(t) \tag{1}$$

where L is a linear operatorof the highest-order derivative which is assumed to be invertible easily, R is the remaining linear operatorof order less than L and N is a nonlinear operator and $g(t)$ is a source term.

Taking the Laplace transform of both sides of the above equation, we get

$$\mathcal{L}[Lu(t)]+\mathcal{L}[Ru(t)]+\mathcal{L}[Nu(t)]=\mathcal{L}[g(t)] \quad (2)$$

Using the differential property of the Laplace transform and using the initial condition, we get

$$s^n\mathcal{L}[u(t)]-s^{n-1}u(0)-s^{n-2}u'(0)-\ldots\ldots\ldots\ldots-u^{n-1}(0)$$
$$+\mathcal{L}[Ru(t)]+\mathcal{L}[Nu(t)]=\mathcal{L}[g(t)]$$

or,

$$\mathcal{L}[u(t)]=\frac{u(0)}{s}+\frac{u'(0)}{s^2}+\ldots\ldots\ldots\ldots\ldots+\frac{u^{n-1}(0)}{s^n}$$
$$-\frac{1}{s^n}\mathcal{L}[Ru(t)]-\frac{1}{s^n}\mathcal{L}[Nu(t)]+\frac{1}{s^n}\mathcal{L}[g(t)]. \quad (3)$$

Now we representing the unknown functions $u(t)$ by an infinite series of the form

$$u(t)=\sum_{n=0}^{\infty}u_n(t). \quad (4)$$

here the components $u_n(t)$ are usually determined recurrently and the nonlinear operator $N(u)$ can be decomposed into an infinite series of polynomials given by

$$N(u)=\sum_{n=0}^{\infty}A_n$$

where A_n are Adomian polynomials of u_0, u_1, … … … u_n defined by

$$A_n=\frac{1}{n!}\frac{d^n}{d\lambda^n}\left[N\left(\sum_{i=0}^{\infty}\lambda^i u_i\right)\right]_{\lambda=0},n=0,1,2,\ldots\ldots\ldots$$

Therefore,

$$\mathcal{L}\left[\sum_{n=1}^{\infty}u_n(t)\right]=\frac{u(0)}{s}+\frac{u'(0)}{s^2}+\ldots\ldots\ldots\ldots\ldots+\frac{u^{n-1}(0)}{s^n}$$
$$-\frac{1}{s^n}\mathcal{L}\left[R\left\{\sum_{n=1}^{\infty}u_n(t)\right\}\right]-\frac{1}{s^n}\mathcal{L}\left[\sum_{n=1}^{\infty}A_n\right]+\frac{1}{s^n}\mathcal{L}[g(t)].$$

In general, the recursive relation is given by

$$\mathcal{L}[u_0(t)]=\frac{u(0)}{s}+\frac{u'(0)}{s^2}+\ldots\ldots\ldots\ldots\ldots+\frac{u^{n-1}(0)}{s^n}$$
$$+\frac{1}{s^n}\mathcal{L}[g(t)] \quad (5)$$

and

$$\mathcal{L}[u_{n+1}(t)]=-\frac{1}{s^n}\mathcal{L}[R(u_n(t))]-\frac{1}{s^n}\mathcal{L}[A_n] \quad (6)$$

Applying the inverse Laplace transform of both sides of (5) and (6), we obtain $u_n,(n\geq 0)$, which is then substituted into (4).

For numerical computation, we get the expression as

$$\phi_n(t)=\sum_{k=0}^{n}u_k(t)$$

which is the nth term approximation of $u(t)$.

3 FOURTH ORDER RUNGE-KUTTA METHOD (RK4)

This method is most commonly used and is generally called as Runge-Kutta Method.

The initial value problem,

$$\frac{dy}{dx}=f(x,y),y(x_0)=y_0.$$

The approximate value of y is given as,

$$y_1=y_0+k.$$

where, $k=\frac{1}{6}(k_1+2k_2+2k_3+k_4)$ in which

$$k_1=hf(x_0,y_0),$$

$$k_2=hf\left(x_0+\frac{h}{2},y_0+\frac{k_1}{2}\right),$$

$$k_3=hf\left(x_0+\frac{h}{2},y_0+\frac{k_2}{2}\right),$$

$$k_4=hf(x_0+h,y_0+k_3).$$

4 POPULATION MODEL

Mathematical models of population growth have been formed to provide an inconceivable significant angle of the true ecological situation. The meaning of each parameter in the models has been defined biologically (Pearl et al., 1920, Ruan et al., 2007). In case of insect population, birth and death rate of insect typically are not constant; instead, they vary periodically with the passage of seasons.

4.1 *Insect population model*

Suppose that an insect population P shows the seasonal growth model. The differential equation of the insect population model is given by –

$$\frac{dP}{dt}=kP\cos\lambda t \quad (7)$$

where k and λ are positive constants.

5 SOLUTION OF THE MODELS

5.1 *Solution of insect population model*

We solve the differential equation (7) by using matlab and we find out the exact solution for $P(0) = 1000, k = 2, \lambda = \pi, h = 0.1.$

a. Solution by LADM:

We now solve Eq. (7) by LADM with initial condition $P(0)$ using the Laplace transform and Adomian Decomposition Method.

$$\sum_{n=0}^{\infty} P_n(t) = P(0) + kL^{-1}\left[\frac{1}{s}L\left(\sum_{n=0}^{\infty} P_n(t)\cos\lambda t\right)\right]$$

The components $P_n(t)$ can be computed as

$$P_0 = P(0),$$

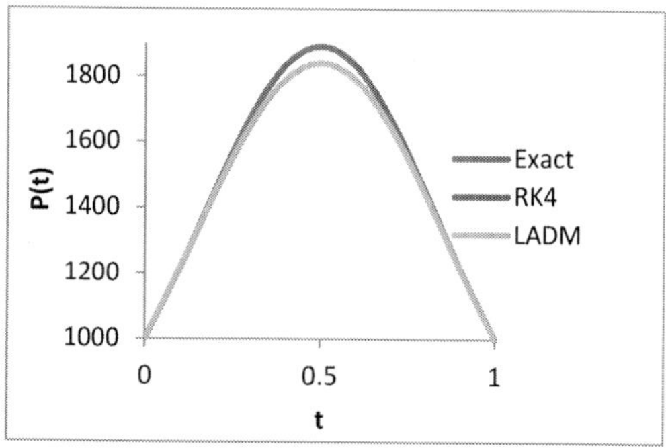

Figure 1. Evaluation among the exact solution and the solutions obtained by using RK4 and LADM methods for Model I.

$$P_{n+1}(t) = kL^{-1}\left[\frac{1}{s}L(P_n(t)\cos\lambda t)\right], n \geq 0$$

Therefore, the 3rd iterative solution is

$$P(t) = P(0) + \frac{kP_0}{\lambda}\sin\lambda t + \frac{kP_1}{2\lambda}\sin\lambda t.$$

b. Solution by RK4:

Consider the initial value problem

$$P'(t) = f(t, P(t)) = kP\cos\lambda t,$$
$$P(t_0) = P_0 = 1000 \textit{ and } h = 0.1.$$

First we define:

$$k_1 = hf(x_0, y_0),$$

$$k_2 = hf\left(x_0 + \frac{h}{2}, y_0 + \frac{k_1}{2}\right)$$

$$k_3 = hf\left(x_0 + \frac{h}{2}, y_0 + \frac{k_2}{2}\right)$$

$$k_4 = hf(x_0 + h, y_0 + k_3).$$

Then an approximation to the solution of initial value problem is made by using Runge-Kutta Method of order 4:

$$P_{n+1} = P_n + k.$$

where, $k = \frac{1}{6}(k_1 + 2k_2 + 2k_3 + k_4)$

Table 1. Numerical Comparison when initially we have $P(0) = 1000, k = 2, \lambda = \pi, h = 0.1.$

T	Exact Sol.	RK4	LADM (3-term)	ERK4	ELADM
0	1000.00	1000.00	1000.000	0.00E+00	0.00E+00
0.1	1217.407	1217.44	1216.073	3.71E–02	1.33E+00
0.2	1453.792	1453.82	1444.178	3.54E–02	9.61E+00
0.3	1673.589	1673.60	1647.568	1.87E–02	2.60E+01
0.4	1831.833	1831.81	1788.521	2.07E–02	4.33E+01
0.5	1889.596	1889.51	1838.842	8.10E–02	5.08E+01
0.6	1831.377	1831.29	1788.122	8.68E–02	4.33E+01
0.7	1672.7981	1672.78	1646.851	1.72E–02	2.59E+01
0.8	1452.8457	1452.91	1443.284	7.25E–02	9.56E+00
0.9	1216.4762	1216.58	1215.158	1.04E–01	1.32E+00
1	999.19568	999.260	999.1956	6.48E–02	3.86E-05

ELADM → Error term of LADM.
ERK4 → Error term of RK4.

c. Numerical Results and Discussion:

The numerical solutions obtained by using the RK4 and LADM are compared with the exact solution (for model-1). Then comparison is shown in Table 1. The result depicts the similar type of error between exact solution and LADM as the amount of error in the calculation between exact solution and RK4.

Once again, the numerical results show that LADM is of good accuracy. The graphical representations of the 3-models reveal that LADM is a better numerical technique for solving such kind of nonlinear equations. So from the three graphs, we can also say that LADM is a reliable numerical technique. Also table-2 and table-3 are rated of population before and after introducing the interspecific competition term in the insect population with respect to the same parameter value. And we perceive that the results in both table are almost same. So we can assume that after introducing interspecific competition term its balanced the ecological system and gives us a good population rate of insect population.

6 CONCLUSION

In this study, we describe a numerical solution of two insect populations. Here we introduce a new method called multistep LADM for the solution of the insect population model. In this approach, the solution is found in the form of a convergent power series with easily computed components. Here Suheil A. Khuri's LADM is successfully applied to the linear and nonlinear insect population models. It is apparent that the Laplace Adomian Decomposition method yields very accurate approximate solutions using only a few iterates. The above figure shows that, the exact solution, fourth-order Runge–Kutta. solutions for Eqs. (7) is compared with the solutions obtained by the LADM using only three iterates. In this paper, our objective has been to show that approximate solutions of nonlinear differential equation systems can be easily obtained without linearization, perturbation or discretization. The above method can solve the nonlinear cases without any complexity. This indicates that the LADM is a suitable method for use in solving nonlinear systems. So we can say that LADM is an accurate and reliable numerical technique for the solution of the linear and nonlinear population model. As can be seen clearly from the graph, LADM gives quite good results after a considerable time interval. This method is a very supporting method, which will undoubtedly be found applicable in broad applications. So it is clear that this method avoids linearization and biologically unrealistic assumptions, and provides an efficient numerical solution. Some simulation results are given as illustrations.

REFERENCES

Adomian G., 1993. *Solving Frontier Problems of Physics: The Decomposition Method*, Springer, New York.

Adomian G. 1992. Differential coefficients with singular coefficients, *Appl. Math. Comput.*, 47: 179–184.

Blumberg, A.A. 1968. Logistic growth rate functions, *J. Theor. Biol.*, 21: 42–44.

Chaudhuri, K. S. 1988. Dynamic optimization of combined harvesting of two species fishery. *Ecol. Model.*, 41, 17–25.

Erbe, L.H., Rao, V.S.H. 1986. Freedman, H.I.: Three-species food chain models with mutual interference and time delays. *Math. Biosci.*, 80, 57–80.

Jana, D., Bairagi, N. 2014. Habitat complexcity dispersal and metapopulations: Macroscopic study of a predator-prey system, *Eco. Complexity*, 17, 131–139.

Khuri, S.A. 2001. A Laplace decomposition algorithm applied to class of nonlinear differential equations, *J. Math. Appl.*, **4:** 141–155.

Kuang, Y., Fagan, W., Loladze, I. 2003. Biodiversity, habitat area, resource growth rate and interference competition. *Bulletin of Mathematical Biology* 65, 497–518.

Kiymaz, O. 2009. An algorithm for solving initial value problems using Laplace Adomian Decomposition Method, *Applied Mathematical Sciences*, 3(29–32): 1453–1459.

Lotka, A.J. 1925. *Elements of Physical Biology. The Williams and Wilkins Co., Baltimore.*

Malthus, T. R. 1798. An Essay on The Principle of Population (1st Edition, plus excerpts, 1803 2nd Edition), Introduction by Philip Appleman, and assorted commentary on Malthus edited by Appleman, *Norton Critical Editions*, ISBN 0-393-09202-X.

Pearl, R., Reed, L.J. 1920. On the rate of growth of the population of the United States since 1790 and its mathematical representation, *Proc. Natl. Acad. Sci.* USA, 6: 275–288.

Ruan, S., Ardito, A., Ricciardi, P., De. Angelis, D.L. 2007. Coexistence in competition models with density-dependent mortality, *Comptes Rendus Biologies*, 845–854.

Verhulst, P.F. 1838. Notice sur la loi que la population suit dans son accroissement, *Corr. Math. Phys.*, 10: 113–121.

Computational Science and Engineering – Deyasi et al. (Eds)
© 2017 Taylor & Francis Group, ISBN 978-1-138-02983-5

Multi stage fuzzy logic based Var compensator for enhanced reactive power and voltage stability in dispersed generation

Asit Mohanty & Meera Viswavandya
CET Bhubaneswar, Odisha, India

Prakash K. Ray
IIIT Bhubaneswar, Odisha, India

Sthita Pragyan Mohanty
CET Bhubaneswar, Odisha, India

ABSTRACT: This paper discusses Reactive Power regulation and stability analysis study in an isolated wind type Micro grid. It proposes a Multi stage fuzzy logic based SVC compensator for compensating Reactive energy in a wind based dispersed hybrid power system. The transfer function small signal analysed hybrid model of the isolated hybrid model has been considered which also consists of a solar cell, a double fed Induction generator based wind generator and a Synchronous generator based DG as a backup. The system performance is watched with variation of reactive power loads, variable wind energy input and non linear solar insolation input variation and the impact upon voltage stabilization and Reactive energy management is studied.

1 INTRODUCTION

Renewable energy resources are often available in a scattered way and are used in an independent manner. Sometimes they operate in grid inter connection mode (Hunter). Centralized management of power system is a complex phenomena while in the inter connected mode and therefore decentralized control methods are quite preferable as it improves the system performance. In order achieve the transient performance and dynamics of power system thyristor based FACTS devices have been considered for improved power management and stability. FACTS devices play an important role and are utilised for overall compensation of the hybrid power network for voltage management and angle stability issues (Bansal, 2007). Stability management and Reactive power control are the most important features in isolated hybrid power system

Wide variation of load and uncontrolled renewable source inputs lead to a unwanted voltage variations with fluctuations in system parameters (Bansal, 2002) Initially mechanically switched capacitors and reactors were used to enhance the steady-state energy transmission, by regulating the voltage associated along the transmission lines but for achieving dynamic compensation as well as real time compensation of power flow FACTS compensation have been implemented (Dong, Hingorani, 2000, Kaldellis).

Equations have been modified into linearised equations by linearising around an operating point and transfer function based on small signal equations have been developed for this particular problem. During connection with the conventional Proportional-Integral (PI) controllers in the linearised hybrid power system model. The PI controllers are increasingly incorporated in case of FACTS compensator for designing internal controllers for distribution and to mitigate voltage flicker.

PI and PID controllers are generally used for performance improvement (Petrov, 2002). In this particular work, a novel multi stage fuzzy logic approach is used to tune the parameters of PID controller for better system performance (Chaung, 2000). The multi stage Fuzzy logic PID controller is thus designed to enhance the system performance in a multiple ways.

2 MATHEMATICAL MODELLING

The stand alone hybrid system is supported by a small signal based transfer function model of SVC which manages and adjusts the reactive power need of the isolated hybrid system and improves system stability.

In this standalone hybrid power model, the total generations like Wind and PV and Diesel generator combinely act together near to the load and the total power output is connected to the common bus i.e. point of common coupling (pcc) with all the loads are connected to the bus bar as receivers of power as shown in Fig. 1. The remote system often faces unbalanced power exchange between the power sources and the main aim is to achieve reactive power management. During a small perturbation there happens a small change in system load bus voltage and the final equation of Reactive power exchange among (Synch Generator, SVC, IG,PV and Load) is expressed as shown in Fig. 3.

$$\Delta Q_{PV} + \Delta Q_{SG} + \Delta Q_{COM} = \Delta Q_L + \Delta Q_{IG}$$

The terminal voltage value gets deviated to a new value after a minute change in input load variation which indirectly affects the reactive power values associated with other elements of the isolated hybrid system. Reactive balanced equation of Micro grid

$$\Delta V(S) = \frac{K_V}{1+ST_V}[\Delta Q_{SG}(S) + \Delta Q_{COM}(S) + \Delta Q_{PV}(S) - \Delta Q_L(S) - \Delta Q_{IG}(S)] \tag{1}$$

$$\Delta Q_{SG} = \frac{V\cos\delta}{X'd\Delta E'q} + \frac{E'q\cos\delta\text{-}2V}{X'd\Delta V} \tag{2}$$

$$\Delta Q_{SG}(s) = K_a \Delta E'_q(s) + K_b \Delta V(s)$$

$$K_a = V\cos\delta / X'd, \quad K_b = E'q\cos\delta\text{-}2V / X'd \tag{3}$$

$$(1+ST_G)\Delta E'q(s) = K_e \Delta E_{fd}(s) + K_f \Delta V(s)$$

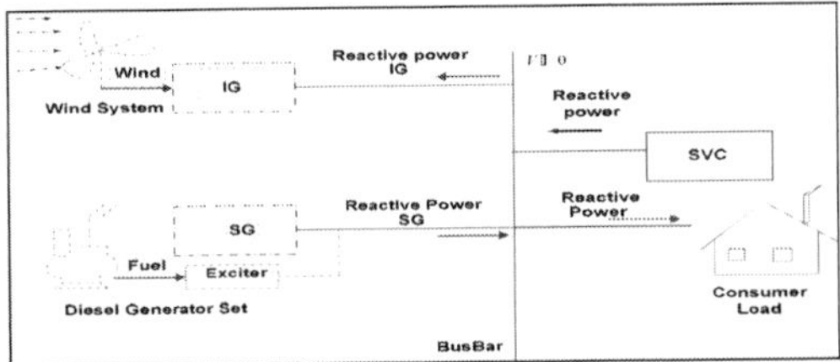

Figure 1. Wind based micro grid showing power exchange.

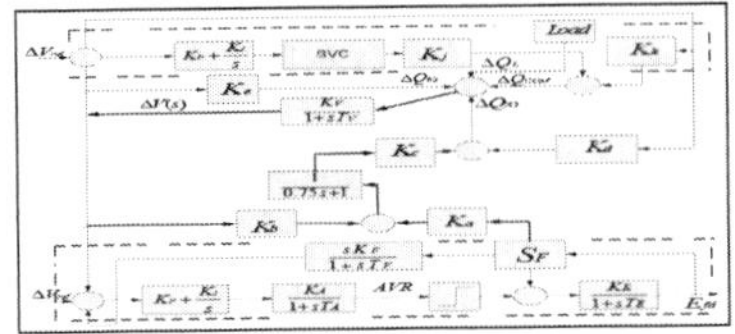

Figure 2. Linearised wind-diesel system with SVC.

$$T_G = \frac{X'_d T_{do}}{X_d}, K_e = \frac{X'_d}{X_d} \text{ and } K_f = \frac{(X_d - X'_d)\cos\delta}{X_d} \tag{4}$$

The reactive energy associated with the Induction generator varies with the input wind speed or wind energy input.

2.1 *A small signal induction generator*

The reactive energy delivered by the Induction generator based wind system at constant slip is

$$\Delta Q_{IG} = \frac{2V_t^0 X_T}{((1\text{-}s)R'_2/S\text{-}R_T)^2 + X_T^2} + \frac{-2(V_t^0)^2 X_T R_Y}{\{2R_Y(P_{1W}\text{-}P) + (V_t^0)^2\}(R_Y^2 + X_T^2)} \Delta P_{1W}$$

$$X_T = X_1 + X'_2 \quad \text{and} \quad R_T = R_1 + R'_2 \tag{5}$$

$$\underline{x} = [\Delta I_{dr}^{ref}, \Delta I_{dr}, \Delta V, \Delta\delta, \Delta E_{fd}, \Delta V_a, \Delta V_f, \Delta E'_q]^T$$

$$\underline{u} = [\Delta V_{ref}] \quad \underline{w} = [\Delta Q_L]$$

The state space representation of the wind-diesel system is written as

$$\dot{x} = Ax + Bu + Cw$$

3A *Fuzzy Logic Controller (FLC)*

Fuzzy control is a control mechanism based on Fuzzy Logic. A Fuzzy controller can work in three stages that are fuzzification, rule based and defuzzification.

3B *Multi stage fuzzy logic PID controller*

The proposed system implements a novel intelligent Multi stage Fuzzy Logic PID Controller for reactive power control of an isolated wind diesel hybrid power system. It consists of a pre-compensator in the first stage with two controllers (FLC), which tunes the parameters Kp and Ki of PI control block as per the required control of the proposed

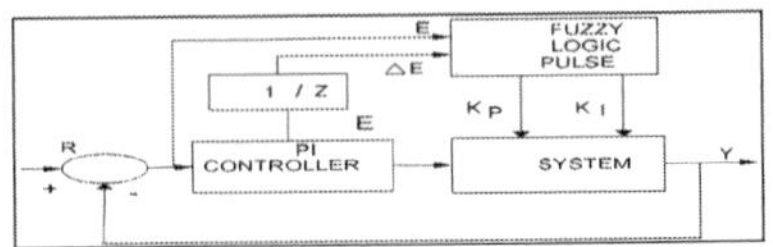

Figure 3. Fuzzy PI controller Block.

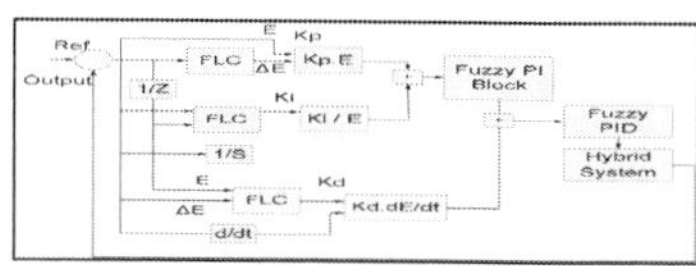

Figure 4. Block diagram multi stage fuzzy.

hybrid system. Still to improve the performance of the hybrid system, another FLC is included to tune the parameter Kd of derivative control block in the next stage. There are three Fuzzy Logic Controllers in this model, which forms the structure of Multi stage Fuzzy logic multi stage controller for improved control performance of the isolated hybrid system.

3C *Design of multi stage fuzzy logic PID controller*

Each Fuzzy logic controller in the proposed Multi stage Fuzzy logic PID Controller has two input variables (E & ΔE) and one output variable. The control signal from the pre-compensator of the Fuzzy logic PID Controller is $F_{P1} = K_p E + K_i \int E$. Control signal from the next stage of this proposed controller is $K_D = K_d \frac{dE}{dt}$. The output control signal from Multi stage Fuzzy logic PID Controller is $U = F_{PI} + F_D$

The control signal F_D is added to the previous stage for better performance. Where K_p K_i and K_d are the fuzzy tuned parameters of PID controller using Mamdani fuzzy inference model. The Multi stage Fuzzy logic PID Controller files (K_P.*fis*, K_i.*fis* and K_d.*fis*) for three Fuzzy Logic Controllers are created in fuzzy tool box. The FIS editor of the K_P K_i and K_d fis files are shown.

Two inputs are error $E(\Delta V)$ and $\Delta E(\Delta V)$. The rules for the proposed Multi stage Fuzzy Logic PID Controller are derived by considering the required performance of the wind-diesel hybrid power system and the properties of the PID controller. Fuzzy rules are framed for every combination of input state variables to tune the parameters K_p K_i and K_d by multi stage fuzzy logic technique. 49 rules are framed with seven linguistic labels for each input variable (E & ΔE). The Multi stage Fuzzy Logic PID Controller parameters K_p K_i and K_d are tuned by Mamdani fuzzy inference, which provide a non-linear mapping of PID parameters.

$$\mu(K_P) = \min(\mu(E), \mu(\Delta E))$$

$$\mu(K_i) = \min(\mu(E), \mu(\Delta E))$$

$$\mu(K_d) = \min(\mu(E), \mu(\Delta E))$$

Rules for K_p

$E/\Delta E$	NL	NM	NS	Z	PS	PM	PL
NL	VL	VL	VB	VB	MB	M	M
NM	VL	VL	VB	MB	MB	M	MS
NS	VB	VB	VB	MB	M	M	MS
Z	VB	VB	MB	M	MS	VS	VS
PS	MB	MB	M	MS	MS	VS	VS
PM	VS	MB	M	MS	VS	VS	Z
PL	M	M	VS	VS	VS	Z	Z

Rules for K_i

$E/\Delta E$	NL	NM	NS	Z	PS	PM	PL
NL	Z	Z	VS	VS	MS	M	M
NM	Z	Z	VS	MS	MS	M	M
NS	Z	VS	MS	MS	M	MB	MB
Z	VS	VS	MS	M	MB	VB	VB
PS	VS	MS	M	MB	MB	VB	VL
PM	M	M	MB	MB	VB	VL	VL
PL	M	M	MB	VB	VB	VL	VL

Rules for K_d

$E/\Delta E$	NL	NM	NS	Z	PS	PM	PL
NL	MB	MS	Z	Z	Z	VS	MB
NM	MB	MS	Z	VS	VS	MS	M
NS	M	MS	VS	VS	MS	MS	M
Z	M	MS	MS	MS	MS	MS	M
PS	M	M	M	M	M	M	M
PM	VL	MS	MB	MB	MB	MB	VL
PL	VL	VB	VB	VB	MB	MB	VL

Table 1. Optimal parameters.

Hybrid System	PID	multi stage fuzzy	fuzzy
Kp	65	35	31
Ki	6000	5250	5100
Rise	0.0943	0.0345	0.355
Overshoot	0.0198	0.0180	0.0126

4 SIMULATION AND RESULTS

The hybrid power system model has been simulated with several loads with undesired disturbances and inspected under various operating instances. At the time of simulation, several studies have been undertaken to judge the performance of different controllers under several unexpected circumstances. Participation of different controllers and their share in minimization of voltage deviation in the hybrid system have been judged. The overall system outcome under normal loading condition and random load input like (±5%) is examined to access about load sharing, compensator performance and the change in system terminal voltage

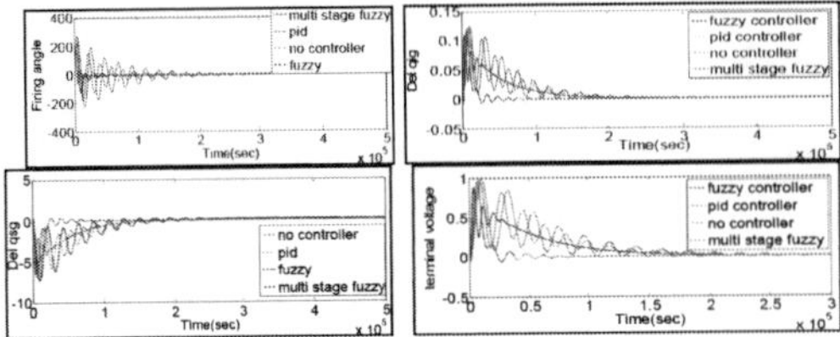

Figure 5(a-d). Comparison of results with multi stage fuzzy and conventional controller.

Table 2. Optimum PID, fuzzy and multi stage fuzzy.

	CONST SLIP		VARIABLE SLIP	
Hybrid system	*Kp*	*Ki*	*Kp*	*Ki*
Multi stage Fuzzy	31	5100	29	4900
Fuzzy controller	35	6000	33	5878
PID Controller	48	7800	45	7890
Without any control	65	13500	60	13120

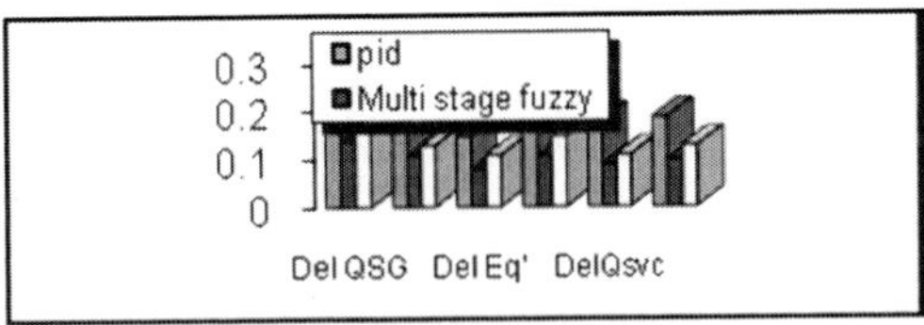

Figure 6. Chart showing maximum peak deviation.

profile. All Improvements have been noted down with the introduction of SVC controller as shown in Fig. 5. Furthermore studies have been carried out with random input of wind and solar insolation to examine the effects upon the system and to testify the robustness. The effectiveness of the controller and reactive power capability are noted. Improvements are done with Fuzzy and Multi stage fuzzy controller as in Fig. 5(a-d).

5 CONCLUSION

The reactive power management with dynamic stability of standalone hybrid power system has been simulated with SVC controller. The hybrid system outcome like settling time duration, peak overshoots magnitude etc. during steady state condition and also during transient condition are judged by the inclusion of SVC compensator. It is noticed that the time taken for settlement and the voltage stability of the hybrid power system get enhanced with the rise in power of wind energy system and the transient stability outcome is improved in addition of fuzzy and multi stage fuzzy based SVC controller.

Appendix-Parameters of hybrid power system.

Hybrid system		*Wind diesel system*	
Wind Turbine		*2 M V A*	
Diesel Engine (Synchronous)		*2 M V A*	
Base Energy		*2.5 M V A*	
Synchronous Generator		*2 MVA*	
X_d' *in pu*	*0.29*	X_d	*1.56 pu*
T_e	*0.55*		*27.8*
K_f	*0.5*	T_a	*0.05*
K_a	*40*	$T_{do'}$	*4.49 sec*

REFERENCES

Bansal, R. C. Bhatti, T. S., Kumar V.: Reactive Power Control of Autonomous Wind Diesel hybrid Power System using ANN. Proceeding of the IEEE Power Engineering Conference (2007).

Bansal, R. C.: Automatic Reactive Power Control of Autonomous Hybrid Power System PhD Thesis, Centre for Energy Studies, IIT Delhi December 2002.

Chaung, F. L., Duon, J. C.: Multi Stage Fuzzy Neural Network Modeling, IEEE Trans. of Fuzzy System 8 No. 2 (2000), 125–142.

Dong, Lee., Wang Li.: Small Signal Stability Analysis of an Autonomous Hybrid Renewable Energy Power Generation/Energy storage system Time domain simulations "IEEE Transaction for Energy Conversion, Vol 23, No 1, March Hingorani, NG., Gyugyi N,: Understanding FACTS, Concepts and Technology of Flexible AC Transmission System, New York, IEEE Power Engineering Society, 2000.

Hunter, R., Elliot, G.: Wind Diesel Systems. Cambridge University Press, Cambridge.

Kaldellis, J.: Autonomous energy system for remote island based on renewable energy sources., in proceeding of EWEC 99, Nice.

Petrov, M., Ganchev, I., Taneva, A.: Fuzzy PID Control of Nonlinear Plants, 2002 First International IEEE Symposium Intelligent Systems, Sep. 2002, pp. 30–35.

Computational Science and Engineering – Deyasi et al. (Eds)
© 2017 Taylor & Francis Group, ISBN 978-1-138-02983-5

Web based prediction and recommendation of products in electronic commerce using association rule learning and genetic algorithm

S. Thakur
Department of Computer Science and Engineering, MCKVIE, Howrah, India

J.K. Sing
Department of Computer Science and Engineering, Jadavpur University, Kolkata, India

ABSTRACT: Recommendation systems make suggestions about artifacts or products to a user. For instance, they may predict whether a user would be interested in purchasing a particular product. Conventional social recommendation systems collect ratings of artifacts from many individuals, and use nearest-neighbor techniques to make recommendations to a user concerning new artifacts. This paper presents a new method for recommending items/products to users based on customer's opinions. The proposed method is a variation of traditional collaborative technique, where predictions and recommendations are computed using a set of customer's opinions from an independent database. The prediction and recommendation is done by using weighted opinions according to their similarity to the user and association rule learning. To achieve better result, the database is optimized by excluding the weak entities using Genetic Algorithm (GA). The final prediction and recommendation is done using the association rule learning on the optimized database. The efficiency of the proposed method is evaluated on a product database of 10 products having 4,500 customer's opinions. The experimental results show that the proposed method overcomes some of the problems associated with collaborative filtering methods, as reported in the literature.

1 INTRODUCTION

E-commerce (Prasad 2007) in any form of business transaction in which the parties interact electronically rather than by physical exchanges or direct contact. It is one of those rare cases where changing needs and new technologies come together to revolutionize the way in which business is conducted. It enables companies to be more efficient and flexible in their internal operations, work more closely with their suppliers, be more responsive to the needs and expectations of their customers, select the best suppliers regardless of their geographical location and sell to a global market.

Recommendations are part of everyday life (Basu *et al.*, 1998). We usually rely on some external knowledge to make informed decisions about an artifact of interest or a course of action, when we are going to see a movie or going to see a doctor. This knowledge can be derived from social processes. At other times, our judgments may be based on available information about the artifact itself and our known preferences.

There are many factors which may influence a person in making these choices, and ideally one would like to model as many of these factors as possible in a recommendation system. There are some general approaches to this problem. In one approach, the user of the system provides ratings of some artifacts or items and the system make informed guesses about what other items the user may like. It bases these decisions on the ratings other users have provided. This is the framework for social-filtering methods (Hill, Stead, Rosenstein & Furnas 1995; Shardanand *et al,* 1995). In a second approach, the system accepts information describing the nature of an item, and based on a sample of the user's preferences, learns to predict which items the user will like (Lang 1995; Pazzani *et al.*, 1996). We call this approach content-based filtering, as it does not rely on social information (in the form of other user's ratings). Both social and content-based filtering can be cast as learning problems: in both cases, the objective is to learn a function that can take a description of a user and an artifact and predict the user's preferences concerning the artifact. Social-filtering systems perform well using only numeric assessments of worth, i.e., ratings. Information concerning the content of each artifact. Social filtering methods leave open the question of what role content can play in the recommendation process.

In order to draw user's attention and to increase their satisfaction towards online information search results, search engine developers and vendors try to predict user preference based on the user behaviour. Recommendation systems are implemented in commercial and non-profit web sites to predict the user preference. For commercial websites, accurate predictions may result in high selling rates.

The main functions of recommendation systems include analyzing user data and extracting useful information for further predictions. Recommendation systems are designed to allow users to locate the preferable items quickly and to avoid the possible information overloads. Recommendation systems apply data mining techniques to determine the similarity among thousands or even millions of data.

Collaborative filtering techniques have been successful in enabling the preferences in the recommendation systems (Hill *et al,* 1995, Shardanand *et al,* 1995). There are three major processes in the recommendation systems: object data collections and representations, similarity decisions, and recommendation computations. The challenge of conventional collaborative filtering algorithms is the scalability issue (Sarwar *et al.* 2000). Conventional algorithms explore the relationship among system users in large datasets. User's data are dynamic, which means the data vary within a short time period. Current users may change their behaviour patterns, and new users may enter the system at any moment. Millions of user data, which are called neighbours, are to be examined in real time in order to provide recommendations (Herlocker *et al.* 1999). Searching among millions of neighbours is a time-consuming process. To solve this, item-based collaborating filtering algorithms are proposed to enable reductions of computations because properties of items are relatively static (Sarwar *et al.*, 2001). Suggest is a Top-N recommendation engine implemented with item-based recommendation algorithms (Karypis. 2000, Deshpande *et al.*, 2004). Meanwhile, the amount of items is usually less than the number of users. In early 2004, Amazon Investor Relations (2004) states that Amazon.com Apparel & Accessories Store provides about one hundred and fifty thousand of items but has more than one million customer accounts that have ordered from this store. Amazon.com employs item-based algorithm for collaborative-filtering based recommendations (Linden *et al.* 2003) to avoid the disadvantages of conventional collaborative filtering algorithms.

Other technologies have also been applied to recommender systems, including Bayesian networks, Clustering, and Horting. Bayesian networks create a model based on a training set with a decision tree at each node and edges representing user information. The model can be built off-line over a matter of hours or days. The resulting model is very small, very fast, and essentially as accurate as nearest neighbour methods (Niu *et al.*, 2002). Bayesian networks may prove practical for environments in which knowledge of user preferences changes slowly with respect to the time needed to build the model but are not suitable for environments in which user preferences model must be updated frequently (Shapiro *et al.*, 1991).

Clustering techniques works by identifying groups of users who appear to have similar preferences. Once the clusters are created, predictions for an individual can be made by averaging the opinions of other users in that cluster. Clustering techniques usually produce less-personal recommendations than other methods, and in some cases, the clustering have worst accuracy than nearest neighbour algorithms (Han *et al.*, 2005).

In this paper, we present a novel method for recommending items/products to users based on customer's opinions. In the proposed method the predictions and recommendations are made using a set of customer's opinions from an independent database by using weighted opinions according to their similarity to the user and association rule learning. To achieve better result, the database is optimized by excluding the weak entities using Genetic Algorithm (GA). The final prediction and recommendation is done using the association rule learning on the optimized database. The remaining part of the paper is organized as follows. Section 2 describes Association Rule Learning, a well known research method used in the proposed prediction system. Section 3 presents the proposed prediction and recommendation system along with the idea of applying Genetic Algorithm (GA). The experimental results on the databases are presented in Section 4. Finally, Section 5 draws the concluding remarks.

2 ASSOCIATION RULE LEARNING

In data mining, association rule learning is a popular method for discovering interesting relations between variables in large databases. Piatetsky-Shapiro describes analyzing and presenting strong rules discovered in database using different measures of interestingness (Han *et al.*, 2005, Agarwal *et al.*, 1993, Jochen *et al.*, 2000). Based on the concept of strong rules, Agarwal (Agarwal *et al.*, 1993) introduced association rules for discovering regularities between products in large scale transaction data recorded by Point-of-Sale (POS) systems in supermarkets.

Association rules are usually required to satisfy a user-specified minimum support and a user-specified minimum confidence at the same time [18]. Association rule generation is usually split up into two separate steps:

1. First, minimum support is applied to find all *frequent item sets* in a database.
2. Second, these *frequent item sets* and the minimum confidence constraint are used to form rules.

Table 1. Database with 5 items with 5 records.

Transaction ID	Shirt	Trouser	Saree	Shoe	Tie
1	1	1	0	0	1
2	0	1	1	0	0
3	0	0	0	1	0
4	1	1	1	0	0
5	0	1	0	0	1

While the second step is straight forward, the first step needs more attention. Finding all *frequent item sets* in a database is difficult since it involves searching all possible item sets (item combinations) (Agarwal *et al.*, 1993, Jochen *et al.*, 2000). Efficient algorithms e.g., Apriori and Elcat can find all frequent item sets.

3 PROPOSED PREDICTION AND RECOMMENDATION SYSTEM

The objective of the proposed Prediction and Recommender system is to present the customers a list of customized products, and allow the customers to choose items/products from the stores online as shown in Fig. 1. For this, we follow a mathematical approach based on Association Rule learning [16, 18, 19], and Genetic Algorithm (GA) [20, 21, 22, 23, 24], which is used to optimize the product database to deliver better prediction and recommendation about the products.

The information pertai*ning to the products are stored o*n an RDBMS at the server side. *The* application which is deployed at the customer database, the details of the items are brought forward from the database for the customer view, based on the selection through the menus and the database of all products are updated at the end of each transaction. The developed system consists of two separate databases as follows:

Product database—This database stores real data that are collected from various garments shops for statistical analysis. It contains the following tables namely *customer_info*, *customer_visit_info*, *item_details*, *place_info* and *purchase_details*. The *product_view* is theview of product database. The view of the product database is shown in Fig. 2.

E_Product database—This database is used to store *user_login_info*, *user_reg_info,* which stores registration data provided by the user at the time of registration along with *user_choices,* i.e. the choice the user has input for the products. The view of the E_product database is shown in Fig. 3.

The database Product database is the main source of comparison of our records with the user choices from E_product database. The comparison of products results in statistical outcome of a particular rule. The rule is calculated by the choices and the

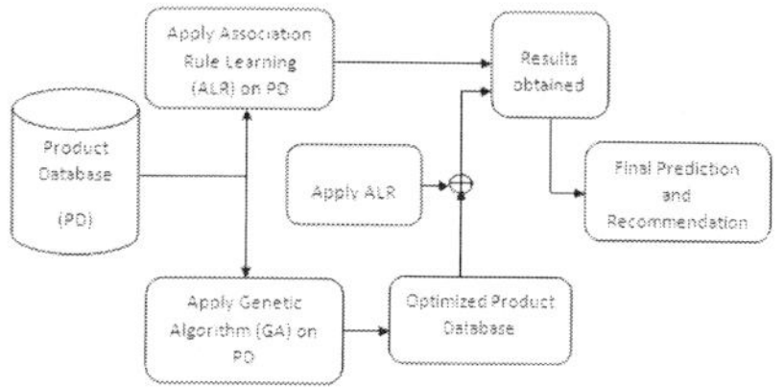

Figure 1. Block Diagram of the Proposed Prediction and Recommendation System.

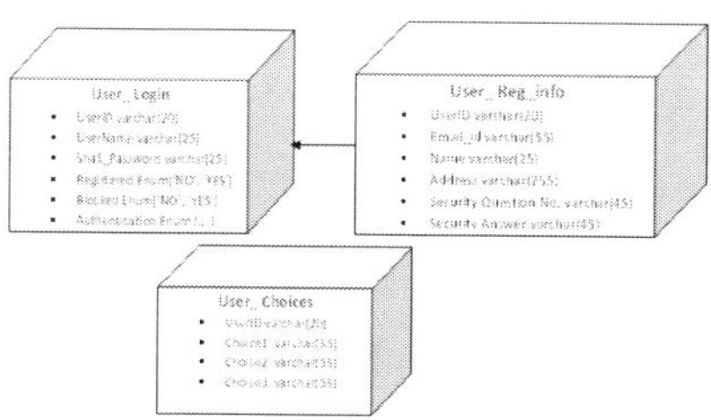

Figure 2. View of User Login and Registration for E_Product database.

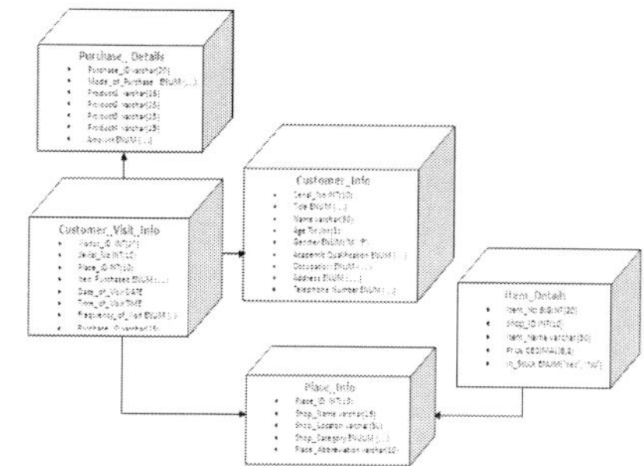

Figure 3. View of E_Product database.

number of customers falls in the selected category is estimated, expressed in probabilistic decimal form, which can be changed later to percentage.

Computing Predictions—The proposed online Prediction and Recommender system uses the Association Rule Learning. First, we have collected datasets of customers from various shopping malls and stored them in a database called product database. This product database acts as a knowledgebase to the rest of the work. Whenever a new customer makes registration in this system, he/she is asked to enter his choice about the product they like to purchase along with two options of prediction rules.

Algorithm 1. Main steps of the proposed prediction and recommendation technique

1. Create a Product Database (PD) for all the items.

2. Computation of Prediction Using Association Rule Learning.
3. Apply Genetic Algorithms (GA) on product database (PD)
   a. Generate an initial population.
   b. Calculation of fitness value—X^* for every chromosome.
   c. Apply Fitness function and use it as a threshold value.
   d. Chromosomes whose fitness value is below threshold value are to be removed from the population.
   e. Generate the next population by per forming selection, reproduction, and crossover operations.
   f. Select the mate randomly and apply crossover operation, generate new population and calculate fitness value X.
   g. Compare the best string X of the current population with X^*.
   h. Apply mutation operation and calculate expected value of mutation and determine its effect on chromosomes.
4. Optimized database is obtained after applying genetic algorithm on Product Database (PD)
5. Go to Step 2 and apply Association Rule Learning on optimized Product database.
6. Compute predictions and recommendations for items closely related.
7. End.

3.1 *Computation of prediction using association rule learning*

In this option, we calculate the probability or chances that the item or item set selected by the customer appear in the database and can also gauge a measure about items, which are popular among the customers. This option calculates the prediction (probability) on the basis of partial matching system, and we also calculates *support*, *confidence*, *conviction* and *lift* for different item sets using Association Rule Learning (ARL) as discussed in Section 2.

Table 2. Chromosome representation and calculation of fitness function.

No. of chromosomes	Mating pool after reproduction (Crossover site shown)	Mate Randomly selected	New population	Fitness value
1	1 1 1 | 1 0	2	1 1 1 0 0	28
2	1 1 1 | 0 0	1	1 1 1 1 0	30
3	1 1 | 0 0 0	4	1 1 0 1 0	26
4	1 1 | 0 1 0	3	1 1 0 0 0	24
5	1 1 0 | 1 1	6	1 1 0 1 1	27
6	1 0 0 | 1 1	5	1 0 0 1 1	19
7	1 0 | 1 0 0	8	1 0 1 0 1	21
8	1 1 | 1 0 1	7	1 1 1 0 0	28

3.2 *Computation of prediction using Genetic Algorithms (GA) and association rule learning*

In this option, we calculate the prediction using association rule learning, where *support*, *confidence*, *conviction* and *lift* have been calculated for different item sets. To achieve better prediction result, the database is optimized by excluding the weak entities using Genetic Algorithm (GA).

4 EXPERIMENTAL RESULTS

The Product prediction and recommendation has been done using the Product database and applying association rule learning and genetic algorithm. Initially the user/customer is requested to input the items in which the user is interested in for e.g. the customer has given choices as shirt, trouser and saree. The task of Prediction option 1 is to search the database whether the items requested by the user is available in our Product database. Further we also calculate the probability based on partial matching system, that the customer will buy the items our system recommends. Finally, we also determine the support i.e. the exact number of full match items of the given user choices and the results are shown in Table 4.

The support, confidence, lift and conviction are calculated, using ARL and applying GA, as defined in section 2 and section 4 respectively. The

Table 3. Fitness value, new population after cross over applied.

Number of chromosomes	Initial Population	Calculation of fitness value—X	Fitness function $f(x)=\sum(X)/N$
1	1 1 1 1 0	30	
2	1 1 1 0 0	28	
3	1 1 0 0 0	24	
4	0 0 0 1 1	3	
5	1 0 0 0 0	16	
6	0 1 1 1 1	15	
7	0 0 0 0 1	1	
8	1 1 0 1 0	26	
9	1 1 0 1 1	27	18.125
10	1 0 0 1 0	18	
11	1 0 0 1 1	19	
12	1 0 1 0 0	20	
13	0 0 1 0 0	4	
14	0 0 1 0 1	5	
15	1 1 1 0 1	29	
16	1 1 0 0 1	25	
		$\sum(X)=290$	

Table 4. Results of partial and full match items from product database.

Item set (X)	Items matched	Total number of items matched	Probability (X)	Support (X)
(Shirt/T Shirt, Trouser /Jeans, Saree)	Full match	290	1.460	290
	2 items matched	1520		
	1 item matched	2630		
	No items matched	80		
(Shirt/T Shirt, Trouser /Jeans, Shoe)	Full match	230	1.790	230
	2 items matched	2420		
	1 item matched	2510		
	No items matched	160		
(Shirt/T Shirt, Trouser /Jeans, Tie)	Full match	380	1.626	380
	2 items matched	1700		
	1 item matched	2750		
	No items matched	105		

Table 5. Results after applying ARL in Row 1 and with ARL in Optimised database on Row 2 for items (Shirt/T-Shirt, Trouser/Jeans, and Saree).

Support (Shirt /T-Shirt, Trouser/Jeans, and Saree) (X U Y)	Support (Shirt/T-Shirt, Trouser/Jeans) (X)	Support (Saree) (Y)	Confidence X => Y	Lift X => Y	Conviction X => Y
0.186	0.533	0.420	0.348	0.834	0.889
0.276	0.723	0.383	0.381	1.00	0.996

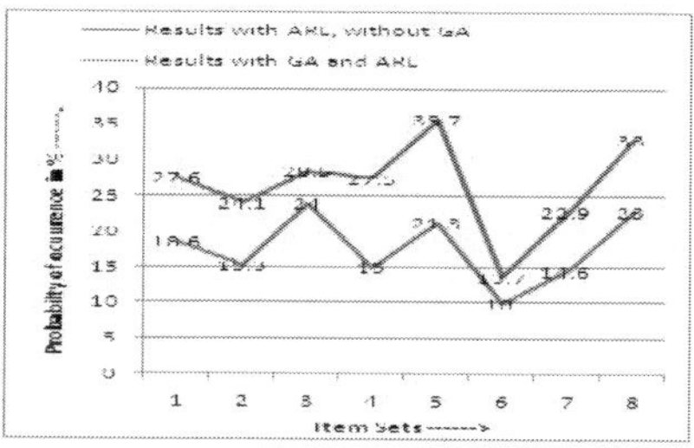

Figure 4. Comparative results of recommendation.

results for the same for different combinations of item sets are shown in Table 5. The results of the conviction shows 24 that for different combinations of item set the value should not exceed 1.0. If the results of conviction exceeds above 1.0 then the rule makes incorrect prediction, as the association between the items sets are based on purely random choice.

It has been observed that the results obtained for support (X) after applying GA and ARL on item sets is better than the results obtained after applying simply ARL. The percentage support (X) or probability of occurrence for the item sets applying GA and ARL shows that the combination (Shirt/T Shirt, Trousers/Jeans, Belt), (Shirt/T Shirt, Trousers/Jeans, Cosmetics) and (Shirt/T Shirt, Trousers/Jeans, Tie) are the best three recommended item sets and (Shirt/T Shirt, Trousers/Jeans, Socks) is the least recommended item sets.

Similarly the percentage support (X) or probability of occurrence for the item sets after applying ARL without GA shows that the combination (Shirt/T Shirt, Trousers/Jeans, Tie), (Shirt/T Shirt, Trousers/Jeans, Cosmetics) and (Shirt/T Shirt, Trousers/Jeans, Belt) are the best three recommended item sets and (Shirt/T Shirt, Trousers/ Jeans, Socks) is the least recommended item sets. In this paper, a unique structure of Work Function Engineered Gate (WFEG) Recessed S/D SOI-MOSFETs with high k dielectric region in between channel and drain end has been developed. Due to the combined benefits of both WFEG recessed S/D structure and high k dielectric material in IR region the proposed structure confirmed the better device performance in terms Drain Induced Barrier Lowering (DIBL) and Hot Carrier Effects (HCE). Comparative performance analysis of SOI and SON in terms of surface potential distribution, threshold voltage behavior and electric field profile reveals that SON has better immunity against various SCE's than its counterpart. Therefore, the proposed recessed S/D IR SOI/SON structure can be a potential tool to optimize the desired performance of the device parameter in the nanometer regime.

5 CONCLUDING REMARKS

In the proposed Product Prediction and Recommendation system implementation is done in three levels. These are (i) building the JSP pages, (ii) creating database and connectivity, (iii) creating the MVC (Model View Controller) design using Servlets and Beans. The complete system is developed in Java 1.6.0_16 using Java as front end as it could run only on Java 1.5 and onward versions correctly. In the backend database is designed using Oracle 11 g Release 1. Application Server used is Apache Tomcat 6.0, the Web browser is Internet Explorer 8.0, database connectivity is done using ODBC, and Java EE libraries is used using Apache Tomcat J2EE library. In the end, we conduct experiments to validate our methods. The support signifies that after buying Shirt/T Shirt and Trouser/Jeans, a customer is likely to buy belt, cosmetics or socks comes to *be 0.357, 0.330 and 0.285* respectively, and then we can make it a rule if ***probability*** exceeds ***0.5.,*** and the same is the value of ***t*** (i.e. threshold). **Probability** that a customer purchase the items the developed system recommended = **0.357 or 35.7%.** The entire work is done on private database and in future work the authors will make the database available in public domain.

ACKNOWLEDGEMENTS

The author S. S. Thakur likes to thanks the Director and Principal, MCKV Institute of Engineering, Liluah for giving permission to use the Laboratory during the proposed work. The author also like to thanks the students of CSE Final Year, for collection of customers purchase data from different shopping complexes, which was really a tough task and the same has been used for the creation of private database.

REFERENCES

Abdelmaguid, T. F., Dessouky, M. M. (2006) A genetic algorithm approach to the integrated inventory-distribution problem, International Journal of Production Research, 44, 4445–4464.

Agrawal, Imielinski, T., Swami, A. (1993) Mining Association Rules between Sets of Items in Large Databases in: Proceedings of the SIGMOD Conference, 207–216.

Basu, C., Hirsh, H., Cohen, W. W. (1998) Recommendation as classification: Using social and Content-based information in recommendation, in: Proceedings of the 15th National Conference on Artificial Intelligence (AAAI98), Madison, WI

Booker, L. B., Goldberg, D. E., Holland, J. F. (1990) Classifier systems and Genetic algorithms in Machine Learning, Paradigms and Methods, Carbonell, J. G. (ed.), The MIT Press, Boston, MA, 235–282.

Breese, J. S., Heckerman, D., Kadie, C. (1998) Empirical Analysis of Predictive Algorithms of Predictive Algorithms for Collaborative Filtering, in Proceedings of the 14th Conference on Uncertainty in Artificial Intelligence, 43–52.

Deshpande, M., Karypis, G. (2004) Item-based top-N recommendation algorithms, ACM Transactions on Information Systems (TOIS), 22(1), 143–177.

Goldberg, D. E. (1989) Genetic Algorithms in Search Optimization, and machine Learning, Addison—Wesley.

Han, J., Kamber, M. (2005) Data Mining Concepts and Techniques, Elsevier.

Herlocker, J., Konstan, J., Borchers, A., Reidl, J. T. (1999) An algorithmic framework for performing collaborating, in: Proceedings of ACM SIGIR 99, ACM, xvi + 339, 230–237.

Hill, W., Stead, L., Rosenstein, M., Furnas, G. (1995) Recommending and Evaluating Choices in a Virtual Community of Use, in: Proceedings of the CHI-95 Conference, Denver, CO.

Holland, J. H. (1975) Adaptation in natural and artificial systems, University of Michigan Press.

Jochen, H., Ulrich, G., Gholamreza, N. (2000) Algorithms for association rule mining—A general survey and comparison. SIGKDD Explorations, 2(2), 1–58.

Karypis, G (2000) Evaluation of item-based top-N recommendation algorithms, Technical Report 00–046, University of Minnesota.

Lang, K. NewsWeeder: Learning to filter netnews, in Proceedings of the Twelfth International Conference in Machine Learning, Lake Taho, California: Morgan Kaufmann.

Lawrence, D. (1991) Handbook of Genetic Algorithms. Van Nostrand Reinhold.

Mitchell. M. (1998) An introduction to genetic algorithms, MIT Press.

Niu, L., Yan, X. W., Zhang, C. Q., Zhang, S. C. (2002) Product Hierarchy-Based Customer Profiles for Electronic Commerce Recommendation, in: Proceedings of the 1st International Conference on Machine Learning and Cybernetics, Beijing, 1075–1080.

Pazzani, M., Muramatsu, J., Billsus, D., Syskill, Webert (1996) Identifying interesting web sites, in: Proceedings of the Thirteenth National Conference on Artificial Intelligence.

Piatetsky-Shapiro, G., Frawley, W. J. (1991) Discovery, analysis, and presentation of strong rules, Knowledge Discovery in Databases, AAAI/MIT Press, Cambridge, MA.

Prasad, B. (2007) A knowledge-based product recommendation system for e-commerce, International Journal of Intelligent Information and Database Systems, 1(1), 1751–5858.

Sarwar, B. M., Karypis, G., Konstan, A. J., Riedl, J. T. (2000) Analysis of recommendation algorithms for E-commerce, in: Proceedings of the 2nd ACM Conference on Electronic Commerce, ACM, vii+271, 158–167.

Sarwar, B. M., Karypis, G., Konstan, A. J., Riedl, J. T. (2001) Item-based collaborative filtering recommendation algorithms, in: Proceedings of the 10th International World Wide Web Conference, 285–295.

Shardanand, U., Maes, P. (1995) Social Information Filtering: Algorithms for Automating "Word of Mouth", in Proceedings of the CHI-95 Conference, Denver, CO.

Stahl, A. (2008) Learning feature weights from case order feedback, Lecture notes in Artificial Intelligence, Springer 78–84.

Computational Science and Engineering – Deyasi et al. (Eds)
© 2017 Taylor & Francis Group, ISBN 978-1-138-02983-5

The challenges of I-voting and its remedy

Kazy Noor E. Alam Siddiquee
Department of Computer Science and Engineering, University of Science and Technology Chittagong, Chittagong, Bangladesh

ABSTRACT: Voting is an important part for any democratic country. Citizen elects their leader through the voting process to lead the country. This vote capturing process can be manually or electronically. Manually means traditional ballot paper voting to go in the voting center to cast the vote. Electronically means voting using electronic devices such as computers. We have introduced here such a voting system that is called I-voting. I-voting that stands for Internet Voting therefore it means participate in voting through Internet from home and abroad by the citizen of any country. This paper identifies the challenges for the growth of an I-voting concept and prefaces starting place and describes the requirements behind the challenges presented and finally a proposed solution and implementation techniques of I-voting system.

1 INTRODUCTION

Voting is the most essential program in democratic options. Therefore, elections and referenda should be available to as many people as possible. It is especially complex for people living abroad to get involved in elections. Thus, digital elections, known as E-voting, is acquiring many more general's attraction (Organization for the Advancement of Structured Information Standards (OASIS), January 2005). The term I-voting is a typical term that represents any kind of voting in electronically kind via Internet. Thus, E-voting includes voting by cell phone as well as digital voting devices in voting workplaces. This perform provides with E-voting in the feeling of voting with the use of a typical pc via the Internet (Organization for the Advancement of Structured Information Standards (OASIS), January 2005 (OSCE/ODIHR, March 2007)). Regardless of how the E-voting is performed and what kind of technological innovation is used, maintaining the voter's option represented by the voter's selection an inviolable key is most essential.

Security is another concern here as it is considered as the driving issue to alter polling results. A threat tree (Pardue, Yasinsac, & Landry, 2010) was proposed and it was vetted by a panel of officials, experts, attorneys, academicians and other representatives of governments. The Canton of Zurich government investigated development of a system (Beroggi, 2014) for using in polling system to start internet voting system and that intended for inclusion of more cantons with federal government to provide a safety standard in i-voting in Switzerland. Another research in Canada showed that i-voting was a patchwork which proved a solid concentration in some areas and no other penetration in other places of federal structure in Canada (Goodman & Jon H. Pammett, 2014). Kevin and Xukai (Butterfield & Zou) traced out several problems on e-voting and kept a solution on internet voting concept.

2 BACKGROUND OF THE RESEARCH

E-Government is about the modernization of team control providing it better group society and companies through the use of details and relationships among technological issues (National Election Committee (Estonia), 2005). In other circumstances, E-Government is developed to decrease the limitations for people getting team companies. E-Government has had a lengthy effective record. E-governmental activities like electronic voting (M¨uller & Prosser), e-taxation, e-procurement etc. was excelled after the pioneering concepts of electronic processing at all aspect of governance. From a European viewpoint, to be able to create E-Government within the European Cooperation (Council of Europe Committee of Ministers, 2004), the European Quantity has released activity plans—the very first E-Government action strategy was released at the Seville the best possible 2000—and several collaborative initiatives and tasks (Council of Europe, 1950). Consequently, the E-Government functions are very developed in many individual states of the European Cooperation. Norway (Council of Europe) in particular has definitely followed its E-Government strategy since the beginning and thus is these days one of essential nations in European nations with regard to E-Government (Council of Europe, March 1952).

Past and persistent E-Government tasks in European nations have had a lot of success at all appropriate levels, at particular, organizational and lawful levels, not only within individual declares but also in a Europe-wide dimension. Consider the success in the position of recognition control, for example, recognition management is one of the most essential E-Government subjects since almost any E-Government application dealing with people or companies attracts on details. Thus, most of the individual declares introduced electronic recognition control ideas developed to their nationwide needs and best for their national legislation and organizational requirements. Furthermore, at the European stage, the individual declares collaboratively addressed the interoperability of current nationwide recognition control solutions to be able to achieve smooth E-Government in a cross-border way. The quantity incredibly allows these developments by releasing financed tasks in digital recognition appropriate control and interoperability issues. Consequently, many nations run amazing recognition control techniques and various studies and ideas on how to accomplish interoperability at a worldwide stage have been developed. The example of recognition control definitely shows the current position of E-Government developments: E-Government is much developed already and interoperability is topic of current improvements. Thus, E-Government these days seems to provide an amazing foundation for programs. The possibility of E-Government is to carry team company's better people. From this factor of view, the next sensible stage forward is to provide people effective contribution in democratic techniques by electronic means. E-Government can be seen as one of the very first E-developments, but apart from E-Government, other professions appropriate to the connection between a nation and its people are available. Therefore, the term "E-Governance" has been offered as applying the going above, with E-Government as one of its sub-disciplines. The self-discipline appropriate to effective contribution is denoted as the e-democracy.

3 CHALLENGES BEHIND THIS SYSTEM

Following challenges need focus

1. Implementation of a remote electronic voting (Müller & Prosser) solution must be preceded by a holistic and integrated project planning including e.g. feasibility studies, cost benefit analysis, system testing and user acceptance.
2. There is a need to set up instruments to support efficient capacity building and knowledge sharing among practitioners in the field. In some cases, emails or faxes may be considered for internet voting (Pardue, Yasinsac, & Landry, 2010 (Beroggi, 2014)); but these may lead threats. The best form of Internet voting is to cast votes through an official electronic ballot across internet from computers (Pardue, Yasinsac, & Landry, 2010). SMS was suggested in some cases but in Switzerland (Beroggi, 2014) it was later on removed as a technique of i-voting. Researchers were rather concerned (in 3 years study) on how to create basement of trust by making the system verifiable individually and universally, and how to ensure transparency at all stages of polling system (Beroggi, 2014).
3. More resources should be allocated for professional training and education for members of election commissions and authorities as well as other personnel involved in e-voting projects (Butterfield & Zou) as it was discovered that traditional voting systems do not provide safety to voters and it is not transparent (Butterfield & Zou). As voting ballots are not made public and thus its acceptance and reliance is got lessened by the system itself (Butterfield & Zou).
4. Campaigning on I-voting telling by its benefits and necessities.
5. Remote e-voting (Butterfield & Zou) based on internet can benefit from closer integration with other electronic services due to increased cost-efficiency, learning effects, and better acceptance. Some of the problems regarding electronic voting referred (Butterfield & Zou) were stationary booth systems, balloting machines where citizens had to go physically to vote. In some cases, compromising of the privacy of voters, altering of potential tallies, threats to voters and other relevant security issues were identified (Butterfield & Zou).
6. It is a good practice that the development process of e-voting systems adapts to the target user with regards to demands, skills, and attitudes (Butterfield & Zou).
7. Fair risk assessment and rational concessions can be helpful in mitigating security concerns pertinent to remote electronic voting (Pardue, Yasinsac, & Landry, 2010). A secret sharing based protocol INFOCOM, Digital Vote-by-Mail Service (DVMS), Helios (Butterfield & Zou) were proposed for maintaining security issues.
8. The level of political maturity in the society needs to be accounted for and actions to expose the value of electronic votes should be taken up. In Canada, I-voting casted in 2014 was surrounded by activities led by local communities (Goodman & Jon H. Pammett, 2014) where the nature of the divided jurisdiction and division of electoral powers has prohibited the use of Internet voting, but the manifestation of supportive legislation and local autonomy

has allowed its implementation (Goodman & Jon H. Pammett, 2014).

9. The organizational routines in e-voting projects need to be improved with the view of enhancing their transparency and good management practices.

4 CONCEPT OF THE PROPOSED SOLUTION

In this study we have proposed an online software solution through which anyone can cast his/her vote to the desired object using his mailing account. Java is preferred as it was preferred in earlier systems (Butterfield & Zou). The user needs to register first having created a mail account where he/she will be confirmed with an acknowledgement notice via the software generated message. During the registration process a user needs a valid national id number or passport number to identify as a unique. The basic principle is that any user with a certain email account can cast his/her vote once. More than once for that session is not allowed and the system is powered to detect that issue. Thus an online ballot (Goodman & Jon H. Pammett, 2014 (Butterfield & Zou)) can be confirmed.

5 IMPLEMENTTION TECHNIQUES AND MEASUREMENTS

From a notional viewpoint, e-voting can be shared into 3 phases:

- *Pre-Voting part*
- *Voting or Balloting part*
- *Post-Voting part*

5.1 *Pre-voting part*

As depicted within the person perspective of the OASIS advanced stage vogue, the numerous comes during this phase are:

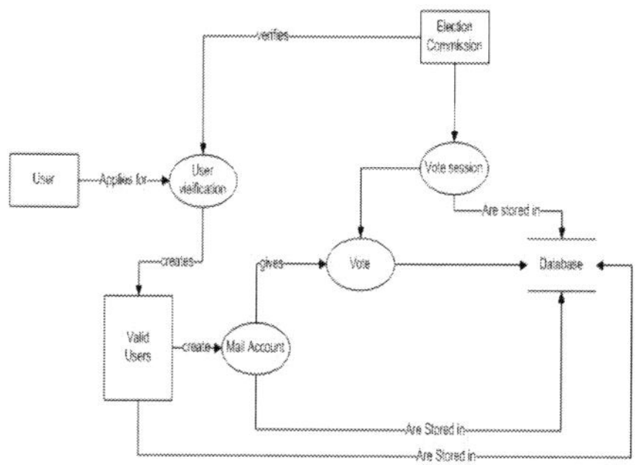

Figure 1. Voting process at a glance.

1. *Individual Nomination Method*
   - Individual Nomination
   - Individual Response
   - Creation of the candidates List
2. *Citizen Language up Method*
   - Citizen Registration
   - Creation of the choice Record

There could also be numerous strategies to become hand-picked as associate individual for election counting on the national legislative. A candidate has got to fulfill some lawful limitations, e.g. she should be sufficiently old, etc. The proposed candidate might need to require her nomination. She has got to select whether or not to require or decline her nomination. Lastly, the nomination procedure outcomes in a very listing containing all candidates, called the candidate list.

5.2 *Voting part*

Based on the outcomes of the pre-voting stage, the vote stage permits all qualified voters to form their decisions and throw their ballots. Thus, by the utilization of the election list, the citizen has got to verify her as associate qualified citizen and her or he has got to throw her personal elect. The fashion does not limit vote to digital vote solely. It's the voter's option to understand with that route she prefers to throw her elect. However, the first chance during this treatise is that the on-line because the digital vote route. In this stage the voter can register through online against her identity number and become eligible as a voter.

There are some criteria to become an eligible as a voter. The identity number can be national identification number or passport number and there will have an age matters. The age of 18 or above only will be eligible for voting. Voting confirmation will be generated automatically through checking the conditions set by the election authority or commission.

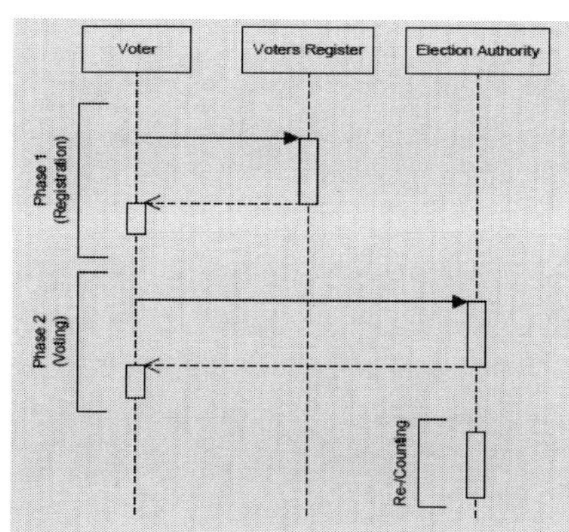

Figure 2. Voter's registration in two phases.

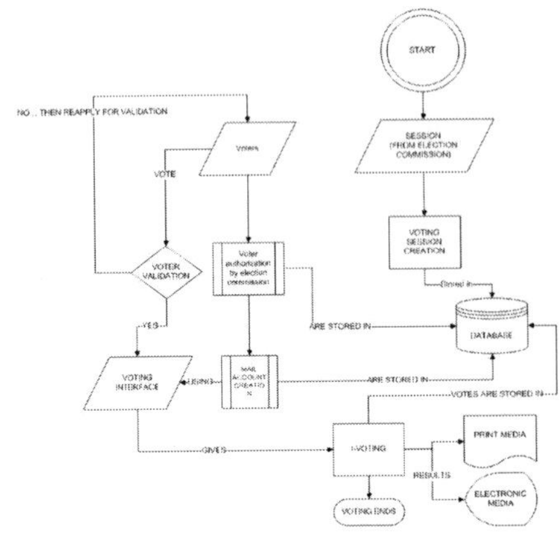

Figure 3. Flow chart of an i-voting system.

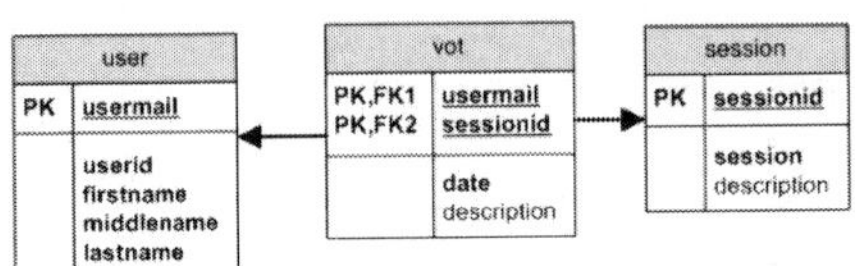

Figure 4. A sample relational framework of the database system for I-voting system.

5.3 *Post-voting part*

The post-voting stage offers with the luscious areas of the e-voting procedure. This stage primarily covers counting and outcome confirming. In inclusion to maintaining track of systems, associate analysis program is awfully needed for the system. Such a software system provides the auditing class and also the election authorities with numerous reviews.

6 CONCLUSION

The presented work is a review on I-voting system in any country (both in developed and under developed countries). This is an electronic way to capture a vote from citizens for E-Government. Still we need more research and consciousness to get its services. Therefore it is an ongoing process and largely its successful will depend on public consciousness and its acceptance of the user. Our future work may include special filters on selection, segregation or partitioning of voters (for setting constraints) for more securing the poling system. Classifiers of support vector machine can aid us a lot on creation of disjoint classes and setting enablers for being a voter. We are focusing on historical contents and our future intention is on primary data and sampling a vast around in both developed and under developed countries.

REFERENCES

Australian Parliamentary Joint Committee on Corporations and Financial Services. (2005). *Inquiry into Corporate Responsibility and Triple-Bottom-Line reporting for incorporated entities in Australia.* Australian Parliamentary Joint Committee on Corporations and Financial Services.

Beroggi, G. E. (2014). *Internet Voting: An Empirical Evaluation,.* Spring Analytica and the Zurich Business School, IEEE Computer Society 0018–9162/14 © 2014 IEEE.

Butterfield, K., & Zou, X. (n.d.). Analysis and Implementation of Internet Based Remote Voting. *11th International Conference on Mobile Ad Hoc and Sensor Systems, 2014 IEEE.*

Council of Europe. (1950). *Convention for the Protection of Human Rights and Fundamental Freedoms.* Council of Europe.

Council of Europe Committee of Ministers. (2004). *Recommendation Record (2004) of the Committee of Ministers to member states on legal, operational and technical standards for e-voting.* Council of Europe, September 2004.

Council of Europe. (n.d.). *Council of Europe, E-Governance-Website.* Council of Europe, http://www.coe.int/T/E/Com/Files/Themes/E-Voting, 2007.

Council of Europe. (March 1952). *Protocol to the Convention for the Protection of Human Rights and Fundamental Freedoms, as amended by Protocol No. 11.* Council of Europe.

Goodman, N. J., & Jon H. Pammett. (2014). The Patchwork of Internet Voting in Canada. *International Conference on Electronic Voting EVOTE2014, E-Voting. CC GmbH © 2014 IEEE.*

Müller, R., & Prosser, A. (n.d.). Electronic voting via the internet. *3rd International Conference on Enterprise Information Systems (ICEIS).* Setubal, Portugal, page 1061 to 1066, 2001.

National Election Committee (Estonia). (2005). *E-Voting System—Overview.* http://www.vvk.ee/engindex.html.

Organization for the Advancement of Structured Information Standards (OASIS). (January 2005). *Election Markup Language (EML) 4.0.* http://www.oasis-open.org.

OSCE/ODIHR. (March 2007). *OSCE/ODIHR Election Assessment Mission Report—PARLIAMENTARY ELECTIONS 2006.* OSCE/ODIHR.

Pardue, H., Yasinsac, A., & Landry, J. (2010). *Towards Internet Voting Security: A Threat Tree for Risk Assessment.* School of Computer and Information Sciences, University of South Alabama, USA, 978-1-4244-8642-7/10, ©2010 IEEE.

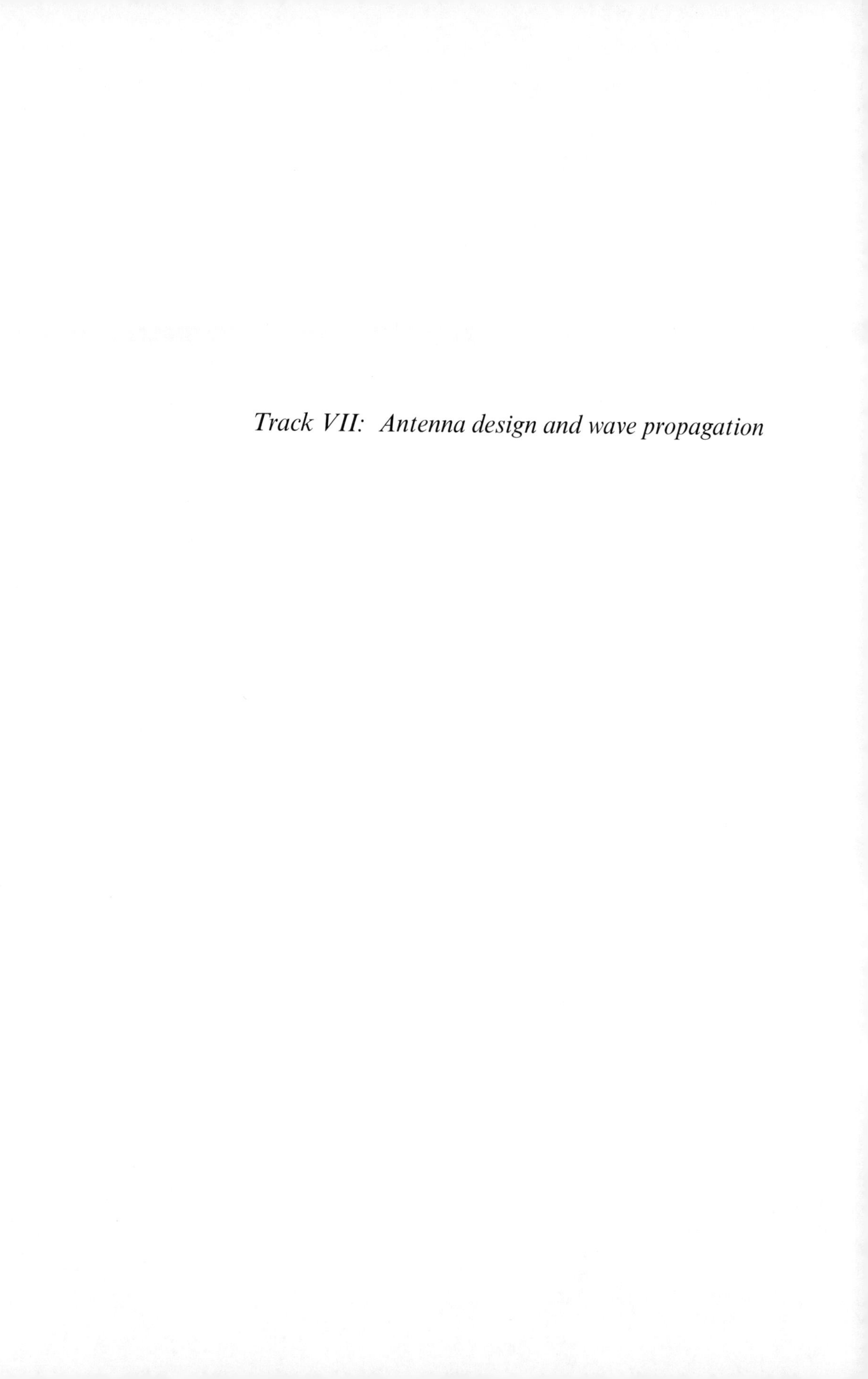

Track VII: Antenna design and wave propagation

Computational Science and Engineering – Deyasi et al. (Eds)
© 2017 Taylor & Francis Group, ISBN 978-1-138-02983-5

Performance analysis of microstrip patch antenna array with 4 different shapes for gain and bandwidth enhancement

Pampa Debnath & Ananga Paul
RCC Institute of Information Technology, Kolkata, India

ABSTRACT: In the current years the progress in communication systems requires the development of cost effective, low weight and low profile antennas that are proficient of maintaining high performance over a wide band of frequencies. This technical development has paying attention into the design of a microstrip patch antenna. The aim of this paper is to design, coaxial probe fed microstrip patch antenna array for high performance application. Desired patch antenna design was simulated by HFSS simulator software. Initially we set our antenna as a single patch and after evaluating the parameters; resonating frequency, radiation patterns, directivity and antenna gain, we transformed it to a 2 × 1 linear array. Finally, we investigated the 3 × 1 linear antenna arrays to increase directivity, gain, efficiency and have improved radiation patterns.

Keywords: microstrip antenna array, rectangular microstrip antennas, triangular microstrip antenna, pentagonal microstrip antennas, circular microstrip antenna, gain, bandwidth, directivity, return loss, resonant frequency, radiation pattern, HFSS

1 INTRODUCTION

Microstrip antennas are presently one of the fastest rising segments in the wireless communications industry. Microstrip antenna technology began its swift development in the late 1970 s. Basic microstrip antenna elements and arrays were fairly well recognized in term of design and modeling by the early 1980s (Pozar *et al.*, 1995). In past few years printed antennas have been mostly studied due to their advantages over other radiating systems, such as low weight, reduction in size, low cost, conformability and possibility of incorporation with active devices.

Wireless communication has experienced an enormous development since it allows users to access network services without being tethered to a wired infrastructure. The wireless systems that have experienced the most fast evolution and wide reputation are the standard developed by IEEE for Wireless Local Area Networks (WLANs), identified as IEEE 802.11 and the Bluetooth technology. Besides being able to specify high-quality signal to noise ratio and immunity to noise, the antennas in microwave communication system will depict compact structures and simplicity of construction to be mounted on various devices. In high performance point to point communication where weight, size, cost, ease of installation, performance, are constraints, low profile antenna is highly demanded. To achieve these requirements, microstrip antenna is preferred.

Although microstrip antenna has several advantages like light weight, low profile and easy to construct, it also has several demerits such as low gain, narrow bandwidth with low efficiency. These disadvantages can however be overcome with incorporated array configuration.

The patch of micro strip antenna can be of different shapes. In this paper triangular, rectangular, pentagonal and circular shaped patches are designed and analyzed. To increase the gain of the system instead of one element, we develop antenna array. In this context, 3 by 1, 2 by 1, single element of all the shapes are configured and analyzed. These structures are simulated using the HFSS software, and all the results are compared based on the simulated result. In this research analysis coaxial feeding technique is incorporated.

2 SINGLE ELEMENT DESIGN

2.1 *Introduction*

In microstrip antenna a patch can be used as an antenna, or can be used as an element of an antenna array. The aim is to study the behavior of patches with different shape and same surface area and then select the appropriate geometrical shape for particular application. In this paper, triangular, rectangular, pentagonal and circular antenna arrays are designed and their characteristics are analyzed (Garg *et al.*, 2001).

2.2 *Patch dimensions and resonant frequency*

Figure 1 shows the geometry of various patch shapes that were designed and simulated. These patches were so designed that each geometrical shaped patch has the same surface area.

Based on the cavity model, the resonant frequency of the rectangular patch antenna having length L can be calculated using the following formula:

$$f_0 = \frac{c}{2L\sqrt{\varepsilon_r}} \tag{1}$$

where c is the speed of light and ε_r is the dielectric constant of the substrate (Pozar *et al.*, 1995, Sidhu et al., 1987).

But in case of equilateral triangle having side length a can be calculated by the following formula:

$$f_0 = \frac{2c}{3a\sqrt{\varepsilon_r}} \tag{2}$$

For circular patch having radius r the formula for resonant frequency for T_{nm} mode is given by:

$$f_0 = \frac{k_{mn}c}{2\pi r\sqrt{\varepsilon_r}} \tag{3}$$

where k_{nm} is the mth zero of derivative of the Bessel function of order n, c is the speed of light and ε_r is the dielectric constant of the substrate (Khraisat et al., 2012).

The resonant frequency of the pentagonal patch can be calculated from the formula given below (Gangwar et al., 2008).

$$f_0 = \frac{X'_{np}c}{2\pi b\sqrt{\varepsilon_r}} \tag{4}$$

Where X'_{np} are the zeros of the derivative of the Bessel function $J_n(n)$ of the order n, as is true for TE mode circular waveguides, however for the lowest order modes

$$X'_{np} = 1.84118$$

2.3 *Substrate selection*

One of the basic steps when designing a micro strip antenna is to select the thickness of the substrate. A thicker substrate will naturally make the antenna mechanically robust and the weight will be more. Apart from this there will be changes in many antenna parameters like radiation power, bandwidth will increase with increase in substrate thickness [2]. On the other hand dielectric loss, surface wave loss will increase with the increase in substrate thickness which will reduce the antenna performance (Pozar *et al.*, 1995, Sidhu et al., 1987). Like the thickness of the substrate the dielectric constant of the substrate plays an important role in the antenna performance. Fringing field, which is responsible for radiation, will increase for a low dielectric constant value of the substrate. So a dielectric material with $\epsilon_r < 2.5$ is always preferable as a substrate. As a result a substrate having $\epsilon_r = 2.33$ and height is 1.575 mm has been used in our design.

2.4 *Design of single element*

In Figure 2 shows all the patches those have been designed using HFSS. The surface area of all patches has been kept as 630 mm^2. The substrate thickness for all the 4 antennas is 1.575 mm. In case of equilateral triangle side length is 38.13 mm, where as for the pentagonal patch the side length is 19.14 mm. The radius of the circular patch is 14.16 mm and for rectangular patch L = 27 mm and W = 23.33 mm.

From Figure 3 it is clear that though the surface area of all the patches is same, the resonant

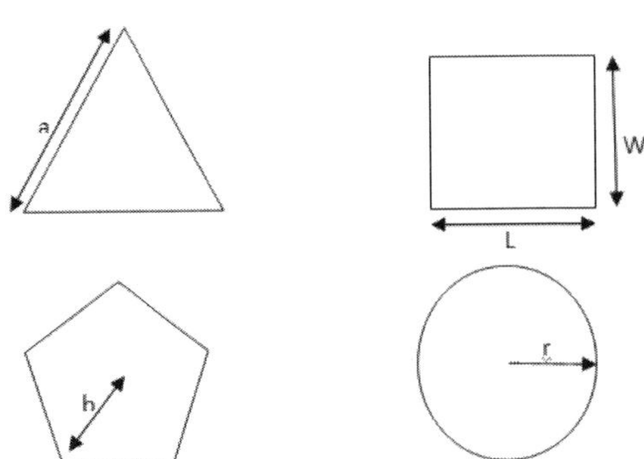

Figure 1. Geometry of equilateral triangle, rectangular, pentagonal and circular patches.

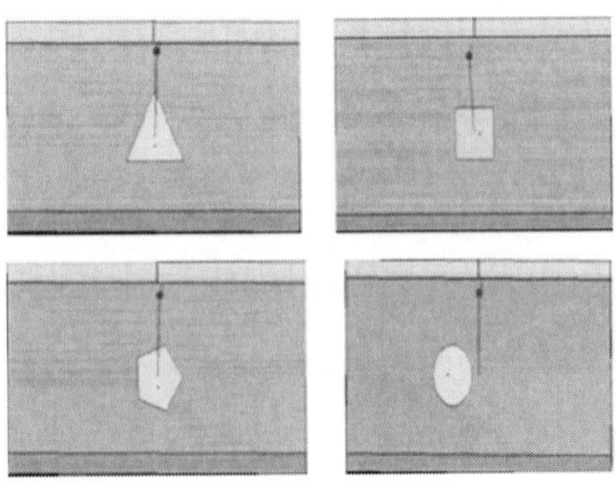

Figure 2. Design of single element equilateral triangular, rectangular, pentagonal and circular patch antenna in HFSS.

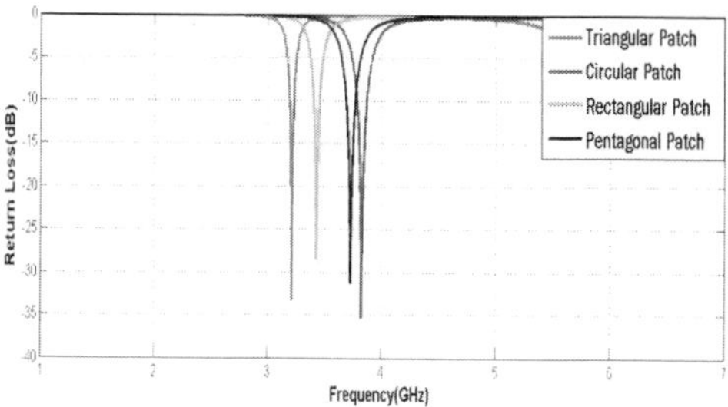

Figure 3. Simulated Return loss (in dB) of all the designs shown in Figure 2.

frequencies are slightly varied for with changing the shape of the patch (Dharsandiya et al., 2014). In case of triangular patch the resonant frequency is lowest and it is 3.213 GHz, for rectangular patch it is 3.433 GHz and for pentagonal and circular patch return losses are 3.739 GHz and 3.832 GHz respectively.

3 ARRAY DESIGN

3.1 *Introduction*

Microstrip antennas are very flexible and are used to synthesize a required pattern that cannot be achieved with a single element. In addition, the beam of an antenna system scanned by them, simultaneously they are used to increase the gain, efficiency, and perform various other functions which would be difficult with any one single element. So in this paper we used an array to develop the performance of this antenna.

Linear, planar or volume array can be constructed by spatially distributing the element of the array. Here a linear antenna array is created for each shape by placing the elements at finite distance apart from each other along a straight line (Sidhu et al., 1987). In this paper coaxial feeding method is used to feed the micro strip antenna array.

3.2 *Design of linear 2 × 1 array*

In order to observe changes in antenna performance for a particular dielectric material of fixed thickness, the substrate used for 2 × 1 antenna array is same as the single element antenna (thickness = 1.575 mm and ϵ_r = 2.33). Figure 4 shows the design of all the arrays which were observed. Surface area of all the patches remains unchanged (630 mm^2) where the distance between two elements of an array is not fixed.

Comparing Figure 3 and Figure 5 it is observed that the resonant frequency of single element and antenna array for a particular geometrical shape remains almost unchanged.

3.3 *Design of linear 3 × 1 array*

Here also to observe changes in antenna performance for a particular dielectric material of fixed thickness, the substrate used for 3 × 1antenna array is same as the single element antenna (thickness = 1.575 mm and ϵ_r = 2.33). Figure 6 shows the design of all the 3 × 1 arrays which were observed.

Also for the 3 × 1 antenna array the resonant frequencies remain almost unchanged. From these three observations it is clear that there is no effect of number of elements in the array on the resonant frequency.

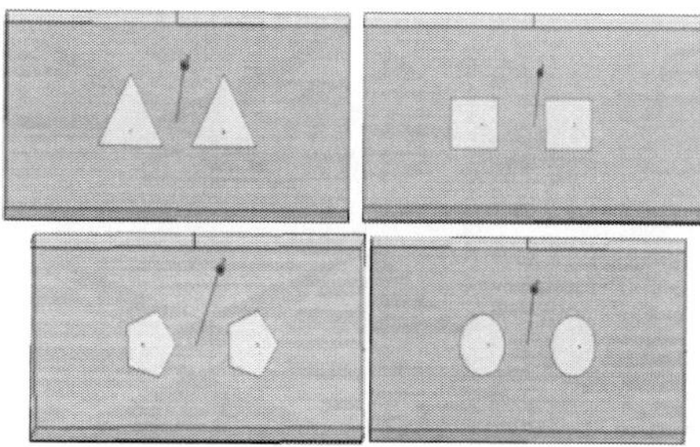

Figure 4. Design of equilateral triangular, rectangular, pentagonal and circular 2 × 1 Antenna array using HFSS.

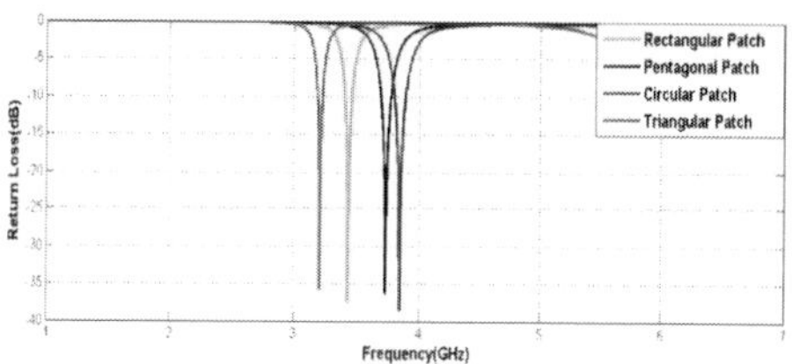

Figure 5. Simulated Return loss (in dB) of all the designs shown in Figure 4.

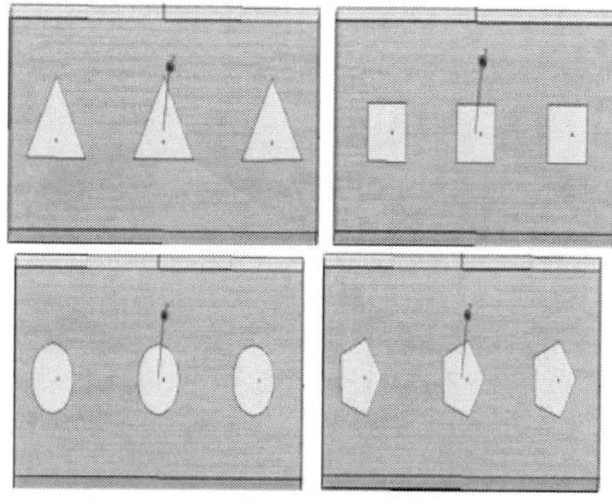

Figure 6. Design of equilateral triangular, rectangular, pentagonal and circular 3 × 1 Antenna array using HFSS.

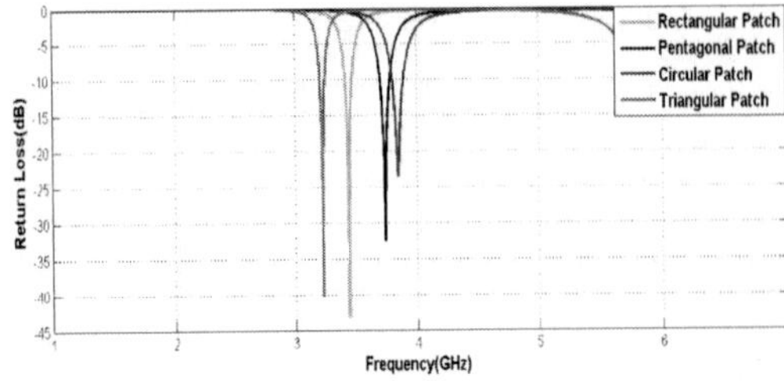

Figure 7. Simulated Return loss (in dB) of all the designs shown in Figure 6.

4 COMPARISON OF SIMULATED RESULTS

Table 1 shows the result obtained after simulation of all the designs. From the obtained data the gain & directivity of antenna increases with the increase in the number of elements as expected. It is also observed that the bandwidth remains constant with the change in number of elements, but there is an effect of patch shape on the bandwidth. For triangular patch the bandwidth is smallest and

Table 1. Comparison among triangular (tri), rectangular (rect), pentagonal (pent) and circular (cir) antenna arrays.

Number of elements	Gain (dB)				Bandwidth (MHz)				Directivity (dB)			
	Tri	Rect	Pent	Cir	Tri	Rect	Pent	Cir	Tri	Rect	Pent	Cir
Single	8.08	7.01	7.45	7.52	42	56	82	93	08.35	07.20	07.54	07.61
2 × 1 array	10.18	10.41	10.96	11.05	43	58	82	91	10.35	10.62	11.05	11.04
3 × 1 array	12.07	12.74	13.17	13.19	42	58	82	93	12.29	12.97	13.30	13.31

Table 2. Comparison of resonant frequencies and return loss.

No. of elements	Return loss (dB)				Resonant frequency (GHz)			
	Tri	Rect	Pent	Cir	Tri	Rect	Pent	Cir
Single	−32.6570	−28.0078	−31.3980	−41.2185	3.213	3.433	3.739	3.832
2 × 1 array	−35.7038	−37.4095	−36.2516	−38.5704	3.207	3.434	3.734	3.845
3 × 1 array	−40.4019	−43.1527	−32.2640	−23.4801	3.222	3.435	3.733	3.845

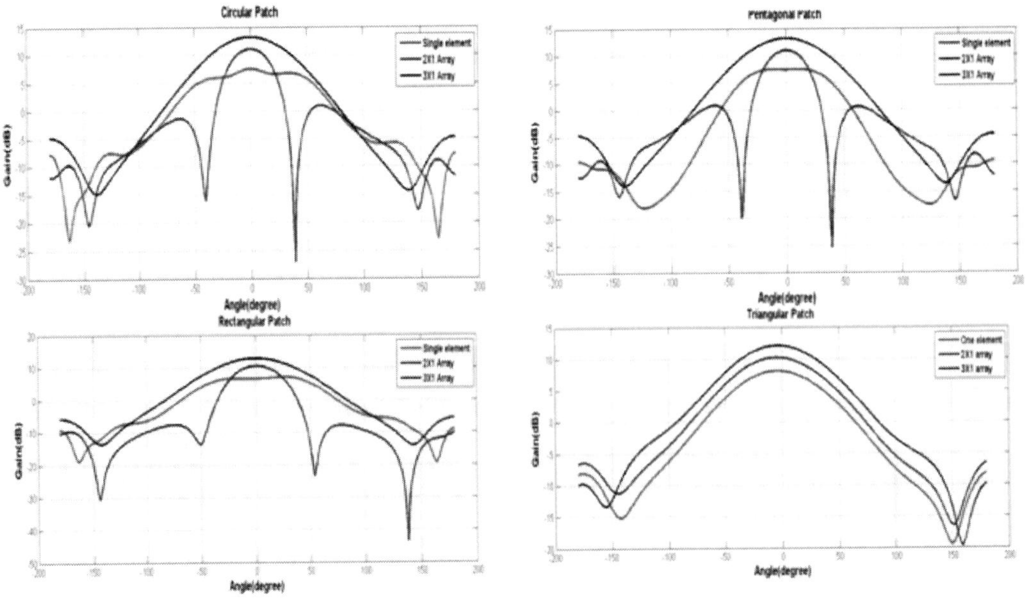

Figure 8. Gain comparison of single element antenna, 2 × 1 & 3 × 1 antenna array.

Table 3. Comparison of radiation pattern of various antenna and antenna array.

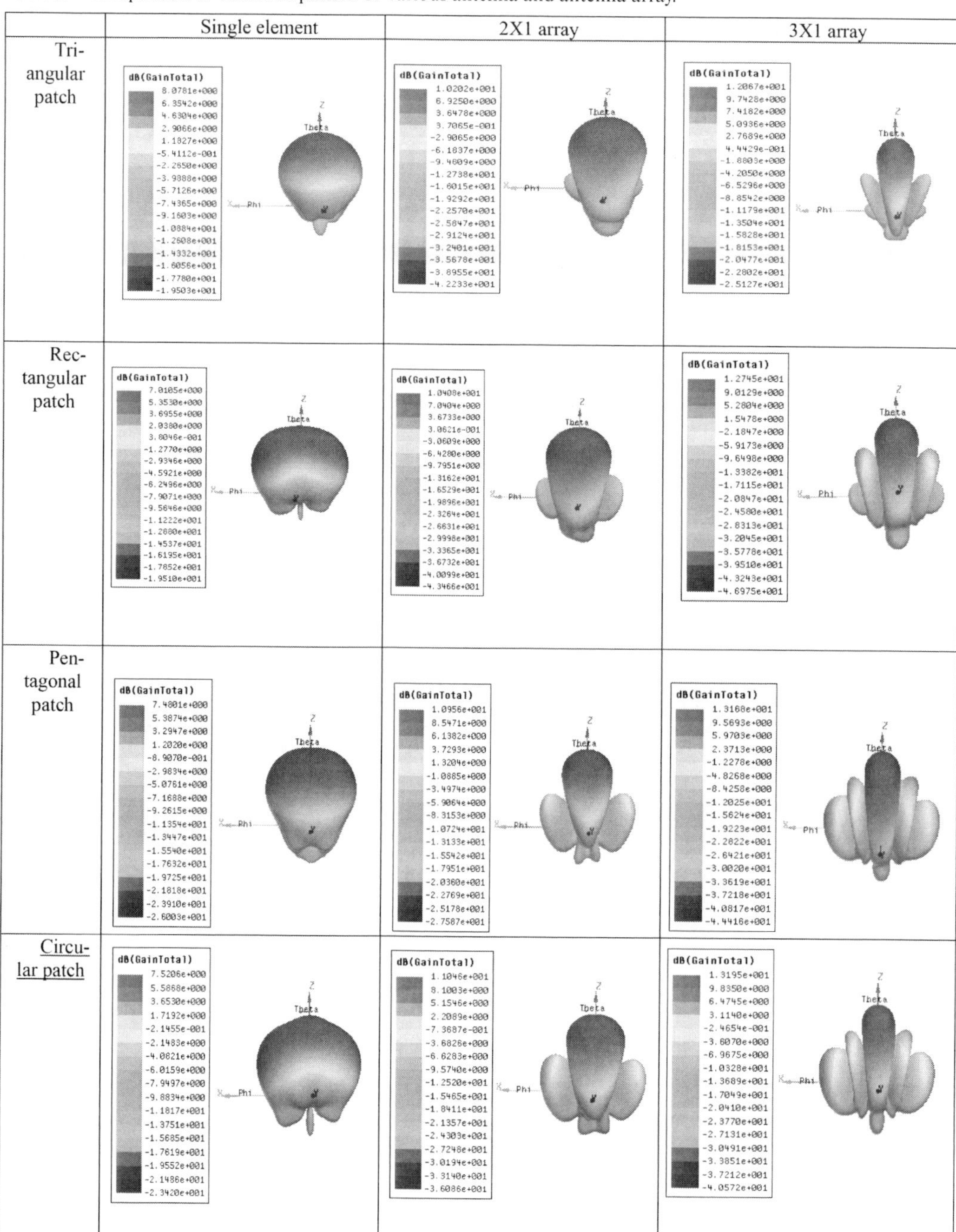

bandwidth is largest for the circular patch antenna. From the below table it can be concluded that the performance of 3 element array antenna is better than single and two element antenna array.

It is observed that, with the increase in number of element side lobes of the radiation pattern increases, as well as the radiation pattern becomes more directive and gain also increases.

5 CONCLUSION

To improve the gain of different antennas and to nullify the side lobes we can go for Phased array. While increasing the number of elements the band width, gain and directivity of circular patch antenna array is quite better compare to other patch antenna array. The gain, bandwidth and directivity obtained for circular patch antenna array is 13.19 dB, 93 MHz and 13.31 dB respectively.

REFERENCES

Foram N. Dharsandiya, Ila D. Parmar," Optimization of Antenna Design for Gain Enhancement Using Array" International Journal of Advanced Research in Computer Science and Software Engineering, Volume 4, Issue 1, January 2014.

Garg, R., P. Bhartia, I. J. Bahl, and P. Ittipiboon, "Microstrip Antenna Design Handbook", Artech House, Boston. London, 2001.

Pozar, D.M., D.H. Schaubert, 'Microstrip Antenna, The Analysis and Design of Microstrip Antennas and Array'. New York, IEEE Press, 1995.

Ravindra Kumar Yadav1, Jugul Kishor2, R. L. Yadava3, "Dielectric Loading on Multi-Band Behaviors of Pentagonal Fractal Patch Antennas", Open Journal of Antennas and Propagation, 2013, 1, 49–56.

Som Pal Gangwar, R P S Gangwar, B K Kanaujia, Paras, "Resonant frequency of circular microstrip antenna using artificial neural networks", Indian Journal of Radio & Space Physics, volume 37, June 2008.

Sumanpreet Kaur Sidhu, Jagtar Singh Sivia, "Comparison of Different Types of Microstrip Patch Antennas", *International Journal of Computer Applications (0975–8887)*.

Vasujadevi Midasala, Dr. P. Siddaiah, S Nagakishore Bhavanam,"Rextangular Patch Antenna Array at 13GHz Frequency Using HFSS", 2014 IEEE International Conference on Computational Intelligence and Computing Research. 978-1-4799-3975-6/14.

Yahya S. H. Khraisat, Melad M. Olaimat, Sharief N. Abdel-Razeq, "Comparison between Rectangular and Triangular Patch Antennas Arrays" Vol. 4, No. 2; 2012.

Computational Science and Engineering – Deyasi et al. (Eds)
© 2017 Taylor & Francis Group, ISBN 978-1-138-02983-5

Design and optimization of HWDA for improvement of gain at Ka band using EBG reflector

Pampa Debnath, Sneha Agarwal, Anirban Pal, Ankana Ashis Saha & Medha Chowdhury
RCC Institute of Information Technology, Kolkata, India

ABSTRACT: In this paper, a simple half wave dipole antenna has been designed for wireless applications. Different types of reflectors have been added to the original design to observe the behavioral change in antenna parameters in the presence and absence of reflectors. The simulation of the aforementioned antenna with and without reflector is simulated using HFSS software. The different reflectors we have used are EBG reflector, rectangular reflector and rectangular slot reflector to observe the change in antenna parameters. Simulation has been carried out using in Ka band using Ansoft HFSS simulator tool to develop planar dipole antenna and to examine the antenna characteristics as return loss, gain and radiation patterns. Comparisons between simulation results of antennas with and without reflector are carried out in order to predict the suggested design. The gain increment has been observes almost doubled from 2.34 dB to 4.34 dB.

Keywords: planar dipole antenna, Ka band, reflector, EBG, gain, radiation pattern, ansoft HFSS

1 INTRODUCTION

With the advent of technology a major growth in the use of wireless sensor technology has been observed. It has been used in various applications like battlefield surveillance, environmental monitoring, intelligent buildings, imaging, safety, medical and health care, logistic and telemetric (Rajalakshmi *et al.*, 2013).

To operate a reliable and low-cost wireless sensor device under stiff constraints of size, data range and energy (Debnath *et al.*, 2016; Hasegawa *et al.*, 2009; Sompan *et al.*, 2011), is the need of the hour. Thus in this proposed paper a half wave dipole antenna has been designed with operating frequency 29 GHz. Due to such high operating frequency this type of antenna finds its application mainly in Ultra-Wideband Range (UWB).To further improve the results different types of reflectors have been introduced in the antenna which has made the antenna very cost efficient and easy to use due to its low size. These designs are easy to fabricate and can be used indoors as well as outdoors.

2 ANTENNA CHARACTERISTICS

Antenna is a sensor device which acts as a specialized transducer to convert RF (Radio Frequency) fields to Alternating Currents and vice versa. It is solely responsible for transferring energy from the transmitter, into free space. The energy is transmitted in the form of electromagnetic Radiations. Now, electromagnetic radiation is a form of energy emitted and absorbed by charged particles. They exhibit wave like behaviour as they transmit through space. The antenna we have chosen to work with is a simple half-wave dipole operating at a design frequency of 29 GHz. For successful analysis, we will be examining the following factors:

2.1 *Bandwidth*

Bandwidth of the dipole defines the range of frequencies over which the antenna can be used (Balanis 2005).

In the proposed paper the central frequency has been taken as 29 GHz. The formula used to calculate the percentage increase in bandwidth with and without reflectors is represented by:

$$B_P = \frac{f_u - f_1}{f_c} \times 100\%$$

where B_p is the bandwidth, f_u is the upper frequency, f_l is the lower frequency and f_c is the centre frequency.

2.2 *Gain and directivity*

An antenna with high gain gives a good range but limits the angular coverage. For this reason, an antenna with optimal gain or low gain is required.

An omnidirectional antenna is one which has low gain; hence it will provide more coverage. The doughnut shaped radiation pattern of an omni-directional dipole, is symmetrical around the axis and maximum in the middle (Balanis 2005).

2.3 *Impedance and matching*

Optimum power transfer between the transmitter and receiver will occur at the point where the impedances match will exist (Balanis 2005).

2.4 *Return loss*

It is a parameter which indicates the amount of power that is "lost" to the load and does not return as a reflection. This parameter is of crucial importance to our project because low return loss will portray lesser mismatch of impedances. Hence, there will be more power transmitted if the mismatch is kept to minimum (Balanis 2005).

3 ANTENNA STRUCTURE AND DESIGN

In the proposed paper a planar dipole antenna operating at a frequency of 29.89 GHz has been designed and simulated using HFSS. Half wave dipole of this design has two identical dipole arms of length 2.3125 mm each and radius of 0.05 mm. The feed gap between the two dipole arms is 0.0625 mm (Debnath *et al.*, 2016; Simon *et al.*, 1994).

3.1 *Design parameters*

The initial condition for the design of half-wave dipole antenna operation is at the design frequency of 29 GHZ. Dimensions of the antenna change according to the design frequency. By taking this into consideration several antenna dimensions have been calculated. The following are discussed below. Design Frequency, f = 29 GHz (Tareq *et al.*, 2014).

Wavelength:

$$\lambda = \frac{c}{f} = 10 \text{ mm} \tag{1}$$

Length of Half—wave dipole antenna:

$$L = \frac{143}{f} = 2.3125 \text{ mm} \tag{2}$$

Feeding Gap of the antenna:

$$G = \frac{L}{200} = 0.125 \text{ mm} \tag{3}$$

Radius of the wire:

$$R = \frac{\lambda}{1000} = 0.05 \text{ mm} \tag{4}$$

From the equation 1 we have calculated the wavelength using the design frequency, based on which the length of the half-wave dipole antenna is calculated along with the Feeding Gap. The Radius of the wire also determined from the equation 4.

To boost the simulation speed and to acquire more accuracy optimization of global mesh properties has been done. Copper (annealed) has been used as the antenna material and between the two antenna arms a sheet has been selected. The designed antenna must have a power source or excitation between its two conducting arms. Hence, the Lumped gap source has been created (Debnath *et al.*, 2016).

Some improvements in the geometric structure of the antenna have been done to improve the gain, directivity and bandwidth of the antenna (Rajalakshmi *et al.*, 2013). A copper reflector has been introduced inside the radiating cylinder beside the antenna arms. The dimensions of the copper reflector have been determined from trial and error method resulting in length = 5.99 mm, width 0.79 mm and thickness = 0.01 mm.The position of reflector has been determined from HFSS for the

Table 1. Shows the parameter table of the proposed dipole antenna (Singh *et. al.*, 2012).

Parameter	Value	Unit
Design frequency (f)	29.89×10^9	GHZ
Wavelength (λ)	10	mm
Length of the dipole (L)	2.3125	mm
Radius of the wire (R)	0.05	mm

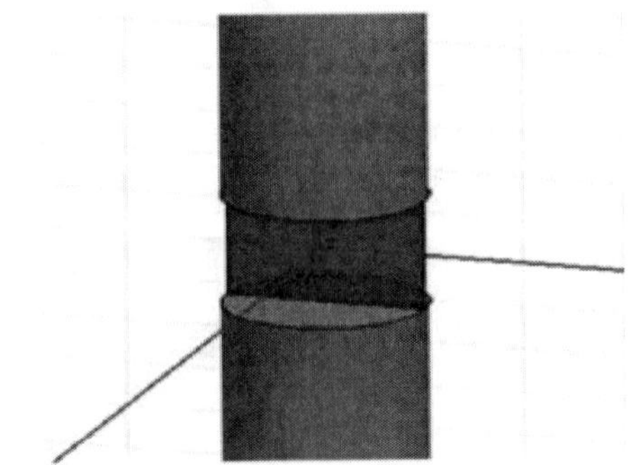

Figure 1. Shows the lumped gap source introduced in the proposed antenna.

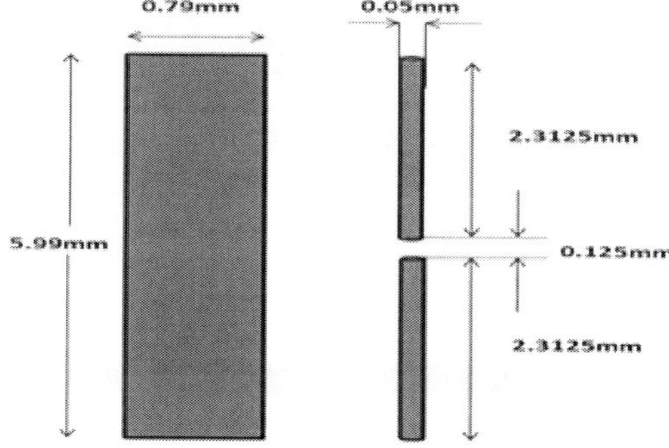

Figure 2. Structure of half wave dipole antenna in the presence of reflector.

best possible case and it is taken as x = −1.5 mm, y = −1.7 mm and z = −3.1 mm.

Fig. 2 shows the structure of half wave dipole antenna in the presence of reflector. The following structure has been simulated using HFSS and finally the results and improvements have been obtained.

4 SIMULATION RESULTS

The comparison between the return losses and VSWR values of dipole antenna without reflector, dipole antenna with rectangular reflector, dipole antenna with EBG reflector and dipole antenna with rectangular slot reflector has been shown in a tabulated form in Table 2.

The comparison between various other antenna parameters like gain, bandwidth and directivity has also been represented in a tabular form in Table 3.

These two tables are followed by graphs representing the various parameters.

The table above clearly shows that the maximum gain is 4.39 dB which was obtained by using EBG as a reflector. Thus the value of gain almost doubles itself in the presence of rectangular or EBG reflectors. However for the rectangular slot reflector the gain remains same but the bandwidth has increased.

From the table above we can see that the lowest return loss has been obtained for rectangular copper reflector at operating frequency 28.75 GHz. There is almost 40.8% reduction in return loss when rectangular reflector is used. Whereas 38.61% and 34.95% reduction in return losses have been obtained with the use of EBG reflector and rectangular slot reflector. But in all cases VSWR will remain more or less same as 1.0.

Fig 3.Represents the Return loss characteristics obtained for a dipole antenna with and without reflectors. It shows the comparison between the return losses of different types of reflectors.

The graph shows that for an antenna without reflector operating at a frequency between 27.5 GHz and 30.0 GHz, the return loss is obtained at 29 GHz and is equal to –74.6 dB (Debnath *et al.*, 2016) and for an antenna with rectangular reflector return loss calculated between operating frequencies 27.5 GHz and 30.0 GHz is –44.16 dB at 28.75 GHz.For antennas with EBG reflector and rectangular slot reflector the return loss obtained at an operating frequency of 28.76 GHz and 29.77 GHz is –45.79 dB and –48.52 dB respectively. Thus the highest decrease in return loss has been obtained by using rectangular reflector.

The table for radiation pattern and gain plots for dipole antenna with and without reflectors has been shown in Table 4.

Table 2. Simulated results of antenna gain, directivity and bandwidth in the presence and absence of different types of reflectors.

Antenna	Reflector	Gain (dB)	Directivity (dB)	Bandwidth (GHz)
Dipole Antenna	Without Reflector	2.34	2.34	4.54
Dipole Antenna	With Rectangular copper Reflector	4.34	4.35	4.4
Dipole Antenna	**With EBG as a Reflector**	**4.39**	**4.41**	**4.3**
Dipole Antenna	With Rectangular slot Reflector	2.34	2.33	5.08

Table 3. The following table shows the effect of different types of reflectors on the return loss and VSWR of the dipole antenna.

Antenna	Reflector	Operating frequency (GHz)	Return loss (dB)	VSWR
Dipole Antenna	Without Reflector	29	−74.6	1.00
Dipole Antenna	With Rectangular copper Reflector	28.75	−44.16	1.00
Dipole Antenna	With EBG as a Reflector	28.76	−45.79	1.010
Dipole Antenna	With Rectangular slot Reflector	29.77	−48.52	1.007

Table 4. Radiation patterns and 3D polar graphs of dipole antenna with different types of reflector.

ANTENNA	REFLECTOR TYPE	RADIATIONPATTERN	3D POLAR GRAPH
DIPOLE ANTENNA	WITHOUT REFLECTOR		
DIPOLE ANTENNA	COPPER REFLECTOR		
DIPOLE ANTENNA	EBG REFLECTOR		
DIPOLE ANTENNA	RECTANGULAR SLOT REFLECTOR		

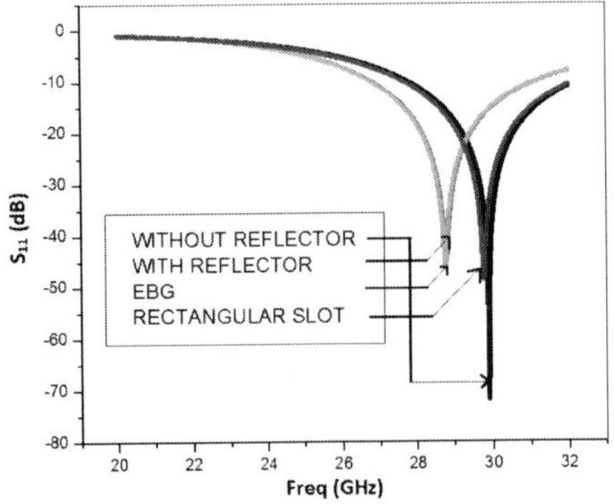

Figure 3. S parameter graphs for dipole antenna in presence of without reflector, with reflector, EBG reflector and rectangular slot reflector.

5 CONCLUSION

In this paper the design, optimization and characteristics of a simple half wave dipole antenna has been examined. Different types of reflectors have been added to the model to improve the performance. A clear comparison between the parameters of the antenna with and without the reflectors has been analyzed with the HFSS software. It is seen that in the presence of reflectors, there is a substantial decrease in return loss and a subsequent increase in gain, in contrast to the reference dipole. The maximum decrease in return loss has been obtained by using rectangular reflector, which is almost 40.8%. It has been reduced from –74.6 dB to –44.16 dB with and without reflector; whereas 38.61% and 34.95% reduction in return losses have been obtained with the use of EBG reflector and rectangular slot reflector. The gain has also become almost double increasing from 2.34 dB to 4.34 dB without and with rectangular reflector and with the same antenna parameters. The gain 2.34 dB remains almost the same with the use of rectangular slot reflector but the bandwidth has increased by 6.7%.The bandwidth without reflector was 4.54 GHZ and that with the rectangular slot reflector has changed to 5.08 GHz, which is the maximum obtained bandwidth among all the reflectors used. Maximum gain is obtained by using EBG reflector which is 4.39 dB.Thus the proposed antenna parameter comparison of the dipole antennas with the three types of reflectors and the dipole antenna without reflector can been used to design an antenna according to the operational requirements.

REFERENCES

Balanis. C. A., 2005, Antenna Theory Analysis and Design, John Wiley & Sons, 3rd Edition, USA, p.151.

Debnath. P., Agarwal. S., Pal. A., Chowdhury. M., Saha. A., 2016 Design and Analysis of Half-Wave Dipole Antenna for Wireless Application, National conference on Materials, Devices and Circuits in Communication Technology, India.

Hasegawa. D., Yasuzumi. T., Suda. T., Kazama. Y., Hashimoto. O., 2009 A Study on Dipole Antenna Having Conical Radiation Pattern with EBG as a Reflector, Conference on Antenna and Propagation, pp 305–308.

Rajalakshmi D., Sanoj. C., Vijayaraj. N., 2013. Design And Optimization of Printed Dipole Antenna Forwireless Sensor Communication At 2.4 GHz, International Conference on Computer Communication and Informatics, India.

Sompan. A., Seewattanapon. S., Mahatthanajatuphat. C., Akkaraekthalin. P., 2011. An Elliptical Dipole Antenna with Rectangular Slot Reflector for Wideband Applications, IEEE International Conference on Electrical Engineering/Electronics Computer, Telecommunications and Information Technology, pp 200–203.

Simon. R., Whinnery. J., Duzer. T., 1994, Fields and Waves in Communication Electronics, John Wiley & Sons, 3rd Edition, Canada.

Singh. P., Sharma. A., Uniyal. N., Kala. R., 2012, Half-Wave Dipole Antenna for GSM Applications, International Journal of Advanced Computer Research, vol. 2(4).

Tareq. M., Alam. D., Islam. M., 2014, Simple Half-Wave Dipole Antenna Analysis for Wireless Applications by CST Microwave Studio", International Journal of Computer Applications, vol. 94(7).

Yu. Y., Ni. J., Xu. Z., 2015, Dual Band Dipole Antenna for 2.45 GHz and 5.8 GHz RFID Tag Application, IJWCMC.

Computational Science and Engineering – Deyasi et al. (Eds)

A bow-tie shaped dielectric resonator antenna excited by a monopole for ultra wideband applications

Rudraishwarya Banerjee, Susanta Kumar Parui & Biswarup Rana
Indian Institute of Engineering Science and Technology, Howrah, India

ABSTRACT: Here, a new 'Bow-Tie' shaped Dielectric Resonator Antenna (DRA) excited by a monopole has been proposed. It offers 105% impedance bandwidth (5.7–17.3 GHz), with uniform monopole like radiation pattern over the entire bandwidth, with the peak gain varying between 5.5–6.5 dBi depending on frequency. The most important aspect of this study is the fact that, this new monopole antenna configuration consumes lesser amount of costly dielectric material when compared with other ultra wideband hybrid monopole antenna available in literature.

1 INTRODUCTION

The enhancement of bandwidth of a monopole antenna by the addition of a ring Dielectric Resonator Antenna (DRA) was initially proposed in (Ittipiboon *et al.* 2002), and a design guideline was derived there. Then a number of DRA-loaded monopole antenna was investigated in (Lapierre *et al.* 2002, Ittipiboon *et al.* 2005, Guha *et al.* 2006) and so forth. Here, a new 'Bow-Tie' shaped DRA, fed by a monopole, is proffered. Unlike other hybrid monopole-dielectric resonator antenna, the DRA used here is not a ring resonator. In this work, two shaped small pieces of Dielectric Resonator (DR) are symmetrically placed on two sides of a monopole, and the configuration looks like a 'Bow-Tie' of ancient English male formal suits. The monopole in turn excites the DR pieces, and 105% impedance bandwidth is achieved, with uniform monopole like radiation pattern over the entire bandwidth. The most interesting part of this work is the fact that much lesser amount of dielectric material is used in this configuration, when compared with other DRA-loaded monopole antenna available till date.

2 ANTENNA CONFIGURATION

The derivation of the monopole fed new 'Bow-Tie' shaped DRA, using a small conical piece of dielectric material is shown in Figure 1. Here, a small conical piece of dielectric resonator, with lower radius, denoted by rL = 7.5 mm, and upper radius rU = 2.5 mm, and height H = 5 mm, is cut into four equal segments, and any two of these four segments are placed symmetrically on both sides

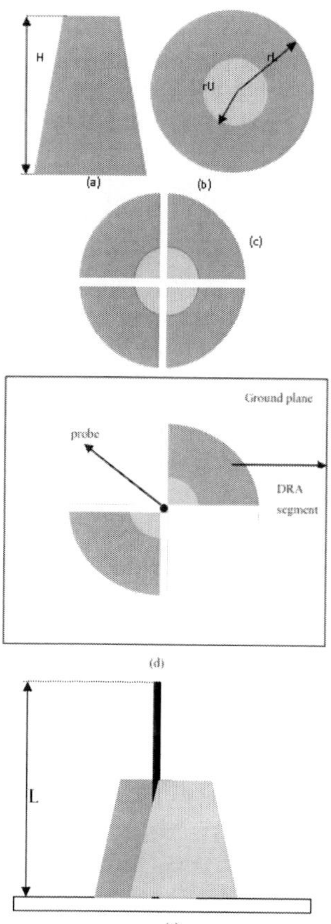

Figure 1. (a) Cross-sectional view of the conical DRA (b) Top view of the conical DRA (c) Top view after segmentation of the conical DRA (d) Top view and (d) Cross-sectional view of the DRA-loaded monopole antenna.

of a monopole, whose height is designated by 'L' in the diagram, and the configuration resembles a 'Bow-Tie' of English formal suits. Various DRA-loaded monopole antennas have been investigated till date, all of which are monopole fed dielectric ring resonator, and have much larger volume.

Undoubtedly, these small pieces of dielectric resonator are much smaller in size than those dielectric ring resonators, available in literature, and hence consume lesser amount of costly dielectric material. Moreover, to be more specific, as only two segments among the four equal segments of a small conical piece of dielectric material serves the purpose of wideband performance, this small piece yields two monopole antennas, which is a commercially important fact. Also this configuration proves that only two segments are sufficient for ultra wideband response, which clearly indicates a new loading style of monopole by dielectric resonator, to achieve ultra wide impedance bandwidth.

3 SIMULATION RESULTS AND DISCUSSIONS

Figure 2 shows the return loss characteristics of the antenna for different values of probe length, which indicates that the best result is obtained for L = 12 mm. The S_{11} characteristics of the only monopole for L = 12 mm is given in Figure 3, from which it is clear that the first resonance of the monopole is at 5.8 GHz. Then, the radiation pattern of the 'Bow-Tie' shaped DRA-loaded monopole is studied at three distinct frequencies, 7.9 GHz, 14 GHz and 16.8 GHz, in Figure 4, Figure 5 and Figure 6, respectively, which shows that the peak gain occurs near ± 50° and varies between 5.5–6.5 dBi, with the maximum peak gain of 6.5 dBi at 7.9 GHz, where the monopole radiates at its dominant mode. Thus, it is observed that excellent monopole like radiation pattern is obtained over the entire bandwidth.

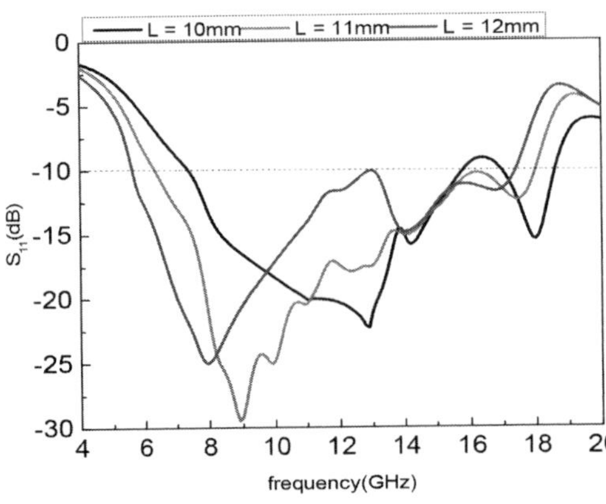

Figure 2. Simulated return loss for different values of probe length [rL = 7.5 mm, rU = 2 mm, H = 5 mm].

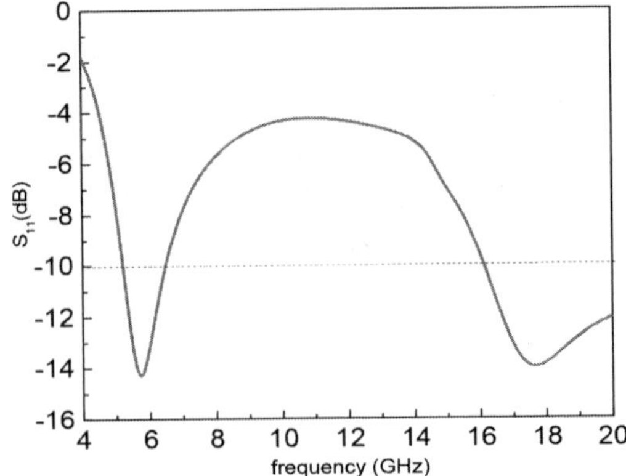

Figure 3. Simulated return loss for probe length L = 12 mm.

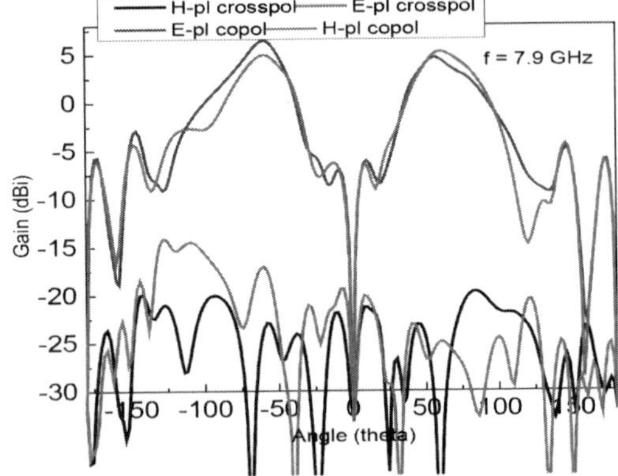

Figure 4. Simulated radiation pattern at the frequency = 7.9 GHz [rL = 7.5 mm, rU = 2 mm, H = 5 mm].

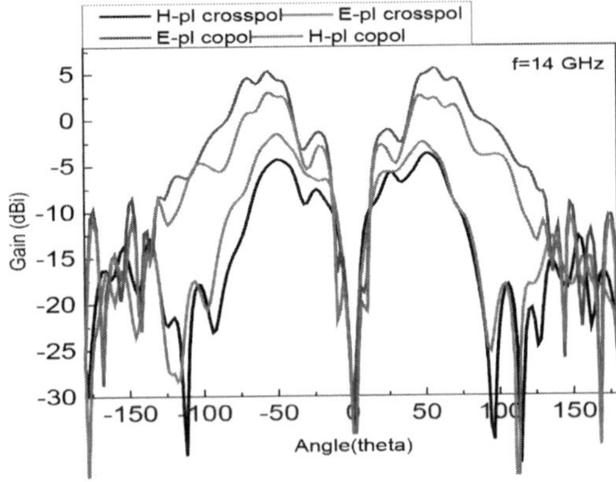

Figure 5. Simulated radiation pattern at the frequency = 14 GHz [rL = 7.5 mm, rU = 2 mm, H = 5 mm].

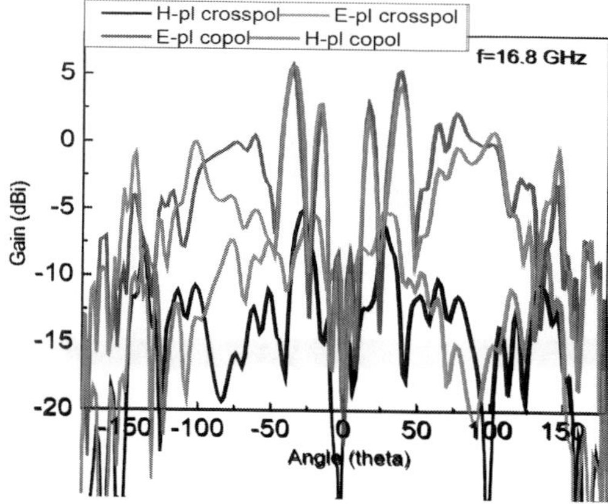

Figure 6. Simulated radiation pattern at the frequency = 16.8 GHz [rL = 7.5 mm, rU = 2 mm, H = 5 mm].

4 CONCLUSION

Thus, this work proffers a different style of loading a monopole by a dielectric resonator, to achieve ultra wide impedance bandwidth. More interestingly, here lesser amount of dielectric material is used to design this UWB antenna, when compared with other UWB antenna proposed till date. Also, here, much higher gain is achieved with excellent monopole like radiation pattern over the entire bandwidth. Another aspect that is noteworthy is the fact that in all the previous works of UWB antenna, usually dielectric ring resonator is used, whereas, here only two small shaped pieces of dielectric resonator serves the purpose, which is certainly easier to fabricate.

ACKNOWLEDGEMENT

Authors are thankful to CSIR, Govt. of India and UGC, Govt. of India for necessary support.

REFERENCES

Guha, D., Antar, Y. M. M., Ittipiboon, A., Petosa, A., & Lee. D., (2006) Improved design guidelines for the ultra wideband monopole-dielectric resonator antenna. IEEE Antennas and Wireless Propagation Letters. 5, 373–377.

Ittipiboon, A., Petosa, A., & Thirakoune, S. (2002) Bandwidth enhancement of a monopole using dielectric antenna resonator loading. Proc. ANTEM Conference, Montreal, QC, Canada, 387–390.

Ittipiboon, A., Petosa, A., Thirakoune, S., Lee, D., Lapierre, M., & Antar. Y.M.M., (2005) Ultra wideband antenna. U.S. Patent no. 6940463 B2.

Lapierre, M., Antar, Y. M. M., Ittipiboon, A., & Petosa, A. (2005) Ultra wideband monopole/dielectric resonator antenna. IEEE Microwave and Wireless Components Letters. 15, 7–9.

Computational Science and Engineering – Deyasi et al. (Eds)
© 2017 Taylor & Francis Group, ISBN 978-1-138-02983-5

Performance of periodic grooves on harmonic rejection in C band folded edge coupled microstrip band pass filters

Tarun Kumar Das
Department of ECE, Future Institute of Engineering and Management, Kolkata, India

Sayan Chatterjee
Department of ETCE, Jadavpur University, Kolkata, India

ABSTRACT: Design of a miniaturized and compact V-shaped half-wavelength edge-coupled microstrip bandpass filter with improved harmonic suppression is presented. The proposed filter is designed at a center frequency of 5.25 GHz with 20% fractional bandwidth (4.75–5.80 GHz) applicable in WLAN. The conventional parallel-coupled bandpass filter structure has been folded symmetrically about horizontal plane and consecutive square and rectangular shaped periodic grooves are inserted to the parallel lines. Thus, the 2nd harmonic is suppressed significantly to –44.38 dB and –43.71 dB and a size reduction of 59.05% and 59.67% is achieved for square and rectangular grooves respectively with respect to the conventional filter.

1 INTRODUCTION

In modern wireless communication system Broadband Wireless Access (BWA) is an important issue and to meet this, the bandpass filters with relatively wide bandwidth are frequently required in the RF front ends of both transmitters and receivers. The primary objectives of a bandpass filter are to achieve minimum insertion loss within the passband and maximum attenuation in the stopbands. There are different topologies for designing a band pass filter at microwave frequency. The complexity of the design belongs to the compactness and comparable structural tolerances. Microstrip type bandpass filters designed from parallel coupled lines have been widely used in traditional RF systems due its planar structure, easy trimming, tightened capacitive coupling and easy integration [8]. In contrary, there has been design trade-off due to the mismatch between the even and the odd-mode phase velocities. Accordingly, due to the asymmetrical characteristics of phase velocities, there is a variation in edge current subject to even—and odd-mode. Traditionally, edge-coupled filters suffer from significant spurious harmonics, near to the passband [8]. Numerous techniques have been employed for 2nd harmonic rejection by matching the phase velocities of even and odd modes or by compensating different electrical length of both Eigen modes. Reactive components such as Photonic Band Gap (PBG), Electronic Band Gap (EBG) and Defect Ground Structures (DGS) have been introduced in [6 and 9] for suppressing the harmonics. Defected Ground Structure (DGS) with PBG, has been proposed in [6] in which by increasing the capacitance and inductance of transmission lines on the defected ground a Lowpass Filter (LPF) has been investigated. However, in [6] leaky wave problem arises and the components become complicated due to the incorporation of discontinuities of current line in ground plane. In [9] a Uniplanar Compact Photonic Band Gap (UC-PBG) structure has been employed to reject the harmonic signals, but the physical and electrical parameters in relation with size reduction is not highlighted. Recently, optimum line structures have been invoked by inserting periodic shapes, such as wiggly, grooved and inter-digitized slots into conventional coupled-line filters [1,5 and11]. Such a kind of periodic structures are used to suppress the second harmonic by introducing the principle of Bragg reflections. High dielectric constant substrate, split ring resonators, Stepped-Impedance Resonator (SIR) structure and over coupled structure effectively achieve compact size and suppression of spurious harmonics [2, 4 and 12]. In [7] over-coupled structure, insufficient-coupled structure, and dividing-feed-in/out structures are highlighted to design a triple-band parallel coupled microstrip bandpass filter having a relatively good suppression ratio. In [2] a compact parallel-coupled line bandpass filter is designed with periodic sawtooth grooves and the 2nd harmonic suppression has been suppressed significantly.

In present article, a folded parallel coupled bandpass filter is designed by folding the symmetrical half of the edge coupled lines w.r.t. the center of the filter length. Along with the earlier method of size reduction [2], the periodic square and rectangular grooves are introduced to the edges of the coupled line, which leads to a better reduction in spurious harmonics.

2 FILTER DESIGN METHODOLOGY

Design parameters of a Chebyshev edge-coupled band pass filter applicable in WLAN have been described in Table 1. Calculated order of the filter is n = 3 and the element values of the Chebyshev prototype low-pass filter with pass-band ripple of 0.1dB, are selected as $g_0 = g_4 = 1.0000$, $g_1 = g_3 = 1.0315$, $g_2 = 1.1474$. The design equations for a bandpass filter with n+1 parallel-coupled line sections are shown in [8]. Table 2 provides the design parameters in microstrip line. Layout of the conventional half-wavelength parallel coupled band pass filter is shown Figure 1. The proposed filter structure is designed on FR4/epoxy substrate material having dielectric constant, $\varepsilon_r = 4.4$, loss tangent, $\tan\delta = 0.016$ and thickness, h = 1.6 mm.

Table 1. Design parameters specifications.

Parameters	Notation	Specifications
Midband Frequency	f_C	5.25 GHz
Fractional Bandwidth	FBW	20%
3 dB Bandwidth	BW	1.05 GHz
Feed line characteristic impedance	Z_0	50 ohms
Insertion Loss	IL	<3 dB
Return Loss	RL	>10 dB
Passband Ripple	L_{Ar}	0.1 dB

Table 2. Microstrip line design parameters.

j	$J_{j,j+1}/Y_0$	$(Z_{0e})_{j,j+1}$ (ohms)	$(Z_{0e})_{j,j+1}$ (ohms)	$w_{j,j+1}$ (mm)	$s_{j,j+1}$ (mm)	$l_{j,j+1}$ (mm)
0	0.55	92.83	37.64	0.80	0.37	8.03
1	0.29	68.61	39.73	1.15	0.94	7.96
2	0.29	68.61	39.73	1.15	0.94	7.96
3	0.55	92.83	37.64	0.80	0.37	8.03

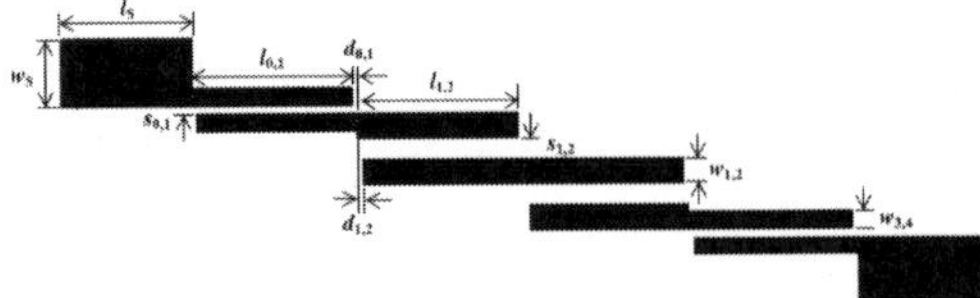

Figure 1. Microstrip layout diagram of conventional parallel-coupled BPF.

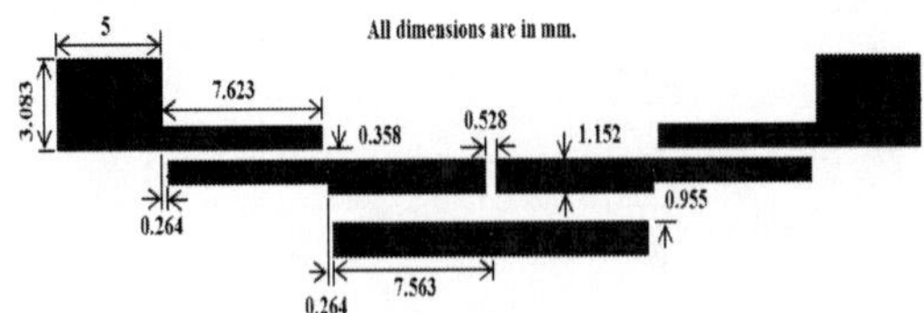

Figure 2. Microstrip layout diagram of folded symmetric parallel-coupled band pass filter.

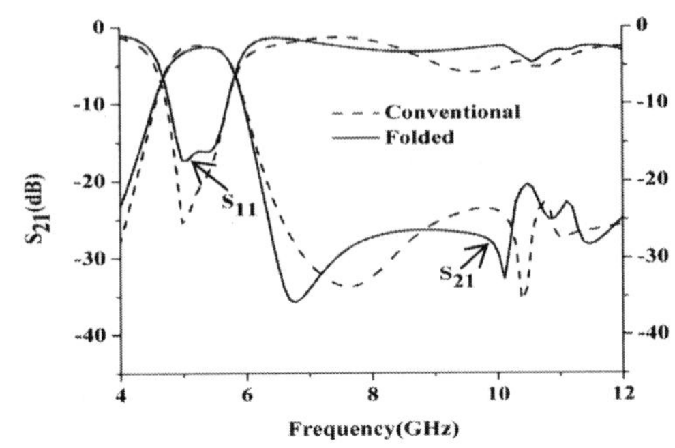

Figure 3. Frequency responses of S-parameters of conventional and compact bandpass filters.

In Figure 1, $d_{j,j+1}$ is the correction of the length of the resonators due to fringing capacitance effects at the open ends and its value is selected as 0.264 mm [8]. Dimensions of the input and output ports are optimized to have 50 ohms characteristic impedance. Accordingly, length, l_s = 11 mm and width, w_s = 3.083 mm are selected. The size of the conventional band pass filter structure has been reduced further by the compact V-shaped symmetric folded structure [2] as shown in Figure 2. Figure 3 shows the simulated results of conventional and compact filters. From Figure 3 it has been observed that the folded symmetrical filter improves skirt characteristics but increases the return loss within the desired passband.

At the same time, there has been degradation in the magnitude of second harmonic. The above phenomenon can be anticipated due to the unwanted open-end capacitive coupling at the symmetry plane of the folded structure. A significant size reduction of 58% has been obtained as compared to conventional filter.

3 SECOND HARMONIC SUPPRESSION

3.1 *Incorporation of periodic grooves*

In this section the problem of harmonic suppression has been addressed by incorporating periodic grooves in folded structure. Traditionally, symmetric microstrip edge-coupled structure supports two quasi TEM modes, i.e., even-mode and odd-mode. For an even-mode excitation, both microstrip lines are induced by the same potentials and magnetic wall symmetry has been observed. Along with this in an odd-mode excitation, both microstrip lines operate at the opposite potentials and electric wall symmetry has been observed. Figure 4 illustrates the even and odd mode symmetry planes [4]. The electromagnetic energy for the odd mode accumulates around the center gap, while for the even mode, it gathers around the outer metallic edges. Hence, the travelling path of the odd mode is to be increased if a coupled section with identical even— and odd—mode electrical lengths is required. Accordingly, due to unequal phase constant i.e., $\beta_o < \beta_e$, odd mode propagates faster than even mode with higher phase velocity. Hence they are characterized by different characteristic impedances. To suppress the second harmonic and to improve the stopband characteristics, periodic square grooves have been introduced symmetrically at the parallel coupled lines of the compact filter structure.

Figure 5 shows the general structure of a parallel-coupled strip line resonator with five square periodic grooves. Here, L_{GR} and W_{GR} are the length and width of the grooves respectively and d is the open end length correction. For 2,3,4 and 5 grooves the periodicity of the location of grooves is L/3, L/4, L/5 and L/6 respectively where L is the length of the resonator.

3.2 *Normalization of Eigen modes*

The symmetric behavior of the phase velocity in the even—and odd-mode of parallel coupled line structure can be observed by inserting periodic grooves, such that the concept of Bragg reflections can be employed in the unit cell for periodic characteristics [1]. Introduction of periodic square grooves in the parallel coupled lines provide an additional electrical path length for the odd-mode. Thus reduction in phase velocity of odd mode has been observed without affecting the electrical path for the even mode. Figure 6 provides the comparison of the group delay plot for different number of square periodic grooves placed symmetrically to

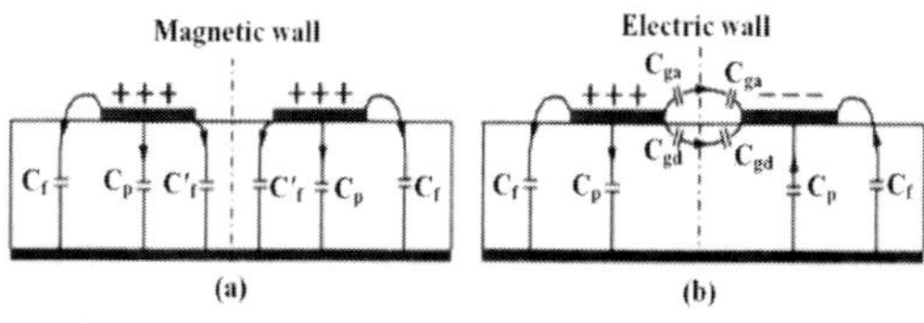

Figure 4. Quasi-TEM modes of a pair of coupled microstrip lines: (a) even mode (b) odd mode [4].

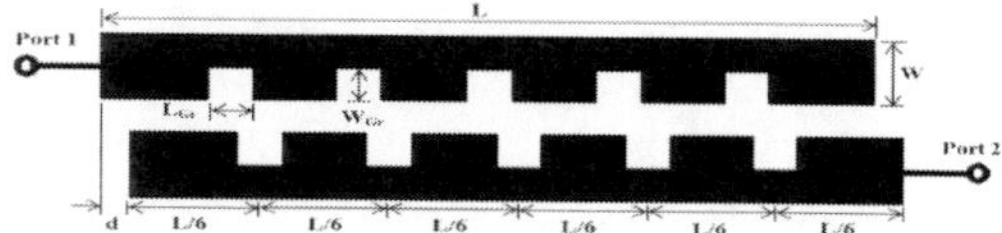

Figure 5. Structure of a unit cell of coupled resonators with square periodic grooves.

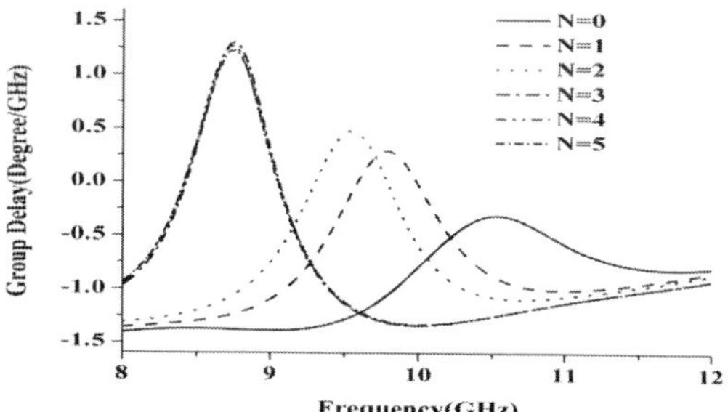

Figure 6. Variation of group delay for a unit cell of coupled lines with square periodic grooves.

the inner edges of the parallel-coupled resonators of a unit cell of band pass filter. From Figure 6 it may be observed that as the number of grooves increases the variation of transmission group delay becomes narrow in bandwidth. Also, the group delay peak value increases and its maxima has been shifted to lower frequency. The Eigen modes normalization and hence the second harmonic suppression has been achieved by optimizing the location of the group delay peak. The most favorable value of the number of grooves (N) has been selected as 3 because of the trivial observation of increased N.

4 PARAMETRIC STUDY

Optimization of length and width of the square grooves are performed by using IE3D EM simulator. Results are shown in Figure 7 and Figure 8 having one groove, located at middle of the resonator. From Figure 7 and Figure 8 it is observed that the periodic square grooves with $L_{Gr} = W_{Gr} = W/2$, provides the satisfactory result of second harmonic suppression. Simulated results of S_{21} are plotted in Figure 9 in which optimization of number of groves (N) has been carried out for folded filter structure. It is observed from Figure 9 that the filter structure with two square periodic grooves located with periodicity L/3 gives the best suppression of second harmonic having suppression level of −44.38 dB. Figure 10 provides the layout of optimized periodic square grooves type filter with two grooves.

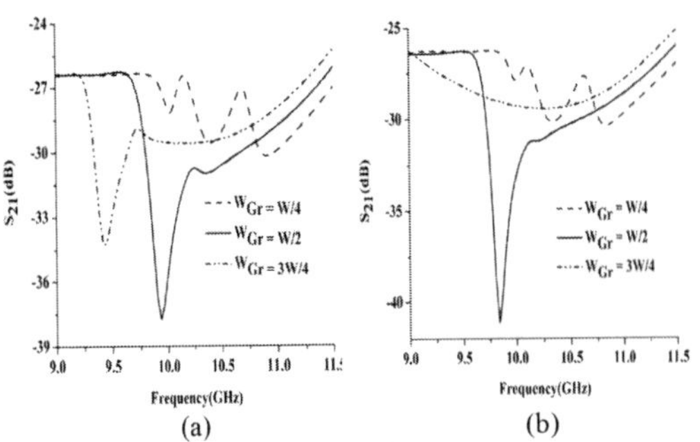

Figure 7. Variation of width of groove with variation in length (a) $L_{Gr} = W/4$, (b) $L_{Gr} = W/2$.

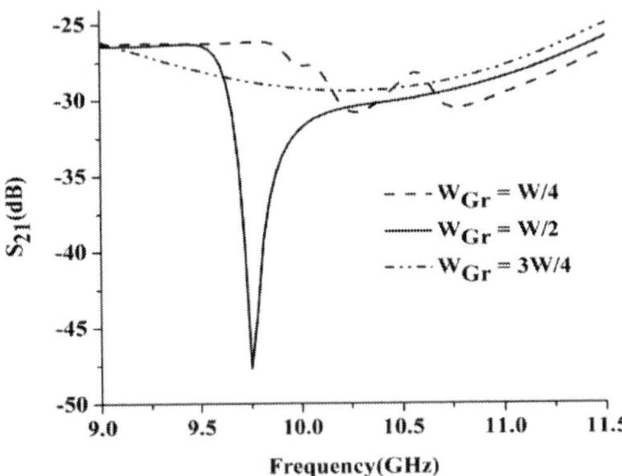

Figure 8. Variation of width of groove with length fixed to 3 W/4.

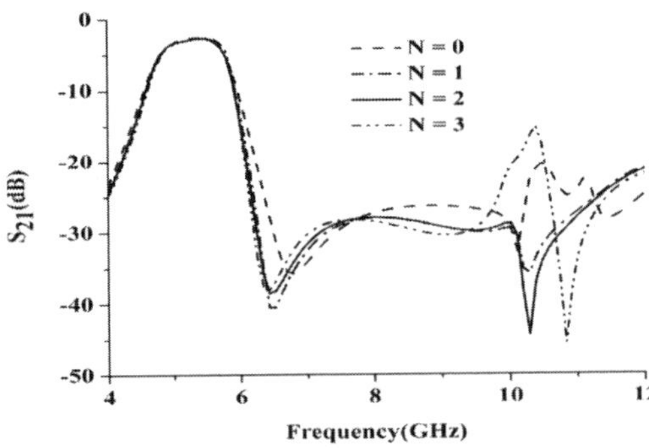

Figure 9. Comparison of simulated S_{21}(dB) versus frequency (GHz) plots of different optimized folded bandpass filters with square grooves.

In Figure 11 the filter characteristics have been investigated by using rectangular grooves with $L_{Gr} = 3$ W/4 and $W_{Gr} = W/2$ in replacement of square grooves. Harmonic rejection level of -43.71dB has been obtained by two rectangular grooves. The size of the optimized filter is 266.27 mm^2. Figure 12 and Figure 13 compare the S-parameters

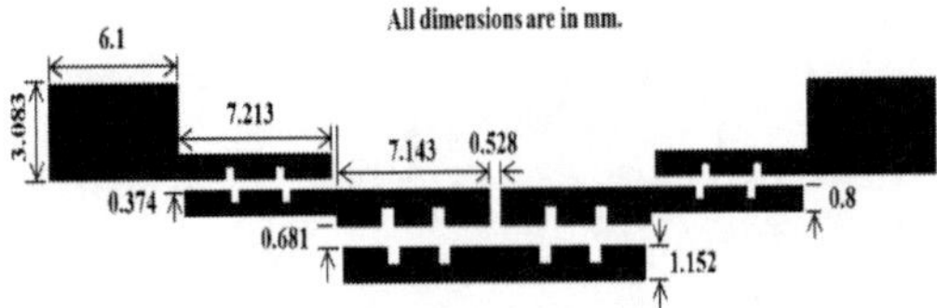

Figure 10. Microstrip layout diagram of a compact parallel-coupled line bandpass filter with two periodic square grooves.

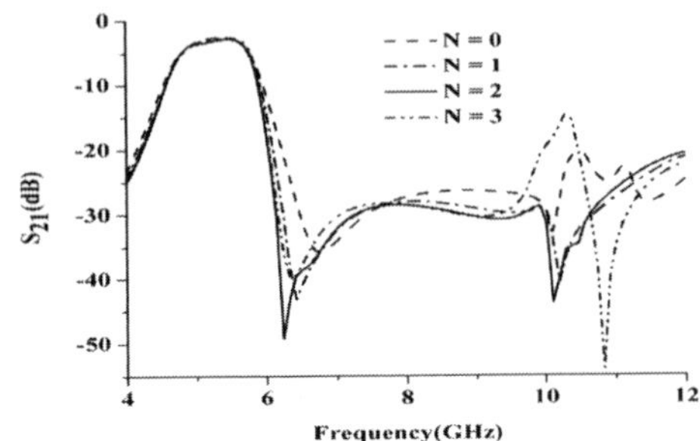

Figure 11. Comparison of simulated S_{21}(dB) versus frequency (GHz) plots of different optimized folded bandpass filters with rectangular grooves.

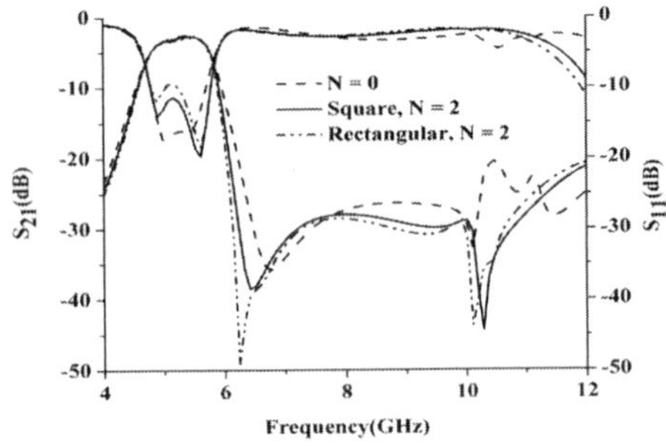

Figure 12. Comparison of simulated S_{21}(dB) versus frequency (GHz) plots of optimized folded bandpass filters with two square and rectangular grooves.

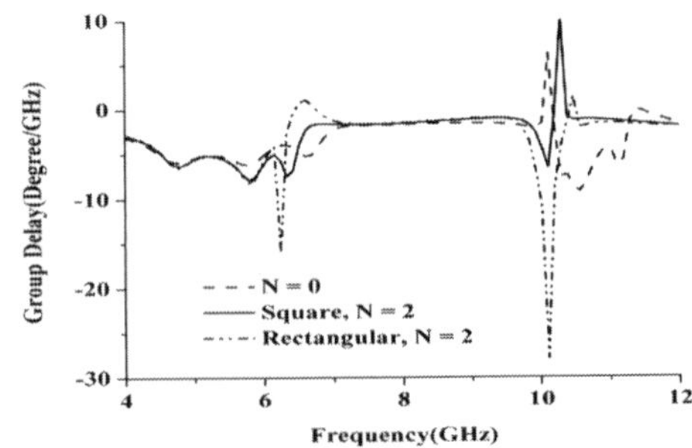

Figure 13. Comparison of group delay of S_{21}(dB) of optimized folded bandpass filters with two square and rectangular grooves.

frequency responses and group delay of S_{21} of folded filter structures with two square and rectangular grooves located at L/3 of each resonator.

5 RESULTS AND ANALYSIS

It is observed from Figure 12 that the filter design with two periodic square grooves gives the best suppression level of -44.38 dB compared to filter design with two rectangular grooves (suppression level of -43.71 dB). The square grooves based design also improves the return loss within the desired passband less than -10 dB, but stopband skirt characteristics are improved for rectangular grooves based BPF.

The Group delay of S_{21} is improved at the 2nd harmonic for square groves based folded filter as shown in Figure 13. The size of the periodic square grooved filter is reduced by 59.05% and that of the periodic rectangular grooved by 59.67% with compared to conventional filter.

6 CONCLUSION

Present paper proposes the design of a compact parallel-coupled microstrip band pass filter design and investigates the effects of second harmonic suppression by incorporating periodic square grooves in the tightly coupled region of coupled line. Asymmetry in phase velocity between even and odd modes has been minimized by the said periodic grooves. Thus an improvement in suppression level of -44.38 dB and -43.71 dB has been achieved using two periodic square and rectangular grooves respectively.

REFERENCES

Bong S. Kim, Jae W. Lee, and Myung S. Song. 2004. An Implementation of Harmonic-Suppression Microstrip Filters with Periodic Grooves. *IEEE Microwave and Wireless Components Letters* 14(9):413–415.

Das, T.K., Chatterjee, S,.2016. Design of a compact symmetrical C-band microstrip bandpass filter with periodic sawtooth grooves. *IAET-2016, Jaipur, India.*

Garcia, J. G., F. Martin, F. Falcone, J. Bonache, I. Gil, T. Lopetegi, M. A. G. Laso, M. Sorolla, and R. Marques. 2004. Spurious passband suppression in microstrip coupled line band pass filters by means of split ring resonators. *IEEE Microwave Wireless Compon. Lett.* 14: 416–418.

Hong, J. S. & Lancaster, M. J. 2001. Microstrip Filters for RF/Microwave Applications. *John Willey & Sons*, Inc.

Jen-Tsai Kuo, Wei-Hsiu Hsu, and Wei-Ting Huang. 2002. Parallel Coupled Microstrip Filters With Suppression of Harmonic Response. *IEEE Microwave and Wireless Components Letters* 12(10): 383–385.

Jun-Seok Park, Jun-sik Yun, and Dal Ahn. 2002. A Design of the Novel Coupled-Line Bandpass Filter using Defected Ground Structure with Wide Stopband Performance. *IEEE Trans on Microwave Theory and Techniques* 50(9): 2037–2043.

Kung, C. Y., Y. C. Chen, C. F. Yang, and C. Y. Huang. 2009. Triple—band parallel coupled microstrip bandpass filter with dual coupled length. *Microwave Opt. Technology Letter* 51: 995–997.

Mattaei, G., L. Young, and E. M. T. Jones, Microwave Filters.1980. Impedance-matching Networks and Coupling Structures. *Artech House, Norwood, MA.*

Mollah, M. N., N. C. Karmakar, and J. S. Fu.2008. Uniform circular Photonic Bandgap Structures (PBGSs) for harmonic suppression of a bandpass filter. *International Journal of Electronics & Communication* 62: 717–724.

Txema Lopetegi, Miguel A. G. Laso, Jorge Hernandez, Miguel Bacaicoa, David Benito, Maria J. Garde, Mario Sorolla, and Marco Guglielmi. 2001. New Microstrip "Wiggly—Line" Filters With Spurious Passband Suppression. *IEEE trans on Microwave Theory and Techniques* 49(9): 1593–1598.

Wu, C. H., C. H. Wang, and C. H. Chen. 2007. Stopband—extended balanced bandpass filter using coupled stepped—impedance resonators. *IEEE Microwave Wireless Compon. Lett.* 17: 507–509.

Computational Science and Engineering – Deyasi et al. (Eds)

Simulation and design of electron gun and focusing system for THz backward wave oscillator

Abhay Shankar
The University of Burdwan, West Bengal, India

Krishna Kumar Belwanshi
Devi Ahilya Vishwavidyalaya, Indore, Madhya Pradesh, India

A. Roy Choudhury & R.K. Sharma
CSIR-Central Electronics Engineering Research Institute, Rajasthan, India

ABSTRACT: This paper presents the simulation and design of electron gun for THz Backward wave oscillator using EGUN, CST-PS and Omnitrak. The 13.5 kV Backward Wave Oscillator electron Gun based on pierce type electron gun consisting of cathode, beam focusing electrode, and anode is under development at CSIR-CEERI. A solenoid magnet focusing is used under confined flow techniques have been designed to achieve desired electron beam focusing with ripples within 10% tolerance.

1 INTRODUCTION

The Terahertz (THz) waves (100 GHz to 10 THz) are of great interest for the development of vacuum electronics devices as it finds enormous applications for wideband communication, imaging radar and spectroscopy (Shur 2005). The BACKWARD WAVE Oscillator (BWO) is the suitable candidate for THz source owing to compact and wide electronics tuning. The design of electron gun for the generation of high density electron beam at THz range is a critical issue of research. The dimensions of THz devices are so small that fabrications are done with special techniques called Micro ElectroMechanical System (MEMs). The practical limits on cathode current density, space charge effect and large magnetic field for beam confinement further add up the complexity to its design. The Pierce type electron gun design with thermionic M-type condenser cathode is the unique choice for BWO Electron Gun. A 13.5 kV convergent electron gun which is capable of delivering beam current ~ 40 mA and having beam radius 0.125 mm is simulated and designed using EGUN, CST-PS and Omnitrak. Approximate gun geometry is drawn consisting of three region cathode, anode and focusing electrode on the basis of synthesis of pierce type electron gun (Gilmour 1986). The optimization of electron gun parameters such as shape of the focusing electrode, cathode and anode are carried out by the simulation tools to get the desired electron beam flow. To minimize the space charge forces a solenoid beam focusing system is designed using CST-PS to achieve a confided beam flow to the entire interaction region.

In this paper, authors have presented a technique to design the electron gun of BWO to be employed for the THz continuous wave generation using commercial software such as EGUN, CST-PS and Omnitrak. Comparisons of theoretical and simulated results have been presented. To achieve the confined flow along with the desired beam waist, solenoid magnetic focusing system has been used.

2 DESIGN OF ELECTRON GUN

The four required beam parameters to design a Pierce type electron gun are as follows: Beam voltage (V_0 = 13.5 kV), beam current (Io = 40 mA), beam waist diameter (2 rw = 0.25 mm) and cathode emission current density ($J_c \cong 0.2$ A/cm^2). Using these beam parameters, the basics gun geometry parameters have been synthesized (Vaughan 1981). The rough gun geometry is drawn from the synthesized parameter like cathode disk and spherical radius, anode-cathode distance, Convergence half angle, which are listed in Table 1.

Table 1. Electron Gun Parameter.

Parameter	Values
Beam Voltage	13.5 kV
Cathode disk radius	0.8 mm
Cathode spherical radius	9 mm
Anode-cathode distance	10.48 mm
Beam Throw	17.25 mm
Convergence half angle	5.09 degree

The above parameters are given as an input to the commercial software like EGUN (Hermannsfeldt 1988), which is based on Finite Difference Method (FDM), along with the required potentials like 13.5 kV at cathode while anode is kept at ground potential. The gun geometry is optimized iteratively by making changes in the shapes of the cathode, Focusing Electrode (BFE) and anode. The process of optimization has been continued until the desired beam parameters like perveance, waist radius and laminarity are obtained, and optimized result is shown in Fig 1. Expanded view of the simulated electron gun is shown in Fig. 2 to show the beam laminarity. Further, the validation of the elecron gun parameter are carried out using CST-PS (https://www.cst.com). It is a 3D electromagnetic simulation tools based on finite integration techniques. The space charge tracking emission model which is based on Child-Langmuir equation is used to solve the space charge limited electron flow.

To achieve the desired current density of beam, the mesh size of the cathode emitting surface is optimized to have the minimum virtual cathode distance. After approximately 30 gun iteration, beam current of ~40.6 mA has been obtained as shown in Fig. 4. The simulated electron gun using CST-PS is shown in Fig. 3.

The Gun design is also re-examined with the FEM code of Omnitark (Humphries 1994) which is shown in Fig. 5. A comparison of the parameters obtained using EGUN, CST, Omnitark are listed in Table 2.

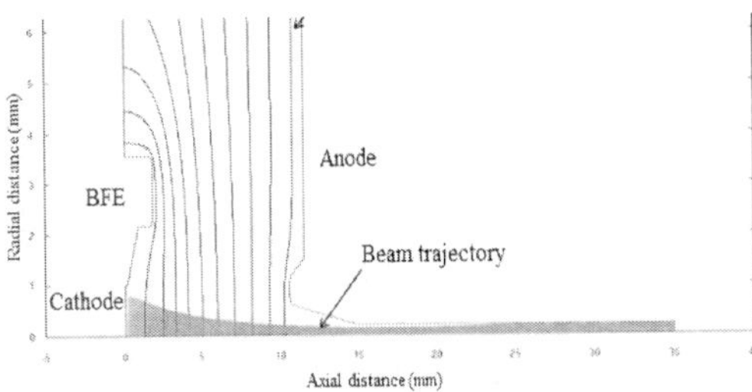

Figure 1. EGUN simulated electron gun.

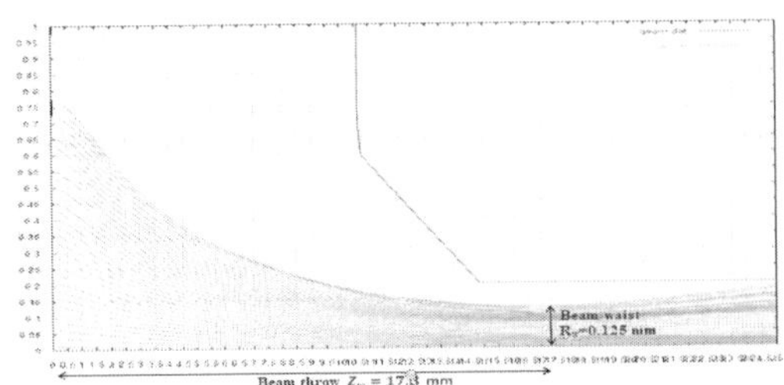

Figure 2. Expanded view of the above-simulated electron gun to show laminarity.

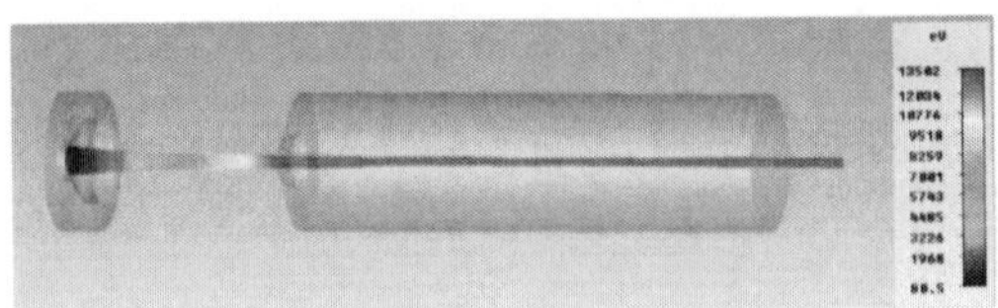

Figure 3. Electron simulated in CST-PS.

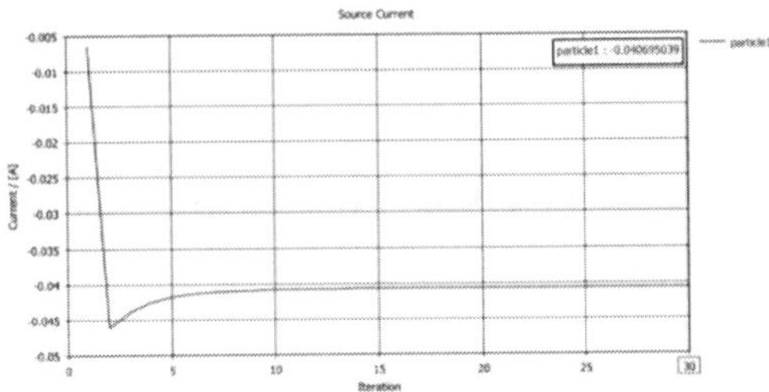

Figure 4. Beam current of electron gun in CST-PS.

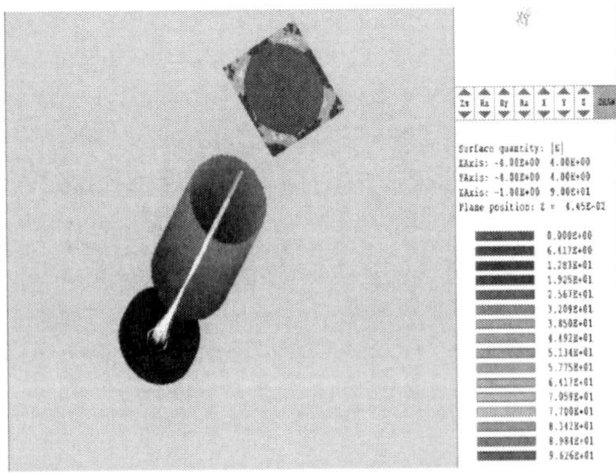

Figure 5. Electron Gun simulated in Omnitrak.

Table 2. The comparison of theoretical and simulated beam parameter

		Simulated		
Parameter	Theoretical	EGUN	CST	Omnitrak
Beam Current (mm)	40	40.4	40.6	39.9
Beam Waist Radius (mm)	0.125	0.125	0.130	0.125

A comparison of different simulator performances in terms of beam perveance is shown in Fig. 6. This shows the close agreement (within

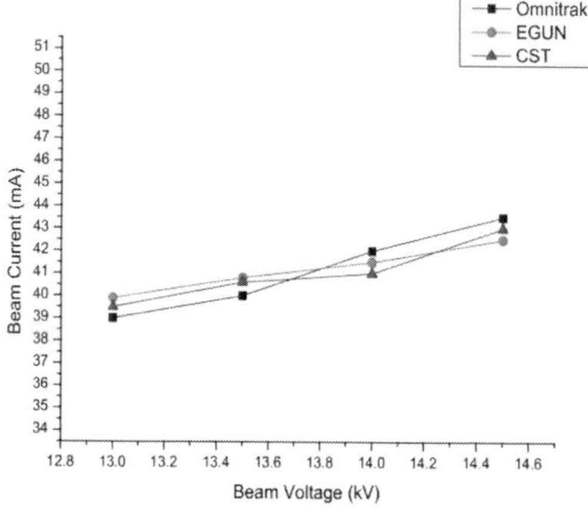

Figure 6. Comparison with different code in terms of perveance with experimental values.

Table 3. Sensitivity Analysis of the Proposed Electron Gun.

Gun	Change by	Change in beam parameter		
Parameter	(mm)	I_0 (mm)	r_w (mm)	Z_w(mm)
Z_{cb}	±0.05	±2.0	±0.06	–
R_c	±0.05	±4.0	±0.005	±0.05
Z_{ca}	±0.05	±2.0	±0.09	±0.35
r_{BFE}	±0.05	±0.5	±0.04	±0.09

Z_{cb} = Axial distance between cathode and BFE.
R_{cb} = Radial distance between cathode and BFE.
Z_{ca} = Axial distance between cathode and anode
r_{BFE} = BFE hat height.

5%) between results obtained from different available commercial software. A sensitivity analysis of the electron gun with respect to the sensitivity of the beam parameters to the different geometrical parameters has been carried out using CST-PS and highlighted in Table 3.

3 DESIGN OF SOLINOIDE FOCUSING

The generation of electron beam in this type of devices gets diverse as it travels away due to space charge forces acting on it. Beam divergence can be prevented with equal and opposite forces and that can be done by using magnetic field aligned with the axis of the beam (Gilmour 1986). The magnet of permanent type or solenoid can be opted having diameter ~30 mm. The focusing system is capable of guiding the beam through the beam tunnel having length of around 75 mm.

The Design of solenoid magnet includes the design of magnetic pole pieces (soft iron) and the non–magnetic (Cu-Ni) spacer for generating the axial magnetic field profile which can be used to focused the electron beam with minimum tolerable limit (i.e. about 10%). Though the both confined (immersed) flow and Brillouin flow focusing are in practices, but for the generation of magnetic field of about 2500 Gauss using confined flow techniques are used. The desired magnetic flux at the cathode for proper launching of the electron beam in the interaction region is calculated from the Busch's Theorem [6]. A confined flow solenoid focusing has been accomplished using CST-PS. Fig. 7 show the simulated electron gun beam flow in the present of optimized axial magnetic field. The beam ripple in this case is ~10% tolerance.

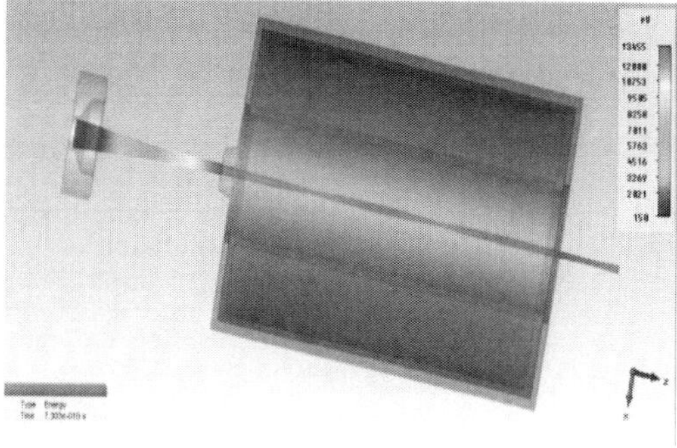

Figure 7(a). Simulated electron gun beam trajectory with magnetic focusing.

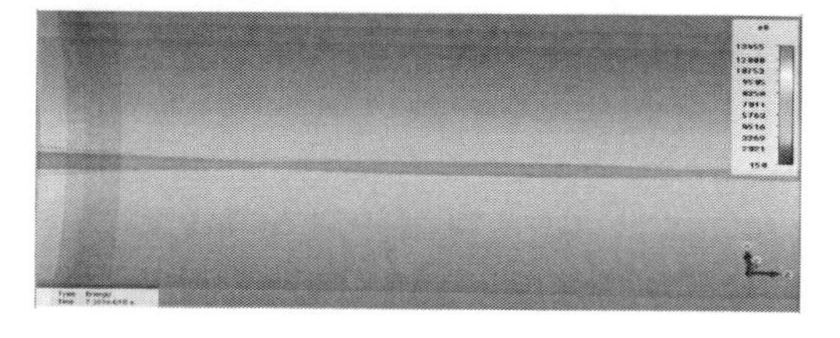

Figure 7(b). Expanded view of above simulated gun to show the scalloping effect.

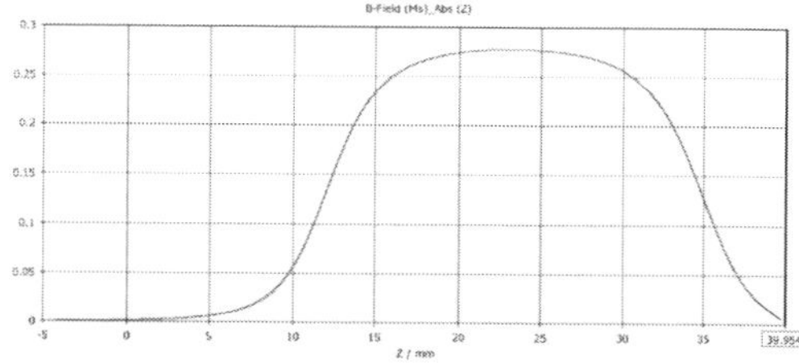

Figure 8. Peak magnetic field profile obtained in CST-PS.

The absolute value of peak magnetic field profile over the interaction region is shown in Fig. 8.

4 CONCLUSION

A pierce type convergent gun for THz Backward wave oscillator has been designed for 40 mA beam current at 13.5 kV beam voltage with 0.125 mm waist radius using EGUN, CST-PS and omnitrak. The beam parameters from different simulation code are in common agreement. Solenoid magnet focusing has been designed using CST-PS for confined flow focusing, keeping beam ripples around 10% tolerance and beam waist radius ~0.125 mm.

ACKNOWLEDGMENT

Authors are thankful to Director CEERI for granting permission to present this work.

REFERENCES

Basu, B. N. 1996. Electromagnetic Theory and Application in Beam-Wave Electronics, World Scientific.

Gilmour, A.S. Jr. 1986. Microwave Tubes. Artech House Inc.

Hermannsfeldt, W. B. 1988. EGUN-An electron optics and gun design Program. SLAC-331, Standford, CA.

Humphries S. Jr., 1994. "OMNITRAK, charged particle Traking in electric and magnetic fields," *Computational Accelerator Physics, AIP Conference Proceeding* 297, pp. 597–604.

https://www.cst.com. CST (computer Simulation Tech nology) microwave software studio 2011.

Sharma, R. K. & Srivastava, V. 2004. Low convergent confined-flow Pierce electron gun for a space TWT. *International Journal of Electronics (IJE)*, vol.91, pp. 97–105.

Shur, Michael. 2005. Terahertz technology: devices and ap plications. *Proceedings of ESSDERC.*, Grenoble, France.

Vaughan.J. R. M. 1981. Synthesis of the pierce gun. IEEE Trans. ED., Vol. 28, pp. 37–41.

Computational Science and Engineering – Deyasi et al. (Eds)
© 2017 Taylor & Francis Group, ISBN 978-1-138-02983-5

VHF to L band scintillation around the EIA crest of Indian longitude zone

S. Chatterjee
Department of Applied Science, RCC Institute of Information Technology, Kolkata, West Bengal, India

D. Jana, S. Pal & S.K. Chakraborty
Department of Physics, Raja Peary Mohan College, Hooghly, India

ABSTRACT: The EIA crest region is most endanger to transionospheric navigation link interruption due to scintillation in the post sunset period of equinoctial months of high solar activity years. Results of studies of ionospheric scintillation at VHF (250.650 MHz), L5 (1176.45 MHz) and L1 (1575.42 MHz) frequency band from Raja Peary Mohan College center (RPMC: 22.66°N, 88.4°E) situated near the Equatorial Ionization Anomaly (EIA) crest reveal remarkable differences in the occurrence features of multifrequency scintillation. The results are discussed in terms of existing theory of evolution, structure and dynamics of electron density irregularities in the low latitude region and corresponding variability.

1 INTRODUCTION

The scientific interest for ionospheric physics, research on ionospheric irregularities has become increasingly important because these phenomena can cause fluctuation in the signal strength of trans-ionospheric links, such as GNSS (Global Navigation Satellite System), IRNSS (Indian Regional Navigational Satellite System) navigation. Ionospheric irregularities related to plasma bubble over the equatorial and low latitude region is a normal occurred phenomenon. The plasma bubble is generated in the bottom side of F-region near the magnetic equator through Rayleigh-Taylor (RT) plasma instability mechanism. Previous studies have found that various scale sizes of irregularity exist inside the plasma bubble, causing scintillation from VHF up to the C band. Ionospheric scintillation refers to random fluctuations in amplitude and phase of radio signals traversing a region of turbulence in the ionosphere. Ionospheric scintillation not only affects satellite communication, but also satellite measurement of ionospheric parameters (*Knepp, 2004*). In this study, we investigate low-latitude ionospheric scintillations. Our study aims to show the characteristics of scintillation associated with low-latitude ionospheric disturbances.

2 RESULTS

One of the major problem to fruitful implementation of the Satellite Based navigation System is the ionospheric scintillation. The most severe effect is observed around the Equatorial Ionization Anomaly (EIA) crest. The prolonged and severe scintillation activity in the equinoctial months of high solar activity years, particularly in the postsunset period, may introduce great difficulties in the operation and in tracking appreciable number of GNSS satellites to achieve the expected accuracy in navigation. This necessitated scintillation studies around the EIA crest in the context of i) feasibility study of SBAS and other GNSS satellites operations, ii) estimating the number of available GNSS tracking during the period of scintillation and iii) deciding the robustness of the receiving system that may accommodate largely the impairments produced in the signal level. GNSS and IRNSS scintillations data recorded at Raja Peary Mohan College Centre (RPMC: geographic 22.66^{0}N, 88.4^{0}E, dip 33.5^{0}N) using GNSS/IRNSS receiver and signal at VHF transmitted from geostationary satellite FSC is recorded using ICOM-7000 receiver at 20/50 Hz sampling rate are used for the present study and S_4 is calculated using usual 3rd peak method.

Ionospheric scintillation data for the vernal equinoctial months of April 2015 are analyzed under present investigation to study multi-frequency, extending from VHF to L band scintillation features near the northern EIA crest of the Indian longitude zone.

Figure 1 shows the temporal variation of S_4 variations at L5, L1 and VHF frequencies showing scintillation on 7th April, 2015. Scintillation Index (S_4) = (P_{max} – P_{min}/P_{max} + P_{min}), where P_{max} is the power amplitude of the third peak down from- the maximum max excursion of the scintillations, and P_{min} is the power amplitude of the third level up from the minimum excursion [*Whitney et al.*, 1969]. S_4 index, normalized standard deviation of signal intensity, is being used to express scintillation strength here. Figure 2 shows percentage occurrence of simultaneous SBAS scintillation during the post sunset period of vernal equinoctial months. The results revealed that most occurrence probability (30–34%) of simultaneous scintillation is observed during 2100–2300 IST hr at L1 frequency band and follows a Gaussian distribution.

A statistical correlation analysis has also been performed between the fade rate at VHF frequency and to those at L5 frequency which shows a statistically significant correlation value (at 99% significant level) of ~ 0.71 (Figure 3). A correspondence between fade rates (below saturation level) at multi frequency band is thus evident.

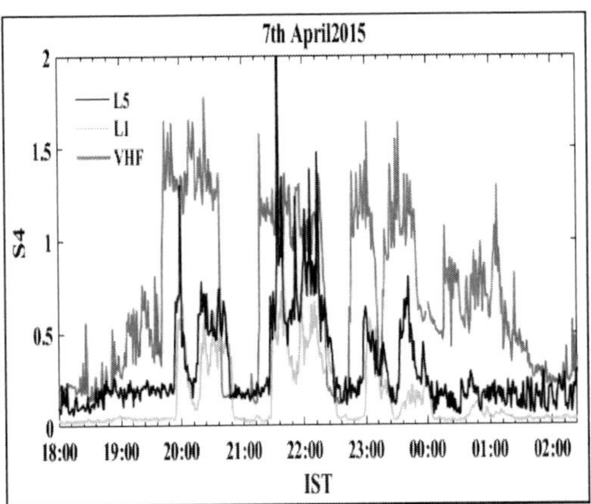

Figure 1. Temporal variation of S4 on 7th April, 2015.

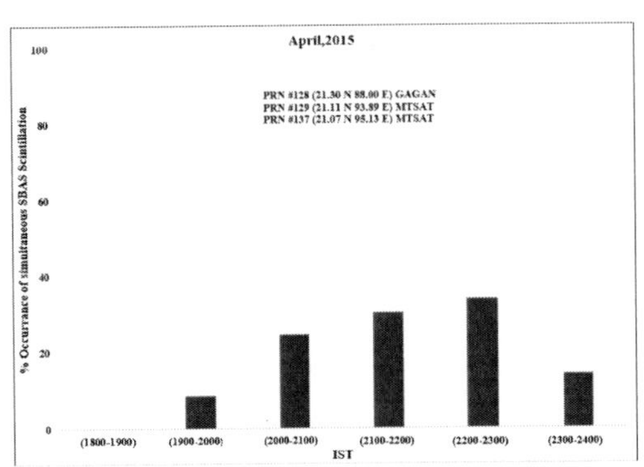

Figure 2. Percentage distribution of simultaneous scintillation in the links of GAGAN and MTSAT.

Recently, a method of measuring the effects of ionospheric irregularities on the GPS signals is being adopted by several research groups [*Wanninger, 1993; Doherty et al., 1994; Aarons et al., 1996*]. It characterizes the GPS phase fluctuations by measuring the time rate of differential phase of dual frequency GPS signals, known as rate of TEC (ROT) in the unit of TECU/min. To identify the present of smaller scale irregularities, a rate of TEC index (ROTI) based on the standard deviation of ROT, i.e., ROTI = $\sqrt{[(ROT^2)-(ROT)^2]}$ [*Pi et al. 1997*] is estimated. We compute the ROTI value for each 1-minute time interval. Figure (4) shows the temporal variation of ROTI in the GPS link #5 on the different days of April, 2015. The median values of ROTI for the scintillating days follow a Gaussian (normal) distribution.

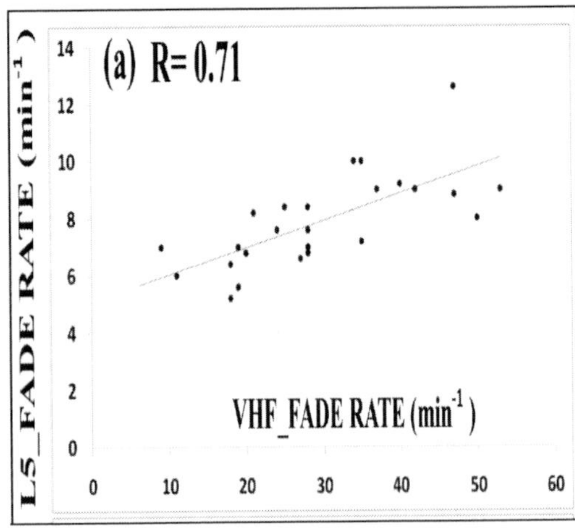

Figure 3. Variation of fade rate VHF vs L5. Statistically significant correlation coefficient (R) are shown.

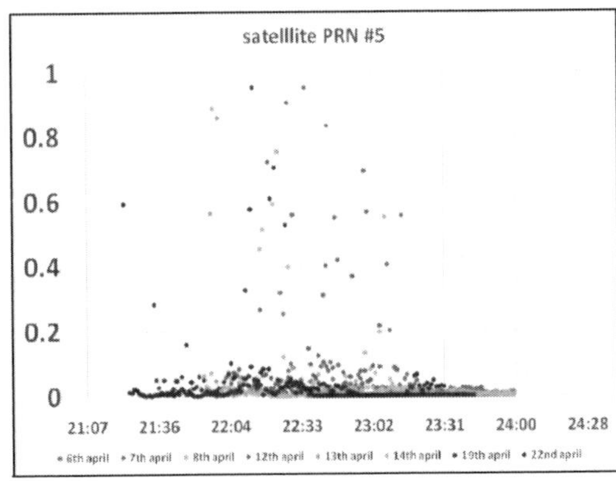

Figure 4. Temporal variation of ROTI on the scintillation day.

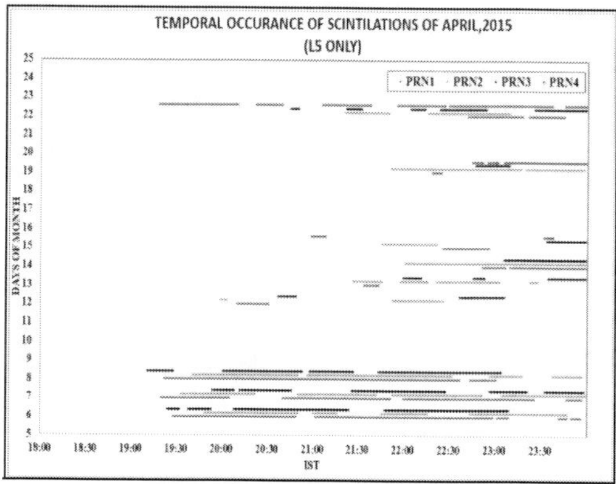

Figure 5. Date wise distribution of scintillation time during the month of April 2015 at L5 for IRNSS links.

Figure 5 shows the bar plot showing temporal distribution of scintillation occurrences at L5 band on the daily basis for the month of April 2015. PRN #1 and #2 and #4 are in Geo-Synchronous orbit while #3 is in Geo-Stationary orbit. Obviously there are several days in the equinoctial months when multi satellite and multi frequency links of IRNSS are plagued by ionospheric scintillation.

3 CONCLUSION

In the present investigation, efforts are made to identify the suitable precursors in post sunset scintillation occurrence near the anomaly crest (Calcutta) during equinoctial months of 2015. The results of multi-frequency scintillation observations at VHF (250 MHz) and L band frequencies reveal distinguishing traits in the fine structures of scintillation occurrence, such as onset time and duration, fade rates and depth, decorrelation time etc. A good correspondence between the coherence time and scintillation strength parameter at L band reflects the fact that perturbation strength is controlling parameter for intense scintillation. Simultaneous occurrence of scintillation at L5 and L1 band at SI > 3dB in all the available IRNSS/GNSS links may call for feasibility study of augmentation facility. Around the EIA crest both the latitudinal and longitudinal span of the Indian longitude zone is much vulnerable to severe scintillation activities producing perturbation in GNSS services. The major concern is simultaneous occurrence of scintillation with $S_4 > 0.5$ at different SBAS as well as IRNSS links leading to failure of the augmentation systems. The extent of ionospheric variability as revealed through studies of scintillation near the equatorial anomaly crest may form an important database for modelling of ionospheric scintillation, which is a major concern of present-day space research activities. To develop a fruitful model of multi-frequency scintillation further study, under varying solar geophysical condition are to be made.

REFERENCES

Aarons, J., M. Mendillo, and R. Yantosca, (1996) GPS phase fluctuations in the equatorial region during the MISETA 1994 campaign, J. Geophys. Res., 101, 26851.

Doherty, P., E. Raffi, J. Klobuchar, M. B. El-Arini, (1994) Statistics of time rate of change of ionospheric range delay, Proc. of ION GPS-94, part 2, 1589.

Franke, S. J. and Liu, C. H., (1983) Observations and modeling of multifrequency VHF and GHz scintillations in the equatorial region, J. Geophys. Res., 88, 7075–7085.

Kinter, P. M., Kil, H., Beach, T. L., and de Paula, E. R., (2001) Fading timescales associated with GPS signals and potential consequences, Radio Sci., 36, 731–743.

Knepp, D. L. (2004) Effects of ionospheric scintillation on Transit satellite measurement of total electron content, Radio Sci., 39, RS1S11.

Pi, X., Mannucci, A. J., Lindqwister, U. J., and Ho, C. M., (1997) Monitoring of global ionospheric irregularities using the worldwide GPS network, Geophys. Res. Lett., 24, 2283.

Rufenach, C. L., (1975) Ionospheric scintillation by a random phase screen: Spectral approach, Radio Sci., 10, 155–165.

Whitney, H. E., Aarons, J., and Malik, C., (1969) Proposed index for measuring ionospheric scintillation. Planet Space sci., 17, 1069.

Whitney, H. E., Basu, S., (1977) The effect of ionospheric scintillation on VHF/UHF satellite communications. Radio Sci., 12, 123–133.

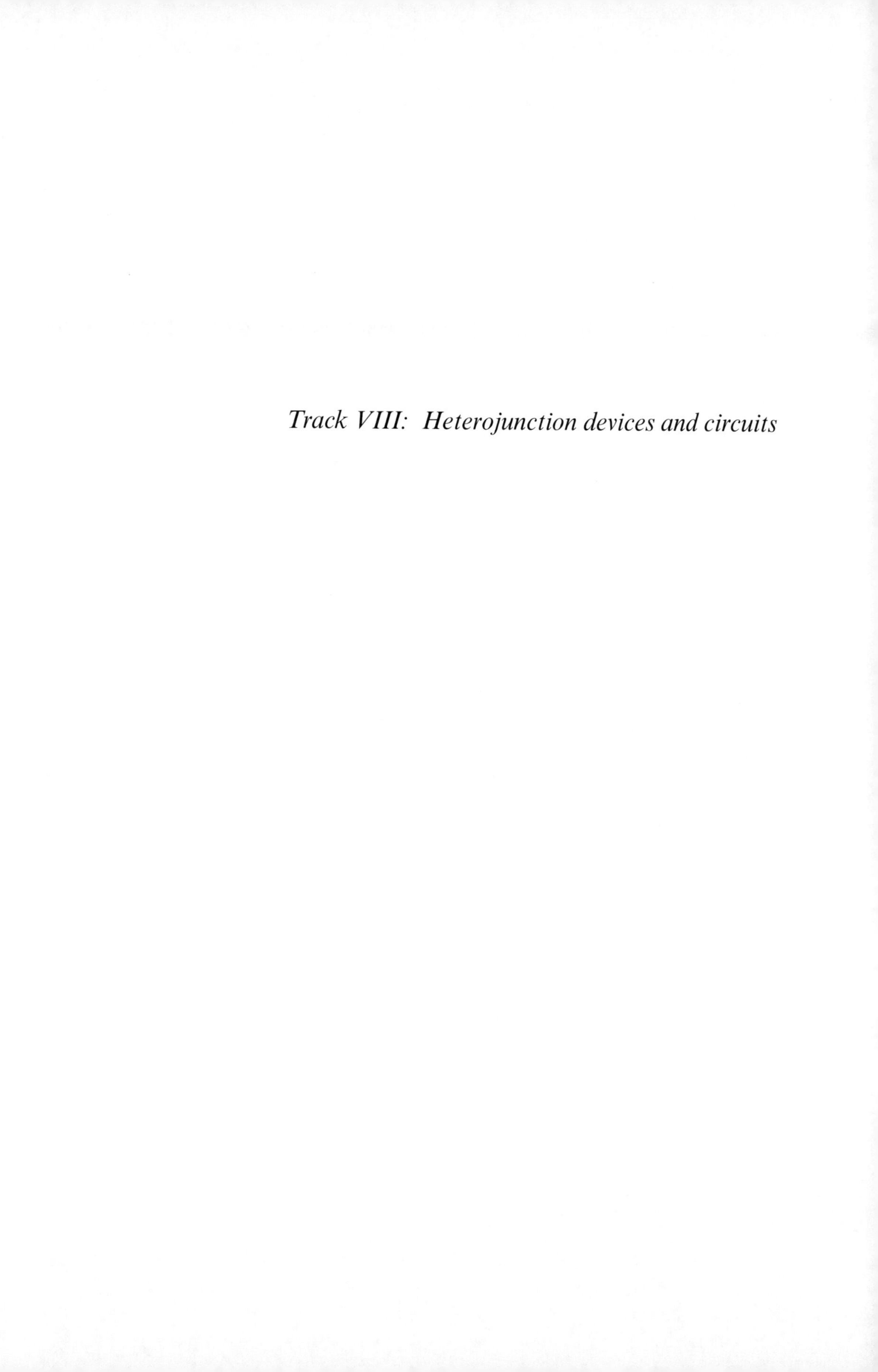

Track VIII: Heterojunction devices and circuits

Computational Science and Engineering – Deyasi et al. (Eds)
© 2017 Taylor & Francis Group, ISBN 978-1-138-02983-5

Studying the comparative performance of p-CuO/n-Si thin film hetero-junction solar cells grown by chemical bath deposition and vapor liquid solid processes

J. Sultana & S. Paul
Centre for Research in Nanoscience and Nanotechnology (CRNN), Kolkata, India

A. Karmakar & S. Chattopadhyay
Department of Electronic Science, University of Calcutta, Kolkata, India

ABSTRACT: This work investigates the comparative performance analysis of CuO thin films grown by employing chemical bath deposition and vapor liquid solid methods on n-Si <100> substrates for the fabrication of p-CuO/n-Si hetero-junction diodes. It is worthy to note that for the growth of CuO film by VLS method, CBD grown CuO powder has been used. Studies are performed in terms of the film morphology, chemical composition, crystallite structure and energy conversion efficiency. The potential for photovoltaic applications of such films are investigated by measuring the relevant junction current-voltage characteristics and by extracting the relevant photovoltaic parameters including open circuit photo-generated voltage, short circuit current density, fill-factor and energy conversion efficiency.

1 INTRODUCTION

Semiconductor materials used in solar cell technology are principally governed by their energy band gap, optical properties, and charge carrier diffusion length (Serin et al. 2005) Copper oxides (CuO) are the transition metal oxides, intrinsically a p-type semiconductor predominately due to copper vacancies. The features of copper oxide semiconductors such as high optical absorption coefficient, non toxicity and low cost fabrication have drawn considerable attention in recent days due to its multitude of applications in the domain of electronics, optoelectronics, photovoltaic's, sensing, catalysis, and magnetic storage media (Paul et al. 2015) CuO thin films can be grown by employing different techniques (Liao et al.2009; Nagase et al. 1999). In the current work, CuO thin films are grown by employing Chemical Bath Deposition (CBD) and vapor liquid solid (VLS) processes. The impact of difference in quality of VLS and CBD grown films on the performance of p-CuO/n-Si hetero-junction in terms of its electrical and photovoltaic characteristics is investigated. The p-CuO films are physically and optically characterized by employing Field Emission Scanning Electron Microscopy (FESEM), X-Ray Diffraction (XRD) and Spectroscopic Ellipsometer (SE) measurements. Electrical performance of the p-CuO/n-Si hetero-junction diode is investigated by measuring its current-voltage (I-V) characteristics and its potential for photovoltaic applications is investigated by extracting the values of open circuit photo-generated voltage (Voc), short circuit current density (Jsc), Fill-Factor (FF) and power conversion efficiency from the relevant electrical measurements in both dark and illuminated conditions.

2 EXPERIMENTAL DETAILS

The CuO thin films are grown on Si (100) substrates. Prior to the growth process, Si wafers are subjected to the standard RCA–I and RCA-II cleaning, followed by a dip in 20% HF solution for native oxide removal. The CuO thin films are grown by following two different techniques: CBD and VLS. For the synthesis of CuO thin films by CBD technique, 0.1 M, 100 ml solution of CuCl2.2H2O is prepared and heated with constant stirring. When the bath temperature rises to 60 OC, 90 drops of ammonia solution are added to form the reaction mixture and the deposition is continued for 25 min for the formation of CuO thin films. After the deposition, the substrate is taken out of the bath and the reaction mixture is further heated until CuO precipitates and is finally separated out and dried up for the extraction of CuO powder. This CBD grown CuO powder is further employed for the synthesis of CuO thin films by VLS technique. The cleaned Si wafers are initially coated with a thin film (~ 7 nm) of gold (Au) by DC sputtering. The growth is performed at 600°C for 40 min in a 2-zone furnace with Ar as the carrier gas at a deposition pressure of 183.9 Torr. The films grown are finally utilized t0 develop p-CuO/n-Si hetero-junction to investigate their comparative photovoltaic performance. For the realization of the fabricated heterojunction, the back contact is taken by evaporating Au and the top CuO layer is covered by an ITO layer of thickness 100 nm and then Al dots of radius 10 μm is deposited to take electrical contact by the W-probe of radius 10 μm. The surface morphology and thickness of the grown samples are characterized by using field-emission scanning electron microscopy (FESEM) (Zeiss Auriga 39–63) and spectroscopic ellipsometric (SENTECH SE850) techniques. The crystallographic analysis and orientation of the grown samples is conducted by employing X-Ray Diffraction (XRD) method. The electrical characterizations of p-CuO/n-Si hetero-junction are performed by using Keithley 4200-SCS, under both the dark and illuminated condition. The photovoltaic parameters are measured under white light illumination of 0.037 mW/cm^2 input power.

3 RESULTS AND DISCUSSION

3.1 *Structural, material and optical characterization of p-CuO thin film grown on n-Si*

Figure 1(a), (b) and (c) shows the SEM micrographs of CuO powder, synthesized by CBD method, and the as-deposited CuO thin film on n-Si substrate prepared by CBD and VLS methods, respectively.

It is apparent from Figure 1(a) that the CBD grown CuO powder acquires a spindle like shape, however, the film grown by VLS technique, using

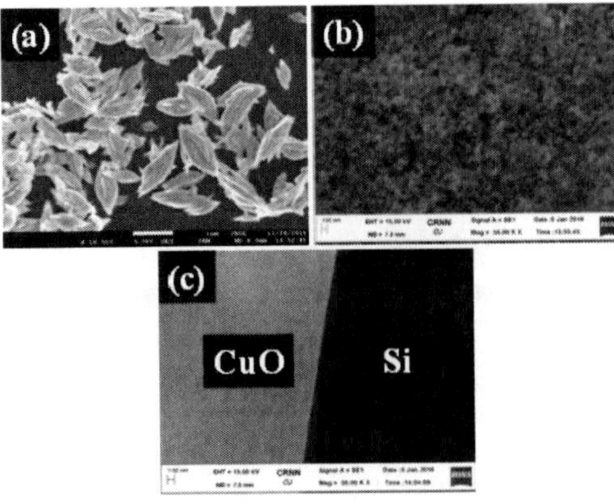

Figure 1. SEM images of (a) CuO powder; (b) CBD grown film and (c) VLS grown film.

the same powder shows a uniform morphology without any visible grain boundary as in Figure 1(c). Also, there is an obvious difference in the morphology of the films grown by CBD and VLS techniques as confirmed from Figure 1(b) and (c). This disparity in the film morphology can be attributed to the growth mechanism of the two methods i.e. an ion assisted growth in CBD, whereas layer assisted growth in VLS method. The average thickness of the films for the CBD and VLS techniques are measured to be ~110 nm and ~34 nm respectively, from the tilted SEM image. Furthermore it is interesting to note that the thickness of the VLS deposited film is much less than the CBD deposited film, however the deposition time of the CBD grown film (25 min) is less than the VLS grown film (40 min). This inconsistency may also be attributed to the different transport mechanism of the formed CuO particles in the respective two methods. In case of CBD technique, the reaction mixture itself acts as the medium for the transport of the CuO species, whereas Argon (Ar) acts as the carrier gas for the transport of CuO vapor in case of VLS technique.

The crystallographic analysis of the grown powder and the films are performed by X-ray diffraction measurements and are shown in Figure 2(a). The grown CuO powder and films exhibited diffraction peaks corresponding to monoclinic CuO which have been verified by JCPDS card no. 80–1917. It is apparent from the XRD patterns that the CBD grown CuO powder and film both comprises of polycrystalline nature, however, the VLS grown CuO film exhibits single crystalline nature, which is attributed to the preferred [111] phase at a growth temperature of 600°C in VLS grown film. The average crystallite size is calculated by using Scherrer formula, $D = 0.9\ \lambda/\beta\cos\theta$, where λ is the wavelength of X-ray radiation used, β is the full-width-at-half-maximum (FWHM) of the peaks at the diffracting angle, θ. The average crystallite size for the dominating [111] phase are calculated to be 14.66 nm, 21.99 nm and 35.32 nm for the CuO powder, CBD grown film and the VLS grown film, respectively. The largest crystallite size of the VLS grown CuO film indicates the highest amount of CuO phase formation and hence the maximum absorption as compared to the CBD grown film.

The spectroscopic ellipsometric (SE) measurements have been performed on the CBD and VLS grown films in the wavelength range of 300–800 nm and the plots of relative amplitude change (Ψ) and phase change (Δ) are shown in Figure 2(b). Such results are simulated to fit the experimental data to extract the thickness of the CuO films and the relevant model structure is considered to be Air/Au/CuO/Si and Air/CuO/Si for VLS and CBD grown films respectively. The optical constants of

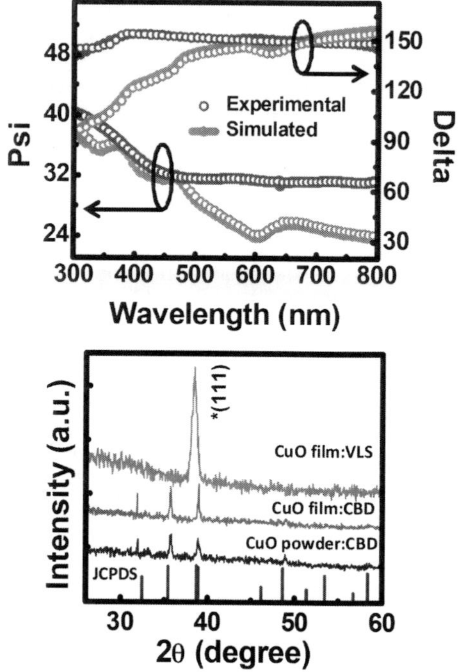

Figure 2. (a) XRD spectra of the grown CuO powder and films and (b) Measured (symbols) and model fit (lines) ellipsometer parameters, Ψ and Δ, for the CuO film on Si substrate in the range of 300–800 nm at 60° (blue for CBD grown sample and red for VLS grown sample).

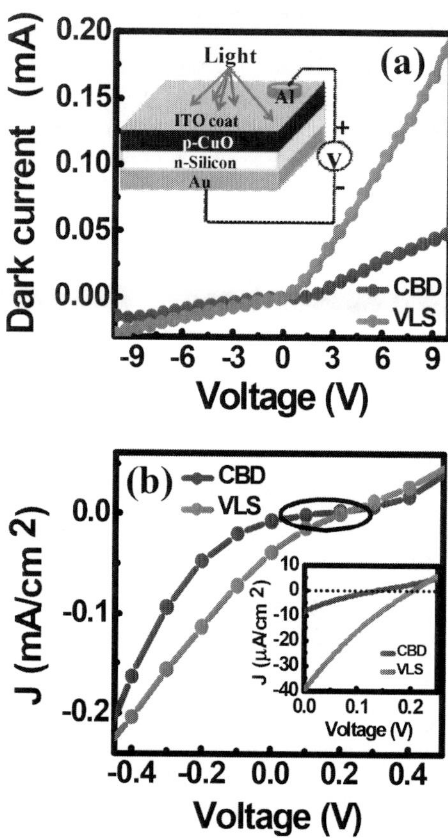

Figure 3. (a) Comparative I-V plots under dark condition in linear scale (inset shows schematic of the fabricated device) and J-V plots under illumination (inset shows the area marked by circle) of the hetero-junction diodes fabricated the CuO films grown by CBD and VLS techniques.

air and the wavelength of incident light have been considered to remain constant in the fitting process. The modeled values of Ψ and Δ of the sample is also plotted in Figure 2(b). It is apparent from such plots that the best-fit data match well with the measured data, thereby, suggesting the model to be extensive over a wide range of wavelength. The thicknesses of the grown CuO films are obtained to be 110.67 nm and 34.14 nm for the growth time of CBD and VLS grown films respectively, which corroborate with the value obtained from tilted SEM images.

3.2 *Electrical characterization of p-CuO/n-Si heterojunction diode*

It is seen from Figure 3(a) that under dark condition, the current for the heterojunction made by VLS grown CuO film is much higher in comparison to the heterojunction fabricated on CBD grown CuO film. This is attributed to the reduction of series resistance and formation of films with superior morphology in VLS method in comparison to the ion-assisted CBD grown process.

The respective values of rectification factor, ideality factor and barrier height are calculated from the I-V plots and are obtained to be 2.53, 5.54; 1.5, 1.44 and 0.5, 0.15 for the hetero-junctions fabricated on CBD and VLS grown CuO films, respectively. The schematic of the fabricated device is shown in the inset of Figure 3(a). Figure 3(b) shows the J-V plots under illumination (inset shows the area marked by circle) of the hetero-junction diodes fabricated by VLS and CBD processes. An obvious photocurrent was observed under illumination and both structures showed characteristic curves with open-circuit voltage and short-circuit current. The negative photocurrent at 0 V bias is due to the photo-generated carriers which overcome the heterojunction barrier height, resulting to a reverse photo-generated current. The VLS assisted p-CuO/n-Si hetero-

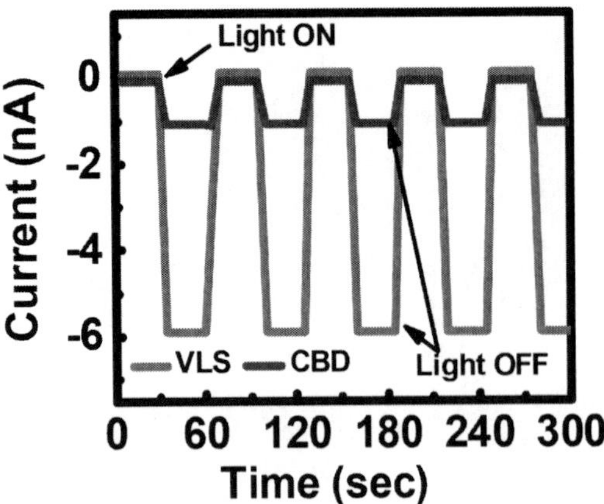

Figure 4. Plot of light 'ON'/ 'OFF' switching characteristics of the hetero-junction photodiodes at zero-bias (0V).

junction exhibits superior performance in comparison to the CBD assisted heterojunction in terms of rectification factor, ideality factor and photovoltaic parameters, which is attributed to the superior crystalline quality, morphology and enhanced absorption owing to high surface to volume ratio. The short circuit current density, open circuit voltage and power conversion efficiency of the CBD and VLS assisted hetero-junctions are obtained to be 0.007, 0.038 mA/cm2; 133, 201 mV and 0.29%, 4.04% respectively, at a power input of 0.037 mW/cm^2.

The time-dependent photocurrent response of the sample is measured periodically by switching 'ON' and 'OFF' the light source for a cycle of 60 s each and the relevant rise- and fall- time, as shown in Figure 4. From Figure 4 the rise and fall time obtained to be 3.8 s, 2.94 s and 4.4 s, 3.2 s respectively for the CBD and VLS employed heterojunctions.

The light 'ON'/ 'OFF' characteristics indicate that such CuO thin film/Si hetero-junction can also be used for photo-detection. The comparatively slow response of CBD grown samples can be attributed to the high built-in potential and series resistance which prevents the transport of carriers thereby, introduces delay in the response time.

4 CONCLUSIONS

CuO thin films have been grown by employing two different techniques: VLS and CBD. Further, for the growth of CuO film by VLS technique the CBD grown CuO powder has been used. The thickness, morphology, crystalline and photovoltaic properties of the grown CuO thin films have been systematically studied for the grown samples by employing different techniques. It is observed that the growth mechanism has a significant impact on the thickness, crystallite size, and photovoltaic properties of the films. It is observed that a good stoichiometric CuO film with an average thickness of 34 nm has been possible to grow by 40 min of growth time using VLS technique. The grown films has been subsequently used for the fabrication of p-CuO/n-Si heterojunction diodes and is observed to deliver the best photovoltaic properties in comparison to the CBD method assisted heterojunction.

ACKNOWLEDGEMENT

Ms. Paul would like to thank University Grants Commission for providing National Fellowship to pursue her research work. The authors would also like to acknowledge the Centre of Excellence (CoE), TEQIP-phase II, World Bank and the Department of Electronic Science (C.U) for providing necessary infrastructure to conduct this work.

REFERENCES

Liao, L., Zhang, Z., Yan, B., Zheng, Z., Bao, Q.L., Wu, T., Li, C.M., Shen, Z.X., Zhang, J.X., Gong, H., Li, J.C., & Yu, T. 2009, Multifunctional CuO nanowire devices: p-type field effect transistors and CO gas sensors, *Nanotechnology* 20: 085203.

Nagase, K.., Y, Zhang., Kodama, Y. & Kakuta, J. 1999, Dynamic study of the oxidation state of copper in the course of carbon monoxide oxidation over powdered CuO and Cu2O, *J. Catal.*187: 123–130.

Paul, S., Das, A., Sultana, J., Karmakar, A., Chattopadhyay, S. & Bhattacharyya, A. 2015, Performance investigation of n-ZnO nanowire / p-CuO thin film heterojunction solar cell grown by chemical bath deposition and vapour liquid solid technique, *Proc. IEEE* 978–1–4673–9513–7/15.

Serin, N., Serin, T., Horzum, S. & Yasemin, C. 2005. Annealing effects on the properties of copper oxide thin films prepared by chemical deposition, *Semicond. Sci. Technol.* 20: 398–401.

Computational Science and Engineering – Deyasi et al. (Eds)
© 2017 Taylor & Francis Group, ISBN 978-1-138-02983-5

Investigation of oxygen vacancy induced resistive switching memory behavior in low-temperature grown n-ZnO/p-Si heterojunction diode

R. Saha, A. Das & A. Karmakar
Department of Electronic Science, University of Calcutta, Kolkata, West Bengal, India

N.R. Saha
Department of Polymer Science and Technology, University of Calcutta, Kolkata, West Bengal, India

S. Chattopadhyay
Department of Electronic Science, University of Calcutta, Kolkata, West Bengal, India

ABSTRACT: The current reports the unipolar bistable resistive switching behavior observed in n-ZnO/p-Si heterojunction diodes. The vertically aligned n-ZnO nanowires/p-Si heterojunction diodes are fabricated by low-temperature chemical bath deposition technique. The growth of ZnO nanowires are confirmed by SEM imaging and their crystalline structure is investigated by X-ray diffraction measurement. Photoluminescence spectra confirm the energy band gap of ZnO nanowires and also the presence of oxygen vacancy related defects and interstitials. The HRS to LRS resistance ratio of the heterojunction is quite high and endurance of the device is also superior compared to other resistive switching devices. Conductance spectroscopy and capacitance-voltage measurements are also performed to reveal the conduction mechanism.

Keywords: ZnO nanowires, resistive switching, oxygen vacancy, conductance spectroscopy

1 INTRODUCTION

Nonvolatile memories based on resistive switching have attracted a great deal of attention for last few decades. Currently, metal oxides like ZnO, HfO_x, TiO_2, NiO and Al_2O_3 based resistive switching memory have been extensively studied as one of the most promising candidates for future nonvolatile memory applications due to its fast switching speed, superior scalability, high ON-OFF ratios, high endurance and compatibility with silicon Complementary Metal-Oxide Semiconductor (CMOS) technology. Owing to its direct bandgap, intrinsic n-type doping and high excitonic binding energy of 60 meV, ZnO is an important material for electronic and optoelectronic applications [Das et al, 2014 & Das et al 2016]. Recently researchers have already shown that ZnO Nanowires (NWs) are the most promising candidate for unipolar and bipolar nonvolatile memory applications. Several methods have been reported for the growth of high quality ZnO NWs including chemical vapor deposition, vapor-liquid-solid method, sputtering, sol-gel and chemical bath deposition process. Among all, Chemical Bath Deposition (CBD) technique is extensively used for NWs growth, which is most effective since it enables the cost effective large area growth. For the wide application of unipolar and bipolar resistive switching based memories in digital circuits, current-voltage (I-V) characteristics of n-ZnO NWs/p-Si heterojunction has been performed. Different mechanism have been proposed to investigate the actual switching phenomena, including the formation of conductive filaments along the NWs by positively charged oxygen vacancy related sites and pure electronic switching, which originates from the trapping and detrapping processes through oxygen vacancies or Zn interstitials defects in ZnO. In this work, resistive switching performance of n-type ZnO/p-Si heterojunction diode is studied by measuring current-voltage, capacitance- voltage and conductance spectroscopy characteristics. The oxygen related defect and vacancy sites are considered to be responsible for such resistive switching behavior and its existence is confirmed by Photoluminescence (PL).

2 EXPERIMENTAL DETAILS

2.1 *Synthesis of ZnO nanowires on p-Si substrate*

For ZnO seed growth, a CBD/CBD technique is employed [Paul et al], where the ZnO seeds are prepared using CBD technique. Standard boron-doped p-type <100> silicon (Si) substrate of carrier concentration $N_A = 4.9 \times 10^{18}$ cm^{-3} is used. Prior to growth, the substrate is cleaned with trichloroethylene, acetone and iso-propyl alcohol sequentially for 5 min under ultrasonication followed by HF oxide etching. The cleaned substrate is sensitized with ZnO nanoparticle seed prepared by CBD method. The substrate is dipped in 100 ml aqueous solution containing 50 mM of Zinc nitrate hexahydrate ($Zn(NO_3)_2.6H_2O$). The solution is heated till the bath temperature reaches 60 °C followed by addition of 90 drops of ammonia drop wise and then again heated for 30 min for the seeds to grow. The samples are taken out, rinsed in running deionised water and dried in N_2 gas flow. For the ZnO nanowire growth, equimolar aqueous solutions, each consisting of 50 mM of zinc nitrate hexahydrate and hexamethyltetramine (C_6H_{12} N_4) are prepared separately using 50 ml of DI water and then mixed together. The ZnO seed sensitized Si substrate is dipped vertically in the respective bath solution using a sample holder with heating up to 90°C for 2 hours. After the nanowire growth, the substrate is rinsed with de-ionized water and dried by blowing N_2 gas.

2.2 *Material characterisation and electrical measurements*

The structural morphology of the grown sample is characterized by using Field Emission Scanning Electron Microscopy (FESEM) (Zeiss Auriga 39–63). The x-ray diffraction (PAN analytical X'Pert Pro X-ray Diffractometer) is employed to study the crystallographic orientation of the grown samples. Energy Dispersive X-ray spectrometry (EDAX, JEOL-JSM 7600F) is used to compare both the compositional as well as the elemental analysis of the grown nanowire. Photoluminescence spectra of ZnO NWs are acquired for the wavelength range 350 nm to 700 nm. The electrical characterizationsare carried out using Keithley 4200 SCS semiconductor characterization system connected to probe station (Cascade Microtech-1200M). Electrical bias is applied to the p-Si side and n-ZnO side is electrically grounded throughout the experiment. The behavior of junction capacitance and conductance-spectroscopy has been performed in detail to investigate the relevant electrical properties and defect states.

3 RESULTS AND DISCUSSIONS

3.1 *Structural and material characteristics of ZnO nanowires on p-Si*

Fig. 1 shows FESEM image of the sample, confirming the growth of vertical ZnO nanowires on

Figure 1. FESEM image of top view of ZnO NWs on p-Si substrate and inset shows the tilted cross sectional image.

a p-Si substrate. It is apparent from the FESEM image that the areal density of grown nanowires is very high. The average diameter and height of the grown nanowires are measured to be 225 nm and 1.45 μm, respectively.

Fig. 2(a) shows the plot of X-Ray Diffraction (XRD) pattern of ZnO NWs. The X-ray pattern confirms the formation of hexagonal wurtzite structure with a dominant peak at 34.42°, which corresponds to <002> plane. Fig. 2(b) represents the EDAX spectroscopy, which confirms the presence of Zn and O atoms in the grown nanowires. Some elemental traces of Silicon are also found which arise from the substrate. The grown ZnO nanowires are of appropriate stoichiometry.

The photoluminescence spectra in Fig. 3 also reveals a strong UV peak at 380 nm corresponds to the Near-Band Edge emission (NBE), where broad green-yellow luminescence ~550 nm, is attributed to oxygen vacancy and Zn interstitials in the ZnO NWs [4]. The broad peak is also attributed to Deep Level Emission (DLE) due to high density of native, interface defects and Zn interstitials.

3.2 *Current-voltage characteristics of n-ZnO/p-Si heterojunction diode*

Fig. 4 shows the typical current-voltage (*I–V*) characteristics of n-ZnO NW/p-Si hetero-junction diode in the voltage range of 0 V to 14 V. The contact area is ~2.1×10^{-3} cm^2 which covered approximately ~10^6 Zn Onanowires (NWs) under it. The unipolar I-V switching characteristics of the n-ZnO NWs/p-Si device are shown in Fig. 4 where the sweeping voltage is 0 V→+14 V→0 V, and the Current Compliance (CC) is 0.1 A. In fast (0.1 V/7.8 ms) voltage sweep cycle, when the applied voltage reaches to ~10.5 V, the current

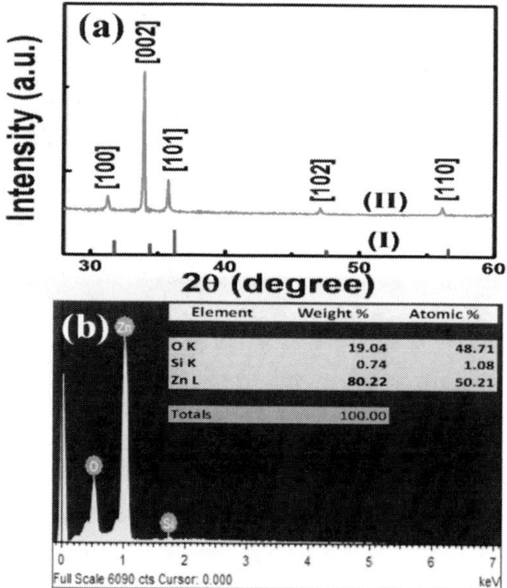

Figure 2. (a) XRD plots for (I) JCPDS Ref. No.-36-1451 and (II) ZnO NW/ZnO seeds/Si, (b) EDAX analysis of the grown ZnO nanowires on Silicon. Inset shows the result data of EDAX.

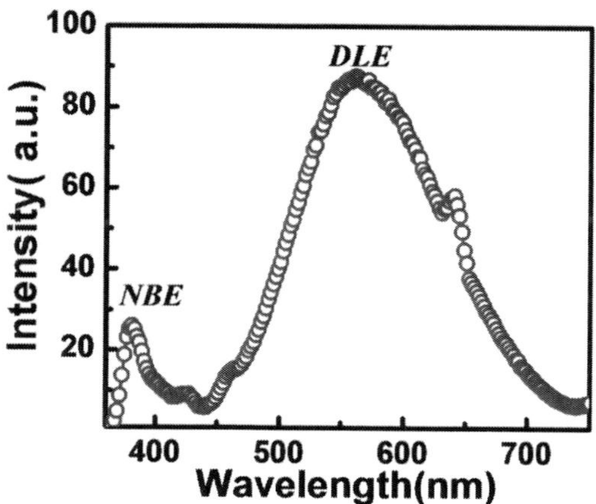

Figure 3. Photoluminescence (PL) spectra of ZnO NWs.

of the devices increase rapidly from 0.39 mA to 1.45 mA, thereby indicating that the diode achieves electrical transition from High Resistance State (HRS) To Low Resistance State (LRS) leading to a "ON" state. Such abrupt increase of current at 10.5 V is attributed to the electroforming process of filament. The hetero-junction nanowire devices switched to LRS (ON) state immediately after the initiation of filament formation process and the arrows indicate the sweeping direction of applied

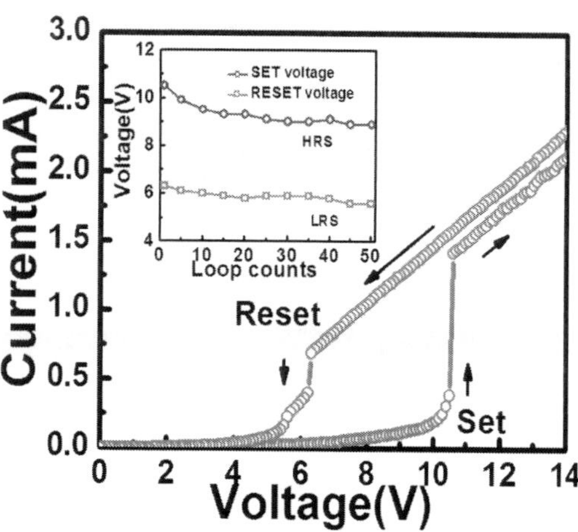

Figure 4. Current- Voltage characteristics of n-ZnO/p-Si heterojunction diode and inset show the plot of set- and reset voltages for 50 voltage sweep cycles.

voltage. When the voltage is swept back to a particular value (reset), a sudden decrease in current occurs and the conductance state switches back to HRS from LRS and a nonvolatile "OFF" state is achieved. Inset of Fig. 4 shows the plots of set and reset voltages for 50 voltage sweep cycles. Moreover, a small dispersion in the value of set-voltage is measured along with a clear and non-overlapped window of operating voltages between the two resistance states. In such a metal oxide nanowire, switching mechanism is quite interesting because oxygen vacancy related ions are more mobile than cations under the external bias. Under the high electric field, oxygen ions are retracted from ZnO and such vacancy related defects gather in individual nanowires to from a filamentary conducting bridge within the nanowire from ZnO nanowire to p-Si/ZnO NWs interface [Chang et al & Qil et al]. Consequently, under the high electric field, current increases rapidly at low resistance state from high resistance state. Further, oxygen ions at p-Si/ZnO NWs interface might be pushed forward into the ZnO NWs under backward voltage sweep. This phenomenon results in a gap between the SET and RESET voltage profile, which is the proof of existence of memory in the ZnO nanowire.

3.3 *Conductance-spectrocopy characteristics of n-ZnO/p-Si heterojunction diode*

The conductance-spectroscopy measurements are performed for investigating the switching properties and also with the switching of n-ZnONW/p-Si diodes by using Keithley 4200 SCS parameter analyzer with a 10 mV ac superimposition. Fig. 5 shows the plots of conductance-spectroscopy for

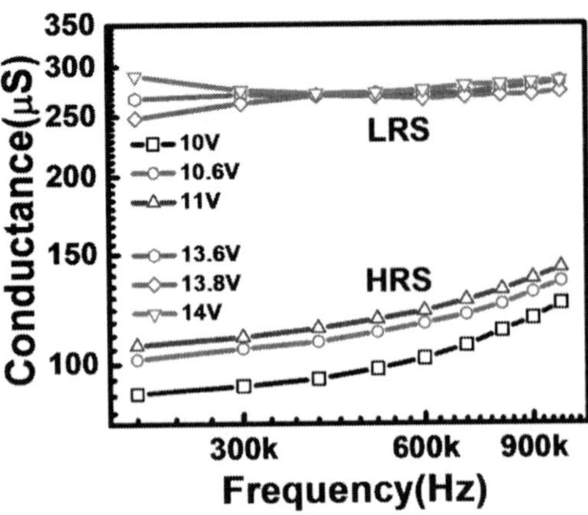

Figure 5. Conductance—frequency characteristics of n-ZnO/p-Si heterojunction diode.

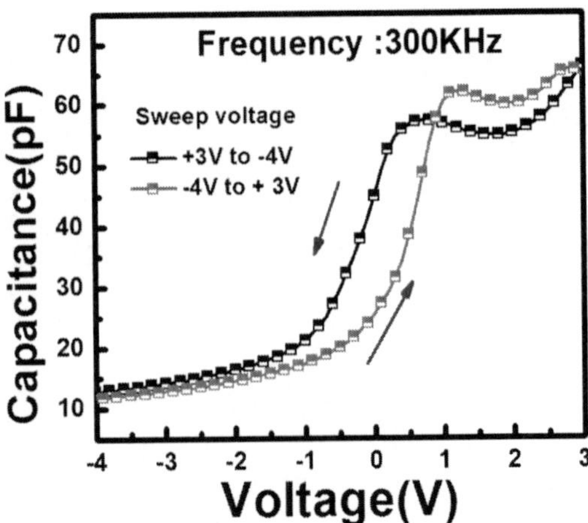

Figure 6. Capacitance-voltage characteristics of n-ZnO/p-Si heterojunction diode.

different voltages related to high resistance state and low resistance state with frequency. The plots indicate that the corner frequency to decrease with the increase of resistance in high resistance state.

However, the increase of conductance with AC small signal frequency is attributed to the dielectric relaxation process, which has several origins including electronic polarization, ionic polarization, and dipolar polarization, however, all of such processes occur at very high frequency mostly above1 GHz. There is one possibility that can cause the increase of AC conductance in the medium frequency range (kHz-MHz) is the hopping through the oxygen vacancy related interface traps. A lower corner frequency in high resistance state connects with a longer transition time for electron from the metal/ZnO NWs interface to vacancy related trap sites within the NWs for DC conduction, which corresponds to a larger tunneling gap in high resistance state [Yu et al]. Therefore, this leads to the gap between SET and RESET currents in a voltage sweep cycle which is also known as '*memory window*'.

3.4 *Capacitance-voltage characteristics of n-ZnO/p-Si heterojunction diode*

The Capacitance-Voltage (C-V) characterization technique is an important tool to understand the effect of oxygen vacancy related interface traps in different frequency variation of n-ZnO-NWs/p-Si heterojunction diodes. Therefore, the C-V measurements have been performed in the range 50 kHz to 500 kHz. The donor concentration and built in potential are calculated from the slope and intercept on x-axis of $1/C^2$-V plots and are obtained to be 4.56×10^{13} cm^{-3} and 0.61 eV respectively, for 300 KHz at room temperature. The C-V curve for frequency 300 KHz signifies the hysteresis between the forward voltage sweep (–4 V to +3 V) and backward sweep (+3 V to –4 V) as shown in Fig. 6.

It is apparent that the C-V curve is shifted towards positive voltage with the increasing capacitance in backward sweep due to the gathering of charged oxygen vacancies near the depletion region. With the accumulation of these charged oxygen vacancies near the depletion region, donor concentration is increased and corresponding depletion width and the built-in voltage in the junction are reduced.

4 CONCLUSIONS

Vertically aligned n-ZnO nanowires are grown on p-Si substrate by Chemical Bath Deposition method (CBD). DLE band in PL spectra confirms the existence of oxygen vacancy related sites in ZnO NWs. The n-ZnO NW/p-Si diodes are electrically characterized by employing the current-voltage, conductance- spectroscopy and capacitance- voltage measurements. I-V characteristics reveal excellent unipolar resistive switching behavior of the hetero-junction diode with quite high ON-OFF ratios and also confirm the high endurance for 50 constant loop measurements. The rise of conductance above a certain corner frequency in high resistance state is attributed to the hopping conduction between the nearest neighboring traps in the conductive filaments. C-V curve for 300 KHz frequency reveals hysteresis between forward and backward sweeps due to modulation of depletion region by oxygen vacancies and space charges. All such results establish the promising application of n-ZnO NWs/p-Si heterojunction diode for unipolar bistable resistive switching memory devices.

ACKNOWLEDGMENT

Rajib Saha and Avishek Das like to acknowledge the Centre of Excellence (COE), TEQIP sponsored by World Bank and University Grants Commission (UGC), India respectively, for providing financial support to pursue their research. The authors would also like to acknowledge the DST PURSE program for providing some necessary infrastructure.

REFERENCES

Chang, W.Y., Lin, C.A., He, H.J. & Wu, T.B. 2010. Resistive switching behaviors of ZnO nanorod layers, *Appl. Phys. Lett.* 96: 242109.

Das, A., Kushwaha, A., Sivasayan, R.K., Chakraborty, S., Dutta, H.S., Karmakar, A., Chattopadhyay, S., Chi, D. and Dalapati, G.K., 2016. Temperature-dependent electrical characteristics of CBD/CBD grown n-ZnO nanowire/p-Si heterojunction diodes. *Journal of Physics D: Applied Physics*, *49*: 145105.

Das, A., Palit, M., Paul, S., Chowdhury, B.N., Dutta, H.S., Karmakar, A. & Chattopadhyay, S. 2014. Investigation of the electrical switching and rectification characteristics of a single standalone n-type ZnO nanowire/p-Si junction diode. *Appl. Phys. Lett.* 105: 083106.

Paul, S., Das A., Chakraborty, A., Chakraborty, B., Palit, M., Chattopadhyay, D. & Chattopadhyay, S. 2014. Growth and characterization of high quality ZnO nanowires by novel CBD/CBD technique. *Proceedings of 1st International Science & Technology Congress*: 467–471.

Qi1, J., Olmedo, M., Zheng, J.G. & Liu, J. 2013. Multimode Resistive Switching in Single ZnO Nano island System, *Scientific Report* 3: 2405.

Yu, S., Jeyasingh, R., Wu, Y. & Wong Philip, H-S. 2011. AC conductance measurement and analysis of the conduction processes in HfOx based resistive switching memory, *Appl. Phys. Lett.* 99: 232105.

Computational Science and Engineering – Deyasi et al. (Eds)
© 2017 Taylor & Francis Group, ISBN 978-1-138-02983-5

Straintronics-MTJ based 3-input universal logic gate (NOR) with high efficiency

Anup Sarkar
Women's Polytechnic, Kolkata, India

Sk. Halim
Calcutta Institute of Technology, Howrah, India

Ankush Ghosh
Dream Institute of Technology, Kolkata, India

Sudhabindu Roy & Subir Kumar Sarkar
Jadavpur University, Kolkata, India

ABSTRACT: Straintronics—Magnetic Tunnel Junction (MTJ) based Nanomagnetic Logic (NML) gates can process the Boolean operation, that can store the output data in the form of magnetisation state. Therefore, it can be used as both logic circuits as well as memory elements. Proposed 3-input NML based NOR gate comprised of a single domain ferromagnet, magneto elastically coupled to the piezoelectric layer PMN-PT with the required MTJ structure. The non-volatile voltage controlled logic gate can fulfill all the logic requirements of a gate, while dissipating ultra-low energy during switching of bits, with extremely fast response including high density logic functionality per unit area.

1 INTRODUCTION

In spintronics, the property of an electron's spin rather than its charge is used as a novel state variable for information storing, processing and communicating. The underlying physics behind these devices is that spins are stored in a ferromagnetic body through interaction and then transfer the angular momentum through an electric but non-magnetic conductor into another ferromagnet. So, it employs the flow of current which eventually minimize the energy advantages of the MTJ. To maximize the energy efficiency, the amount of charge required to switch the MTJ state should be minimized. Straintronics is an alternative approach, which follow the energy efficient method to switch the MTJ state and has recently proposed in (Bandyopadhyay, 2012 & Barangi, 2014).

Straintronics based logic gate are gaining significant interest as it can process and store information simultaneously. It provides flexible designing of computer architecture by eliminating the need of refresh clock cycles, ultimately reducing energy dissipation. Due to its non-volatility it can improve the system reliability and eliminate the delay of a computer.

However the digital logic gate should require satisfying all the requirements as mentioned in (Srinivasan, 2011 & Wasir, 2003). For this reason universal logic gate proposed in (Behin, 2010 & Ney, 2003) may not be useable in all circumstances. Moreover, all the logic gates based on non-volatile scheme supposed to be inferior to CMOS based logic in terms of energy-delay product. Straintronics based two inputs universal NOR gate with ultra-low energy dissipation have been proposed in recent paper (Sarkar, 2016). However, in a digital IC 3-input logic gates are preferred to avoid the complex connectivity between the 2-input gates for difference purposes (Rabacy, 2003). The proposed 3-input gate structure is based on principle of straintronics, which are very energy efficient as well as fast and hence can provide more functionality on a chip area.

2 THEORY

Straintronics is designed with the combination of a magnetostrictive material with a piezoelectric layer PMN-PT. An applied stress can cause the rotation of magnetization vector and thus changes the state of the nanomagnet. Recently, switching of magnetization vector in single magnet using this approach has been proposed in (Roy, 2011).

Straintronics device as depicted in Fig. 1(a) is a combination of 4-layers. The first layer is a thicker layer of piezoelectric PMN-PT. It is more than 4 times thicker than nanomagneticlayer. The second layer is a thin layer of soft magnet, whose magnetization will rotate and decide the logic '0' or logic '1' state. Third layer is very thin layer of insulator MgO (1 nm) required for Tunnelling Magneto Resistance (TMR). The bottom layer is a Synthetic Antiferromagnet (SAF) with very high anisotropy energy. It is permanently magnetized in one direction of the easy axis, say, –Y direction. TMR is high or '1' when magnetization of soft layer, i.e., anti-parallel with the hard layer and is '0' for the parallel. The device is a cylindrical ellipse with its major and minor axis lying on X-Y plane in Y and X direction respectively as shown in Fig. 1(a). In the zero stress condition magnetization vector align itself along easy axis (Y-axis), since this is the minimum energy position. An applied voltage across the PMN-PT generate an electric field that converts into strain, 'S', where S = (ΔL/L) and ΔL = change of the length L. This physical change of length in PMN-PT layer transfers a stress (mechanical energy) to the magnet. Now, depending on the polarity of the input voltage, an energy minimum is created along minor X-axis due to Villari-effect. This will allow the magnetization to rotate towards minor X-axis. Whenever magnetization rotates towards minor axis, then after judicial withdrawal of stress, magnetization can settle towards opposite state.

The potential profile of a 3-input NOR gate based on straintronics MTJ is shown in Fig. 2 and its corresponding logic in Table 1. The plot shows that the anisotropy energy barrier of a nanomagnet can be inverted by the applied voltages (stress) at the input terminal, so that the new minimum energy position comes at $\theta = 90^0$ from $\theta = 0^0$ or 180^0.

Now, for any one or more combination of input voltage, if the input layer voltages is greater than critical voltage then the switching of soft layer will be occurred to get our desired logic operation as shown in logic Table 1. The dimension of the cylindrical elliptical nanomagnet is taken as (120 × 80 × 6) nm, so that Barkhausen effect is negligible and it can behave as a single domain nanomagnet. PMN-PT is taken as piezoelectric layer and the E field along Z-axis, converted to

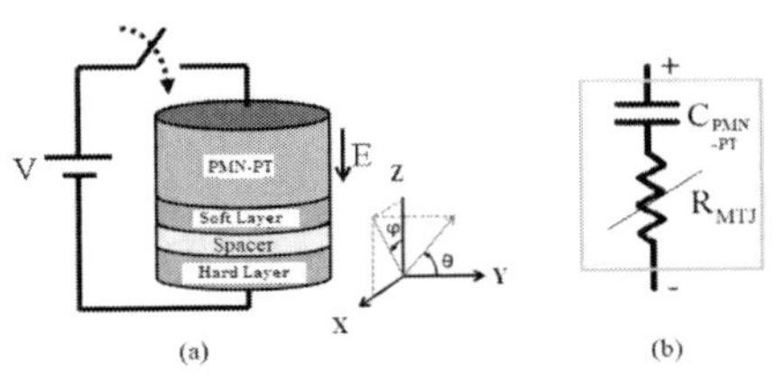

Figure 1. (a) STR devices with reference coordinate specified. (b) Equivalent electrical model of the device.

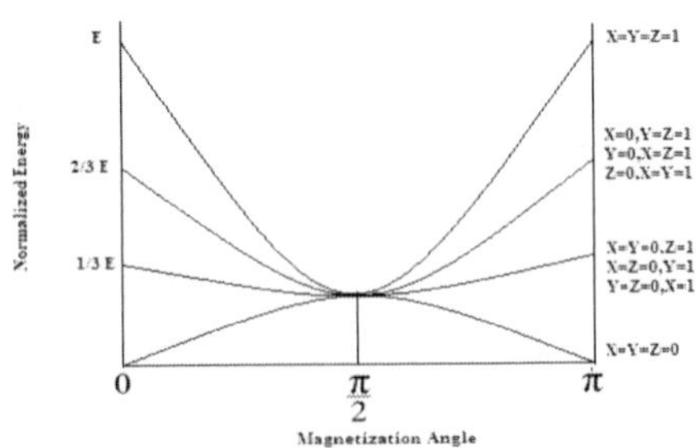

Figure 2. Energy profile of a 3 input NOR Gate.

Table 1. Logic Table.

XYZ	TMR	OUT	Swithes?
000	H	1	NO
001	L	0	YES
010	L	0	YES
011	L	0	YES
100	L	0	YES
101	L	0	YES
110	L	0	YES
111	L	0	YES

strain in X-Y plain through d_{31} coefficient. The dimension of nanomagnet ensures that the anisotropy energy barrier = 150 KT. In this energy barrier magnet's spontaneous switching error probability becomes e^{-150}, which is very low. Now the retention time of the magnet for 1THz frequency will be = 2.77×10^{52} years, almost nonvolatile.

3 IMPLEMENTATION

In this paper a 3-input NOR gate is proposed as depicted in Fig. 3. Here 3-inputs A, B and C are applied at the top layer, which can generate stress for each of the inputs. The fourth input terminal, 'Set' is required to switch the soft layer opposite to that of the hard layer, and it is required only when the magnetoresistance of the MTJ is low. By taking the value of read current it can be determined that whether the soft layer is parallel or anti parallel with the hard layer.

Here, PMN-PT is taken as piezoelectric material which is the top layer and its thickness is 50 nm. The next layer is ferromagnetic layer of cobalt of dimension (120 × 80 × 6) nm. The bottom layer is a synthetic anti ferromagnet with large anisotropy energy barrier, which is permanently magnetized in the left direction of the easy axis. The layer between ferromagnet and anti ferromagnet is a spacer layer of MgO of thickness 1 nm. It is used for Tunnellingmagne to Resistance (TMR) in MTJ structure.

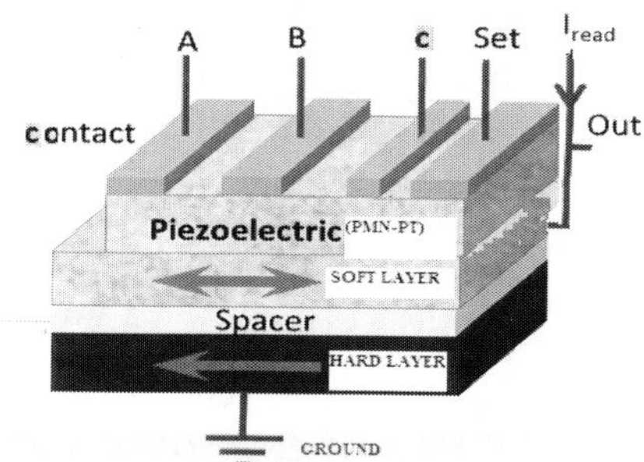

Figure 3. Proposed structure of 3-input NOR Gate.

4 RESULT AND DISCUSSION

It can be shown from magnetic susceptibility model (Barangi, 2014) that critical voltage needed for switching the magnetization of the nanomagnet is

$$\text{Vc} = \frac{\left(\mu_0 / 2Ms^2\left(N_x - N_y\right) + K_u\right)t_{PMN-PT}}{3/2\lambda_s \gamma d_{31}}$$

where $t_{PMN-PT}$is the thickness of piezoelectric layer, γ is the Young's modulus of the soft magnet (cobalt) = 2.09×10^{11} Pa.d_{31} is the piezoelectric effect coefficient = −3000 pm/V (Nikonov, 2013), K_u = uniaxial anisotropy coefficient = 450 J/m³ for cobalt (Barangi, 2014).

Therefore, for the proposed magnet the critical voltage for filliping the magnetization is 38.4 mV.

Now any voltage above the critical voltage will rotate the magnetization vector towards minor axis. The desired result will be obtained if the applied stress can be withdrawn within the time period of successful pulse width.

Capacitance generated at piezoelectric surface is C = 1.7 fF for the proposed dimension.

For our simulation input is taken as 0.1V, so that sufficient noise margin can be achieved during the application of stress.

Total energy of a nanomagnet can be obtained by the Energy equation

$$E = E_{sh} + E_u + E_\sigma$$

E_{sh} = shape anisotropy energy =

$$\frac{\mu_0}{2}VM_s^2\begin{bmatrix} N_z Cos^2\varphi(t)Sin^2\theta(t) + N_x Sin^2\varphi(t) \\ Sin^2\theta(t) + N_y Cos^2\theta(t) \end{bmatrix}$$

where, θ(t) = Polar angle, ϕ(t) = azimuthal angle,

N_x, N_y &N_z = De magnetization factor along X, Y & Z direction respectively.

V = volume of the nanomagnet

E_u = crystal anisotropy energy = $\left[K_u Sin^2\theta(t)\right]$V where K_u = uniaxial crystal anisotropy constant

E_σ = stress anisotropy energy = $\frac{3}{2}\lambda_s\sigma V$ $(1 - Sin^2\theta(t))$

where λ_s = magneto striction co-efficient = 20ppm & σ = applied stress on the nanomagnet

From the energy diagram of Fig. 4, it is understood that energy barrier of the nanomagnet decreases and becomes zero with the application of applied stress increases at the level of critical stress (σ = 31 M).

Magnetization dynamic of a nanomagnet can be best described by LLG equation:

$$\frac{\overrightarrow{dM}}{dt} = -\frac{\gamma_0}{1+\alpha^2}\left(\vec{M}x\vec{H}\right) - \frac{\gamma_0}{M_s x\left(\alpha + \frac{1}{\alpha}\right)}\vec{M}x\left(\vec{M}x\vec{H}\right)$$

where the symbols have their usual meaning.

Although the switching time of magnetization vector can be calculated from LLG equation, but it requires rigorous mathematical computation which is very time consuming. A simplified dynamic model by taking second degree control equation for damping (Barangi, 2014) can be used to analysis the magnetization dynamic of the system by the equation is,

$$\frac{d^2\theta}{dt^2} + 2\zeta\omega_0\frac{d\theta}{dt} + \omega_0^2\theta = 0$$

where ζ = general damping factor =

$$\frac{\alpha(M1+M2)}{\sqrt{4(1+\alpha^2)M_1M_2 - \alpha^2(M1+M2)^2}},$$

ω_0 = natural frequency of oscillation =

$$\frac{\sqrt{4(1+\alpha^2)M_1M_2 - \alpha^2(M1+M2)^2}}{2}$$

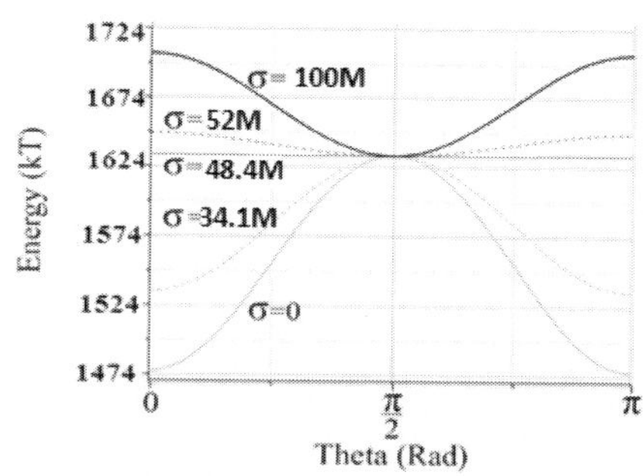

Figure 4. Energy profile of the proposed nanomagnet with the application of stress.

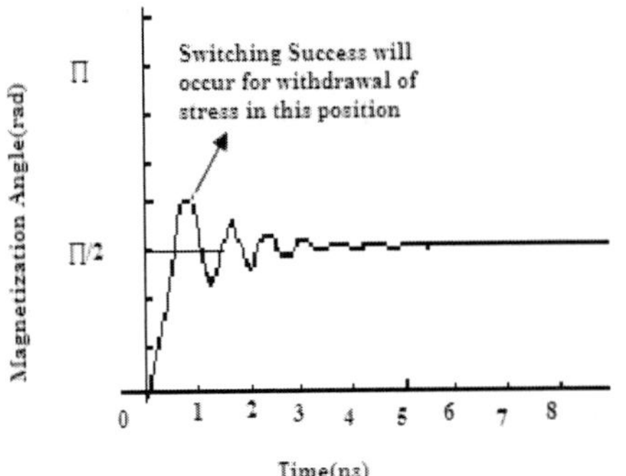

Figure 5. Switching graph of magnetization vector with respect to time with 0.1V input.

Table 2. Performance comparison with CMOS.

Device type	Power dissipation	Propagation delay	Switching success probability
CMOS	pJ	12 ns	100%
Straintronics	aJ	2.76 ns	99.99%

ω_d = damped frequency of oscillation = $\omega_0\sqrt{1-\zeta^2}$

Values of M_1 & M_2 depend on material properties and applied stress

By solving the equation it can be obtained that

$$\theta(t)=\frac{\Pi}{2}-\frac{\Pi}{2}e^{-\zeta\omega_0 t}\cos(\omega_d t)$$

which is depicted in Fig. 5. From the graph it can be interpreted that minimum delay time for magnetization to align at minor axis is = 0.7 ns. Now, relaxation delay for cobalt to settle at the final stage after withdrawal of stress is approx. 2.06 ns (Barangi, 2015).

So total delay time of magnetization to settle at opposite state $\tau = 2.76$ ns.

Now energy delay product = $E\tau = 4.7 \times 10^{-30}$ Js, which is less than CMOS based logic gate.

5 PERFORMANCE COMPARISON WITH CMOS BASED DEVICE

Here ultra-low power dissipation occurs due to absence of physical current flow during switching. During the flipping of a single bit power dissipation can be estimated in the order of Auto Joule. Where as in modern CMOS based device it is of the order of Pico Joule only (Romli, 2015). Moreover switching time for flipping a bit in Straintronics based device is in the order of Nano second with error probability of 10^{-6} (Biswas, 2016), which is also comparable to CMOS device.

6 CONCLUSION

Straintronics based logic gates are less energy dissipative than CMOS based gate. Furthermore, straitronics based device is comparable to CMOS device in terms of propagation delay and switching error probability. But, CMOS device is volatile. So STR—MTJ based device can be used to build-up as a basic building block for non-volatile memory with high frequency applications. Cobalt can be the perfect material for this case due to their fast response and tolerable noise margin. Furthermore, due to low functional voltages, the STR-MTJ allows digital logic gates to operate synchronously with FIMS or STT based logic gates.

REFERENCES

Bandyopadhyay, S. & Atulasimha, J, 2012. Hybrid Straintronics and Spintronics: An ultra energy efficient paradigm for logic and memory, in proc. 2012 70th Annual device research cong. (DOI) June 18–20, 2012, pp. 35–36.

Barangi M. & Mazumder P., 2014, Straintronics based magnetic tunnelling junction: Dynamic and static behaviour analysis and material investigation, Appl. Phys. Lett., Vol.104, No.16, p.p 162403–5.

Barangi & Mazumder P., 2015, IEEE transaction on magnetic, vol. 51, No. 5.

Behin-Aein B, Datta D, Salahuddin S and Datta S, 2010, NatureNanotechnol. 5 266–2699.

Ney, A, Pampuch, C, Koch, R. & Ploog, K.H, 2003, Programmable computing with a single magneto strictiveelement. Nature 425, 485–487.

Nikonov, D.E & Young, I.A, 2013, Overview of beyond CMOS devices and uniform methodology for their benchmarking proc. IEEE 101, 2498.

Rabaey, G.M., Chandrasekhar A.P., & Nikolic, B. 2003, Digital integrated ckts. (Pearson education).

Roy K, Bandyopadhyay S, Atulasimha, 2011, Hybrid spintronics and straintronics: A Magnetic technology for ultra-low energy computing and signal processing, J Appl. Phys. Lett. Vol. 99, p. 063108.

Sarkar, A. Dutta, P.K., Ghosh, A., Ray, S. & Sarkar, S.K., 2016, "Implementation of universal gate (NAND) based on nanomagnetic logic using multiferroic" ASP Quantum matter, vol 1–5.

Srinivasun, S., Sarkar, A. Behin-Aein, B. & Dutta, S. 2011, All spin logic device within built non-reciprocity. IEEE trans., 47 4026–4032.

Waser, R 2003 (ed) Nanoelectronics and information technology, Ch. III (Wiley-VCH).

Biswas, A.K., 2016, PhD Thesis and Dissertation at Virginia Commonwealth University.

Romli, N. B. Minhad, K.N. Reaz, M.B. IAmin, Md. S. 2015, An overview of power dissipation and control techniques in CMOS technology, Journal of Engineering Science and Technology, 10–3, 364–382.

Computational Science and Engineering – Deyasi et al. (Eds)
© 2017 Taylor & Francis Group, ISBN 978-1-138-02983-5

Comparative study of single electron threshold logic based and SET-CMOS hybrid based 1 bit comparator

Arpita Ghosh
RCC Institute of Information Technology, Kolkata, India

Subir Kumar Sarkar
Jadavpur University, Kolkata, India

ABSTRACT: This paper presents the design implementation of the 1 bit comparator circuit using two different design approaches such as single electron threshold logic based approach and the hybrid SET-CMOS based approach. After the basic logic gates the comparator is one of the basic building blocks of any decision making circuit. The further reduction in the size and the power consumption can be done by designing the circuit using other technological approaches. The comparative analysis of the proposed circuit using two approaches is illustrated in this paper.

Keywords: 1 bit comparator, Single electron tunneling, SET-CMOS hybrid, Single electron threshold logic CMOS based circuit, MIB model, SIMON

1 INTRODUCTION

The recent trends in the technological solutions are mainly concentrating in the reduced feature size and the ultra low power consumption. The shrinking dimensions of conventional MOSFET has come to a certain limit due to the problems faced related to scaling. As a solution different other advanced technologies are coming in the scenario. The Single Electron Tunneling Technology (SET) (Likharev, 1999) is one of them. The presence and the absence of a single electron can be encoded as the two basic logic levels. As well as the comparison between the weighted sum and the threshold value of the single electron tunneling (Lageweg, 2003) based gate works as a single electron threshold logic gate. Different logic circuit implementation (Ghosh et. al. 2013) can be performed using this approach. Though SET provides desired results in the nanometer regime with far reduced power consumption and unique coulomb oscillation feature, it suffers from different issues related to room temperature operation, low current drivability and small voltage gain. For getting the benefits of both the technologies the SET can be further combined with the conventional CMOS technology. It leads to a new technological solution called hybrid SET-CMOS technology (Mahapatra et. al, 2006, Ghosh et. al, 2014).

For any logic circuit design apart from the basic logic gates the comparators also play one of the vital roles. Here in this paper the design and implementation of a single bit comparator is explained using both the approaches. The overall comparative analysis of the 1 bit comparator design aspect is the main area of discussion in this paper. The organization of the paper is divided into several sections among which the first one gives the brief introduction to the different technological design approaches. The next two sections deal with the detailed design of 1 bit comparator using the single electron threshold logic based approach and the hybrid approach. The fourth part of the paper gives the elaborate deliberation of the comparative analysis. The overall conclusion of the study is given in the last section.

2 ONE BIT COMPARATOR WITH SINGLE ELECTRON THRESHOLD LOGIC APPROACH

The circuit diagram of 1 bit comparator using conventional logic gates is shown in Figure 1, where mainly two AND gates, 2 NOT gates and a NOR gate is required for the implementation. When the input A is less than B then the output is indicated with L, for A > B the output is G and for A = B the output is E(equal).

The truth table of the one bit comparator is shown in Table 1. The Figure 2 shows the basic threshold logic gate where the weighted sum in the inputs is compared with threshold value.

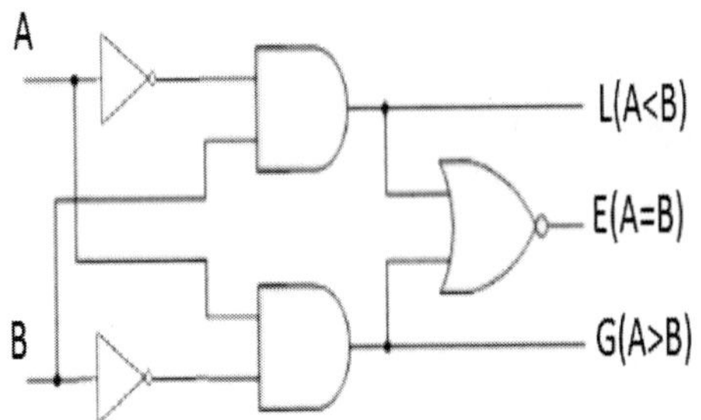

Figure 1. 1 bit Comparator using logic gates.

Table 1. Truth table of 1 bit comparator.

Input		Output		
A	B	G(A > B)	L(A B)	E(A = B)
0	0	0	0	1
0	1	0	1	0
1	0	1	0	0
1	1	0	0	1

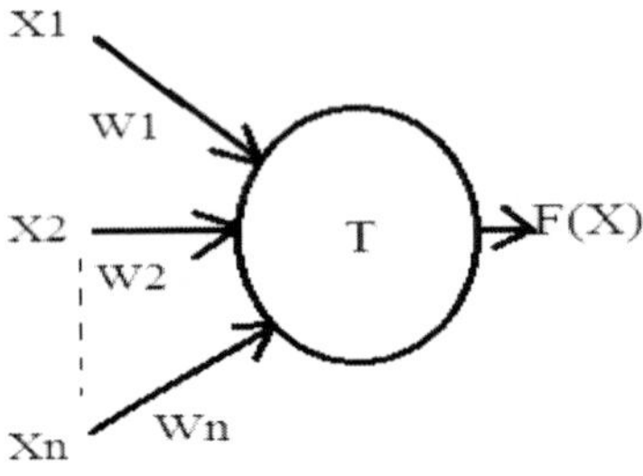

Figure 2. Threshold logic gate with x1, x2,...xn as the input values with their respective weights w1, w2,...wn, the threshold value T and output F(X).

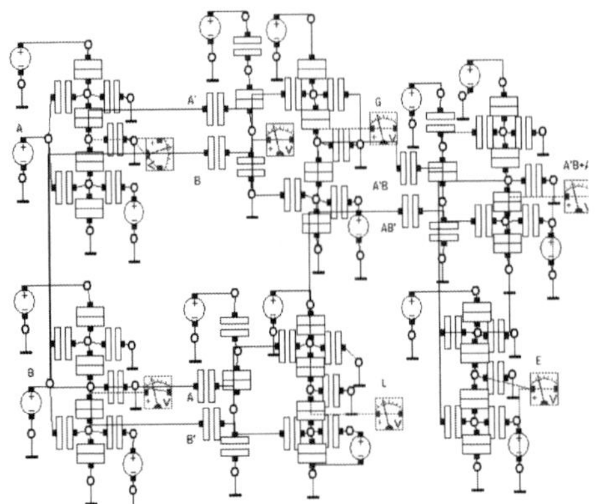

Figure 3. Single electron threshold logic based 1 bit comparator.

If the weighted sum is greater than the threshold then the output of the gate is denoted with logic1 and logic 0 stands for weighted sum below the threshold value.

Figure 3 shows the proposed 1 bit comparator with single electron threshold logic. The complete circuit is designed and simulated in the SIMON

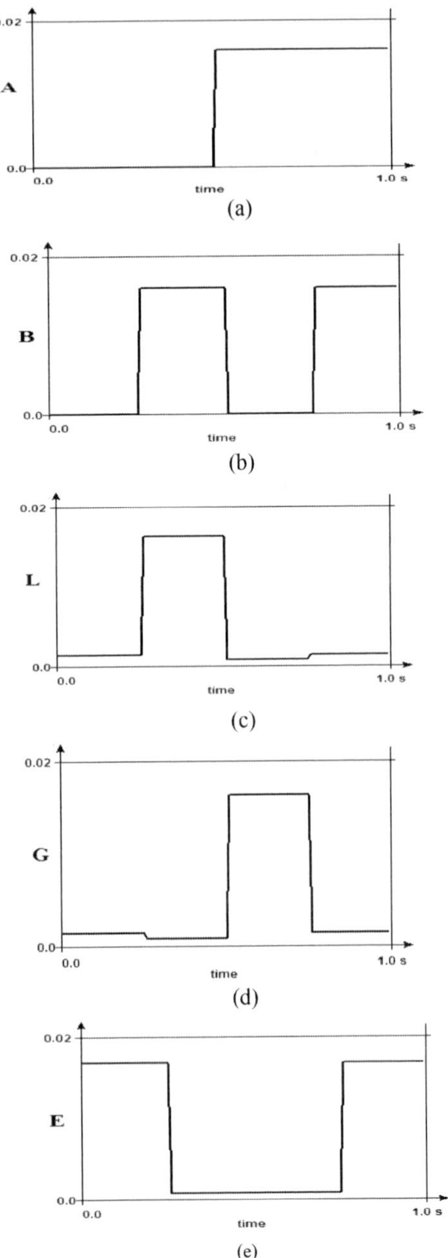

Figure 4. Input and output waveforms (a) A, (b) B, (c) L, (d) G and (e) E.

(Wasshuber 1997) environment, which is a Monte Carlo based simulation tool. The circuit consists of 2 input threshold logic based AND gates, a 2 input OR gates and inverter. The 2 input AND gate and OR gate equations for threshold logic is shown in equation (1) and (2).

$$Y = AND(A,B) = \text{sgn}\{A + B - 1.5\} \quad (1)$$

$$Y = OR(A,B) = \text{sgn}\{A + B - 0.5\} \quad (2)$$

For buffered gates equations (3) and (4) are used.

$$Y = AND(A,B) = \text{sgn}\{-A - B + 1.5\} \quad (3)$$

$$Y = OR(A,B) = \text{sgn}\{-A - B + 0.5\} \quad (4)$$

The simulated results of the circuit shown in Figure 3 is depicted in Figure 4. Figure 4(a), (b), (c), (d), (e) can be justified using the truth table shown in Table 1. The input signals A and B are shown in Figure 4(a) and (b). Similarly the simulated output waveforms for L, G and E are shown in Figure 4(c), (d) and (e) respectively.

3 HYBRID SET-CMOS BASED 1 BIT COMPARA-TOR

The SET-CMOS hybrid approach based design is shown in the Figure 5. The designed circuit is simulated in the Tanner SPICE environment. For the co-simulation of SET and CMOS both of them are replaced by their corresponding equivalent SPICE compatible models.

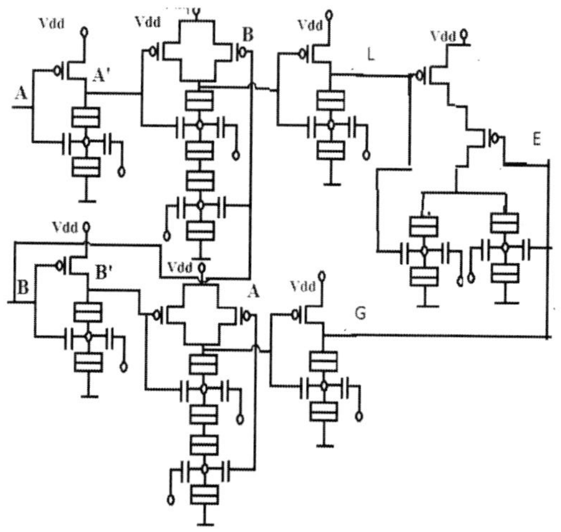

Figure 5. Hybrid SET-CMOS based 1 bit comparator circuit.

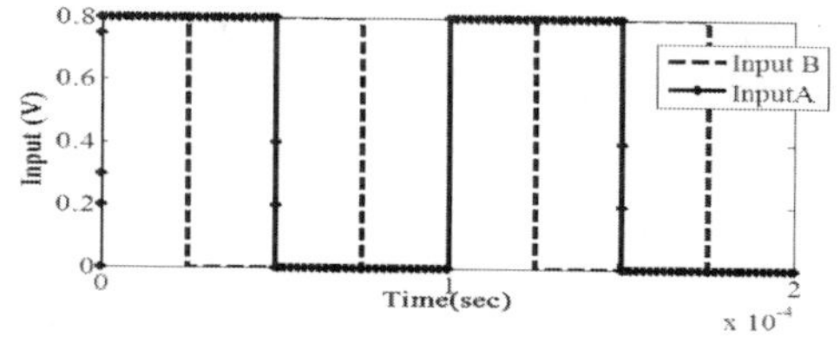

(a)

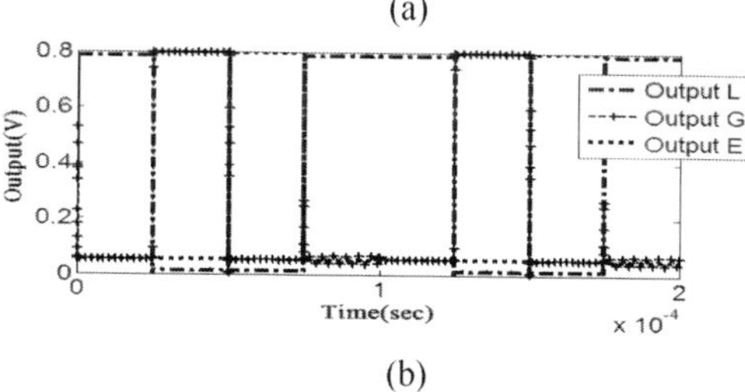

(b)

Figure 6. Input and output Waveforms for (a) Input A and B, (b) Output L, G, E.

Table 2. Different design approach based 1 bit comparator and their properties.

Parameters	CMOS based	Threshold logic	SET-CMOS
Number of MOSFETs	20	NA	10
Number of Tunnel junctions	NA	27	20
Number of Capacitors	7	42	20
Number of Islands	0	24	10
Power Supply	0.8V	0.016V	0.8V
Operating Temperature	300 K	0 K	4 K
Power Consumption	2.21e–05 W	0.22e–09 W	6.05e–07 W

For the MOSFET the BSIM4.6.1 model file is used and for SET MIB model (Mahapatra et.al, 2006) is included. The simulated results of the proposed circuit is shown in Figure 6.

The input signals A and B are shown in Figure 6(a). Here 0.8 V is used to indicate logic 1 and 0V is used for the indication of logic 0. In the similar way the output L, G and E are shown in Figure 6(b).

4 COMPARISON OF DIFFERENT DESIGN APPROACHES

The comparative analysis of the 1 bit comparator design is given in the Table 2 with respect to the different design parameters. From the table it is clear that the conventional CMOS technology for designing the circuit consumes more power than the other two technological approaches. The single electron threshold based approach consumes the lowest power (0.225e-09 watts), whereas the operational temperature is very low (0 K). The power supply requirement of the threshold logic approach (0.016 V) is lowest among the three approaches. The other parameters are the number of MOSFETs, tunnel junctions, capacitors, islands.

5 CONCLUSION

The complete paper discusses the design and implementation prospects of one bit comparator using different design approaches such as single electron threshold logic based and SET-CMOS based. The corresponding input and output waveforms are verified using the truth table. The comparison between the design approaches along with the conventional CMOS approach is also deliberated with respect to the different parameters. The comparative analysis illustrates that the threshold logic based design consumes power in the nano watt range which is the lowest among the three approaches but suffers from the problem related to the operating temperature.

REFERENCES

Ghosh, A., Jain, A, Singh, N.B. & Sarkar, S.K 2014 Design and implementation of SET-CMOS hybrid half subtractor *Annual IEEE India Conference (INDICON)*, Pune, India.

Ghosh, A., Jain, A & Sarkar, S. K. 2013 Design and Simulation of Single Electron Threshold Logic Gate based Programmable Logic Array *Procedia Technology 10 pp.866–874.*

Likharev, K. *1999* Single-Electron Devices and Their Application *Proc. IEEE, Vol. 87(4), pp. 606–632.*

Likharev, K. 1999 Single-Electron Devices and Their Application *Proc. IEEE, Vol. 87(4), pp. 606–632.*

Lageweg, C., Cotofana, S., & Vassiliadis, S. 2003 Evaluation Methodology for Single Electron Encoded Threshold Logic Gates *Proc. of the IFIP International Conference on Very Large Scale Integration of Systems-on-Chip(VLSI-SOC) pp. 258–262*, Darmstadt, Germany.

Mahapatra, S & Ionescu, A.M 2006 Hybrid CMOS Single-Electron-Transistor Device and Circuit Design *Artech House, Inc., ISBN:1596930691.*

Rehan, S. E. 2010 The implementation of 2-bit decoder using single electron linear threshold gates *Int. Conf. On Micro-electronics (ICN), Dec. pp. 180–183* Mansoura, Egypt.

Wasshuber, C. 2001 Computational single-electronics *Springer Verlag ISBN: 321183558X.*

Wasshuber, C. 1997 SIMON-A simulator for single—electron tunnel devices and circuits *IEEE Trans. On Computer Aided Design of Integrated Circuits and Systems* 16, *937–944.*

Computational Science and Engineering – Deyasi et al. (Eds)
© 2017 Taylor & Francis Group, ISBN 978-1-138-02983-5

Surface potential based threshold voltage modeling of work function engineered gate recessed source/drain IR SOI/SON MOSFET for better nanoscale device performance

Tiya Dey Malakar
Department of Electronics and Communication Engineering, RCCIIT, Kolkata, India

Partha Bhattacharyya
Department of Electronics and Tele Communication Engineering, IIEST, Shibpur, Howrah, India

Subir Kumar Sarkar
Department of Electronics and Tele Communication Engineering, Jadavpur University, Kolkata, India

ABSTRACT: In this paper, we represent the threshold voltage modeling along with drain current, based on the surface potential of linearly graded binary metal alloy gate (Work Function Engineered Gate) Recessed Source / Drain (Re S/D) SOI/SON MOSFET with additional insulator region(IR -SOI). The proposed structure is similar to that of the Recessed S/D SOI MOSFET with the exception that there is an insulator region of high k dielectric in between the channel and drain region. Due to the presence of high k dielectric material the electric field is substantially reduced in the channel and drain regions. Therefore the proposed structure suppressed various short channel effects such as Hot Carrier Effect (HCE) which define the device degradation mechanism and hot electron reliability. The analytical surface potential model has been developed by solving two dimensional Poisson's equation in the channel region considering appropriate boundary condition with parabolic potential profile.

Keywords: Work Function Engineered Gate (WFEG), Recessed Source/Drain (Re S/D), high K dielectric, Short Channel Effects (SCEs) etc.

1 INTRODUCTION

Investigation of the nanoscale device become crucial for high density integrated circuit. But this continuous downscaling of the device dimension in submicron regime introduce various Short Channel Effects (SCE's) such as Drain–Induced Barrier Lowering (DIBL), Hot-Carrier Effect (HCE), Sub-threshold conduction and junction leakage etc (Mattausch et al. 2008). Among the several non conventional MOS structure Fully Depleted Silicon on insulator (FD SOI) MOSFET establish as a potential candidate for their lower parasitic capacitances, better immunity against radiation, higher speed of operation (Colinge, 1997). However these FD SOI MOSFET structure is still prone to various Short Channel Effects (SCE's). Hence the researchers has invented Silicon –On – Nothing structure where the silicon active layer is replaced by air or so called "noting" to enhance the device performance (Deb et al, 2010). Although the various short comes of the SOI/SON structure such as threshold voltage instability and DIBL at higher drain bias can be overcome with Work Function Engineered Gate (WFEG) SOI-MOSFETs as reported by researchers (Deb et al, 2012). In order to increase the channel conductivity recessed S/D SOI MOFET structure has been investigated and fabricated by various research group (Zhang et al. 2004).

In this paper, a unique structure of Work Function Engineered Gate (WFEG) Recessed S/D SOI-MOSFETs with high k dielectric region in between channel and drain end has been proposed to enhance the device performance in terms of threshold voltage, surface electric field, HCE and drain current. A Ta–Pt binary alloy system as gate electrode with linearly varying work function from 100% Pt (at source side) to 100% Ta (at drain side) have been considered in our present structure for analytical modeling and a performance comparison has been carried out for the SOI and SON MOSFETs.

2 ANALYTICAL MODELING

Figure 1 shows the cross-sectional view of layered WFEG Re-S/D IR SOI/ SON MOSFET. The thicknesses of front gate oxide, buried layer, silicon substrate and channel silicon film are represented by t_f, $t_{box/air}$, t_{sub} and t_{si} respectively. L is the device channel length, t_{rsd} and $d_{box/air}$ are the Source/ Drain recessed thickness and length of the Source/Drain overlap region over Buried Layer.

The effective work function of Pt-Ta binary alloy system with linearly varying mole fraction can be correlated with the horizontal channel position (along x coordinate) and be expressed as-

$$\phi_{meff}(x) = (x/L)\phi_b + (1 - x/L)\phi_a \quad (1)$$

Work functions of Pt and Ta are denoted by ϕ_a and ϕ_b respectively so that $\phi_{mef}(x) = \phi_a$ at source side (x = 0) and $\phi_{mef}(x) = \phi_b$ at the drain side (x = L) (Manna et al., 2012).

2.1 *Surface potential distribution*

To find out the surface potential distribution of our proposed structure Two dimensional Poisson's equation has been solved considering uniform charge density in the thin silicon film region (Reddy et al, 2005).

$$\frac{\partial^2 \phi_i(x,y)}{\partial x^2} + \frac{\partial^2 \phi_i(x,y)}{\partial y^2} = \frac{qN_a}{\varepsilon_{Si}} \quad i = 1,2 \quad (2)$$

This equation (2) has been solved by using parabolic potential approximation as per Young (Young et al, 1989) and can be expressed as-

$$\phi_1(x,y) = \phi_{s1}(x) + a_{11}(x)y + a_{12}(x)y^2 \quad (3)$$

For$(0 \le x \le L - L_{CIR},\ 0 \le y \le t_{Si})$

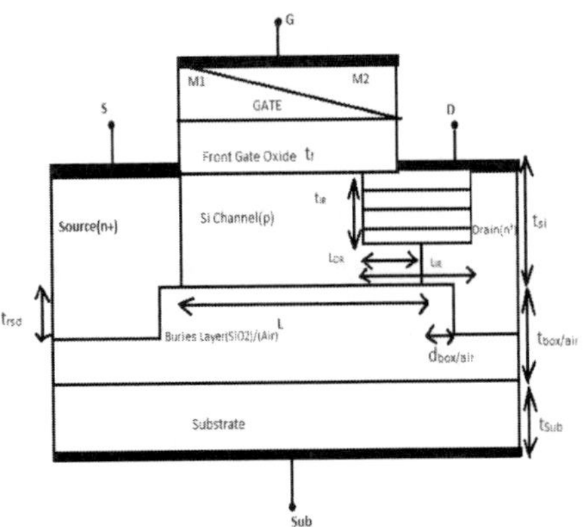

Figure 1. A schematic cross sectional view of layered Recessed S/D SOI/SON structure.

$$\phi_2(x,y) = \phi_{S2}(x) + a_{21}(x)y + a_{22}(x)y^2 \quad (4)$$

For $(0 \le x \le L_{CIR},\ 0 \le y \le t_{Si})$

Here the front interface surface potential in the region I region II are represented by $\phi_{s1}(x)$ & $\phi_{s2}(x)$ respectively. And $a_{11}(x)$, $a_{12}(x)$, $a_{21}(x)$ & $a_{22}(x)$ are the arbitrary coefficient of function x. With these constants equation (2) can be solved with some proper boundary conditions as follows

i. Electric Field continuity at the front gate oxide-Silicon channel interface therefore we have

$$\left.\frac{d\phi_1(x,y)}{dy}\right|_{y=0} = \frac{\varepsilon_{ox}}{\varepsilon_{-si}} \frac{\phi_{s1}(x) - (V_{gs} - V_{fbf}(x))}{t_f} \text{ for region I} \quad (5)$$

$$\left.\frac{d\phi_2(x,y)}{dy}\right|_{y=0} = \frac{\varepsilon_{ox}}{\varepsilon_{-effective}} \frac{\phi_{s2}(x) - (V_{gs} - V_{fbf}(x))}{t_f} \text{ for region II} \quad (6)$$

where $V_{fbf}(x) = \phi_{meff} - \phi_S$ is the front channel interface flat band voltage, ε_{ox} and ε_{-Si} are the relative permittivity of Silicon di-oxide and Silicon respectively. $\varepsilon_{-effective}$ is the effective dielectric constant of the region II.

ii. Electric field continuity at the Silicon Channel-Buried layer interface in region I and region II:

$$\left.\frac{d\phi_1(x,y)}{dy}\right|_{y=t_{si}} = \frac{C_{box1}(V_{Sub1} - \phi_{b1}(x))}{\varepsilon_{-si}} + \frac{C_{rsd1}(V_{S1} - \phi_{b1}(x))}{\varepsilon_{si}} + \frac{C_{rsd2}(V_{D1} - \phi_{b1}(x))}{\varepsilon_{si}} \quad (7)$$

$$\left.\frac{d\phi_2(x,y)}{dy}\right|_{y=t_{si}} = \frac{C_{box2}(V_{Sub2} - \phi_{b2}(x))}{\varepsilon_{-effective}} + \frac{C_{rsd3}(V_{S2} - \phi_{b2}(x))}{\varepsilon_{-effective}} + \frac{C_{rsd4}(V_{D2} - \phi_{b2}(x))}{\varepsilon_{-effective}} \quad (8)$$

where $V_{Sub1} = V_{Sub} - V_T \ln(N_{Sub}/N_i)$, $V_{S1} = V_S - V_T \ln(N_A N_D/N_i^2)$ and $V_{D1} = V_D - V_T \ln(N_A N_D/N_i^2)$ represents the effective substrate, Source and Drain biases respectively. $C_{box1} = C_{box2} = \varepsilon_{box/air}/t_{box/air}$ define the buried layer capacitance and other four capacitances are due to the existence of recessed

Source/ Drain regions as proposed by Sviličić et al. (Sviličić et al, 2009) and adapted according to our device geometry.

iii. Potential at the Source end becomes:

$$\phi_1(0,0) = \phi_{s1}(0) = V_{bi} \tag{9}$$

iv. Potential at the drain end becomes:

$$\phi_2(L,y) = \phi_{s2}(L) = V_{bi} + V_{DS} \tag{10}$$

v. Surface potential at the interface of the two dissimilar channel materials is continuous:

$$\left.\frac{d\phi_1(x,y)}{dx}\right|_{x=L-L_{CIR}} = \left.\frac{d\phi_2(x,y)}{dy}\right|_{x=L-L_{CIR}} \tag{11}$$

where V_{bi} is the junction built- in potential. 2D Poisson's equation has been solved by using parabolic potential approximation as in equation (3) and then first two boundary conditions can be used to evaluate the values of coefficients.

Now replacing the values of these coefficients in equation (3) & (4) and then in equation (2) we get the differential equation for front and back surface potentials. These are non-homogeneous differential equation and have been solved by separation of variable method to find the complete solution of front & back surface potential.

In conventional SOI MOSFET, the threshold voltage is taken as that value of V_{GS} at which minimum surface potential becomes twice of the Fermi potential (i.e $\phi_{S,min} = 2\phi_{f,Si}$). The position of the minimum of the interface surface potential ($\phi_{fmin\ or}\ \phi_{b,min}$) can be evaluated considering $\left.\frac{d\phi_f(x)}{dx}\right|_{x=x_0} = 0$ or $\left.\frac{d\phi_b(x)}{dx}\right|_{x=x_0} = 0$ Where x_0 define the position where interface potential is minimum.

3 RESULTS AND DISCUSSIONS

Analytical modeling has been carried out for proposed Work Function Engineered Gate (WFEG) Recessed S/D IR SOI and SON MOSFETs structures. The results obtained from the present structures are compared with the results of both WFEG Recessed SOI/SON MOSFETs and conventional SOI/SON MOSFET. For both the structures we have taken a binary alloy system consisting of Ta–Pt as gate electrode with work function variation in metal alloy for linearly increasing discrete compositions from 100% Pt (from source side) to 100% Ta (at drain side). The simulations parameters used here are given in the following table.

Due to the continuous work function adjustment from source to drain side the Electric field can be reduced in recessed WFEG IR device structure. Additional presence of the insulator region in between channel & Drain creates a depletion region under the gate. This depletion region considerably reduced the electric field in the inserted region by preventing the carrier to enter the gate oxide from the channel. Also the peak value of electric field at the drain side decreases as shown in fig. 2. Hence the proposed model provides better immunity against severe Hot Carrier Effects (HCEs). It is also evident from the figure that SON has Lower values of electric field than its SOI counter parts.

Figure 3 shows the variation of threshold Voltage against the silicon film thicknesses of WFEG Recessed IR SOI/SON structures. It is also compared with the Recessed SOI/SON structure. It reveals that the threshold voltage variation of Recessed IR SOI/SON structures are more sensitive to the silicon film thicknesses than the Recessed SOI/SON.

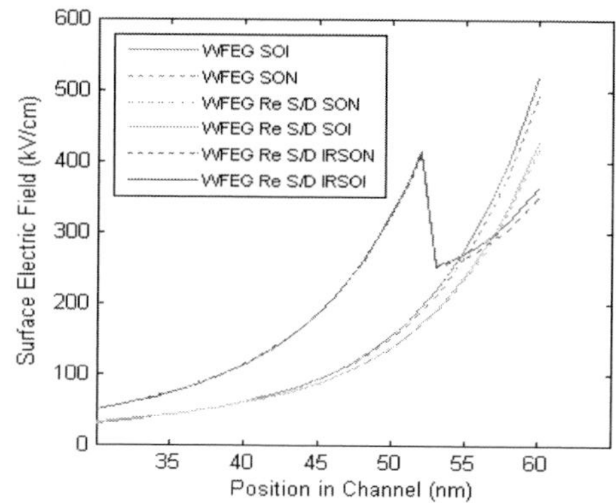

Figure 2. Variation surface electric field along the channel of proposed SOI and SON MOSFETs for $V_{DS} = V_{GS} = 0.1$ V.

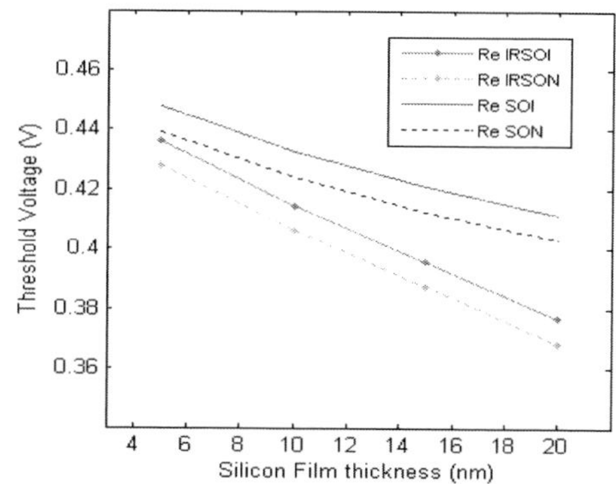

Figure 3. Variation of threshold Voltage versus silicon film thicknesses of WFEG Recessed IR SOI/SON structures.

Table 1. Value of Parameters used for simulation.

Parameters	Values
N_A	$10^{21}\ m^{-3}$
N_{SUB}	$10^{21}\ m^{-3}$
$N_{S/D}$	$10^{26}\ m^{-3}$
t_{Si}	20 nm
t_f	1.5
$t_{box/air}$	100 nm
L_{IR}	20 nm
L_{CIR}	4 nm
$d_{box/air}$	3 nm
t_{sub}	200 nm

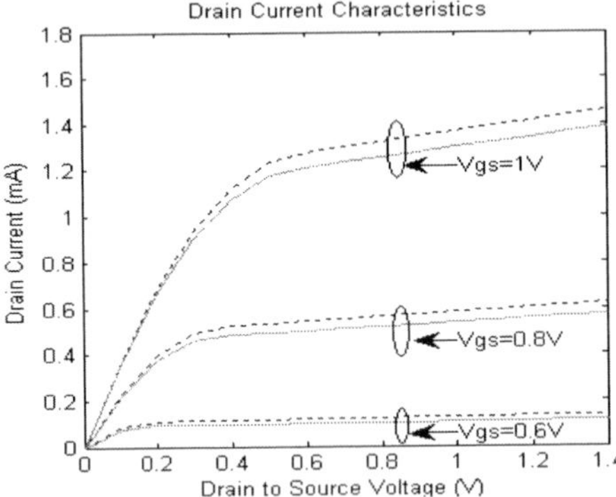

Figure 4. Variation of Drain Current against the Vds of WFEG Recessed IR SOI/SON structures for various gate to source voltages.

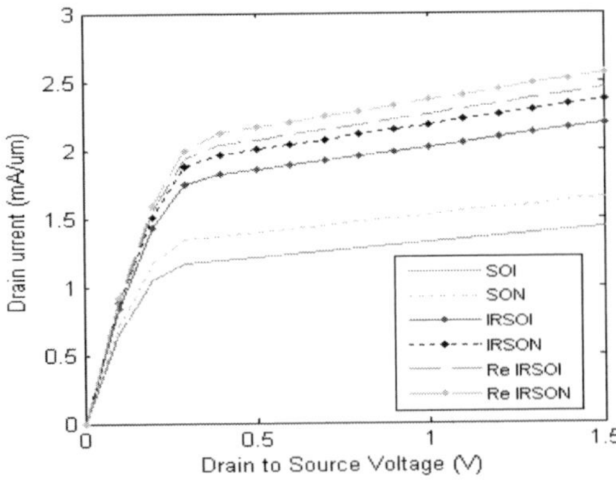

Figure 5. Variation of Drain Current versus voltage of WFEG Recessed IR SOI/SON structures.

Figure 4 show the Variation of Drain Current versus the drain to source voltage of WFEG Recessed IR SOI/SON structures for various gate to source voltages. Due to the lower parasitic capacitances SON structure provides higher drain current than SOI structure at a given gate to source voltage.

Figure 5 depicted the variation of Drain Current against the drain to source voltage of WFEG Recessed IR SOI/SON structures and compared with recessed and conventional SOI/SOI structures. The proposed model shows the highest drain current compared with others due to the presence of additional insulator region which considerably reduce the Hot Carrier Effects.

4 CONCLUSIONS

In this paper, a unique structure of Work Function Engineered Gate (WFEG) Recessed S/D SOI-MOSFETs with high k dielectric region in between channel and drain end has been developed. Due to the combined benefits of both WFEG recessed S/D structure and high k dielectric material in IR region the proposed structure confirmed the better device performance in terms Drain Induced Barrier Lowering (DIBL) and Hot Carrier Effects (HCE). Comparative performance analysis of SOI and SON in terms of surface potential distribution, threshold voltage behavior and electric field profile reveals that SON has better immunity against various SCE's than its counterpart. Therefore, the proposed recessed S/D IR SOI/SON structure can be a potential tool to optimize the desired performance of the device parameter in the nanometer regime.

REFERENCES

Colinge J.P. (1997) Silicon on Insulator Technology: Materials to VLSI, second ed., Kluwer Academic Publishers, Norwell, MA, Kluwer.

Deb. Sanjoy, Singh. N.B., Das. D., Dey. A.K. and Sarkar, S.K. (2010) Analytical model of Threshold voltage and sub-threshold slope of SOI and SON MOSFET: A comparative study, Journal of Electron Devices, vol.8, pp. 300–309.

Deb Sanjoy et al. (2012) Work Function Engineering with Linearly graded Binary Metal Alloy Gate Electrode for Short Channel SOI MOSFET, IEEE Trans. on nanotechnology, vol. 11, issue 3, pp. 472–478.

Manna Bibhas, Sarkhel, S., Islam, N., Sarkar, S., Sarkar, S.K. (2012) Spatial Composition Grading of Binary Metal Alloy Gate Electrode for Short-Channel SOI/SON MOSFET Application, IEEE Transaction on Electron Devices, Vol-59, Issue-12, Pp-3280–3287.

Mattausch M.M., et al. (2008) The Physics and Modeling of MOSFET, World Scientific Publishing Co. Pte. Ltd., Singapore.

Reddy Venkateshwar, M. Jagadesh Kumar, (2005) A new dual-material double-gate (DMDG) nanoscale SOI MOSFET-twodimensional analytical modeling and simulation, IEEE Trans. Electron Dev. 4(2), pp. 260–268.

Svilicic, B. et al. (2009), Analytical models of front- and back-gate potential distribution and threshold voltage for recessed source/drain UTB SOI MOSFETs, Solid State Electron. 53, pp. 540–547. Young, K.K. (1989) Short-channel effects in fully depleted SOI MOSFET's, IEEE Trans. Electron Dev. 36 pp. 399–402.

Zhang. Z. et al. (2004) Self-align recessed source/drain ultrathin body SOI MOSFET, IEEE Electron Dev. Lett. 25, 740–742.

Computational Science and Engineering – Deyasi et al. (Eds)
© 2017 Taylor & Francis Group, ISBN 978-1-138-02983-5

Study the effect of catalytic metal contacts on ethanol vapour sensing performance of WO_3-Si hetero-structure sensor

Subhashis Roy, Bijoy Kantha & Subir Kumar Sarkar
Jadavpur University, Kolkata, West Bengal, India

ABSTRACT: Tungsten Oxide (WO_3) based thin film heterojunction ('n' type WO_3 - 'p' type Si < 100 > substrate) gas sensors with Gold contact and noble catalytic metal contact of Palladium–Silver (70%) have been fabricated by sol gel process to detect ethanol vapor. Surface morphology of thin film WO_3 sensor has been studied by Scanning Electron Microscope (SEM). The sensitivities of the sensors with two different contacts have been studied. Response magnitude of the sensor showed that the linear change of sensitivity with respect to concentration variation from 192 ppm to 3110 ppm. Maximum 78% sensitivity has been achieved at 3110 ppm ethanol vapor with Pd-Ag contact with respect to 66% sensitivity with Au contact for the same concentration.

Keywords: Ethanol, heterojunction, catalytic metal, sensitivity

1 INTRODUCTION

Organic pollutants are broadly used in industries and daily life and are the key effluents released by the industries to the environment. Among various organic pollutants, ethanol is the most common organic pollutant which effects the environment, aquatic system and human health due to their toxicity and hazardous effect. With advancement of time and technology different new techniques have been developed for growing low power, high sensitivity sensors like thin film sensors, modifying sensing layer's surface using novel catalytic metals like Ag, Au, Pd, Pt etc.; catalytic metals as sensor surface contact, metal oxide heterojunction structure type gas sensors by Banergee, N. et al. (2014), Wang, C. et al. (2010), Kantha, B. et al. (2015) and many other authors.

Tungsten Oxide (WO_3) is a versatile n type compound semiconductor with cubic perovskite like structure. The wide band gap of WO_3 (~2.6 eV at room temperature as described by Ottaviano, L. (2001)) and fairly good lattice matching with Si ensures that heterostructure formed by these two materials gives very less amount of defect states at their interfaces/grain boundaries. The synergistic effect between components and heterojunction interfaces between these two components are the key factors for enhancing the sensing characteristics with respect to homojunction sensor.

Little work on WO_3–Si based heterojunction ethanol sensor has been presented in literature. ZnO–Si based heterojunction structure grown by galvanic method as gas sensor with catalytic metal contact Pd-Ag (26%) with improved sensitivity and response time at low concentration of methane gas was reported by Bhattacharyya, P. et al. (2007) and Roy, S. et al. (2013) have shown the development of ethanol sensor on CBD grown ZnO nanorods. Nobel catalytic materials are high effective oxidation catalysts which enhances the reaction on the gas sensor surface when used as surface modifier or surface contact. The grown thin film layer of WO_3 on Si showed less nonporous and nanocrystalline structure in this work. In another work the sol gel process has been taken to grow ZnO on Si by Mishra, G.P. et al. (2010) and effect of catalytic metal contact Pd-Ag (70%) with respect to Au contact has been studied. The result shows device with both side Pd-Ag contact gives better response time with higher response magnitude compared with device with both side Au contact.

Therefore in the present investigation an attempt has been made to fabricate two WO_3 thin film heterojunction (WO_3-Si) sensors with novel material Pd-Ag (70%) and Au as surface contact. The sol gel process has been selected to get extra advantages like low cost, better surface morphology and ease of execution process. The surface morphology of the grown sensing layer has been corroborated via Scanning Electron Microscopy (SEM). The sensor demonstrated high sensing properties in terms of higher sensitivity and lower limit of detection with respect to other works.

2 SENSOR FABRICATION AND CHARACTERIZATION

Nanocrystalline WO_3 thin film deposited on p-Si < 100 > substrate by sol gel method. WCl_6 was used as a precursor to prepare WO_3 thin film. WCl_6 was dissolved in isopropanol at a ratio of 5 g/100 ml and stayed in dry air for 3 days. Then the sol was deposited onto the Si substrate (resistivity 1 Ω cm, 500 μm thick) by dip coating method and the substrate was dried in air at 150° C for 30 minutes. Post deposition heat treatment (annealing) at 600°C for 2 hours was carried out to improve microstructure and crystalline properties of the deposited thin film. The contacts of noble catalytic metal Pd-Ag (70%) and Au have been formed on both side of the sensor. The figure 1 and figure 2 show the pictorial view of the sensor and surface morphology of grown thin film WO_3 layer by Scanning Electron Microscopy process at 100 nm scale.

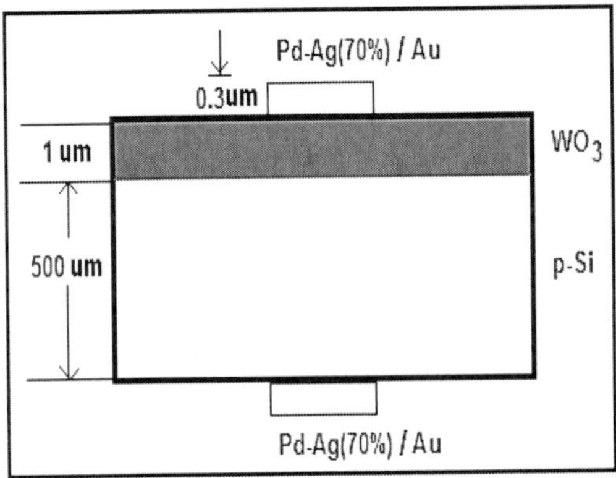

Figure 1. Schematic view of fabricated sensor structure.

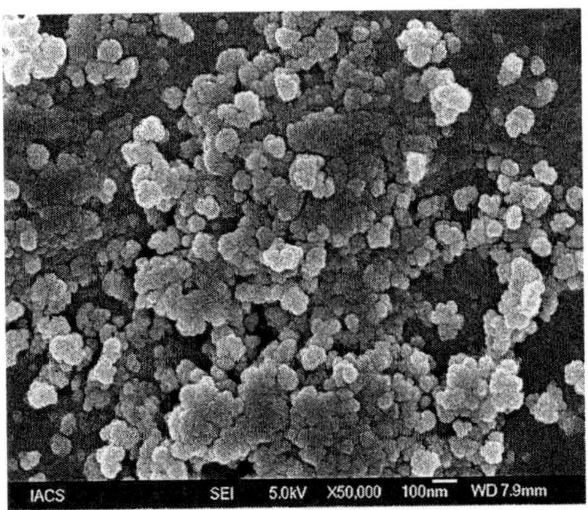

Figure 2. SEM view of thin film sensor surface.

3 EXPERIMENTAL SETUP

Figure showed the schematic of ethanol vapour sensor measurement set up. The cylindrical sensor chamber is made of glass (length 25 cm and diameter 3.5 cm). The thin film sensor is inserted into the chamber (by contact connection). The Temperature Controller (TC) was used to attain the operating temperature. Temperature controller was operated by resistive heating coil (≈8 cm of constant heating zone, with temperature accuracy ±1°C). The gas flow and mixing ratio were precisely monitored and controlled with the help of Mass Flow Controller (MFC) (Alicat scientific, M-50SCCM-D) for air (reference gas). The homogeneous mixture carrying the desired percentage of the target vapour was fed into the chamber with a flexible PVC pipe. During the testing the gas pressure on the sensor was 1 atm.

4 RESULTS AND DISCUSSION

The operating temperature of the sensor was found 200°C and 250°C for Pd-Ag and Au contact based sensors with 0.1% ethanol vapour (shown in figure 4) respectively. Then by varying the concentration of the ethanol vapor from 192 ppm to 3117 ppm the change in conductivity was measured at fixed operating temperature for the two samples (shown in figure 5). The reducing gas ethanol reacted with oxygen adsorbed on the surface of the sensor rather lattice oxygen.

The adsorption of the C_2H_5OH on WO_3 surface can take place by the following route:

$$C_2H_5OH_{(vap)} + O^-_{(ads)} \rightarrow CH_3CHO_{(ads)} + H_2O_{(vap)} + e-$$
$$CH_3CHO_{(ads)} + O_{(lattice)} \rightarrow CH_3CHO_{(vap)} + V_O$$

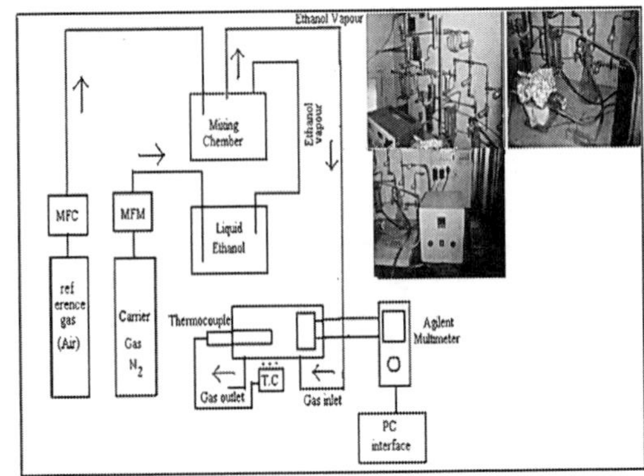

Figure 3. Laboratory experimental setup.

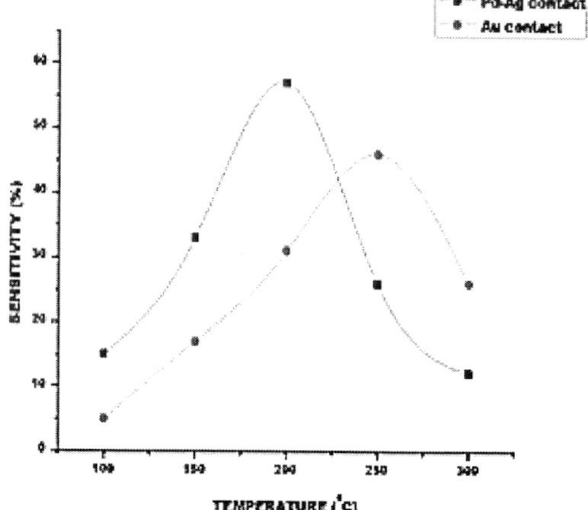

Figure 4. Sensitivity as a function of temperature for two sensors.

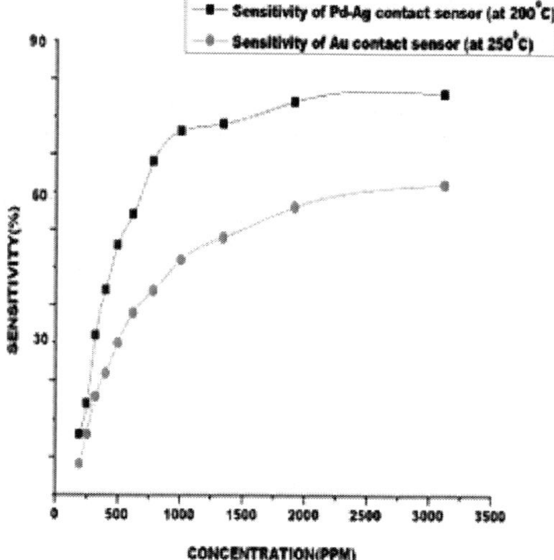

Figure 5. Sensitivity as a function of concentration for two sensors.

Catalytic noble metal Pd-Ag (70%) contact based sensor shows greater sensitivity (78%) than the Au contact based sensor (66%). At an elevated temperature, the oxygen molecules are weakly bounded with the catalytic metal atoms of Pd-Ag which produces oxygen atoms. These atoms then undergo spillover process and finally form negatively charge surface ions by gaining electrons from oxide surface, giving high electrostatic potential in the junction. The space charge region, being depleted of electrons, is more resistive than bulk, enhancing overall conduction process of metal oxide gas sensor with respect to Au contact gas sensor.

5 CONCLUSION

In the present work heterojunction WO_3 thin film metal oxide gas sensor has been fabricated for sensing the ethanol vapour concentration (0.019%–0.31%). Its sensitivity has been studied for Pd-Ag catalytic metal contact and Au contact. The result shows higher sensitivity for Pd-Ag contact than Au contact.

ACKNOWLEDGEMENT

Author thankfully acknowledges the financial support for this research work obtained from UGC UPE Phase II sponsored project.

REFERENCES

Banerjee, N., Roy, S., Sarkar, C. K., Bhattacharyya, P., 2014. Effect of Humidity on Ethanol Sensing Performance of Pd Sensitized ZnO Nanorod based Sensors, 2014. *Journal of Surfaces and Interfaces of Materials, (American Scientific Publishers)*, vol. 2, No. 2, pp. 154–160.

Bhattacharyya, P., Basu, P., Saha, H., Basu, S., 2007. Deposition of Nanocrystalline ZnO thin Films on p-Si by Novel Galvanic Method and application of the Heterojunction as Methane Sensor. *Journal of Materials Science: Materials in Electronics (Springer), vol.18, 8, pp. 823–829*

Kantha, B., Roy, S., Sarkar, S.K., 2015. Implementation of Pd modified WO3 thin film gas sensing system with Bulk-CMOS and SOI-CMOS for monitoring leakage of hydrogen gas, 2015. *Journal of Nanoelectronics and Optoelectronics, American Scientific Publisher, Vol. 10, No. 1.*

Mishra, G. P., Sengupta, A., Maji, S., Sarkar, S. K., Bhattacharyya, P.,2010. The Effect of Catalytic Metal Contact on Methane Sensing Performance of Nanoporous ZnO-Si Heterojunction. *Int. J. on Smart Sensing and Intelligent Systems (ISSN1178-5608), vol. 3(2), pp. 292–303.*

Ottaviano, L., Passacantando, M., Santucci, S., 2001. On the spatially resolved electronic structure of polycrystalline WO3 films investigated with scanning tunneling spectroscopy. *Surface Science, vol. 475, pp. 73–82.*

Roy, S., Banerjee, N., Sarkar, C.K. and Bhattacharyya, P., 2013. Development of an Ethanol Sensor based on CBD Grown ZnO Nanorods,2013. *Solid State Electronics (Elsevier), vol. 87, pp. 43–50.*

Wang, C., 2010. Metal Oxide Gas Sensors: Sensitivity and Influencing Factors, *Sensors, pp. 2088–2106.*

Computational Science and Engineering – Deyasi et al. (Eds)
© 2017 Taylor & Francis Group, ISBN 978-1-138-02983-5

Study the effects of annealing temperatures on sol-gel derived TiO_2 sensing element

Anup Dey, Bijoy Kantha & Subir Kumar Sarkar
Jadavpur University, Kolkata, West Bengal, India

ABSTRACT: The paper presents the sol-gel derived Titanium-di-oxide thin film and study the effect of annealing temperatures on the film for getting appropriate surface morphology on selected annealing temperature. This paper also investigates the suitable operating temperature at170°C. Titanium-di-oxide has been prepared by using Tin (IV) tert-butoxide on p-Si substrate sol-gel deposition technique. The Scanning Electron Microscopy (SEM) is used to investigate structure and morphology of the fabricated thin film. It is observed that Titanium-di-oxide thin film with 750°C has smaller grain size (42 nm) and sharper grain image compared to the other annealed temperatures.

Keywords: TiO_2, sol-gel deposition technique, annealing temperatures, SEM

1 INTRODUCTION

Now a day, different toxic gases (like. H_2, CO_2, NH_3, CH_3OH etc.) is a clean and renewable energy source which is expected to progressively replace the existing fuels and also have wide combustion range of 5–75%, so continuous monitoring is required of toxic gases for safety operation. A gas sensor is a gas detector that detects in the presence of toxic atmosphere. They contain micro-fabricated point-contact gas sensors and are used to locate leaks. They are considered low-cost, compact, durable, and easy to maintain as compared to conventional gas detecting instruments. MEMS based hydrogen sensor mainly the combination of nanotechnology and Micro-Electro-Mechanical Systems (MEMS) technology allows the production of a hydrogen micro sensor that functions properly at room temperature.

A metal-oxide type semiconductor such as, TiO_2, WO_3, SnO_2, ZnO etc. can detect a gas which changes of its conductivity. Among these metal-oxides, TiO_2 has good resistance against photo-corrosion in acidic aqueous solution and its energy band gap can be changed with the variations in crystalline by Sakai, G. et al. (2001). TiO_2 has found applications in many field, such as sensing by Higuchi, T. et al. (2009) and Ito, K. et al. (1982), photo catalysis by Tang, WM. et al. (2005), tsai, TH. et al. (2008) and electro chromic application. TiO_2 has exhibited excellent performance for many target gases like H_2S, H_2, NH_3, NO_2 and CH_3OH by liu, C.C. et al. (1982) and dey, A. et al. (2016). A TiO_2 nanostructure includes nanorods by Sakai, G et al. (2001) have been synthesize by different methods. Some of metal oxides are utilized as sensor materials due to their high mobility of electrons, high surface to volume ratio and good chemicals and thermal stability under different operating condition.

The sol-gel method is simple and low cost technique for deposit of nanostructure film in large scale. Thin TiO_2 film has been prepared via physical and chemical routes, focusing on sol-gel coating, CVD, solution spray and thermal annealing.

In this paper, we have synthesized TiO_2 thin film and study the effect of different annealing temperatures (450°C–750°C) on the film. This paper also investigates the suitable operating temperature at 170°C. The preparation of thin film is explained in Section 2. Results and discussions are reported in Section 3. The concluding remarks appear in Section 4.

2 PREPARATION OF THIN FILM

Sol-gel method is a simple and low cost method to prepare TiO_2 thin film by using Tin (IV) tert-butoxide as a precursor by Dey, A. et al. (2016). Tin (IV) tert-butoxide was mixed to ethanol and HCl at certain percentage and placed in dry air for 48 hours. After that spin coating technique is used to drop the obtained sol onto the p-si substrate. Initially, the substrate was cleaned and dried at 100°C. The gel film was dehydrated in air at 150°C for 20 minutes. To get the crystalline film, the dried film was annealed at 500°C–750°C for 3 hours. The detailed structures are shown in Figure 1. The SEM image of WO_3 thin film after different annealing temperatures is presented in Figure 2-Figure 4.

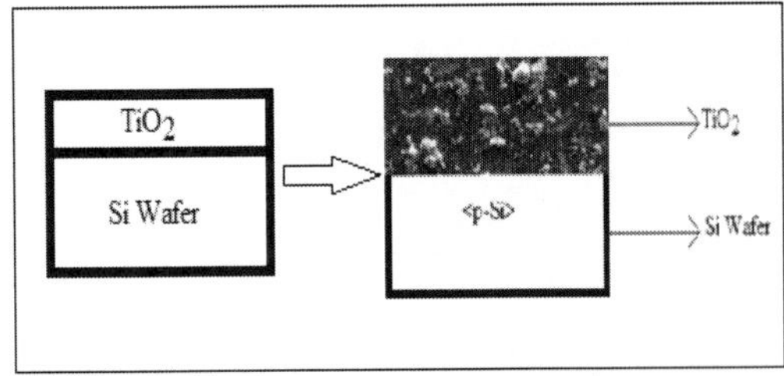

Figure 1. Structure of TiO_2 thin film device.

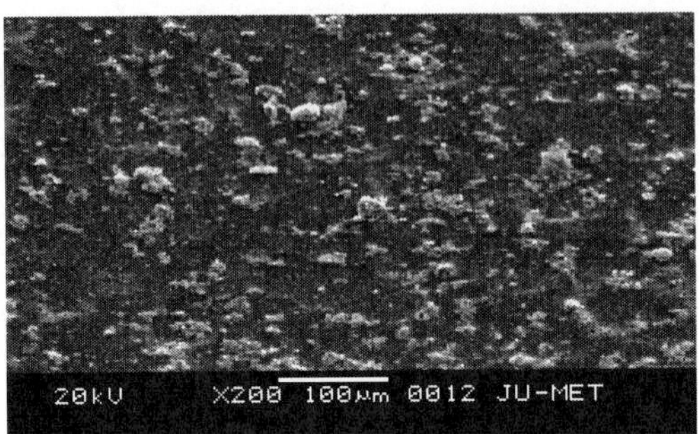

Figure 2. SEM image of TiO_2 thin film sensor at annealing temperature of 550°C.

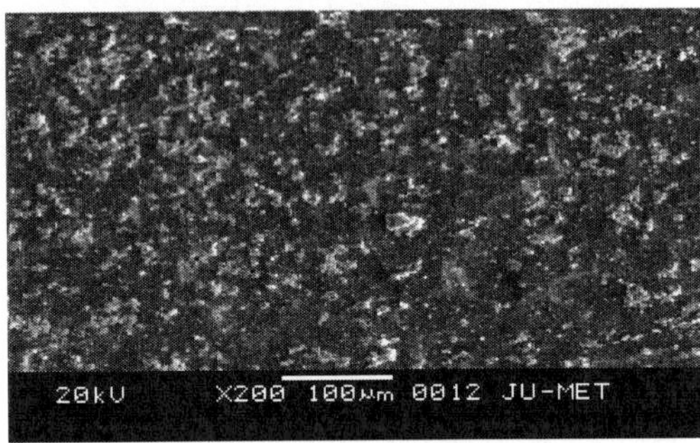

Figure 3. SEM image of TiO_2 thin film sensor at annealing temperature of 650°C.

Figure 4. SEM image of TiO_2 thin film sensor at annealing temperature of 750°C.

Table 1. Different annealing temperature with corresponding to particle grain size and operating temperature.

Annealing temperature (°C)	Particle grain size (nm)	Operating temperature (°C)
550	68	250
650	57	220
750	42	170

3 MICROSTRUCTURE CHARACTERISTICS

Microstructure characteristics measured by SEM (Scanning Electron Microscope) of the TiO_2 thin films with different annealing temperature (550,650 and 750°C) are shown in Figure 2-Figure 4. SEM images propose that sol–gel developed metal oxide thin films have different polycrystalline structure with different grain size (68 nm, 57 nm and 42 nm). Typical SEM characteristics are also found that TiO_2 thin film with annealing temperature of 750°C have sharper grain image and smaller grain size (42 nm) compare to the other annealed temperature.

4 CONCLUSION

In the present work, the design and the effect of annealing temperatures TiO_2 thin film sensor has been studied in the operating temperature range of 150°C to 350°C. It is observed that TiO_2 thin film sensor are suitable at the operating temperature 170°C. It is also found that Titanium-di-oxide thin film with 750°C has smaller grain size and sharper grain image compared to the other annealed temperatures (Table 1).

ACKNOWLEDGEMENT

Author thankfully acknowledges the financial support for this research work obtained from UGC—UPE Phase II sponsored project.

REFERENCES

Dey, A. Kantha, B. & Sarkar, S.K. 2016 Sol–gel grown Pd modified WO_3 thin film based methanol sensor and the effect ofAnnealingtempetures. *MicrosystemTechnologies* DOI10.1007/s00542–016 –2841–3

Higuchi, T. Nakagomi, Sh. & Kokubun, Y. 2009 Field effect hydrogen sensor device with simple structure based on GaN. *Sens Actuators B Chem; 140:79–85.*

Huang, H. Luan, W. Zhang. J.S. Qi, Y.S. & Tu, S.T. 2008 Thermoelectric hydrogen sensor working at room temperature prepared by bismuth–telluride P–N couples and Pt/gAl_2O_3. *Sens Actuators B Chem; 128:581–5.*

Ippolito, S.J. Kandasamy, S. Kalantar, Z.K. & Wlodarski, W. 2005 Layered SAW hydrogen sensor with modified tungsten trioxide selective layer. *Sens Actuators B Chem a; 108:553–7.*

Ippolito, S.J. Kandasamy, S. Kalantarzadeh, K. & Wlodarski, W. 2005 Hydrogen sensing characteristics of WO_3 thin film conductometric activated by Pt and Au catalysts. *Sens Actuators B Chem b; 108:154–8.*

Ito, K. & Kojima, K. 1982 Hydrogen detection by Schottky diodes. *Int J Hydrogen Energy; 7:495–7.*

Liu, C.C. 1992. Development of chemical sensors using microfabrication and micromachining technique, Pro. *Int. Meet. Chemical Sensors.*

Sakai, G. Matsunaga, N. & Shimanoe, K. 2001 Theory of gasdiffusion controlled sensitivity for thin film semiconductor gas sensor. *Sens Actuators B Chem; 80:125–31.*

Suehle, J.S. Cavicchi, R.E. Gaitan, M. & Semancik, S. 1993. Tin oxide gas sensor fabricated using CMOS microhotplates and in situ processing, *IEEE Electron Device Lett.*

Tang, W.M. Lai, P.T. Xu, J.P. & Chan, C.L. 2005 Enhanced hydrogensensing characteristics of MI SiC Schottkydiode hydrogen sensor by trichloroethylene oxidation. *Sens Actuators A Phys; 119:63–7.*

Tong, M. Dai, G. & Gao, D. 2001 WO_3 thin film sensor prepared by sol-gel technique and its low-temperature sensing properties to trimethylamine. *Materials Chemistry and Physics 69:176–179.*

Tsai, T.H. Chen, H.I. Lin, K.W. Hung, C.W. Hsu, C.H. & Chen, LY. 2008 et al. Comprehensive study on hydrogen sensing properties of a Pd–AlGaN based Schottky diode. *Int J Hydrogen Energy; 33:2986–92.*

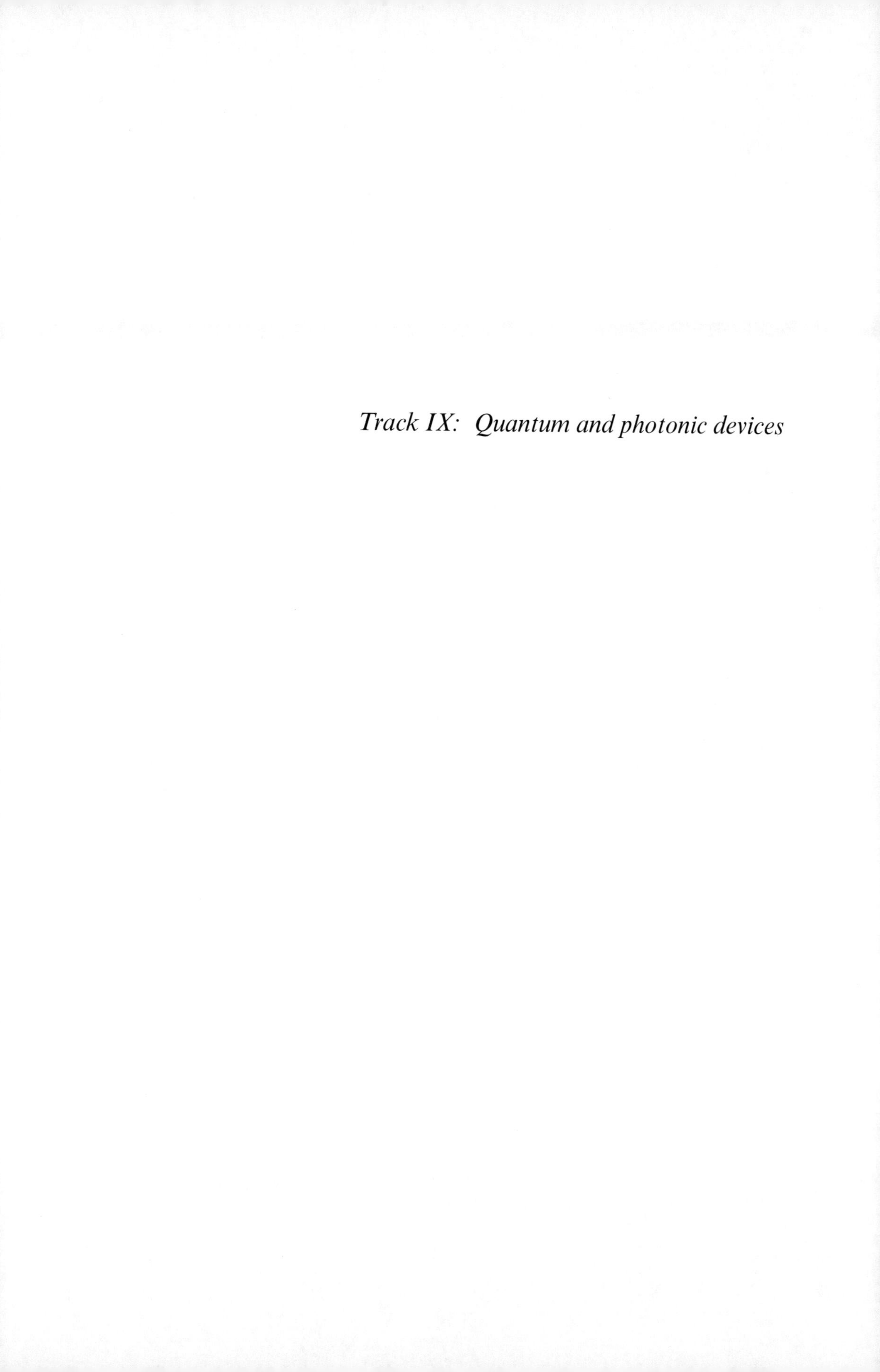

Track IX: Quantum and photonic devices

Computational Science and Engineering – Deyasi et al. (Eds)
© 2017 Taylor & Francis Group, ISBN 978-1-138-02983-5

Effect of structural parameters on tunneling current density for rectangular double quantum well device

Shuvodeep Saha, Aparupa Chakrabarty & Arpan Deyasi
RCC Institute of Information Technology, Kolkata, India

ABSTRACT: Tunneling current density of rectangular double quantum structure is analytically computed by considering propagation of wave function along the quantized direction. Traveling outflow is normalized w. r. t incident wave, and its change is computed at each grid point in the direction of propagation. Ben-Daniel Duke boundary conditions are considered at interfaces for effective mass mismatch. Well and barrier widths are varied within appropriate ranges of forward bias, and material composition is also tuned within type-I limit. Results are important for application of the device as resonant tunneling diode.

1 INTRODUCTION

Rapid hrinkage of MOSFET dimensions led exponential progress in VLSI technology in the past two decades, which becomes realizable owing to development of existing microelectronics technology. But recently, the progress is far deviated from Moore's law due to the limitations exhibited by the growth materials, and the limitations of fabrication set-up. Thus, in order to continue the miniaturization of integrated circuits following Moore's law well in the present day, design of microelectronic devices should be replaced with novel designs that take advantage of the quantum mechanical effects, which will also speak in favor of lower power consumption. Solid-state quantum nanoelectronic devices are the nearest alternative to maintain the continuous trend of increasing packing density and speed of information processing. Among the different low-dimensional quantum-mechanical devices, resonant tunneling devices are found the interest of both theoretical (Goldhaber-Gordon *et al.*, 1997) and experimental researchers (Sen *et al.*, 1987, Maranowski *et al.*, 2000) for the past decade owing to its novel electronic properties (Esaki *et al.*, 1974), and also its less complex development mechanism which is supported by the controlled microelectronic growth techniques with various combination of semiconducting materials (Talale *et al.*, 2008, Yamamoto 1987). Electrical and optical properties of these heterostructure devices can be computed from the knowledge of quantum transport processes, and precise estimation of transmission coefficient is essential for the device with incorporation of physical parameters (Simion et al., 2000, Ghatak *et al.*, 1988). Easki and Tsu first proposed a semiconductor symmetric double barrier structure (Esaki *et al.*, 1970) where electronic transport proceeds via resonant tunneling mechanism. This pioneering work makes the road for future research using quantum-confined devices. They showed that a series of energy levels and associated subbands are produced due to the confinement of carriers along one direction of otherwise bulk structures. Computation for transmission coefficient carried out, (Chanda *et al.*, 1982, Christodoulides *et al.*, 1985) and later Scandella (Scandella *et al.*, 1992) was without effect of material parameters, which was later realized (Chang *et al.* 1974, Read *et al.*, 1986); who computed resonant tunneling probability in semiconductor double barrier structure for different material parameters. They showed that computation of thermal probability is essential to calculate current from quantum devices. Thermal probability was also computed (Wessel *et. al.*, 1989) for thin barrier considering the $GaAs/Al_xGa_{1-x}As$ material composition. Influence of the electron interference effects on the inhomogeneous spatial distribution of the probability current density for the electron waves in semiconductor 2D nanostructures was theoretically investigated (Petrov *et al.*, 2009). Researchers also proposed (Song *et al.*, 1996) a transition layer model used to calculate resonant tunneling in a double-barrier quantum well system. Modified TBDQW structures are used (AlMuhanna *et al.*, 2011) to design long wavelength semiconductor lasers with low threshold current and small beam divergence. Experimental workers showed the importance of structural parameters (Tsai *et al.*, 2014) in design of LED.

In this present paper, current density is analytically computed from the knowledge of tunneling probability, and this calculation is based on the estimation of wave vector and its propagation through different layers. Barrier and well dimensions are changes within range of interest, and composition of barrier material is also varied to see the effect on current density. Occurrence of negative differential resistances is seen from the current density profile. Result will play crucial role for design of resonant tunneling devices.

2 MATHEMATICAL MODELING

For one-dimensional confinement, time-independent Schrödinger equation is given by

$$-\frac{\hbar^2}{2}\frac{\partial}{\partial z}\left[\frac{1}{m^*(z)}\frac{\partial}{\partial z}\psi(z)\right]+V(z)\psi(z) - q\xi(z)z = E(z)\psi(z) \tag{1}$$

where ζ(z) is the applied field along the direction of wave propagation.

Propagation vector can be obtained from Eq. (1) as

$$k_j = \left[\frac{2m^*\left(E - qV_j\right)}{\hbar^2}\right]^{0.5} \tag{2}$$

Thermal equlibrium probability can be calculated from the knowledge of range of wave vector

$$P = \frac{dk}{2\pi\hbar^2 \ln\left[1+\exp\left(E_F - \hbar^2(n_k-1)dk + k_{\min}\right)^2\right]} \tag{3}$$

where n_k denotes the range of 'k' values.

Tunneling current may be assumed as the probability of finding the electron in a particualr saptial region due to the flow of wave vector, either form left or right of the strucutre. This is defined as

$$J_z = \frac{\hbar}{2m^*}\left[\psi'\frac{\partial\psi}{\partial z} - \psi\frac{\partial\psi'}{\partial z}\right] \tag{4}$$

where ψ' denotes the complex conjugate of ψ.

3 RESULTS AND DISCUSSION

Fig. 1 shows the current density profile of the rectangular double quantum well structure as a function of applied voltage. It is seen from the plot that peak of the current density is obtained at 2.775 volt when well width is kept at 4 nm and middle

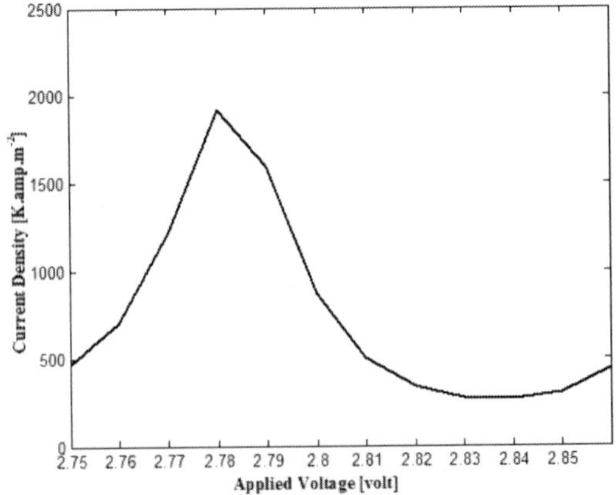

Figure 1. Current density variation with applied voltage for $Al_{0.1}Ga_{0.9}As/GaAs/Al_{0.1}Ga_{0.9}As/GaAs/Al_{0.1}Ga_{0.9}As$ DBQW structure.

barrier width at 6 nm. Negative differential resistance appears in the region which speaks in favor of quantum tunneling phenomenon. By increasing the range of applied voltage, we can see that multiple peaks appear at higher bias values. This is due to that fact that with increasing positive voltage, the net potential value in the Schrodinger equation reduces, which makes transmission easier. The lowermost eigenenergy level becomes negative (at a particular bias), and the first excited state becomes the lowermost state. This causes generation of second peak, and so on. at a given value of applied voltage, transmission occurs very close to the quasi-bound level, and thus current density becomes very high, which is semi-classical phenomenon. This is shown in Fig. 2.

Effect of structural parameters has been investigated on the tunneling phenomenon, and thus on the current density. In Fig. 3, effect of middle barrier thickness is analyzed and plotted. It is observed from the plot that increasing the barrier thickness first increases the tunneling probability, and hence current density increases. But higher value of barrier layer decreases the current. This is because at a particular gap between the adjacent wells, eigenstate of both the wells are aligned, and hence tunneling current becomes maximum. Changing the barrier width in either side reduces tunneling current. Also after attaining the peak, the current becomes oscillatory function of applied voltage.

Similarly, the effect of well width is also analyzed, and graphically represented in Fig. 4. It is seen from the figure that current density becomes very high when well widths are 2 nm and 6 nm respectively, but comparatively low for 4 nm well width. The reason is same as discussed in the previous paragraph. For those well dimensions eigenstates of the

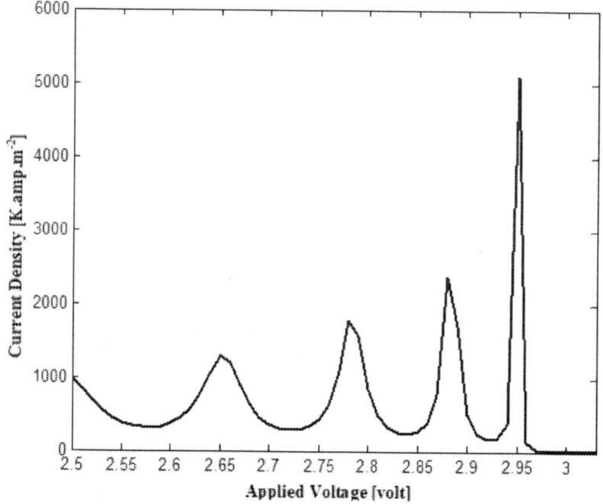

Figure 2. Current density variation with applied voltage for $Al_{0.1}Ga_{0.9}As/GaAs/Al_{0.1}Ga_{0.9}As/GaAs/Al_{0.1}Ga_{0.9}As$ DBQW structure.

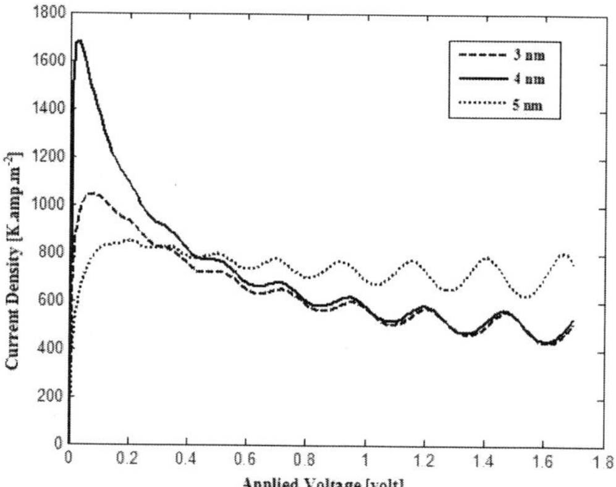

Figure 3. Current density variation with applied voltage for different middle barrier widths.

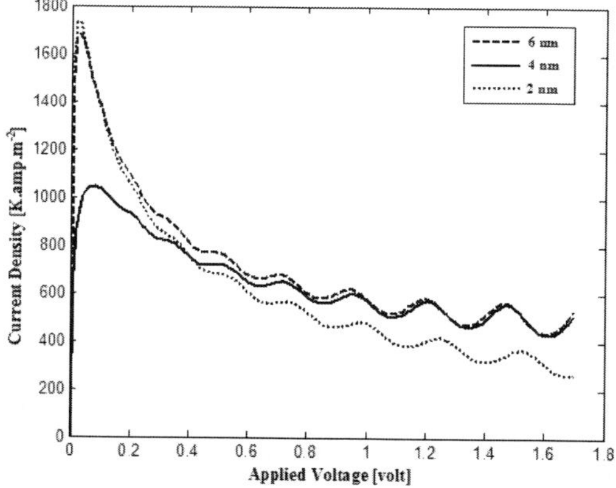

Figure 4. Current density variation with applied voltage for different well widths.

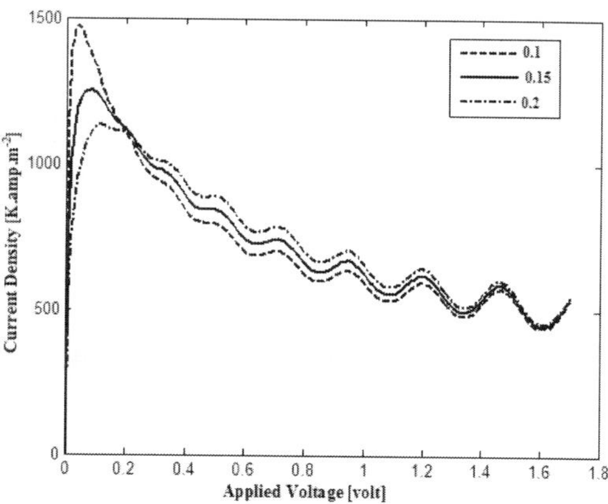

Figure 5. Current density variation with applied voltage for different material compositions of barrier widths.

adjacent wells are equal, then resonant tunneling occurs, and higher tunneling current is obtained. But for the other dimensions, transmission probability is lower, and thus the peak of current.

By varying the material composition of the contact and middle barrier layers, current density is computed and plotted as a function of applied bias, shown in Fig. 5. It is observed that by increasing the percentage of AlAs in barrier layer, peak of current density decreases. This is because higher percentage of Alas increases the potential barrier, which enhances the quantum confinement. This restricts the wave propagation between external contact and the quantum structure. Thus peak of current decreases.

By using this main method, we have created the object of our other classes which is placed in the same package and by using the object we have taken the inputs to perform the calculation and plot the graph. To calculate and plot the graph, we go through the mathematical calculations for calculating transmittivity of photonic multiple quantum well for normal incidence and transmittivity of SiO2-Air is calculated.

4 CONCLUSION

Effect of middle barrier width and well width, as well as composition of barrier widths are calculated on tunneling current density of rectangular double quantum well device. Choice of applied bias plays a crucial role in calculating current. Higher bias may generate multiple peaks which ultimately lead to the quasi-bound state transmission. Choice of structural parameters plays important role for resonant tunneling, which is crucial in designing RTD.

REFERENCES

Al-Muhanna. A, Alharbi. A, Salhi. A, 2011, Waveguide Design Optimization for Long Wavelength Semiconductor Lasers with Low Threshold Current and Small Beam Divergence, Journal of Modern Physics, 2, 225–230.

Chanda. L. A, Eastman. L. F., 1982, Quantum Mechanical Reflection at Triangular Planar-Doped' Potential Barriers for Transistors, Journal of Applied Physics, 53, 9165–9169.

Christodoulides. D. N., Andreou. A. G., Joseph. R. I., Westgate. C .R., 1985, Analytical Calculation of the QuantumMechanical Transmission Coefficient for a Triangular, Planar-Doped Potential Barrier, Solid State Electronics, 28, 821–822.

Chang. L. L, Esaki. L., Tsu. R., 1974, Resonant Tunneling in Semiconductor Double Barriers, Applied Physics Letters, 24, 593–595.

Esaki. L, Chang. L. L, 1974, New Transport Phenomenon in Semiconductor Superlattice, Physical Review Letters, 33, 495–498.

Esaki. L, Tsu. R, 1970, Superlattice and Negativedifferential Conductivity in Semiconductors, IBM Journal Research Division, 14, 61–65.

Goldhaber-Gordon. G, Montemerlo. M. S, Love. J. C, Opiteck. G. J, Ellenbogen. J. C, 1997, Overview of Nanoelectronic Devices, Proceedings of the IEEE, 85, 521–540.

Ghatak. A. K., Thyagarajan. K., Shenoy. M. R, 1988, A Novel Numerical Technique for Solving the One-Dimensional Schrödinger Equation using Matrix Approach—Application to Quantum Well Structures, IEEE Journal of Quantum Electronics, 24, 1524–1531.

Maranowski. K. D, Gossard. A. C, 2000, Far-infrared Electroluminescence from Parabolic Quantum Well Superlattices Excited by Resonant Tunneling Injection, Journal of Applied Physics, 88, 172–177.

Petrov. V. A, Nikitin. A. V, 2009, Penetration of QuantumMechanical Current Density under Semi-Infinite Rectangular Potential Barrier as the Consequence of the Interference of the Electron Waves in Semiconductor 2D Nanostructures, Proceedings of SPIE, 7521.

Reed. M. A., Koestner. R. J., Goodwin. M. W., 1986, Resonant Tunneling Through a HeTe/$Hg_{1-x}Cd_xTe$ Double Barrier, Single Quantum Well Structure, Journal of Vacuum Science and Technology A, 5, 3147–3149.

Sen. S, Capasso. F, Gossard. A. C, Spah. R. A, Hutchinson. A. I, Chu. S. N. G, 1987, Observation of Resonant Tunneling through a Compositionally Graded Parabolic Quantum Well, Applied Physics Letters, 51, 1428–1430.

Simion. C. E, Ciucu. C. I, 2007, Triple–Barrier Resonant Tunneling: A Transfer Matrix Approach, Romanian Reports in Physics, 59, 805–817.

Scandella. L., Güntherodt. H. J, 1992, Field Emission Resonances Studied with dI/ds(V) and dI/dV(V) Curves, Ultramicroscopy, 42, 546–552.

Song. Y, 1996, A Transition Layer Model and its Application to Resonant Tunneling in Heterostructures, Physics Letters A, 216, 183–186.

Tsai. C. L., Wu. W. C, 2014, Effects of Asymmetric Quantum Wells on the Structural and Optical Properties of InGaNBased Light-Emitting Diodes, Materials, 7, 3758–3771.

Talele, K., Patil, D. S. 2008, Analysis of Wavefunction, Energy and Transmission Coefficients in GaN/AlGaN Superlattice Nanostructures", Progress In Electromagnetics Research, 81, 237–252.

Wessel. R., Alterelli. M., 1989, Quasi Stationary Energy Level Calculation for Thin Double Barrier GaAs-$Ga_{1-x}Al_xAs$ Heterostructures, Physical Review B, 39, 10246–10250.

Yamamoto. H, 1987, Resonant Tunneling Condition and Transmission Coefficient in a Symmetrical OneDimensional Rectangular Double-Barrier System, Applied Physics A: Materials Science & Processing, 42, 245–248.

Computational Science and Engineering – Deyasi et al. (Eds)
© 2017 Taylor & Francis Group, ISBN 978-1-138-02983-5

Measurement of transmittivity in 1D photonic crystal under normal incidence implemented by JAVA-GUI application

Pritam Dutta, Dibyaduti Gorai, Rahul Mukherjee & Soumen Santra
Department of Computer Application, Techno India College of Technology, Kolkata, India

Arpan Deyasi
RCC Institute of Information Technology, Kolkata, India

ABSTRACT: In this paper, a package is made based on JAVA-GUI platform to measure the transmittivity in one-dimensional photonic crystal considering electromagnetic wave propagation along the direction of confinement. For a suitable range of structural parameters within practical limit (for fabrication), computation is carried out around 550 nm for the purpose of optical communication. Dimensions of both the constituents, number of layers and refractive indices of the materials are considered as input parameters for development of the package, and simulated results provide typical characteristics of optical bandpass filter. Though the package is developed for normal incidence of e.m wave only, but later can be extended for oblique incidences also.

1 INTRODUCTION

One-dimensional photonic crystal may be looked as repetition of a unit cell with two/three dielectric materials; and thickness and refractive indices of the materials are such that electromagnetic wave can propagate along the confined direction following Bragg's principle (Loudon 1970, Yablonovitch 1987). The alternative lower and higher dielectric constants of the constituent materials and thickness of the layers are chosen in such a way that when spectra of e.m wave is incident on it, only a certain spectrum is allowed to transmit, and other are rejected (Shambat *et al.*, 2009, Maity *et al.*, 2013). The application of this microstructure is not only limited for development of optical filter (Mao *et al.*, 2008), but also in development of photonic crystal fiber (Russell 2006), quantum computation (Azuma *et al.*, 2008), optical transmitter (Limpert *et al.*, 2003), optical receiver (Reininger *et al.*, 2012), optical sensor (Shanthi *et al.*, 2014) etc.

Graphical user interface of java is one of the versatile tools for implementation of user-friendly package. In this paper, we have developed a package using JAVA-GUI application for measurement of transmittivity in 1D photonic crystal under normal incidence condition. In section II, proposed algorithm is discussed, in section III, development of code is mentioned using the application, in section IV, result is highlighted along with conclusion and future possible extensions.

2 ALGORITHM

In order to observe the graphical representation of transmittivity of photonic crystal with respect to wave-length of light, here we introduced a user interface which is developed by Java Frame (JFrame).

The frame consists of five input text fields and an internal output frame or panel.

We set the input values in those input text fields and get the output graph plot in the internal output panel by clicking the 'get plot' button

Entering the inputs, when we click the button the following actions are performed

1. Inputs are scanned and saved the values according their data types.
2. The scanned input values are passed through the parameterized constructor of the class named 'Graph plot'
3. The method 'Graph plot' is invoked in order to plot the graph.
   - 3.1. Calculate theta1, theta2.
   - 3.2. Calculate reflectivity '*r*', transmitivity '*t*' and propagation length '*d*'.
   - 3.3. Generate transfer matrix corresponding to the interface M12, M21.
   - 3.4. Repeat this step until (i < 1001)
     - 3.4.1. Calculate wave vector corresponding to the value of wavelength.
     - 3.4.2. Calculate transfer matrix for the elementary cell as ML1 and ML2.

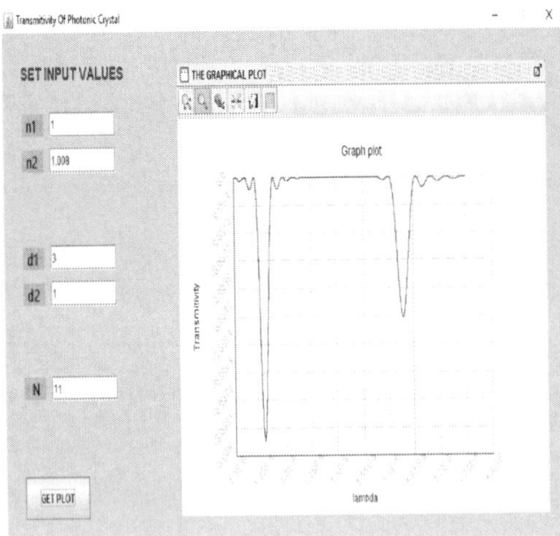

Figure 1. Bandpass filter for given input parameters.

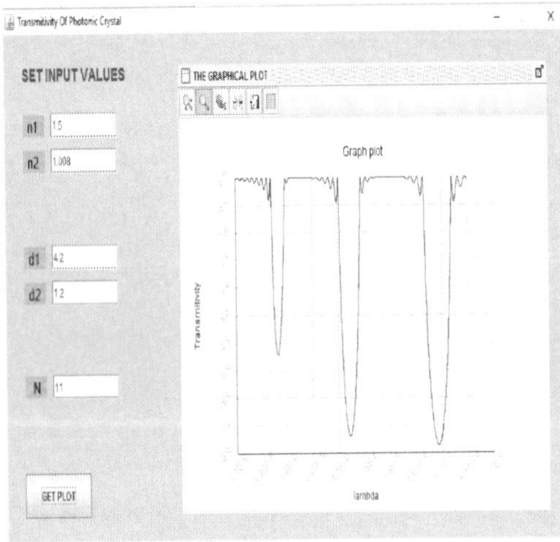

Figure 2. Multiple bandpass filter.

3.4.3. Repeat this step until (Z < = N)
3.4.3.1. Calculate the total transfer matrix of Z such elementary cells.
3.4.3.2. Exit loop (7.3).
3.4.4. Calculate Transmittivity corresponding to the wavelength.
3.4.5. Exit loop (7).

4. The function 'Graph plot' returns the final graphical representation to the internal output panel of the interface frame.

3 CODE DEVELOPMENT

By using this main method, we have created the object of our other classes which is placed in the same package and by using the object we have taken the inputs to perform the calculation and plot the graph. To calculate and plot the graph, we go through the mathematical calculations for calculating transmittivity of photonic multiple quantum well for normal incidence and transmittivity of SiO_2-Air is calculated.

Below is the code for the Main class containing the main method. Executing this, results in the whole GUI to be able to perform correctly.

```
public class Main {
public static void main (String [] args) {
    EventQueue.invokeLater(new Runnable () {
public void run () {
try {
            PhFrame window = new PhFrame ();
            window.frame.setVisible(true);
        } catch (Exception e) {
            e.printStackTrace();
        }
      }
    });
  }
}
```

4 RESULTS

After developing the package, transmittivity is calculated and plotted as a function of wavelength. A few snapshots of the GUI are shown below.

5 CONCLUSION

Transmittance of one-dimensional photonic crystal structure is computed for normal incidence of s-polarized electromagnetic wave with centre wavelength at 1550 nm. Computation reveals the fact that slab thicknesses and material composition are two key design features for desired tuning of filter bandwidth. Design parameters can be optimized in such a way that centre wavelength of the proposed bandpass filter will always be surrounded by two photonic bandgap, represented by sudden increases of reflectance in the simulate profiles. The user friendly package will help the end users to gain knowledge before fabrication. The structure is a suitable candidate for optical communication.

6 FUTURE SCOPE

With the help of this easy-to-use package, one can get the transmittivity vs wavelength graph of a photonic crystal by simply putting the required data within the Text Field of the GUI as shown in both the figures. Further this package can be modified to get comparison output of different compositions within the crystal. It provides higher productivity, while facilitating a lower cognitive input data.

REFERENCES

Azuma. H (2008) Quantum Computation with Kerr-Nonlinear Photonic Crystals, Journal of Physics D: Applied Physics, vol. 41, p. 025102.

Limpert. J, Schreiber. T, Nolte. S, Zellmer. H, Tunnermann. T, Iliew. R, Lederer. F, Broeng. J, Vienne. G, Petersson. A, Jakobsen. C (2003) High Power Air-Clad Large-Mode-Area Photonic Crystal Fiber Laser, Optic Express, vol. 11, pp. 818–823.

Loudon. R (1970) The Propagation of Electromagnetic Energy through an Absorbing Dielectric, Journal of Physics A, vol. 3, pp. 233–245.

Maity. A, Chottopadhyay. B, Banerjee. U, Deyasi. A (2013) Novel Band-Pass Filter Design using Photonic Multiple Quantum Well Structure with p-Polarized Incident Wave at 1550 nm, Journal of Electron Devices, vol. 17, pp. 1400–1405.

Mao. D, Ouyang. Z, Wang. J. C (2008) A Photonic-Crystal Polarizer Integrated with the Functions of Narrow Bandpass and Narrow Transmission Angle Filtering, Applied Physics B, vol. 90, pp. 127–131.

Reininger. P, Kalchmair. S, Gansch. R, Andrews. A. M, Detz. H, Zederbauer. T, Ahn. S. I, Schrenk. W, Strasser. G (2012) Optimized Photonic Crystal Design for Quantum Well Infrared Photodetectors, Proc. of SPIE, vol. 8425, p. 84250A.

Russell. P. S. J (2006) Photonic-Crystal Fibers, Journal of Lightwave Technology, vol. 24, pp. 4729–4749.

Shambat. G, Mirotznik. M. S, Euliss. G, Smolski. V. O, Johnson. E. G, Athale. R. A (2009) Photonic Crystal Filters for Multi-band Optical Filtering on a Monolithic Substrate, Journal of Nanophotonics, vol. 3, p. 031506

Shanthi. K. V, Robinson. S (2014) Two-dimensional Photonic Crystal based Sensor for Pressure Sensing, Photonic Sensors, vol. 4, pp. 248–253.

Yablonovitch. E (1987) Inhibited Spontaneous Emission in Solid-State Physics and Electronics, Physical Review Letters, vol. 58, pp. 2059–2061.

Computational Science and Engineering – Deyasi et al. (Eds)
© 2017 Taylor & Francis Group, ISBN 978-1-138-02983-5

Use of multi-passing technique for reduction of v_π voltage in KDP crystal

R. Maji & S. Mukhopadhyay
Department of Physics, The University of Burdwan, West Bengal, India

ABSTRACT: The multi-passing technique is already established for reduction of V_π voltage in $LiNbO_3$ crystal. In general the V_π voltage of KDP is very high in comparison to that of $LiNbO_3$. Again the modulation of light in KDP is more active and strong. Therefore the reduction of V_π is more necessary for practical application of electro-optic modulation by KDP. In this paper the authors show that the multi-passing technique is more suitable in KDP for electro-optic modulation. If this multi-passing technique is used for reduction of the V_π voltage in KDP, then this V_π is dropped down from several KVs to volt level. This technique, therefore, is very suitable for light modulation with looser electric power.

1 ABOUT THE SYSTEM

The half-wave voltage V_π of an electro-optic modulator is defined as the voltage required to produce a phase shift of 180° in a light beam passing into the modulator (Ghatak 2002, Maji 2011, 2010). There are many applications of electro-optic cell in Q-switching. For exploiting the electro-optic switching character of a Pockels modulator like KDP, ADP, KD*P it requires high voltages in the order of few KVs. This gives a real problem for practical modulation of an electronic message signal by optical one. The authors of this paper here give a proposal for reduction of the V_π voltage in case of KDP crystal, acting as an Pockls cell (Ghatak 2002, Gu 2012, Hu 2011, Maji 2010_a, 2010_b, 2011, Yariv 1985).

2 ANALYSIS OF GETTING LOWER V_Π VOLTAGE IN AN ELECTRO-OPTIC POCKELS CELL BY MULTI PASSING OF A BEAM

In Figure 1 the schematic representation of the V_π reducing process is shown. KDP based electro-optic cell is attached with external biasing voltage V, along its Z axis or its optic axis. The length of the cell is ℓ. Now a beam polarized along 45° (in X-Y plane) to both X and Y axis is taken which is sent through the cell along its Z direction. The polarizer P helps to select the polarization angle of the light along the 45° of X axis. The refractive index of the material for the components of the light polarized along X direction is $n_X = n_0 - \frac{1}{2}n_0{}^3r_{63}E_Z$ and that along Y direction is $n_Y = n_0 + \frac{1}{2}n_0{}^3r_{63}E_Z$.

The mirrors in the Figure 1 gives change the direction of the light as shown.

3 METHOD OF REDUCTION OF V_Π IN KDP

The electric field component of the polarized wave along X direction after traversing the length ℓ along Y direction the KDP electro optic modulator, as.

$$E_X = E_0\cos(\omega t - n_x k_0 \ell)$$
$$E_X = E_0\cos\left(\omega t - n_0 k_0 \ell + \frac{1}{2}n_0{}^3 k_0 r_{63} E_Z \ell\right) \quad (1)$$

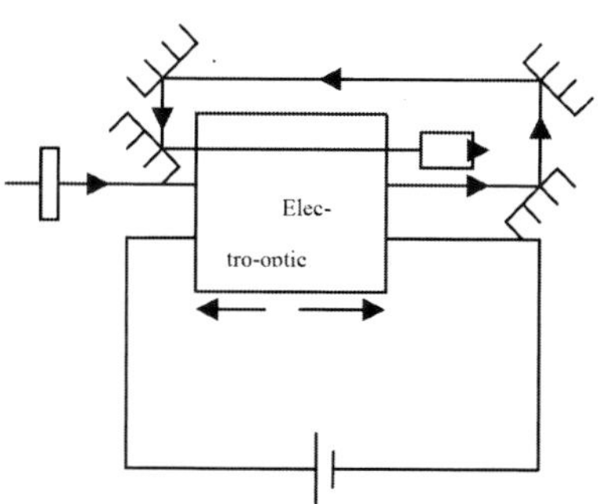

Figure 1. A representation for the reduction of V_π voltage in KDP (M_i, i = 1 to 4 are used mirrors, S denotes the source of light P Denotes the Polarizer, V denotes voltage).

Where n_x is the refractive index of KDP crystal in x direction and E_Z is the external electric field along z direction. After completion of 1st cycle it gets its electric field at the output as,

$$E_{X_1} = E_0 \cos\left(\omega t - 2n_0 k_0 \ell + 2\frac{1}{2} n_0^3 k_0 r_{63} E_Z \ell + \phi_1\right) \quad (2)$$

Similarly after completion of 1st cycle the expression of the electric component of the light wave polarized along the Y direction and passing through the Y_1 direction, is

$$E_{Y_1} = E_0 \cos(\omega t - 2n_0 k_0 \ell - 2\frac{1}{2} n_0^3 k_0 r_{63} E_Z \ell + \phi_2) \quad (3)$$

Thus the phase difference of E_{X1} and E_{Y1} after crossing the electro-optic modulator is

$$\Delta\phi_1 = 2n_0^3 r_{63} k_0 E_Z \ell + (\phi_1 - \phi_2) \quad (4)$$

Now if a compensator is used to remove $(\phi_1 - \phi_2)$ term the V_π voltages becomes,

$$V_{\pi 1} = \frac{\lambda_0}{4n_0^3 r_{63}} \quad (5)$$

And similarly, the phase difference of E_{X2} and E_{Y2} after crossing the electro optic modulator (after 2 times passing) is

$$\Delta\phi_2 = 3n_0^3 r_{63} E_Z \ell \quad (6)$$

Removing the 1st term by a suitable compensator the V_π voltage becomes

$$V_{\pi N} = \frac{\lambda_0}{2(n+1) n_0^3 r_{63}} \quad (7)$$

Thus if the n times rotations done the V_π voltage becomes

$$V_{\pi N} = \frac{\lambda_0}{2(n+1) n_0^3 r_{63}} \quad (8)$$

Here n is +ve integer.

By the method we can reduce the half wave voltage (V_π) of an electro-optic cell (KDP) by the desired amount after the suitable number of rotations.

4 CONCLUSION

From the obtained equation 8 it is seen that the half wave voltage reduces inversely with n. For example if the V_π voltage of KDP is 64kv, t then in a one-time passing system V_π is 32 KV. Similarly for 8 times it comes down to 8 KV. So increasing the number of turns of passing one can decrease the V_π in order of few volts.

REFERENCES

Ghatak A., and Thyagarajan K., Optical Electronics by (Cambridge university press 2002).

Gu Y., Hu J., Li S., Han X., Wang M., Wu P., and Zhao M., "Novel photonic broadband microwave frequency measurement based on intensity-modulated link with output microwave interference detection", Optical Engineering 51(1), 019001, DOI: 10.1117/1.OE.51.1.019001 (2012).

Hu S., Han X., Wu P., Gu Y., Zhao M., "A photonic technique for microwave frequency measurement employing tunable dispersive medium", Microwave Photonics, 2011 International Topical Meeting on & Microwave Photonics Conference, 2011 Asia-Pacific, MWP/APMP, IEEE,165–168, DOI: 10.1109/MWP.2011.6088695.

Maji R. and Mukhopadhyay S., "A method of reducing the half wave voltage (Vπ) of an electro-optic modulator by multi passing a light through the modulator", Optik International Journal Light Electron opt (2011). DOI:10.106/ijleo.2011.07.035.

Maji R. and Mukhopadhyay S., "A new method of controlling the self-focusing length of a bulk nonlinear material using electro-optic material", IUP journal of Physics, Vol III, No 3 (2010_a).

Maji R. and Mukhopadhyay S., "An alternative optical method of determining the unknown microwave frequency by the use of electro-optic materials and semiconductor optical amplifier", Optik International journal Light Electron opt (2010_b) doi:10.1016/ijleo.2010.10.013.

Yariv A. Optical Electronics, Halt Rinehart and Winston, New York (1985).

Computational Science and Engineering – Deyasi et al. (Eds)
© 2017 Taylor & Francis Group, ISBN 978-1-138-02983-5

An alternative approach of all optical frequency encoded DIBIT based latch along with simulated verification

Partha Pratim Sarkar & Bitan Ghosh
Department of Electronics and Communication Engineering, University Institute of Technology, The University of Burdwan, Burdwan, West Bengal, India

Sourangshu Mukhopadhyay
Department of Physics, The University of Burdwan, Burdwan, West Bengal, India

Sankar Narayan Patra
Department of Instrumentation Science, Jadavpur University, Kolkata, West Bengal, India

ABSTRACT: Super fast all optical memory and optical logic gates are the basic building blocks for optical computation and communication systems. Realization of a very fast memory-cell in the optical domain is very challenging. Recently different theoretical papers based on combinational and sequential logic circuits for developing flip flops as well as memory cells have been reported by the different authors. Here, the authors have proposed a new scheme of all optical frequency encoded dibit based latch circuit using reflected semiconductor optical amplifier and add-drop multiplexer with its proper simulation. The use of dibit representation technique along with optical frequency encoding technique makes the system very fast and reduce bit error problem.

1 INTRODUCTION

All optical data processing technique is the most alternative and successful replacement to overcome the speed related problems due to its inherent properties of parallelism. Now, different types of all optical logic gates and memory circuits are the fundamental building blocks for optical data processors and communication systems. Again, dibit representation technique provides the benefits of low bit error problem by increasing high signal to noise ratio. Here, the authors have proposed a new scheme of all optical frequency encoded latch by Reflected Semiconductor Optical Amplifier (RSOA) and Add/Drop Multiplexer (ADM) blocks with dibit representation technique, where digital value '0' is represented as [0] [1] logic states and digital value '1' is represented as [1] [0] respectively. In other way, the presence of the two frequencies side by side [υ_1] [υ_2] represents digital logic state '0' and [υ_2] [υ_1] does the same as digital logic state '1'. To implement the all optical dibit based latch circuit, RSOA and ADM are two important optical switches, discussed bellow.

1.1 *Reflected semiconductor optical amplifier*

Reflected semiconductor optical amplifier is a switch where, if a light of wavelength considered as weak probe beam say, $\lambda_1 = 1550$ nm (in term of frequency say υ_1) and another light of wavelength considered as a strong pump beam say, $\lambda_2 = 1540$ nm (in term of frequency say υ_2) are inserted to the input terminals. The frequency of the output beam is equal to the frequency of probe beam and the power of the output beam is equal to the power of pump beam. If υ_1 and υ_2 frequency of light beam are supplied to the pump beam and probe beam respectively, the RSOA provides υ_2 frequency of light beam at the output terminal. Opposite frequency of light beam comes out at the output terminal if the input frequencies of light beams are made alternate, which is shown in Figure 1. So it could be established as a very suitable optical device for constructing many all optical logical operations.

1.2 *Add-drop multiplexer*

Optical add-drop multiplexer is a frequency selective switch. Now, when it is adjusted with a fixed biasing current in terms of a light beam of a particular frequency then it reflects that particular frequency of light beam and passes all the rest frequencies of light beam through it. If υ_1, υ_2, υ_3...... frequencies of light beams are the input of ADM and it is biased with a current in terms of a light beam of υ_1 frequency, then the ADM reflects only light beam

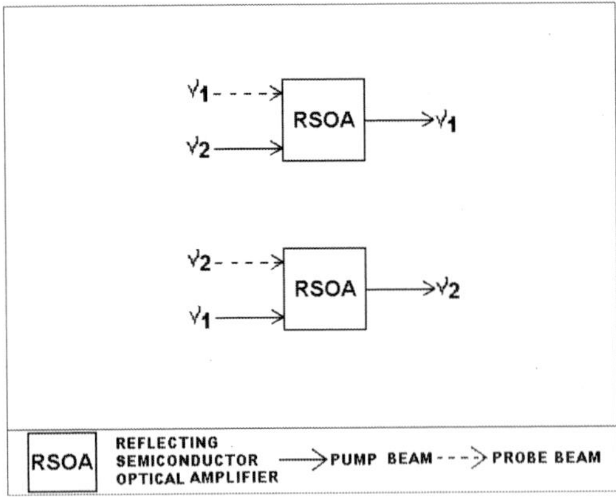

Figure 1. Schematic diagram of Reflected Semiconductor Optical Amplifier (RSOA).

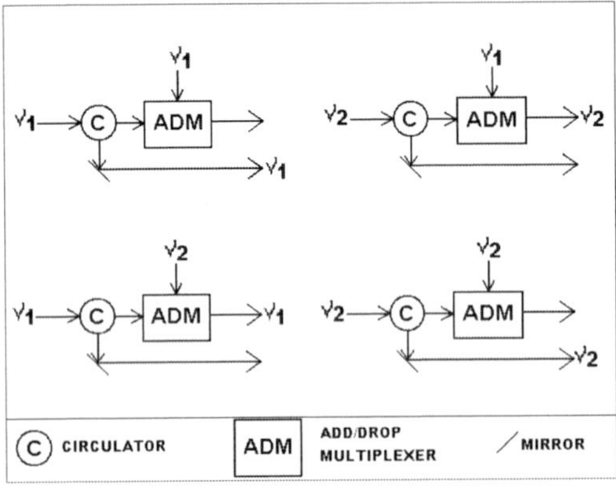

Figure 2. Schematic diagram of Add/Drop Multiplexer (ADM).

of υ_1 frequency and passes all other light beams. Alternative incident happens if the value of biasing current changes. This block is shown in Figure 2.

2 PRINCIPLE OPERATION OF FREQUENCY ENCODED ALL OPTICAL DIBIT NOT BASED LATCH

To prepare a complete unit of frequency encoded all optical memory cell, the first step is to prepare a memory unit or a latch circuit as it can store a dibit. The proposed system which is described here, based on frequency encoding principle, shown in Figure 1. Here we use dibit representation technique. In the dibit representation, I' and I" represents 1st bit and 2nd bit of input and O' and O" are the dibit output terminals respectively. To implement the optical NOT based latch logic with dibit technique Beam Splitters (BS), Mirrors (M), Reflected Semiconductor Optical Amplifier (RSOA) and Add-Drop Multiplexer (ADM) are used at different position of the system.

In Figure 3, the block diagram of optical dibit based latch circuit is shown. Here optical wave beams in form of $I' = \upsilon_1$ and $I'' = \upsilon_2$ are applied at input terminal, now, υ_1 frequency is moved as weak probe beam and υ_2 frequency is sent as strong pump beam to the $RSOA_1$. So the output of $RSOA_1$ provides υ_1 frequency of light beam. Then υ_2 frequency of light beam enters to ADM_1, which is biased by υ_2 frequency. So it passes υ_1 frequency to $RSOA_2$ as a pump beam. But there is a constant probe beam of υ_2 frequency at $RSOA_2$. For this reason output becomes in form of υ_2 frequency of light beam and goes to output terminal O´. Again this output goes to $RSOA_3$ as a pump beam by the feedback path.

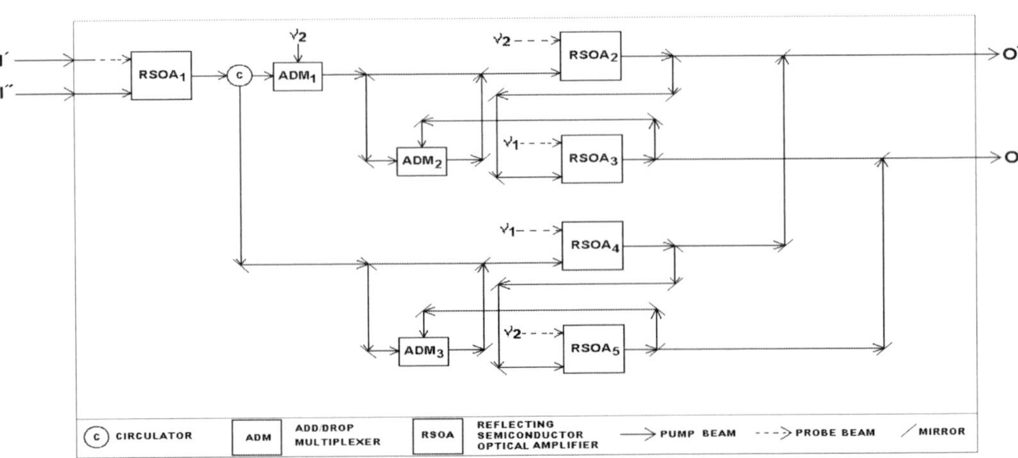

Figure 3. Schematic diagram of frequency encoded DIBIT based Latch.

Since there is a constant source of probe beam of υ_1 frequency, so υ_1 frequency of light beam comes out from $RSOA_3$ then this light beam of υ_1 frequency goes to another output terminal O′′. Now, to sustain the output continuously, there is a feedback process, where the output light beam of υ_1 frequency from O′′ goes to the pump beam terminal of $RSOA_2$. For this reason, we get continuous output of υ_2 and υ_1 at the terminal O′ and O′′ respectively, even if the input frequency of light beam is absent.

Again, if we apply optical light beam in form of I′ = υ_2 frequency and I′′ = υ_1 frequency at input that is opposite of the first case, now it is noticed that υ_2 frequency and υ_1 frequency of light beam are moved as weak probe beam and strong pump beam to the $RSOA_1$ respectively. So υ_2 frequency of light beam is reflected by ADM_1 because it is biased by optical wave beam of υ_2 frequency. So υ_2 frequency of light goes to $RSOA_4$ as a pump beam. And there is a constant probe beam of υ_1 frequency. Then output becomes υ_1 frequency of light beam and this light wave of υ_1 frequency goes to output terminal O′. Now a part of this output goes to $RSOA_5$ as a pump beam by the feedback path. But there is a probe beam of υ_2 frequency. So, υ_2 frequency of light beam goes to output terminal O′′. But this output from O′′ goes to the pump beam terminal of $RSOA_4$. So, the output is maintained incessantly by this feedback process. Here, we get continuous output of υ_1 and υ_2 at the terminal O′ and O′′ respectively, where the input light beam is absent. The output frequencies at both the terminals remain unchanged until or unless the input frequencies of light beams are altered. This is the overall connection of the all optical latch circuit.

When dibit logic state '[0][1]' is applied at the input, the upper portion of the system is activated but lower half does not. So one can ensure the υ_2 and υ_1 frequency of lights come at the output terminal O′ and O′′ respectively i.e. O′ = digital state 1 and O′′ = digital state 0.

So at the output dibit logic state '[1][0]' is obtained. Similarly when dibit logic state '[1][0]' is applied to the input terminal, the upper half does not get active but the lower half of the system is activated. Again we get the output O′ = digital state 0 and O′′ = digital state 1, i. e. dibit logic state '[0][1]' is obtained. The input and output of the optical dibit based latch or memory cell is also shown in the truth table in Table 1, which satisfies the fundamental logic principle of NOT based latch. So, NOT logic is established in this mechanism. If dibit logic state '[0][1]' and '[1][0]' are withdrawn from input side, the system will continue to show the last shown values of O′ and O′′ as the final output due to the feedback mechanisms. So, this system can be used as a frequency encoded all optical dibit based memory cell.

Table 1. Truth Table of dibit based latch.

DIBIT I/P		Digital I/P	DIBIT O/P		Digital O/P
I′	I′′	O′	O′′		
[0]	[1]	0	[1]	[0]	1
[1]	[0]	1	[0]	[1]	0

3 SIMULATION METHOD OF THE DIBIT NOT BASED LATCH

Here for simulation of dibit based latch we have used the MATLAB (2008a) simulink software. First, for designing the dibit sources, we considered mathematical value '8' = digital state '1' and mathematical value '3' = digital state '0'. With the help of simulink tool box library a dibit control box has been designed.

Using this box the mathematical models of RSOA and ADM have been added accordingly in Figure 4 following the schematic diagram shown in Figure 3. These are simulated by constant blocks, whose values may be changed, taken from simulink library. Now RSOA unit is properly programmed using 'C' language for choosing the proper output at the output terminal. In presence of both inputs to the RSOA block, this system provides the wavelength of the probe beam at the output terminal. Again, for ADM unit two embedded MATLAB functions are taken and properly programmed using 'C' language for choosing the proper output at the output terminal. The input optical signal beam with υ_2 frequency will come at the output as drop or reflected signal, if the biasing current is taken as '1' and other frequencies (here υ_1) passes as transmitted beam. Now, the opposite incident happens, if the biasing current is chosen as '0'. This principle fully supports the function of optical ADM.

Now, following the block diagram of Figure 3, we have simulated the DIBIT based optical latch circuit with MATLAB Simulink programming in Figure 4, where there are two output terminals, which provide the changeable output depending on the variation of inputs. Like if we provide the dibit based input [υ_1][υ_2] or [0][1] (here [3][8]) this simulated block provides the output in form of [υ_2][υ_1] or [1][0] (here [8][3]), now if we change the input, output changes accordingly. Again, if there is no input is given at the input terminal, this simulated block holds the previous output. Now, if we apply the input in form of [υ_2][υ_2] or [1][1] (here [8][8]), which is an abnormal input combination for latch circuit, for this reason a toggle is raised in the Figure 5.The combination of inputs and outputs shown in Figure 5, which fully support the truth table of latch circuit in Table 1.

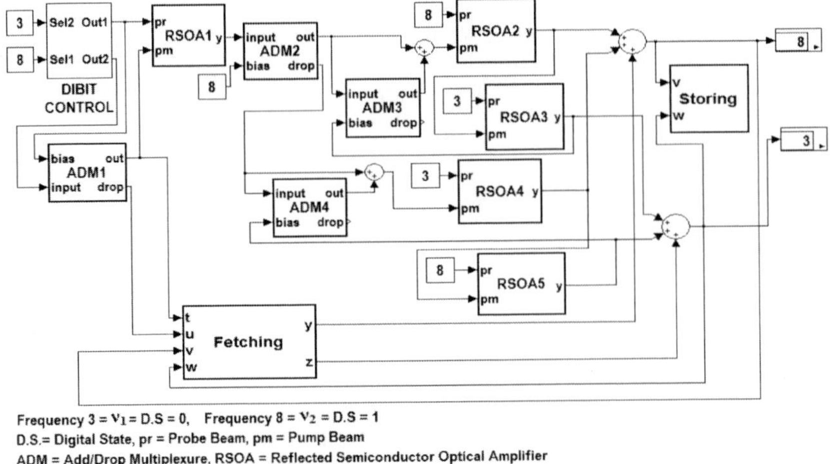

Figure 4. Mathematical model of all optical frequency encoded DIBIT based latch.

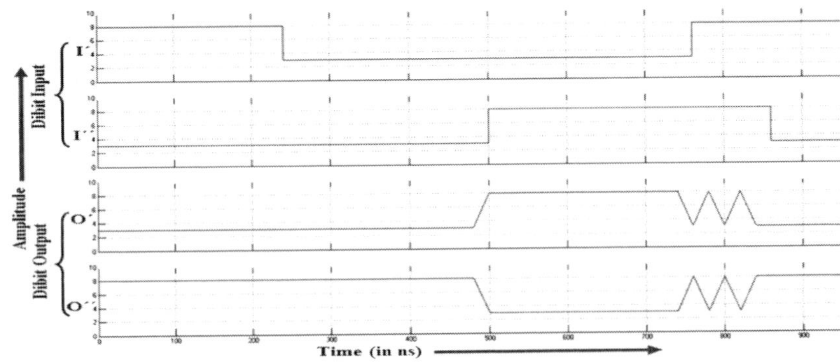

Figure 5. Graphical output of all optical frequency encoded DIBIT based latch.

4 CONCLUSION

This system can exhibit high degree of parallelism and reduces bit error problem. Also truth table satisfies the dibit based latch logic circuit. Therefore its performance can directly be utilized for the development and verification of the performances of different logic devices based on frequency encoding principle. Using this dibit representation one can implement for other sequential and combinational all optical operations like flip-flops, multivibrators, memories etc.

REFERENCES

Chandra, S. K. et al. 2014. All-optical phase encoded 4-to-1 phase multiplexer using four wave mixing in semiconductor optical amplifier. *Optik-International Journal for Light and Electron Optics* 125(23): 6953–6957.

Dutta, S. & Mukhopadhyay, S. 2011. Alternating approach of implementing frequency encoded all-optical logic gates and flipflop using semiconductor optical amplifier. *Optik - Int. J. Light Electron Opt.* 122(2): 125–127.

Ghosh, B. & Mukhopadhyay, S. 2013. A novel realization of all-optical dibit represented frequency encoded Boolean and quaternary inverters without switching device. *Optik-International Journal for Light and Electron Optics* 124(21): 4813–4815.

Guo, L. Q. & Connelly, M. J. 2007. A poincare approach to investigate nonlinear polarization rotation in semiconductor optical amplifiers and its applications to all optical wavelength conversion. *Proc. of SPIE* 6783(678325): 1–5.

Mukhopadhyay, S. 1992. Binary optical data subtraction by using a ternary dibit representation technique in optical arithmetic problems. *Applied Optics* 31(23): 4622–4623.

Sarkar, P. P. & Mukhopadhyay, S. 2014. All optical frequency encoded NAND logic operation along with the simulated result. *Journal of Optics* 43(3): 177–182.

Wang, K.Y. & Foster, A. C. 2012. Ultralow power continuous-wave frequency conversion in hydrogenated amorphous silicon waveguides. *Optics letters* 37(8): 1331–1333.

Computational Science and Engineering – Deyasi et al. (Eds)
 ISBN 978-1-138-02983-5

Quantized conductance characteristics of Nano–MESFET under optical illumination

Kasturi Mukherjee
Heritage Institute of Technology, Kolkata, India

Arpan Deyasi
RCC Institute of Information Technology, Kolkata, India

Deepam Gangopadhyay
Heritage Institute of Technology, Kolkata, India

ABSTRACT: Current-voltage characteristics of a Metal-Semiconductor Field-Effect Transistor under optical illumination are calculated when gate length is reduced into nanometer dimension from micrometric magnitude. In order to calculate the drain current flowing through the device under dark condition, potential distribution and the dimension of depletion region are calculated from the solution of the two-dimensional Poisson's equation. The photoconductive component of the current is computed to estimate total current flowing through the device. Submicron gate length introduced quantized conductance characteristics in presence of incident radiation. The suitability of this unique characteristic of the device for use as optical detector is highlighted in this paper.

Keywords: Current-voltage characteristics, Nanometer gate length, Flux density, Photoconductive current, Drain current

1 INTRODUCTION

Metal-semiconductor field effect transistor has been focused by many researchers because of its novel light sensitivity property and its application in the microwave photonics domain (Vilcot *et al.*, 2003). The higher carrier mobility, current, transconductance and transit frequency being higher for the MESFET has made its electrical characteristics suitable for application in microwave circuits (Allen *et al.*, 1997, Huang *et al.*, 2004). Furthermore, optical illumination on MESFET has been found to be a controlling factor for tuning the frequency range of MESFET microwave oscillator, injection locking, gain and phase control of MESFET amplifiers and oscillators, and operation of MESFET as a switch (Kraemer *et al.*, 2009. Peterson *et al.*, 1992, Gautam *et al.*, 1993). MESFET has been used as an on-chip optical detector in MMICs especially for optically fed phased arrays (Madjar *et al.*, 1993). Researchers also (Bandhawakar *et al.*, 2001) presented the important role of MESFET as switch under optical illumination for designing low voltage, low power VLSI devices. The difference in the spectral response and quantum efficiency of an Optically Controlled Silicon MESFET for the variation in the impurity profile distribution has been analyzed (Chattopadhyay *et al.*, 2010). The experimental results on the optical response of the current-voltage characteristics of GaAs MESFET photodetectors designed for the X—band applications have been reported (Umeda *et al.*, 1987). Optical control of the switching characteristics as well as suitable Schottky configuration of MESFETs for high speed photodetection has been addressed by the researchers (Chaturvedi *et al.*, 1983, Singh *et al.*, 1987). Workers (Zebda *et al.*1997) have also modeled the time-varying characteristics of the MESFET using perturbation technique. The photoconductive and photovoltaic phenomena in the channel and depletion region of the MESFET have also been theoretically analyzed (Gautier *et al.*, 1985). Most of the scientists have focused on the optical characteristics of the MESFETs having gate length in the micron range. Effect of nanometric gate length is recently analyzed where distorted behaviors is explained and consequently its effect on electrical performances (Mukherjee *et al.*, 2014). In this paper, we have focused on the current-voltage characteristics of GaAs MESFET with nanometer gate length in the presence of optical illumination obtained from the solution of the Poisson's equation.

The article is organized as follows. The theoretical background is presented in section II. The results are explained in section III. Finally, a conclusion is given in section IV.

2 MATHEMATICAL FORMULATION

Computation of current requires the calculation of the potential and the depletion layer width distribution throughout the MESFET which further depends on its structural parameters. The GaAs MESFET device under our study consists of a gate of length L and width 'Z', active layer thickness 'a' and constant channel doping density (N_D). The width of the depletion region is non-uniform under the source, gate and drain terminal when the MESFET is subjected to external bias.

The Poisson's equation governing the two-dimensional channel potential distribution is given by Eq. (1) (Sze, 1997)

$$\frac{\partial^2 \psi}{\partial x^2} + \frac{\partial^2 \psi}{\partial y^2} = \frac{-qN_D}{\varepsilon_r \varepsilon_0} \tag{1}$$

where, $\psi(x, y)$, q, ε_0, ε_r represent the potential, electronic charge, and absolute and relative permittivities respectively. The potential distribution obtained from the solution of Eq. (1) is expressed as the superposition of the 1D ($U(y)$) and 2D ($\varphi(x, y)$) potential function forming the Laplace's equation respectively (Khemissi *et al.*, 2012) as shown in Eq. (2–3),

$$\psi(x,y) = U(y) + \phi(x,y) \tag{2}$$

where,

$$\frac{\partial^2 U(y)}{\partial y^2} = \frac{-qN_D}{\varepsilon_r \varepsilon_0} \tag{3.1}$$

$$\frac{\partial^2 \phi(x,y)}{\partial x^2} + \frac{\partial^2 \phi(x,y)}{\partial y^2} = 0 \tag{3.2}$$

The solution of Eq. (3) is expressed in a generalized form as follows

$$\phi(x,y) = (Ae^{kx} + Be^{-kx})(C\cos ky + D\sin ky) + Exy + Fx + Gy + H \tag{4}$$

The constants are evaluated by applying the boundary conditions. Considering the source potential to be zero (reference potential), the constant potential at the gate electrode ($y=0$) is independent of x which gives us,

$$H = -(V_G + V_{bi}) \tag{5}$$

Here, V_G is the gate voltage and V_{bi} is the built-in potential of the Schottky barrier. It is expected that at the edge of the depletion region, the transverse electric field vanishes along the y-direction. The following eqn. is obtained from this boundary condition,

$$\frac{\partial \phi(x,y)}{\partial y} = 2ADk\sinh(kx)\cos(kh_1) + Ex + \frac{qN_D h_1}{\varepsilon} \tag{6}$$

Here, h_l is the maximum width of the depletion layer along the y-direction (considering one-dimension) under the gate terminal. Again, the longitudinal electric field reaches its saturation value (E_S) at $x=L_l$ corresponding to the width (h_l) of the depletion region.

$$\frac{\partial \phi(x,y)}{\partial x}\Big|_{x=L_1} = \frac{\partial \psi(x,y)}{\partial x}\Big|_{x=L} = E_S \tag{7}$$

Using Eq. (2) the potential distribution in the MESFET channel under the gate terminal is given by (Khemissi *et al.*, 2012)

$$\psi(x,y) = \frac{qN_D}{2\varepsilon_r\varepsilon_0}a^2 + \frac{2aE_s}{\pi} \times \left[\frac{\sinh(k(L-x))}{\cosh(kL)-1} + \frac{\sinh(kx)}{\cosh(kL)-1}\right]\sin(ky) \tag{8}$$

Here, $k=\pi/2a$. The two-dimensional depletion layer width at the source and drain ends are calculated from by Eq. (9),

$$w_S = \sqrt{\frac{2\varepsilon_r\varepsilon_0(V_{bi} - V_G - \phi(0,h_S))}{qN_D}} \tag{9.1}$$

$$w_D = \sqrt{\frac{2\varepsilon_r\varepsilon_0(V_{bi} - V_G + V_D - \phi(L,h_D))}{qN_D}} \tag{9.2}$$

and

$$\phi(0,h_S) = \frac{2aE_S}{\pi}\left[\frac{\sinh(kL)}{\cosh(kL)-1}\right]\sin\left(\frac{\pi}{2}\sqrt{\frac{V_b - V_g}{V_p}}\right) \tag{10}$$

$$\phi(L,h_D) = \frac{2aE_S}{\pi}\left[\frac{\sinh(kL)}{\cosh(kL)-1}\right] \times \sin\left(\frac{\pi}{2}\sqrt{\frac{V_d + V_b - V_g}{V_p}}\right) \tag{11}$$

The drain current through the MESFET in the absence of incident light is solely controlled by the applied gate-to-source and drain-to-source voltages. This is expressed in the generalized form as follows (Sze, 1997)

$$I_{DS}{}^{dark} = \frac{q^2 N_D{}^2 z\mu(E)}{\varepsilon_r \varepsilon_0 L} \int_{W_S}^{W_D} \left[(a-W)W\right] dW \tag{12}$$

Here, the mobility μ is a function of the electric field.

When the MESFET structure is exposed to the background radiation the photogenerated carriers i.e. electrons (n^{ch}) and holes (p^{ch}) are expected to contribute to the current through the device. The photoconductive current density is given by

$$J_{DS}{}^{pc} = q(n^{ch} v_n + p^{ch} v_p) \tag{13}$$

Here, v_n and v_p represent electron and hole drift velocities respectively. Electrons and holes have equal recombination rates in the active region leading to Eq. (14)

$$J_{DS}{}^{pc} = q(n^{ch} v_n + p^{ch} v_p) \tag{14}$$

The total photoconductive current can be obtained by integrating the factor J_{DS}^{pc} within the physical boundary limit as shown below,

$$I_{DS}{}^{pc} = 0.5Zq \frac{\alpha^2 \phi_0 \tau_p}{\alpha^2 L_p{}^2 - 1} \left[\frac{\tau_n}{\tau_p} v_n + v_p \right] (a-h)^2 e^{-\alpha h} \tag{15}$$

Here, τ_n, τ_p are electron and hole lifetimes respectively and α is the absorption coefficient.

The net drain current through the MESFET structure under optical illumination is calculated using Eq. (16)

$$I_{DS} = I_{DS}{}^{dark} + I_{DS}{}^{pc} \tag{16}$$

3 RESULTS AND DISCUSSION

Drain current characteristics of the MESFET are plotted for different intensities of optical illumination in Fig. 1. For a given gate-to-source voltage the drain current is found to increase with the increase in the drain-to-source voltage for all the cases. The linear, nonlinear and saturation regime of the drain current is distinctly realizable in Fig. 1(a) in the absence of incident light.

Even when the background radiation is of the order of $9 \times 10^{22}/m^2$, the situation remains almost same [Fig 1(b)]. However, with a subtle change in the intensity of the exposed flux density of the order of $9 \times 10^{23}/m^2$ or $10^{24}/m^2$, the nature of the drain current yields a noticeable kink in the linear region of its characteristics [Fig. 1(c), Fig. 1(d)]. The drain current shoots up for a very small range of drain-to-source voltage and the rest portion of the curve resembles the nature exhibited by Fig. 1(a).

The net current flow in a device in presence of incident radiation is the algebraic sum of the dark current i.e. the current flowing through the device in absence of light and the photocurrent i.e. the current contribution from the photogenerated carriers. Consequently, the drain current increases linearly and with a slightly greater magnitude compared to that yielded under dark conditions at lower range of drain-to-source voltages. However, the photogenerated minority carriers have the nature to drift in the opposite direction to the flow of the majority carriers, precisely when minority carrier concentration is higher.

In MESFET structure, concentration of minority carriers in the depletion region becomes considerable when the negative gate-to-source voltage of the device approaches towards pinch-off con-

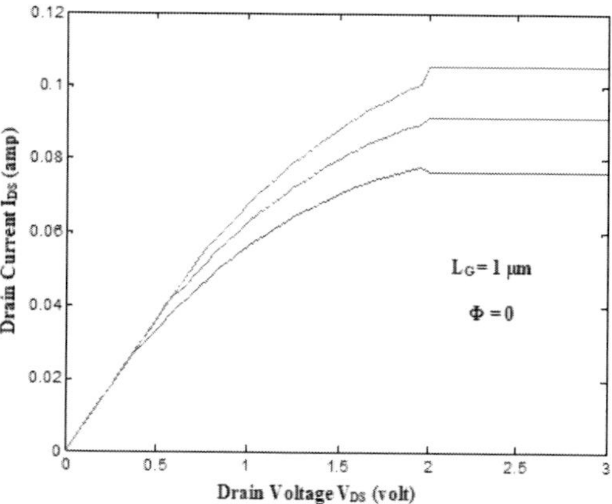

Figure 1a. Drain current as a function of drain voltage at 1 µm gate length in absence of incident radiation.

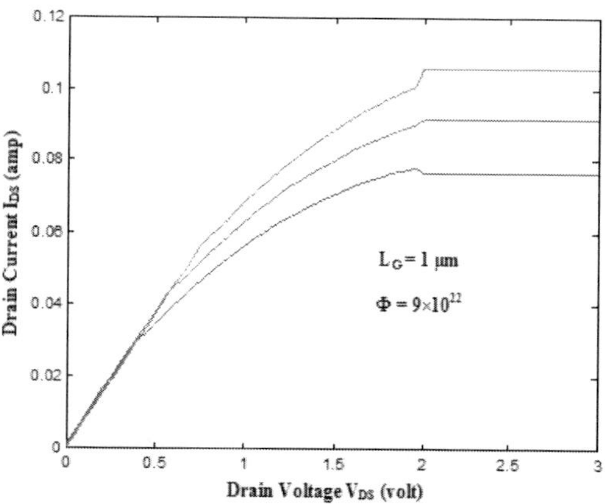

Figure 1b. Drain current as a function of drain voltage at 1 µm gate length in presence of incident radiation (**9×10^{22} w.m^{-2}**).

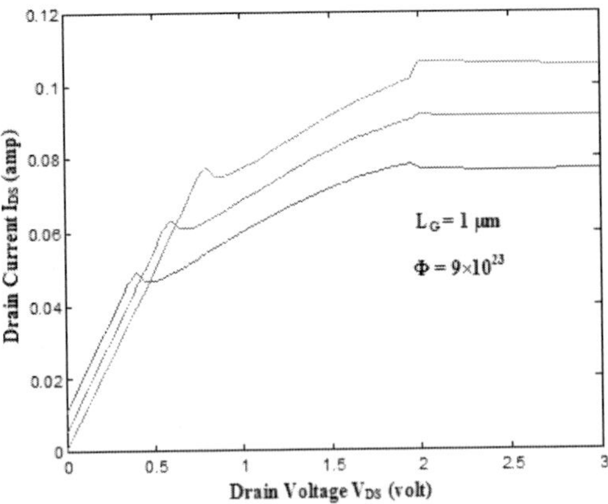

Figure 1c. Drain current as a function of drain voltage at 1 μm gate length in presence of incident radiation **(9×10^{23} w.m^{-2})**.

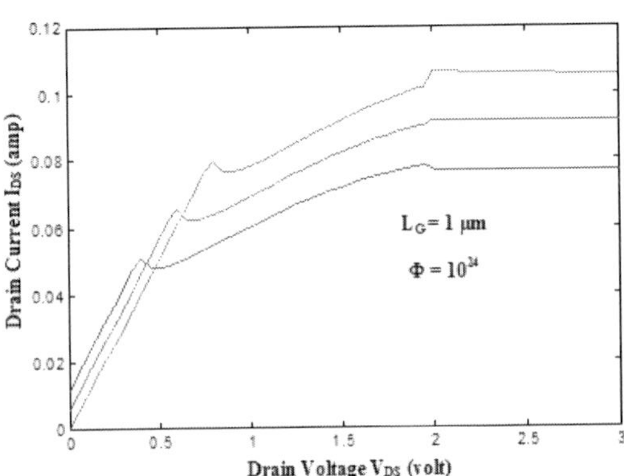

Figure 1d. Drain current as a function of drain voltage at 1 μm gate length in presence of incident radiation **(10^{24} w.m^{-2})**.

dition. This counter-propagating direction of the dark and photo currents tends to reduce the net current transport in the device over a very small range of V_{DS} values. This is manifested in the kink in the graph.

However, the photocurrent magnitude being very small compared to the order of the dark current cannot dominate the suppression of the net current at higher values of drain-to-source voltage. As a result the current rises gradually and saturation is reached at a larger drain-to-source voltage. Further, it is seen that the magnitude of the drain current under illumination at the kink positions can be varied by changing the gate bias. Thus, the response of the MESFET device to the optical illumination through the variation of its drain current characteristics can be utilized for optical detector applications.

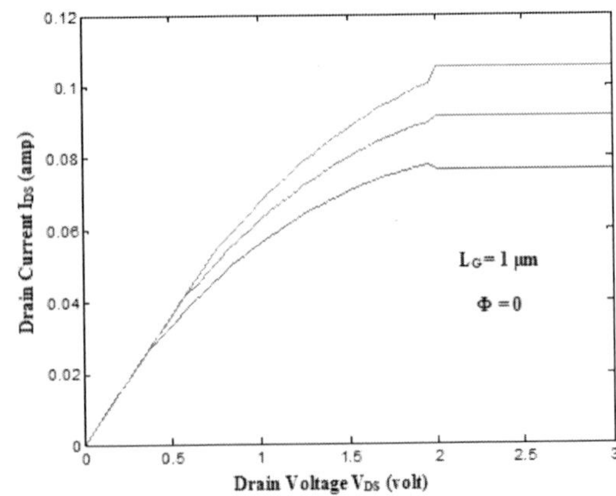

Figure 2a. Drain current as a function of drain voltage at 0.01 μm gate length in absence of incident radiation.

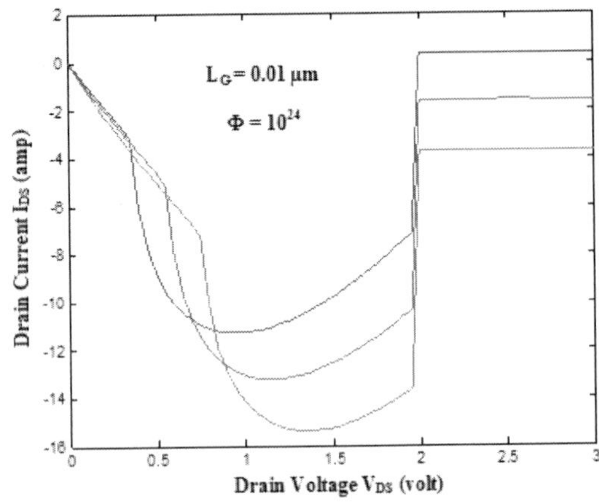

Figure 2b. Drain current as a function of drain voltage at 0.01 μm gate length in presence of incident radiation **(10^{24} w.m^{-2})**.

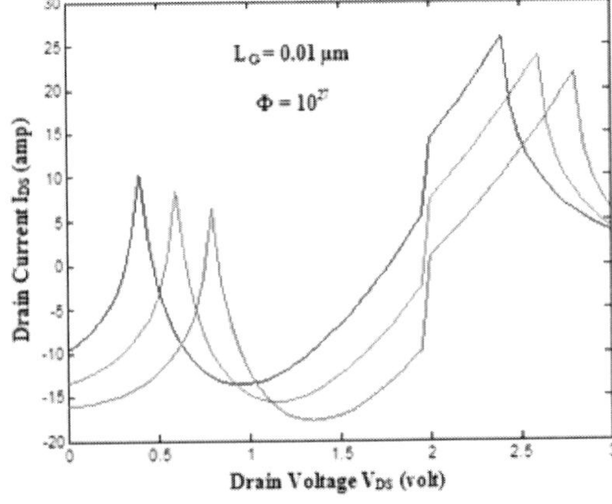

Figure 2c. Drain current as a function of drain voltage at 0.01 μm gate length in presence of incident radiation **(10^{27} w.m^{-2})**.

The static characteristics of nanometer gate length MESFET in presence of optical illumination is shown in Fig. 2. The drain current displays a different behavior for the MESFET with 10 nm gate length in Fig. 2(b) compared to that in Fig. 1(b). The initial reduction in the drain current is attributed to the scattering events obstructing the motion of the carriers by reducing their drift velocity.

But ballistic transport of the electrons with velocity greater than the saturation drift velocity is responsible for the sudden increase of the current. This nature of the drain current shown in the absence of radiation [Fig. 2(a)] remains same even when the incident flux density is of the order of $10^{24}/m^2$ [Fig. 2(b)].

When the optical flux density is increased to the order of $10^{27}/m^2$, the MESFET with nanometer gate length exhibits quantized conductance. Sharp peak is observed at very low bias, which reveals the fact that scattering is very less at lower horizontal field. This helps to grow current due to large number of photon incidence. With increase of field, scattering starts to dominate causes sharp reduction of channel current, and again current is increased when potential barrier is reduced. This feature can be utilized for the use of nano-MESFET as optical modulator applications.

4 CONCLUSION

The bulk MESFET shows kink in the drain current characteristics in presence of incident radiation of high flux density. However, the device with nanometer gate length exhibits quantized conductance characteristics for higher flux density and hence found to be suitable for optically controlled modulator / detector applications.

REFERENCES

Allen. S. T., Sadler. R. A, Alcom. T. S, Palmour. J. W, 1997, Silicon Carbide MESFETs for High Power S-Band Applications, IEEE MTT–S International Proceedings of Microwave Symposium Digest, 1, 57–60.

Bandhawakar. G, Pal. B. B, 2001, Optically Controlled E-MESFET for VLSI Application, Proceedings of SPIE Advances in Microelectronic Device Technology, 4600.

Chattopadhyay. S. N, Overton. C. B, Vetter. S, Azadeh. M, Olson. B. H, Naga. N. E, 2010, Optically Controlled Silicon MESFET Fabrication and Characterizations for Optical Modulator/Demodulator, Journal of Semiconductor Technology & Science, 10(3), 213–224.

Chaturvedi. G. J, Purohit. R. K, Sharma. B. L, Optical Effect on GaAs MESFET's, Infrared Physics, 23(2), 65–68.

Gautam. D. K, Ishida. K, 1993, Carrier Induced MESFET Optical Switches for Photonic Integration, IEEE Proceedings of J Optoelectronics, 10(5), 317–324.

Gautier. J. L, Pasquet. D, Pouvil. P, 1985, Optical Effects on the Static and Dynamic Characteristics of a GaAs MESFET, IEEE Transactions on Microwave Theory and Techniques, 33(9), 819–822S.

Huang. C, Pai. H, Chen. K, 2004, Analysis of Microwave MESFET Power Amplifiers for Digital Wireless Communications Systems, IEEE Transactions on Microwave Theory & Techniques, 52(4) 1284–1291.

Khemissi. S, Azizi. C, 2012, A Two-Dimensional Analytical Modeling of the Current-Voltage Characteristics for Submicron Gate–Length GaAs MESFET's, International Journal of Engineering & Technology, 12(4), 27–33.

Kraemer. R, Katz. M, 2009, Short-Range Wireless Communications: Emerging Technologies and Applications, John Wiley & Sons, UK, 1st edition.

Madjar. A, Paollela. A, Herczfeld. P. R, 1993, The Comprehensive Nature of Optical Detection in GaAs MESFETs and Possible Application as an RF Logarithmic Amplifier, IEEE Transactions on Microwave Theory and Techniques, 41(1), 165–167.

Mukherjee. K, Deyasi. A, 2014, Current-Voltage Characteristics of MESFET with Nanometer Gate Length, 1st International Science and Technology Congress (ELSEVIER), paper id: 50, 347–352.

Peterson. A. K, Jagd. A. M, 1992, Tuned Optical Front-End MMIC Amplifiers for a Coherent Optical Receiver, Proceedings of 22nd European Microwave Conference, 2, 1065–1070.

Singh. V. K, Pal. B. B, 1987, Optically Controlled Switching Characteristics of Silicon MESFET's, Solid State Electronics, 30(3), 267–272.

Sze. S. M., 2000, Physics of Semiconductor Devices, John Wiley & Sons, UK, 2nd edition.

Umeda. T, Members. Y. C, 1987, I-V Characteristics of Photodetector using the GaAs MESFET, Electronics and Communications in Japan (Part II: Electronics), 70(5), 27–33.

Vilcot. A, Cabon. B, Chazelas. J, 2003, Microwave Photonics: From Components to Applications and Systems, Kluwer Academic Publishers, Netherlands.

Zebda. Y, Abu-Helweh. S, 1997, AC Characteristics of Optically Controlled MESFET (OPFET), IEEE Journal of Lightwave Technology, 15(7), 1205–1212.

Computational Science and Engineering – Deyasi et al. (Eds)
© 2017 Taylor & Francis Group, ISBN 978-1-138-02983-5

Analytical modeling and sensitivity analysis of dielectric-modulated junctionless gate all around gate stack—FET as biosensor

A. Chakraborty
ECE Department, Bengal Institute of Technology and Management, Bolpur, India

A. Sarkar
ECE Department, Kalyani Government Engineering College, Kalyani, India

ABSTRACT: An analytical model of dielectric-modulated junctionless gate all around gate stack –FET (JLSRGGS) for application as a biosensor is presented. An expression of channel-center potential is obtained by solving 2-D Poisson's equation using parabolic potential approach. An analytical model of threshold voltage is developed from the minimum channel-center potential to analyze the sensitivity of the biosensor. Moreover, the effects of the variation of the different device parameters were investigated to study the permittivity effects of the biomolecules present within the nanogap cavity. Dependence on the sensitivity for varying dimensional parameters has also been discussed. The analytical model is verified and validated with the help of TCAD device simulation.

1 INTRODUCTION

Biosensor is a device that is used to provide rapid detection of biomolecules. The biosensor for label-free detection of neutral (biotin–streptavidin) and charged biomolecules (DNA) has seen potential application in biomedical field (Bergveld 1989) of science and technology. Nanogap is a gap (<100 nm) between the gate metal and the gate oxide layer. The gap is made by etching the sacrificial material between the gate and gate oxide regions. In absence of the biomolecules, air ($K = 1$) fills the nanogap. Now in the presence of the biomolecules, the region is filled with biomolecules having different dielectric constant ($K > 1$). Due to these higher values of dielectric constant, shift in threshold voltage (V_{Th}) occurs which can be used as a metric to detect the biomolecule knwon as dielectric modulated device. Recently Junctionless (JL) surrounding-gate MOSFET devices emerge as one of the most promising novel structures primarily due to their simple fabrication process with low thermal-budget, their ability to suppress the short channel effects (Chiang et al. 2012), and good electrostatic control of the channel (Gnani et al. 2011). An experimental demonstration of nanogap-embedded biosensor has already been available (Wenga et al. 2013). However, a JLSRGGS-based biosensor has not been reported yet. Therefore, for the first time, the concept of JLSRGGS-based biosensor has been presented in this paper using analytical and simulation-based study.

2 DEVICE STRUCTURE

Figure 1 shows the two-dimensional (2-D) cross-sectional device structure of Junctionless Surrounding Gate (JLSRG) based biosensor.

The typical physical structural parameters used for this JLSRG-based biosensor is listed in Table 1, unless otherwise mentioned and/or varied. Two-dimensional device simulation has been performed using TCAD Silvaco (Atlas 2013) device simulator software in order to verify and validate the correctness of the analytical model developed. The models and methods used for simulating the JLSRGGS-based biosensor are listed in Table 2. In this study,

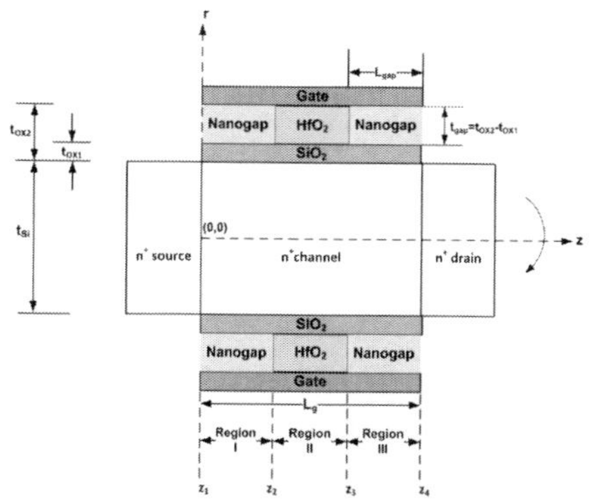

Figure 1. 2-D cross-sectional device structure of junctionless surrounding gate based biosensor.

Table 1. Physical parameters of JLSRGGS -based biosensor device.

Physical parameters	Value
L_g (Metallurgical gate length)	50 nm
Length of Source and Drain	20 nm
t_{OX2} (Thickness of the HfO_2 layer in region II)	11 nm
t_{OX1} and t_{OX3} (Thickness of SiO_2 layer in region I & III)	1.0 nm
t_{Si} (diameter of Silicon channel)	20 nm
N_d (Si channel doping conc.)	10^{18} cm^{-3}
Φ_M (Work function of gate)	5.5 eV

Table 2. Simulation setup with the choice of the models and methods used for simulation of the JLSRGGS -based biosensor.

Physical models	Name of the model used
Carrier statistics	Fermi-Dirac (FERMIDIRAC)
Carrier transport	Drift-diffusion (DD)
Mobility reduction model	Lombardi CVT
Carrier Generation-Recombination	SRH, AUGER
Method	GUMMEL & NEWTON

Quantum-Mechanical Effects (QMEs) and ballistic transport are not considered since they are not significant for devices with film thickness larger than 10 nm (Ortiz-Conde et al. 2005) and device channel length larger than 10 nm (Lundstrom et al. 2002). The presence of a biomolecule in the nanogap of the biosensor is contemplated by substituting the air (K = 1) with a dielectric material (K > 1) for a collection of biomolecules (Ionescu-Zanetti et al. 2006).

It is worth mentioning that non-hybridized single strand DNA have the property of both the dielectric constant (Offenhäusser et al. 2009) and the charge as well (Narang et al. 2012). The consequence of a charged biomolecules in the nanogap cavity is considered by introducing fixed oxide interface trapped charges N_f in the dielectric layer (Kang et al. 2008). The negative charge of the biomolecules in the dielectric layer is attributed by an increase the flatband voltage (V_{FB}) of the region where it is present. The idea is similar to that of a familiar method to model the interface traps and localized charges present in the oxide-semiconductor interface in MOSFETs. As a result, these charged biomolecules causes an increase in the flatband voltage equals to (qN_f/C_1), which results in a decrease in the surface potential and body-center potential in the nanogap cavity regions of the device. The simulation is carried based on the charge values reported for DNA (Dashiell et al. 2002), a negatively charged biomolecules. In this study, the charge density of DNA N_f equals to 10^{-12} cm^{-2} no. of charges is considered (Young et al. 1989).

3 ANALYTICAL MODELING OF CHANNEL CENTER POTENTIAL

The electrostatics of the JLSRG –based biosensor is developed by solving the 2-D Poisson's equation in the cylindrical coordinate system

$$\frac{1}{r}\frac{\partial}{\partial r}\left(r\frac{\partial}{\partial r}\Psi(r,z)\right)+\frac{\partial^2\Psi(r,z)}{\partial z^2}=\frac{-qN_d}{\varepsilon_{Si}} \tag{1}$$

where r is the radius varying from center of the channel (r = 0)to surface of the channel (r = R = $t_{Si}/2$) in radial direction, z is the position along the channel with reference to the source side varying in axial direction from source end (z = 0) to drain end (z = L_g), $\Psi(r,z)$ is the 2-D potential distribution, q is the electronic charge, N_d is the uniform doping concentration used in the JLSRG-based biosensor.

Approximating the variation of $\Psi(r,z)$ in r-direction according to a parabolic potential profile (Razavi et al. 2008)

$$\Psi(r,z)=C_1(z)+C_2(z)r+C_3(z)r^2 \tag{2}$$

where $C_1(z)$, $C_2(z)$ and $C_3(z)$ are constants to be determined from the boundary conditions. The solution is

$$\Psi_C(z)=ae^{\frac{z}{\lambda}}+be^{-\frac{z}{\lambda}}+\varphi_C \tag{3}$$

The channel region is divided into three regions as shown in Figure 1.

For Region I and III, the effective gate capacitance is the series combination of the capacitance of gate oxide and the nanogap region given by

$$C_{eff}=\frac{C_{ox1}C_{gap}}{C_{ox1}+C_{gap}} \tag{4}$$

where C_{ox1} is given by

$$C_{ox1}=C_{ox3}=\frac{\varepsilon_{ox}}{Rln\left(1+\frac{t_{ox1}}{R}\right)} \tag{5}$$

C_{gap} is given by

$$C_{gap}=\frac{\varepsilon_{ox}}{Rln\left(1+\frac{t_{ox2}-t_{ox1}}{R}\right)} \tag{6}$$

For Region II, the effective gate capacitance is the total oxide capacitance per unit area including the gate stack of high-k HfO_2 and SiO_2 layer, given by

$$C_{ox2} = \frac{\varepsilon_{ox}}{R\ln\left(1+\frac{t_{oxeff}}{R}\right)} \quad (7)$$

where t_{oxeff} is the effective oxide thickness (HfO_2+ SiO_2), as given by (Kang et al. 2008)

$$t_{OXeff} = t_{OX1} + t_{OX2}\left(\frac{\varepsilon_{SiO_2}}{\varepsilon_{HfO_2}}\right) \quad (8)$$

where t_{OX1} is the SiO_2 and t_{OX2} is the high-k oxide layer thickness. Flatband voltage V_{FB} is expressed by $V_{FB} = \Phi_M - \Phi_{Si}$, where Φ_{Si} represent the work function of Silicon. In region I and region III, due to the presense of the negatively charged biomolecules flatband voltage decreases, and is given by

$$V_{FB1} = V_{FB3} = \Phi_M - \Phi_{Si} - \left(\frac{qN_f}{C_1}\right) \quad (9)$$

Among the electrostatic potentials V_1 at source end (z = 0) and V_4 at drain end ($z = L_g$) is known and is given by $V_1 = 0$ and $V_4 = V_{DS}$. The other two potentials V_2 and V_3 are unknown and determined from the continuity of the Electric Field at $z = z_2$ and $z = z_3$

4 ANALYTICAL MODEL OF THRESHOLD VOLTAGE

In a junctionless MOSFET, the threshold voltage is determined by equating the minimum channel-center potential to zero. It is observed that minimum channel-center potential is always located in region I. Because nanogap filled-up with biomolecules will effectively deposit negative charges in the oxide layer source side and the drain side also. Negative interface-trapped charges present in the gate oxide layer lowers the channel-center potential (Kang et al. 2008). At the drain side, due to applied drain bias channel-center potential increases. Hence, minimum channel-center potential will always be positioned in the region I. To find the location of the minimum channel-center potential,

$$\left.\frac{d\Psi_{C1}(z)}{dz}\right|_{z=z_{min}} = 0 \quad (10)$$

We obtain the position of the minimum channel-center potential, given by

$$z_{min} = \frac{\lambda_1}{2}\ln\left(\frac{b_1}{a_1}\right) \quad (11)$$

Substituting Z_{min} from (11) in (10), expression of minimum channel-center potential is obtained

$$\Psi_{C,min} = 2\sqrt{a_1 b_1} + \varphi_{C1} \quad (12)$$

Substituting values of a_1 and b_1 in (12) and also substituting $V_{GS} = V_{Th}$ in (12) and then solving for V_{Th}, we obtain an analytical expression of the threshold voltage, given by

$$V_{Th} = \frac{u_2 \pm \sqrt{u_2^2 - 4u_1u_3}}{2u_2} \quad (13)$$

5 RESULTS AND DISCUSSIONS

The variation of the channel-center potential as a function of the position of the channel along z-axis is plotted in Figure 2, obtained with the aid of the analytical model developed and from TCAD simulation. An excellent agreement was observed between the modeled expression and TCAD simulation. Figure 2 clearly shows the distortion of the potential distribution under the nanogap regions. For non-zero length of L_{gap}, channel-center potential lowers in region I and III due to the presence of the biomolecules with k > 1 and negatively charged charges in the dielectric layer. Figure 2 also indicates that as Lgap increases, distortion of the potential distribution increases and potential minimum also increases. The change in potential minimum is the crucial factor for determining the threshold voltage and hence the sensitivity of the device. Therefore, it can be concluded that higher values of L_{gap} is required in order to obtain highly-sensitive biosensor. The sensitivity of a biosensor is defined by the difference of threshold voltage V_{Th} for nanogap filled with air and nanogap filled with biomolecules.

$$\Delta V_{Th} = V_{Th(gap=air)} - V_{Th(gap=filled)} \quad (14)$$

Figure 3 reflects the fact that that for a biomaterial with high dielectric constant (K = 12), as L_{gap}

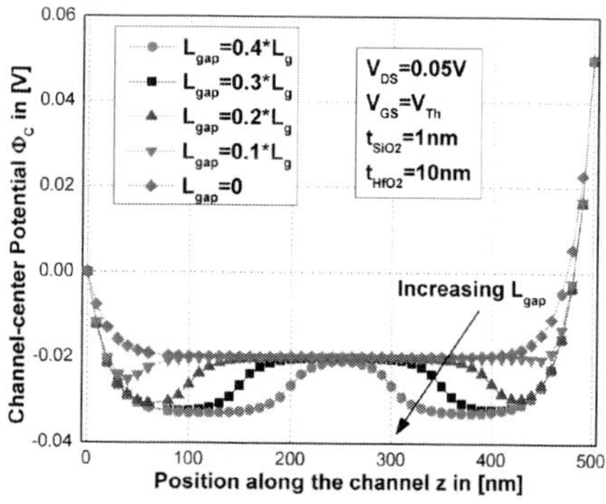

Figure 2. Plot of Channel-center potential Φ_C as a function of position in the channel along z-axis for different length of nano-gap cavity L_{gap}.

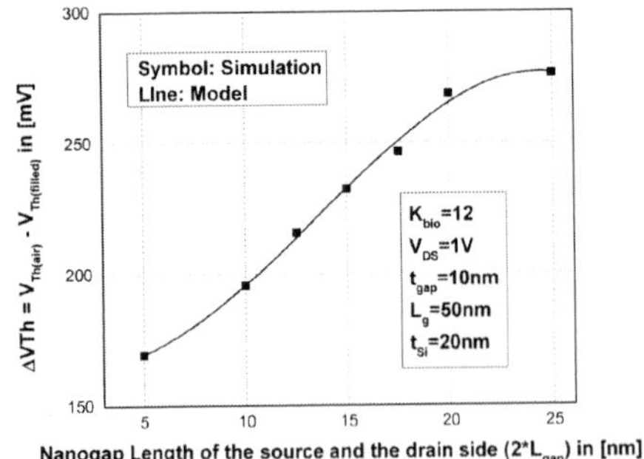

Figure 3. Plot of the variation of sensitivity parameter ΔV_{Th} as a function of nanogap langth L_{gap}.

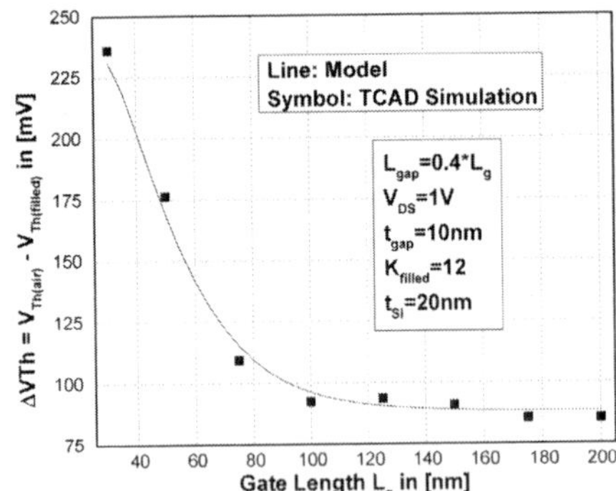

Figure 4. Plot of the variation of sensitivity parameter ΔV_{Th} as a function of gate length Lg.

i.e. the length of the nannogap increases, sensitivity increases. However, it is worth mentioning that after a certain limit, ΔV_{Th} saturates. Highest sensitivity occurs for $2*L_{gap}$/ Lg ratio = 25/50 = 0.5. For a symmetric biosensor with nanogap on both source and drain sides, this corresponds to L_{gap}/L_g ratio equals to 0.4 for which highest sensitivity occurs. This is in accordance with the findings of the previously reported results ((Narang et al. 2012).

In order to determine the optimal size of the device, sensitivity is plotted against gate length L_g for a nanogap filled with high-K biomolecules (K = 12) and for $L_{gap} = 0.4\ L_g$. Figure 4 clearly reveals the fact that as length of the surrounding gate transistor decreases, the sensitivity increases. However, it is worth mentioning that if L_{gap} is fixed at some particular value, then sensitivity will not depend on the length of the device L_g, as reported previously ((Narang et al. 2012)).

6 CONCLUSION

The concept of junctionless gate all around gate stack –FET (JLSRGGS) device for application as a biosensor has been proposed and analyzed with the help of an analytical model. The adequate sensitivity with the ease of fabrication for junctionless device makes JLSRGGS-based biosensors an attractive alternative for FET-based biomolecule sensing applications. Hence, the proposed analytical model can act as useful guideline to provide the optimal device dimensions, favorable design and fabrication of JLSRGGS-based biosensors.

REFERENCES

Bergveld, P. 1986. The development and application of FET- based biosensors. *Biosensors*, 2(1): 15–33.

Chiang, T-K. 2012. A new quasi-2-D threshold voltage model for short-channel junctionless cylindrical surrounding gate (JLCSG) MOSFETs. *IEEE Trans. Elec. Dev.* 59(11); 3127–3129.

Dashiell, M. W. & Kalambur, A. T. & Leeson, R. & Roe, K. J. & Rabolt, J. F. & Kolodzey, J. 2002. The electrical effects of DNA as the gate electrode of MOS transistors," in *Proc. IEEE Lester Eastman Conf.*, 259–264.

Device simulator ATLAS User manual. Silvaco Int., Santa Clara, CA. May 2011 [Online].

Gnani, E. & Gnudi, A. & Reggiani, S. & Baccarani. G. 2011. Theory of the junctionless nanowire FET. *IEEE Trans. Elec. Dev.* 58 (9): 2903–2910.

Ionescu-Zanetti, C. & Nevill, J. T. & Di Carlo, D. & Jeong, K. H. & Lee, L. P. 2006. Nanogap capacitors: Sensitivity to sample permittivity changes. *J. Appl. Phys.*, 99(2): 024 305–5.

Kang, H. & Han, J.-W. & Choi, Y.-K. 2008. Analytical thresh old voltage model for double-gate MOSFETs with localized charges," *IEEE Elec. Dev. Lett.* 29(8): 927–930.

Lundstrom, M. S. & Ren, Z. 2002. Essential Physics of Carrier Transport in Nanoscale MOSFETs. *IEEE Trans. Elec. Dev.*, 49(1): 133–141.

Narang, R. & Saxena, M. & Gupta, R. S. & Gupta, M. 2012. Dielectric modulated tunnel field effect transistor—A bio molecule sensor. *IEEE Elec. Dev. Lett.* 33(2): 266–268.

Offenhäusser, A. & Rinaldi, R. 2009. *Nanobioelectronics—For Electronics, Biology, and Medicine*. New York: Springer-Verlag.

Ortiz-Conde, A. & Garcia-Sanchez, F. J. & Malobabic, S. 2005. Analytic solution of the channel potential in undoped symmetric dual-gate MOSFETs. *IEEE Trans. Elec. Dev.* 52(7): 1669–1672.

Razavi, P. & Orouji, A.A. 2008. Dual material gate oxide stack symmetric double gate MOSFET: improving short channel effects of nanoscale double gate MOSFET, *Electronics Conference, 2008. BEC 2008. 11th International Biennial Baltic*, 83–86.

Wenga, G. & Jacques, E. & Salaun, A.-C. et al. 2013. Step-gate polysilicon nanowires field effect transistor compatible with CMOS technology for label-free DNA biosensor. *Biosensors Bioelectron.*, 40(1): 141–146.

Young, K.K. 1989. Short-channel effects in fully depleted SOI MOSFET's, *IEEE Trans. Elec. Dev.* 36(2): 399–402.

Computational Science and Engineering – Deyasi et al. (Eds)
© 2017 Taylor & Francis Group, ISBN 978-1-138-02983-5

Effect of grating length on wave propagation inside paired DNG/air composition

Bhaswati Das & Arpan Deyasi
RCC Institute of Information Technology, Kolkata, India

ABSTRACT: Replacing conventional SiO_2-air composition by DNG material—air combination and considering magnitude of operating wavelength just below or above the Bragg wavelength (1.55 μm), remarkable oscillating wave is observed inside one-dimensional photonic crystal. Results are observed by varying the grating length for forward and backward propagating waves for different metamaterials which are already practically feasible. Results are playing an important role for guided transmission in optical communication for DNG metamaterial based photonic crystal with negligible loss. In this paper, a package is made based on JAVA-GUI platform to measure the transmittivity in one-dimensional photonic crystal considering electromagnetic wave propagation along the direction of confinement. For a suitable range of structural parameters within practical limit (for fabrication), computation is carried out around 550 nm for the purpose of optical communication. Dimensions of both the constituents, number of layers and refractive indices of the materials are considered as input parameters for development of the package, and simulated results provide typical characteristics of optical bandpass filter. Though the package is developed for normal incidence of e. m wave only, but later can be extended for oblique incidences also.

1 INTRODUCTION

The Photonic crystals are periodic optical nanostructure of different dielectric layers of different refractive indices. In the photonic crystal where photons cannot propagate, is called photonic bandgap (Yablonovitch 1987). Photonic bandpass filter (Robinson et al., 2011) and photonic crystal fibers are also based on this structure (Belhadj et al., 2006) and it is already used in making optical transmitter (Szczepanski 1988), receiver (Fogel et al., 1998), sensor (Shanthi et al., 2014), memory (Lima et al., 2011), quantum device (Jiang et al., 1999) etc. This structure works as building block for the next generation communication system. The DNG material based PhC designs are not reported for those works, though importance of metamaterial is already verified in high-frequency communication application. Due to the novel property of improved signal-tonoise ratio and negative refractive index, DNG metamaterial is used to design the high-frequency antenna using metamaterial is the subject of research in the preset decade (Xiong et al., 2012, Hwang et al., 2009). The noise rejection property in the desired frequency spectrum, i.e. the region of optical communication (with minimum attenuation) is become the choice of interest for the communication engineers. So, the metamaterial array becomes the key area of study in designing photonic integrated circuit.

In the last decade, various research works are published regarding 1D photonic crystal (Xu et al., 2007, Rudziński 2007) because of ease of fabrication and less complexity, and also a few papers are also published involving 2D microstructure (Zhang et al., 2004). These works are based on conventional SiO_2/air composition of metamaterial (Gao et al., 2011), or semiconductor heterostructure (Maity et al., 2013).

But as far the knowledge of the authors, there are very few publications involving DNG materials based research work. Our present work explaining the forward and backward wave nature inside 1D photonic crystal, which shows that the wave can travel for a longer grating length compared to the conventional PhC by setting the operating wavelength slightly less (1.5 μm) or greater (1.6 μm) than Bragg wavelength (1.55 μm). Effect of strong and medium coupling coefficient on wave propagation is studied for paired nanorod ($n = -0.3$), nano-fishnet with rectangular void ($n = -1$), nano-fishnet with elliptical void ($n = -4$) structures using three different type of grating lengths. For analysis of guided transmission inside 1D photonic crystal structure these results are very important.

2 MATHEMATICAL MODELING

Inside a photonic crystal, the forward and backward propagating waves may be represented as

$$B(z.t) = b\exp\left[-j\left(\beta_b z - \omega t\right)\right] \quad (1.1)$$

$$A(z.t) = a\exp\left[j\left(\beta_a z + \omega t\right)\right] \quad (1.2)$$

where β_b and β_a are the propagation constants. Considering the propagation part along with the d. c magnitude, equations may be rewritten as

$$A(z,t) = a(z)\exp[j\omega t] \quad (2.1)$$

$$B(z,t) = b(z)\exp[j\omega t] \quad (2.2)$$

Hence coupling of total power is dependent on a(z) and b(z) only. In absence of any coupling,

$$\frac{da(z)}{dz} = j\beta_a a(z) \quad (3.1)$$

$$\frac{db(z)}{dz} = -j\beta_b b(z) \quad (3.2)$$

When the waves are coupled, the Eq. (3.1) and Eq. (3.2) will be modified as

$$\frac{da(z)}{dz} = j\beta_a a(z) + \kappa_{ab} b(z) \quad (4.1)$$

$$\frac{db(z)}{dz} = -j\beta_b b(z) + \kappa_{ba} a(z) \quad (4.2)$$

Subject to the appropriate boundary condition, Eq (4.1) may be modified as

$$a(z) = a\exp\left[\frac{jz(\beta_a - \beta_b)}{2}\right] \times \exp\left[\frac{\sqrt{4\kappa_{ab}\kappa_{ba} - (\beta_a + \beta_b)^2}}{2} z\right] \quad (5.1)$$

Similarly, we can write

$$b(z) = b\exp\left[\frac{jz(\beta_a - \beta_b)}{2}\right] \times \exp\left[\frac{-\sqrt{4\kappa_{ab}\kappa_{ba} - (\beta_a + \beta_b)^2}}{2} z\right] \quad (5.2)$$

Using Eq. (4), we may write

$$\exp\left[\frac{4\kappa_{ab}\kappa_{ba} - (\beta_a + \beta_b)^2}{2} z\right] = \pm j\frac{b}{a} \quad (6)$$

As the right-hand side of Eq. (6) is imaginary, so the term in the square-root should be less than zero, i.e., we can write this in the mathematical form

$$\kappa_{ab}\kappa_{ba} < \frac{1}{4}(\beta_a + \beta_b)^2 \quad (7)$$

This can be written as

$$\kappa_{ab} = \kappa_{ba}^{*} = \kappa \quad (8)$$

With the boundary conditions $b(0) = b_0$ and $a(L) = 0$ these equations may be solved analytically. Expressions of forward and backward waves thus can be given as

$$a(z) = b_0 \frac{\kappa\exp[-j\Delta\beta z]\sinh[\alpha(z - L)]}{\Delta\beta\sinh\alpha L - j\alpha\cosh\alpha L} \quad (9)$$

$$b(z) = b_0\kappa\exp[j\Delta\beta z] \times \frac{\left[\Delta\beta\sinh\{\alpha(z - L)\} + j\alpha\cosh\{\alpha(z - L)\}\right]}{-\Delta\beta\sinh\alpha L + j\alpha\cosh\alpha L} \quad (10)$$

where $\alpha = \sqrt{\kappa^2 - \Delta\beta^2}$, L is the grating length.

3 RESULTS AND DISCUSSION

Based on the mathematical formulation, simulation is carried out to observe the oscillating wave inside one-dimensional photonic crystal for DNG material air composition.

Results are plotted for three types of operating wavelengths, one is set just below of the Bragg wavelength (1.5 μm), one is just above the Bragg wavelength (1.6 μm) and other is Bragg wavelength (1.55 μm) itself. Results are observed for paired nanorod (–0.3), nano-fishnet with rectangular void (–1) and nano-fishnet with elliptical void (–4) structures respectively.

Considering the finalized forward and backward wave equations, wave profiles are simulated and plotted as a function of grating length for one-dimensional photonic crystal. Fig. 1a and Fig. 1b exhibit the forward and backward wave propagating characteristics for three different grating at Bragg wavelength (1.55 μm), and Fig. 2a and Fig. 2b is for 1.6 μm respectively. For all these results we consider paired nanorod (–0.3 μm)/air composition.

Observing Fig. 1a and Fig. 1b we can see that with increase of grating length, magnitude of

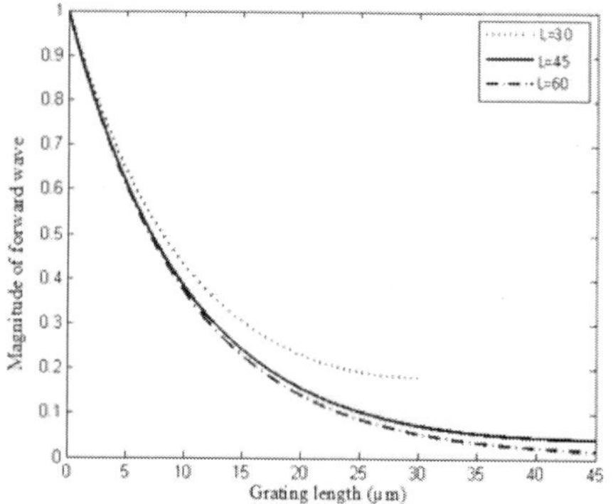

Figure 1a. Profile of forward wave with grating length for paired nanorod/air composition at 1.55 μm for different grating lengths.

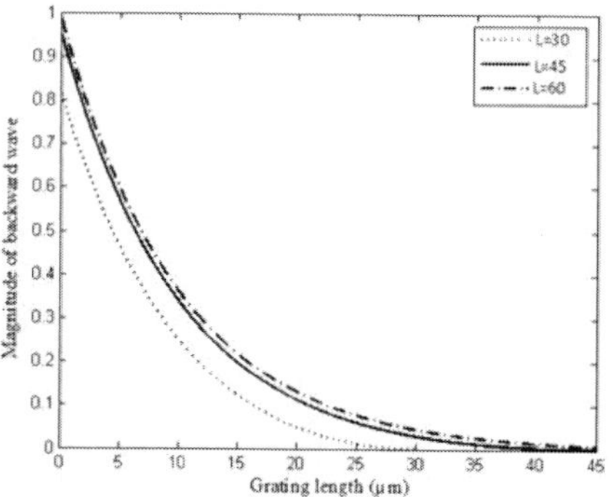

Figure 1b. Profile of backward wave with grating length for paired nanorod/air composition at 1.55 μm for different grating lengths.

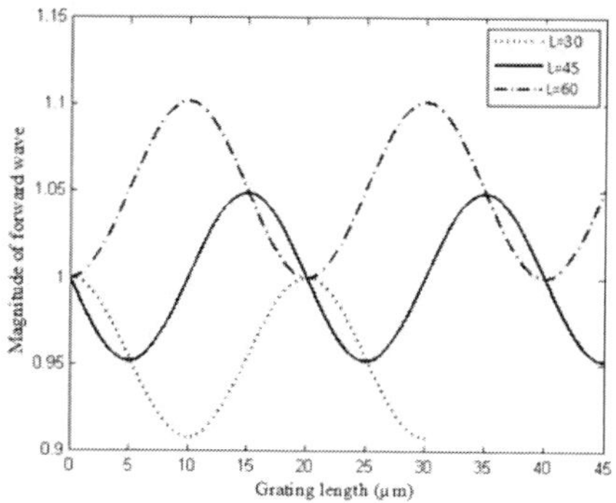

Figure 2a. Profile of forward wave with grating length for paired nanorod/air composition at 1.6 μm for different grating lengths.

the wave decreases exponentially but in Fig. 3 we observe oscillating wave for both the forward and backward wave. For the entire backward wave we get the highest magnitude for grating length 60 μm and lowest for 30 μm.

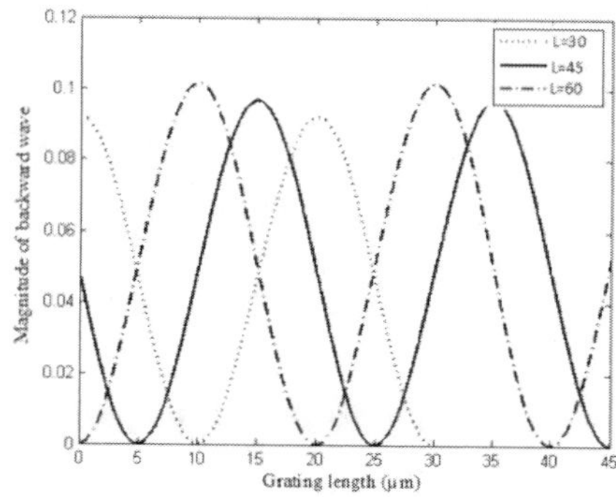

Figure 2b. Profile of backward wave with grating length for paired nanorod/air composition at 1.6 μm for different grating lengths.

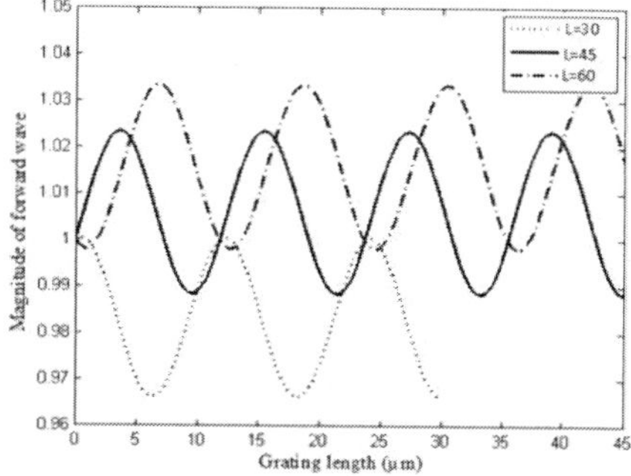

Figure 3a. Profile of forward wave with grating length for paired nano-fishnet with rectangular void at 1.5 μm for different grating lengths.

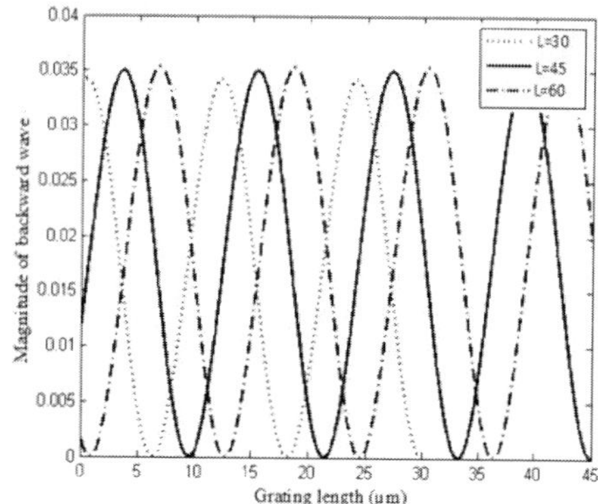

Figure 3b. Profile of backward wave with grating length for paired nano-fishnet with rectangular void at 1.6 μm for different grating lengths.

Similarly, simulations are also carried out for nanofishnet with elliptical void structure. Comparing with the results obtained for rectangular void, it is found out that frequency of oscillation is increased with increase of negative refractive

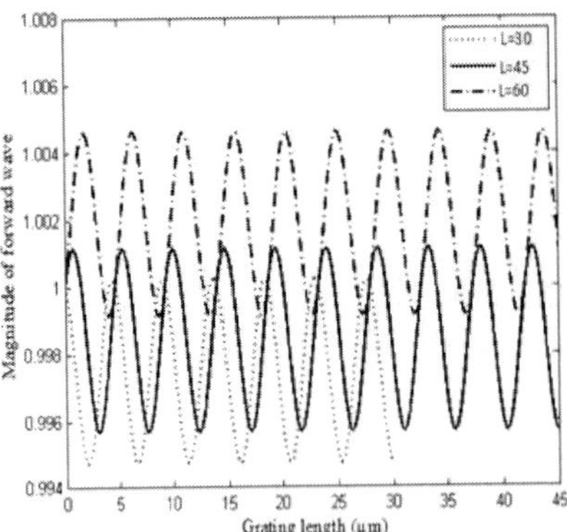

Figure 4a. Profile of forward wave with grating length for paired. nano-fishnet with elliptical void at 1.5 μm for different grating lengths.

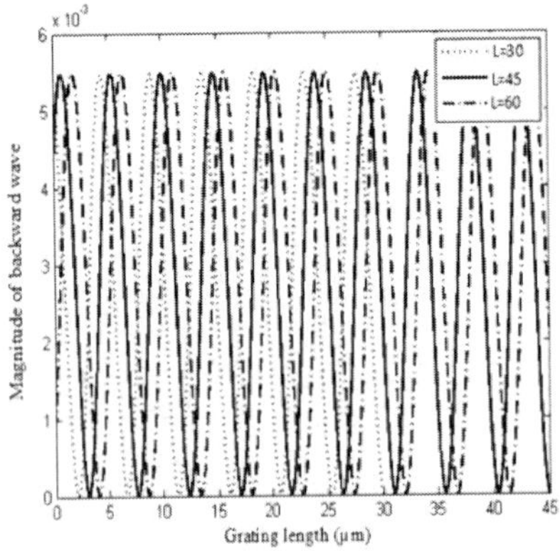

Figure 4b. Profile of backward wave with grating length for paired nano-fishnet with elliptical void at 1.5 μm for different grating lengths.

index. These are reported in Fig. 4a and Fig. 4b for 1.5 μm wavelength. Also oscillation frequency is higher when operating wavelength exceeds predetermined Bragg value, i.e., for 1.6 μm wavelength.

4 CONCLUSION

Both forward and backward waves show monotonic decreasing behavior for conventional PhC. This property suggests that for a longer distance e. m wave propagation is not possible. But DNG material/air composition gives completely different behavior which suggests the longer propagation of e. m wave compared to the conventional one, if operating wavelength is set at slightly higher (1.6 μm) or lower (1.5 μm) than Bragg wavelength (1.55 μm). Result suggests that with the increase of magnitude the curve decreases exponentially and makes attenuated transmission for Bragg wavelength but for other two wavelengths the curves oscillate periodically. Again we can see that the curves oscillate more for the 60 μm grating length and strong coupling condition. So, these structures also exhibit bandpass filter characteristics at the Bragg wavelength (depending on structural parameters), and it can serve the dual purpose.

REFERENCES

Belhadj. W, AbdelMalek. F, Bouchriha. H, (2006) Characterization and Study of Photonic Crystal Fibres with Bends, Material Science and Engineering: C, 26, 578–579

Fogel I. S, Bendickson. J. M, Tocci. M. D, Bloemer. M. J, Scalora. M, Bowden. C. M, Dowling. J. P, (1988) Spontaneous Emission and Nonlinear Effects in Photonic Bandgap Materials, Pure and Applied Optics, 7, 393–408

Gao. Y, Chen. H, Qiu. H, Lu. Q, Huang. C, (2011) Transmission Spectra Characteristics of 1D Photonic Crystals with Complex Dielectric Constant, Rare Metals, 30, 150–154

Hwang. R. B, Liu. H. W, Chin. C. Y, (2009) A metamaterial-based E-plane Horn Antenna, Progress in Electromagnetics Research, 93, 275–289

Jiang. Y, Niu. C, Lin. D. L, (1999) Resonance Tunneling through Photonic Quantum Wells, Physical Review B, 59, 9981–9986

Lima Jr. A. W, Ferreira. A. C, Sombra. A. S. B, (2011) Optical Memory made of Photonic Crystal working over the C-band of ITU, Journal of Optical Fiber Communication Research, 1–16

Maity. A, Chottopadhyay. B, Banerjee. U, Deyasi. A, (2013) Novel Band-Pass Filter Design using Photonic Multiple Quantum Well Structure with p-Polarized Incident Wave at 1550 nm, Journal of Electron Devices, 17, 1400–1405

Robinson. S, Nakkeeran. R, (2011) Two Dimensional Photonic Crystal Ring Resonator based Bandpass Filter for C-Band of CWDM Applications, Nat. Conf. Comm, pp. 1–4 (2011)

Rudziński. A, (2007) Analytic Expressions for Electromagnetic Field Envelopes in a 1D Photonic Crystal, ACTA Physics Polonica A, 111, 323–333

Szczepanski. P, (1988) Semiclassical Theory of Multimode Operation of a Distributed Feedback Laser, IEEE Journal of Quantum Electronics, 24, 1248–1257

Shanthi. K. V, Robinson. S, (2014) Two-Dimensional Photonic crystal based Sensor for Pressure Sensing", Photonic Sensors, 4, 248–253

Xiong. H., Hong. J. S, Peng. Y. H, (2012) Impedance Bandwidth and Gain Improvement for Microstrip Antenna using Metamaterials, Radio Engineering, 21, 993–998

Xu. X, Chen. H, Xiong. Z, Jin. A, Gu. C, Cheng. B, Zhang. D, (2007) Fabrication of Photonic Crystals on Several Kinds of Semiconductor Materials by using Focused-Ion Beam Method, Thin Solid Films, 515, 8297–8300

Yablonovitch. E, (1987) Inhibited Spontaneous Emission in Solid-State Physics and Electronics, Physical Review Letters, 58, 2059–2061

Zhang. Z, Qiu. M, (2004) Small-volume Waveguide-section High Q Microcavities in 2D Photonic Crystal Slabs", Optics Express, 12, 3988–3995

Computational Science and Engineering – Deyasi et al. (Eds)
© 2017 Taylor & Francis Group, ISBN 978-1-138-02983-5

An all-optical feedback loop based frequency encoded data-storing unit by EDFA

Baishali Sarkar
B.N. Mahavidyalaya, Itachuna, Hooghly, West Bengal, India

Sourangshu Mukhopadhyay
The University of Burdwan, Burdwan, West Bengal, India

ABSTRACT: Erbium Doped Fiber Amplifier (EDFA) is an established and potential optical device which can amplify an optical weak probe signal strongly by the use of a suitable pump beam. This fiber has the main advantage of amplifying as well as modulating an optical signal. There are reported many optical switching operations and systems where EDFA has been used massively. Here a new concept is proposed by the current authors of using an optical feedback loop with the use of an EDFA, for the development of an all-optical and frequency encoded 1 bit latch and n bit memory system. The advantage of the whole system is that it is all-optical in nature and thus it gives superfast operation speed.

1 INTRODUCTION

Nonlinear materials are very much established for the development of several optical switches. Different all optical switching based logic operations, optical modulations, demodulations etc. are proposed by the use of such materials (Esmaeilian *et al.*, 2011, Chakraborty, *et al.*, 2009, Pal, *et al.*, 2009, Bonk, *et al.*, 2012, Johnson, *et al.*, 2010, Kusalajeerung, *et al.*, 2011). Again the nonlinearity of Erbium doped optical fiber is also a promising optical system not only for amplifying an optical signal, but also for switching and modulation purposes (Desurvire, *et al.*, 1987, Chakraborty, *et al.*, 2011, Mahad, *et al.*, 2009). EDFA is an optical fiber doped with erbium, which acts as an amplifier. If a high power pump beam (wavelength of 980 nm or 1480 nm) is mixed with a signal beam (in the wavelength region 1530–1570 nm) and passed through the erbium-doped silica fiber, then after a certain distance the pump power is abruptly dropped and the small power of the signal beam is enhanced strongly (Yariv, 1991, Ghatak, *et al.*, 1999, Desurvire, 1990, Desurvire, 1994). In our proposal we report a process of developing a frequency encoded 1 bit all optical memory unit (latch) by the use of the above property of EDFA. The EDFA was earlier used in optical latch (Chakraborty, *et al.*, 2011). There the role of EDFA was only for amplifying an optical signal. Here in this proposed scheme EDFA is used as an optical switch.

2 FREQUENCY ENCODING PRINCIPLE

In optical data processing and parallel computation several encoding/decoding techniques are seen to be used. For example intensity encoding, polarization encoding, phase encoding and frequency encoding (Garai, *et al.*, 2010, Garai, *et al.*, 2009, Garai, *et al.*, 2010, Ghosh, *et al.*, 2011). These encoding techniques uses certain reference level of a signal to indicate the logic state '1' and '0' for Boolean representation of data. Except the frequency encoding principle all other types of encoding changes the reference level of '1' and '0' during transmission and operation of data. In case of intensity encoding (for example) the value of intensity representing the '1' logic state may be dropped down to the '0' state because of absorption/attenuation of data in the medium during transmission. So the bit error problem may come. On the other hand frequency encoding/decoding has the reliability in encoding of bit, as generally frequency of a signal are not changed in the medium during operation/transmission. That is why a faithful and reliable optical operation can be achieved by frequency encoding process. Here if a particular frequency of light is considered as '1', the frequency of another light is considered as '0' logic state. Several frequency encoded optical logic operations are proposed by scientists and technologists.

3 EDFA AS AN OPTICAL SWITCH

Erbium doped fiber amplifier can be used very successfully for developing an optical switch. Its amplifying character, which is described below can be used for optical switching purpose. If a pump beam of wavelength $\lambda_p = 980$ nm and pump power $P_{pi} = 7$ mw is jointly applied with a probe signal

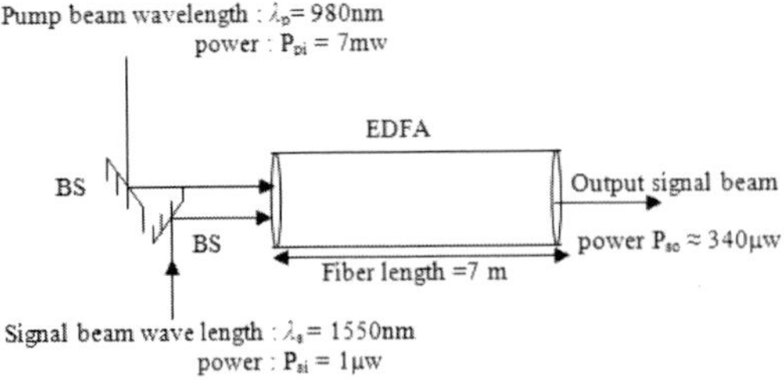

Figure 1. Schematic diagram of a latch.

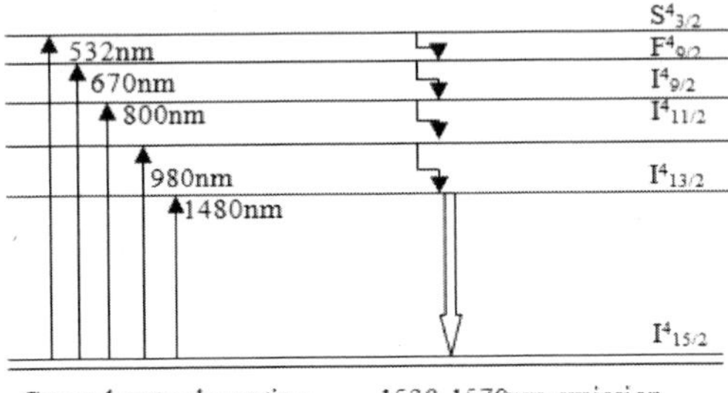

Figure 2. Energy level diagram of EDFA.

beam of wave length λ_s = 1550 nm and signal power P_{si} = 1 μw in the input side of an EDFA of length 7 m long, then at the output the whole pump power is found to be absorbed by the gain of the signal power at the output. The scheme is shown in Figure 1. The approximate output power of the signal is 340 μw strongly (Yariv, 1991, Ghatak, *et al.*, 1999, Desurvire, 1990, Desurvire, 1994).

This operation of EDFA is supported by its energy level diagram given in Figure 2. From this diagram it is clear that if any of pump beams of wavelength 1480 nm, 980 nm, 800 nm, 670 nm, 532 nm is applied, same type of amplification of probe beam in the wavelength region 1530 nm–1570 nm is obtained strongly (Yariv, 1991, Ghatak, *et al.*, 1999, Desurvire, 1990, Desurvire, 1994).

4 THE PROPOSED ALL-OPTICAL FREQUENCY ENCODED LATCH BY EDFA

The operation of latch is described by the Figure 3. Here the input is amplified by an amplifier, and a portion of the amplified output is fed back into the input. The principle of operation is that when the input is withdrawn the output continues and remains functional as the feedback signal energizes the system. The proposed all optical memory is shown in Figure 4 where frequency encoding

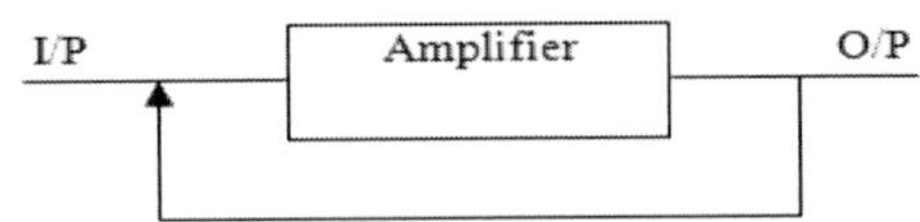

Figure 3. Schematic diagram of a latch.

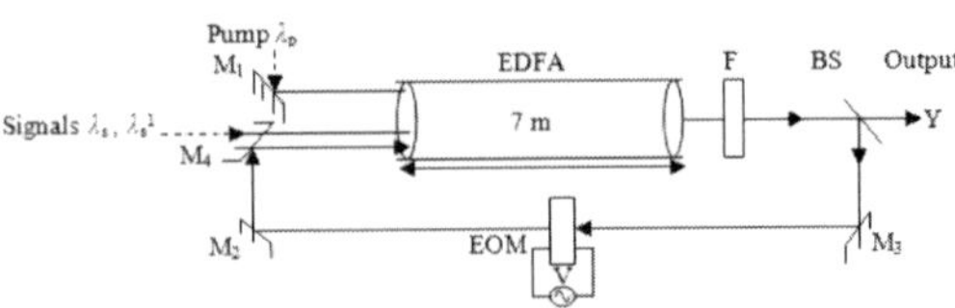

Figure 4. The schematic diagram of the proposed memory unit (EDFA is erbium doped optical fiber of 7 m length, F is a λ_s, $\lambda_s^{\prime}$ pass filter, BS is a beam splitter, M_1, M_2, M_3, M_4 are four mirrors). EOM represents the electrically biased Electro-optic modulator.

operation is used. Here one EDFA of length 7 m is taken and a pump beam of power 7 mw and wavelength λ_p = 980 nm is applied as the biasing signal. Now λ_s and $\lambda_s^{\prime}$ are two wavelengths lying in the range 1530 nm–1570 nm are taken and considered as logical state '1' and logical state '0' respectively. At the output of the fiber an optical filter is kept which passes λ_s, $\lambda_s^{\prime}$ signals and blocks the light of wavelength λ_p if any. After that a beam splitter is used to divide the output light into two parts. One part is connected with input by the feedback shown in Fig. 4. The other part gives the respective output of the memory unit.

Now when λ_s (1) is applied at the input the output is λ_s (1), even if the input is withdrawn, the output continues. Similarly when $\lambda_s^{\prime}$ (0) is applied at the input the output will be $\lambda_s^{\prime}$ (0). Thus $\lambda_s^{\prime}$ (0) will be continued at the output even if the input is stopped. When the operation of obtaining the output at λ_s(1) is changed to the operation of getting $\lambda_s^{\prime}$ (0) at output, the supply of the pump beam (λ_p) should be stopped before the application of $\lambda_s^{\prime}$ (0) at the input side. Then λ_p is to be applied again for continuation of $\lambda_s^{\prime}$ (0). Similarly to change $\lambda_s^{\prime}$ (0) based operation to λ_s (1) operation the pump beam (λ_p) also is to be stopped. Then after application of λ_s (1) the pump beam is to be applied again for continuation of λ_s (1). The feedback mechanism actually sustains the operation. The whole operation can be referred by a truth table as given in Table 1.

In the feedback path one can put an electro-optic Pockels cell, which is biased electrically by some external potential. Varying the potential the refractive index of the crystal can be changed, which ultimately can change the delay in the feedback. This introduction of electro-optic Pockels material can thus control suitably the speed of operation of the system.

Table 1. The truth table of a frequency encoded all optical latch by EDFA.

Input		Output	
Input wavelength	Input logic state	Output wavelength	Output logic state
λ_s	1	λ_s	1
withdrawn	λ_s	1	
λ_s^I	0	λ_s^I	0
withdrawn	λ_s^I	0	

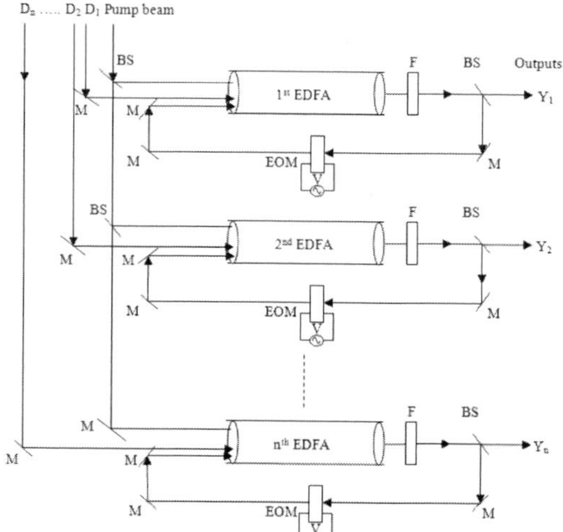

Figure 5. A schematic diagram of EDFA based optical data storing unit (M represents the mirror, BS is beam-splitter).

To get a stable output power, when the input probe beam is withdrawn, the feedback system (here it is the beam splitter BS in Figure 4) has the very important role. The splitting ratio of the beam splitter should be in such order so that a very small part (~1 μw) of the output power is coupled to the input. Again if the feedback power drops below the threshold probe power as required for the amplification of this system will stop its operation.

5 ALL-OPTICAL N-BIT MEMORY SYSTEM USING EDFA

The implementation of the two-bit or any higher bit frequency encoded optical data storing unit can be developed also using the feedback mechanism in n-no. of EDFAs. The scheme is shown in Figure 5.

Here frequency encoded n-bit data are given input at n-different channels marked by D_1, D_2, ..., D_n and the stored output data is received at the output channels marked by Y_1, Y_2, ..., Y_n. now if the frequency encoded Boolean data string is withdrawn from the input, the entire information will be stored at Y_1, Y_2, ..., Y_n channels. To implement this scheme we should select the wavelengths of $\lambda_s(1)$ and $\lambda_s^I(0)$ greater than λ_P (wavelength of the pump beam). The pump beams are drawn from a single source.

6 CONCLUSIONS

The whole system is all-optical as no electronics is involved here to support the operation. This scheme can give a real time operation in case if experiment is performed. Again the system can store a digital memory for very long time unless the pump is not withdrawn. The operation is reliable and faithful as frequency encoding technique is used here. The whole scheme described above is based on the optical feedback mechanism. This feedback can be made positive or negative by making a phase delay, which can be implemented by putting an electro-optic material in the feedback path. The scheme needs a very small light power for the operation. The fan in and fan out of the optical proposed circuits are also high. Here a data comprised by a large no. of bits can be stored in the memory unit.

REFERENCES

Esmaeilian, A. Abas, A.F. Mahdi, M. A. & Samsudin, K. (2011). Ultrafast two-bit all-optical analog-to-digital conversion based on femto-second soliton sequence sampling. Opt. Eng. 50: 125001.

Chakraborty, B. & Mukhopadhyay, S. (2009). Alternative approach of conducting phase-modulated all-optical logic gates. Optical Engineering 48: 035201.

Pal, A. & Mukhopadhyay, S. (2009). Using modified Mah-Zehnder interferometer to get better nonlinear correction term of an isotropic nonlinear material. Chinese Optics Letters (OSA) 7: 624.

Bonk, R. Huber, G. Vallaitis, T. Koenig, S. Schmogrow, R. Hillerkuss, D. Brenot, R. Lelarge, F. Duan, G. H. Sygletos, S. Koos, C. Freude, W. & Leuthold, J. (2012). Linear semiconductor optical amplifiers for amplification of advanced modulation formats. Optic Express 20: 9657.

Johnson, S. & Biswas, A. (2010). Topological soliton perturbation for sine-gordon equation with full nonlinearity. Phy. Lett. A 374: 3437.

Kusalajeerung, C. Chiangga, S. Pitukwongsaporn, S. & Yupapin, P. P. (2011). Nonlinear switching in silicon-based ring resonators. Optical Engineering 50: 024601.

Desurvire, E. Simpson, J. R. Becker, P. C. (1987). High-gain erbium doped traveling wave fiber amplifier. Optics Lett. 12: 888.

Chakraborty, B. & Mukhopadhyay, S. (2011). All-optical method of developing a phase encoded latch by using EDFA. Optik 122: 1844.

Mahad, F. D. B. Supa'at, M. & Sahmah, A. (2009). EDFA Gain optimization for WDM' System. Electrika II: 34.

Yariv, A. (1991). Optical Electronics. USA: Saunders College Publishers.

Ghatak, A. & Thyagarajan, K. (1999). Introduction to Fiber Optics. New Delhi: Cambridge University Press India Pvt. Ltd.

Desurvire, E. (1990). Analysis of noise figure spectral distribution in erbium doped fiber amplifiers pumped near 980 and 1480 nm. Applied Optics 29: 3118.

Desurvire, E. (1994). Erbium Doped Fiber Amplifiers. New York: John Wiley & Sons.

Garai, S.K. & Mukhopadhyay, S. (2010) Analytical approach of developing the expression of output of all-optical frequency encoded different logical units and a way out to implement the logic gates. Optical Fiber Technology 16: 250.

Garai, S. K. & Mukhopadhyay, S. (2009) Method of implementing frequency encoded multiplexer and demultiplexer systems using nonlinear semiconductor optical amplifiers. Optics and Laser Technology 41: 972.

Garai, S. K. Mukhopadhyay, S. (2010). A novel method of developing all-optical frequency encoded memory unit exploiting nonlinear switching character of semiconductor optical amplifiers. Optics and Laser Technology 42: 1122.

Ghosh, B. & Mukhopadhyay S. (2011). All-optical wavelength encoded NAND and NOR operations exploiting semiconductor optical amplifier based Mach-Zehnder interferometer wavelength converter and phase conjugation system. Optics and Photonic Letters (World Scientific) 4: 47.

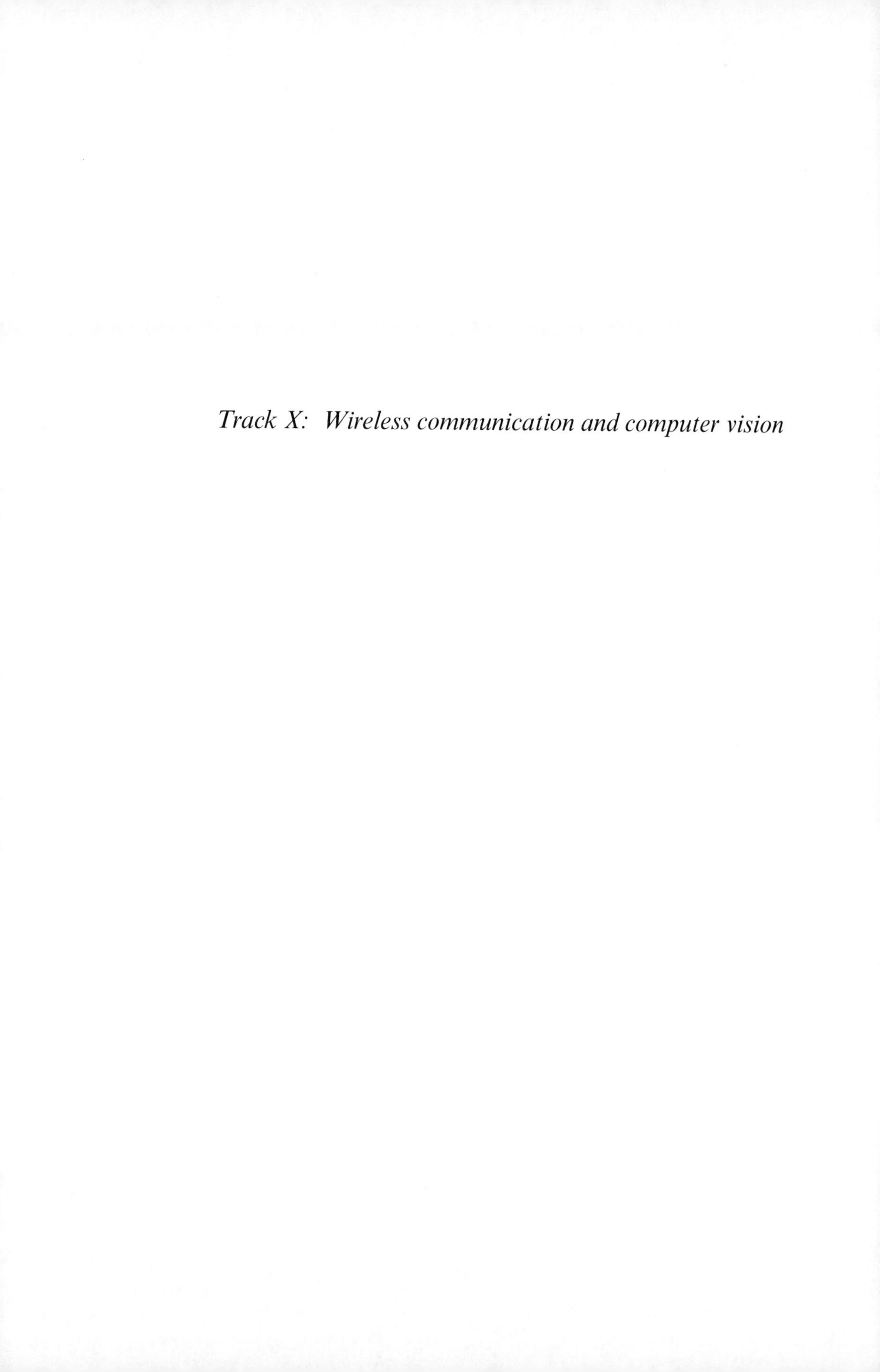

Track X: Wireless communication and computer vision

Computational Science and Engineering – Deyasi et al. (Eds)
© 2017 Taylor & Francis Group, ISBN 978-1-138-02983-5

Some studies on profit analysis in location management for personal communication service network

Manashi De & Srijibendu Bagchi
RCC Institute of Information Technology, Kolkata, India

ABSTRACT: In a Personal Communication Service (PCS) network, Location management is an important issue in cellular networks. In our research, a new approach for profit analysis of the service provider in Location Management of PCS network has been introduced. In our approach, users' movement has been considered as random in this work. The probability of an user's movement from one location to other is taken from a number of distributions through simulation varying within 0 to 1. The profit of the service provider has been calculated with the number of states keeping all other parameters constant. The distributions are Normal, Gamma, Nakagami-*m*, Weibull, Pareto, Burr. However, in our work it has been clearly observed that, with the increase of number of changing states which is denoted by M, the profit of the service provider also increases, which immensely affects in the reduction of overall cost of Location Management.

1 INTRODUCTION

In a Personal Communication Service (PCS) network, the location of a mobile user should be identified prior to the conveyance of an incoming call to the mobile user. the PCS network keeps the trajectory of the location of the mobile user through a mechanism called Location Management (LM) (Keshav, 2006, Zhang, 2003) LM is one of the vital concerns in cellular networks. However, lm possesses two fundamental operations, such as Location Update (LU), and paging. In lu, a mobile user dynamically updates its registration of location in the network, whereas in case of paging, the network pinpoints the exact area of the mobile user, where the mobile user is currently residing in order to route an incoming call (Halawani et al., 2011). Both LU and paging sustain significant cost, which should be minimized. Intuitively, frequent LU of the mobile users indicates the exact location information of the mobile user, and thus paging cost will be reduced. On the other hand, LU cost will increase. Thus, a tradeoff exists between the LU and paging costs. Previously, many algorithms have been proposed to solve the problem of LM cost. In our research, new techniques have been introduced to solve the above issues. In addition, the profit of the service provider has been analyzed using different probability distribution of states such as ormal, Gamma, Pareto, Nakagami-*m*, Weibull, Burr. The paper is organized as follows. Section II enlightens some related works of LM. Section III describes the proposed approach for reducing the LM cost and thereby increasing the profit of the service provider. Section IV provides a discussion of simulation results of our proposed work using MATLAB. Section V describes the conclusion and some future scope.

2 RELATED WORKS

Some scholars provide a discussion about the forwarding and resetting operation. Forwarding operations contains a chain of databases to locate a mobile user. Resetting operation updates the databases to track the current location of the mobile user directly without following the chain of databases (Chen et al., 1998). The PCS network acts as a server and provide services to the mobile user for 'updating the location of the mobile user as the user moves across a database boundary' and 'locating the mobile user'. They also describe the use of Markov chain to describe the behavior of mobile user and optimize the service rate of the PCS network, while the forwarding and resetting operation is performed.

Some researchers discussed the movement based LM with the concept of Home location register (HLR) and Visitor Location Register (VLR) architecture. It has also concluded that among dynamic LM scheme, the movement based approach is the most practical and easy to implement (Wang et al., 2008).

Some scientists enlighten the dynamic intersystem LM technique based on the concept of boundary location area and boundary location register to provide a reliable inter-system roaming scenario. They provide a discussion about the Mobility Application Part (MAP) protocol, which is used to design the intersystem of location registration and call delivery procedures. They further describes about the signaling cost, latency of location registration and call delivery, call losses due to intersystem roaming (Akyildiz et al., 2001).

Some researchers conferred about the development of embedded Markov chain to analyze the signaling cost of the Movement based schemes of LM.

They derived analytical formulas of signaling cost of Movement based schemes by considering the (LA) architecture, which was further manifested by simulation (Wang et al., 2014).

3 PROPOSED SCHEME

As an extension of the previously proposed scheme, a certain section has been modified, which shows an improved performance in some scenarios. In this current study, a new approach for profit analysis of the service provider in LM of PCS network has been proposed. The profit of the service provider has been calculated with the number of states keeping all other parameters constant. In this section, the problem has been treated in a generalized way. Here, instead of considering arbitrary pseudorandom sequence for the probability of user's location change, different probability distributions of states has been adopted by following the aforesaid distributions. The idea behind it is that, a distribution can be fitted from the received data of user's movement. Since, most generalized distributions has been considered (i.e. a number of distributions may be treated as their special cases), a large number of possibilities have been covered. However, the parameters of the distributions are chosen arbitrarily but if they are estimated from a real time scenario, the results would not differ significantly.

3.1 *Proposed algorithm*

Step 1: The signaling cost of the network has been calculated for the following cases:

Case 1: $n \geq 2M\text{-}1$

Case 2: $2M = 3n\text{-}1$

Case 3: $2M \geq 3n$

Case 4: $M \leq 2n\text{-}2$

Case 5: $M \geq 2n\text{-}1$, where M is the changing states within the network, $n = 0,1,\ldots,5,..$ The total signaling cost (*TSC*) is derived as (Wang et al 2014)

$$TSC = N_{VLR}\delta_{VLR} + N_{VLR'}\delta_{VLR'} + N_{HLR}\delta_{HLR} + N_{HLR'}\delta_{HLR'} + N_{paging}\delta_{paging} \quad (1)$$

Step 2: The throughput of the network is calculated which is followed by the following expression (Chen et al 1998)

$$X = \left(\sum_{i=0}^{k-1} P_{(C,i)} \times \mu_i\right) + \left(\sum_{i=1}^{k-1} P_{(I,i)*} + P_{(I,i)**} + P_{(C,i)*} + P_{(C,i)**}\right) \times \mu_p) + \left(P_{(I,k)*} + P_{(C,k)*}\right) \times m_k \quad (2)$$

Step 3: The profit of service provider is obtained by subtracting the throughput of the network from the *TSC*, and it is denoted by the expression, Profit of the service provider = *TSC* – *X*.

Step 4: The profit of the service provider is further analyzed with the help of the following above distributions.

4 SIMULATION RESULTS

Some scenarios of profit analysis of the service provider using different probability distributions and considering the lower limits of n in different scenarios are described by the following graphs:

From the six scenarios as depicted in this section, the profit analysis of service provider for the different cases of n has been identified by using the following above mentioned distribution. It can be concluded that broadly for most of the distribu-

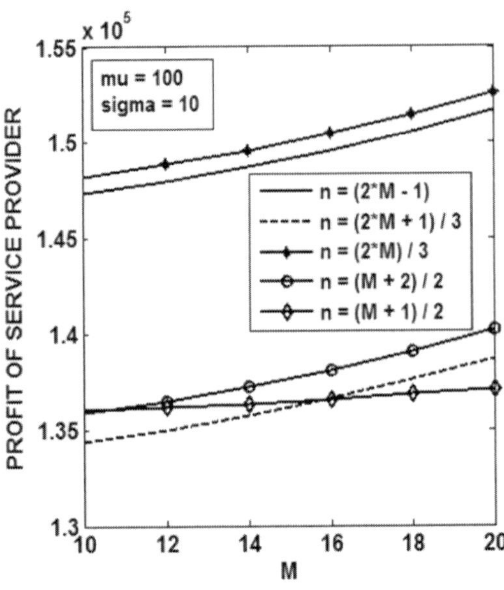

Figure 1. Profit of service provider using Normal disrtribution.

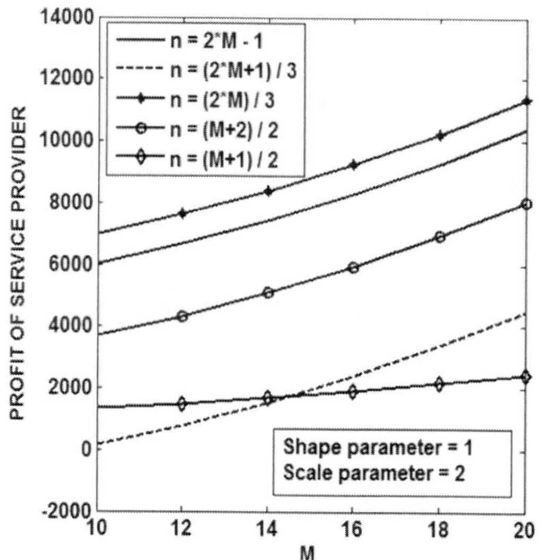

Figure 2. Profit of service provider using Gamma distribution.

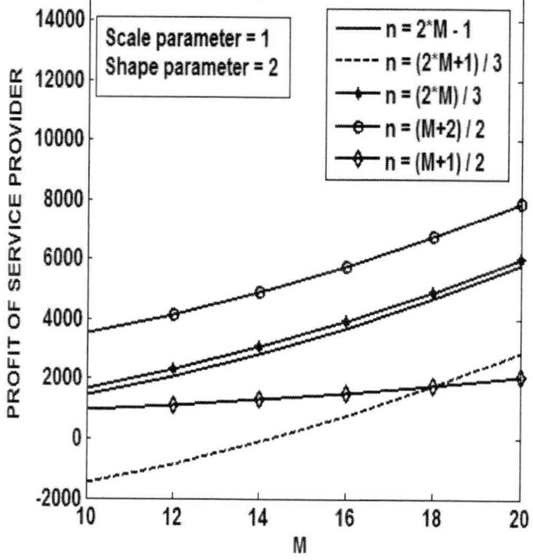

Figure 3. Profit of service provider using Nakagami-m distribution.

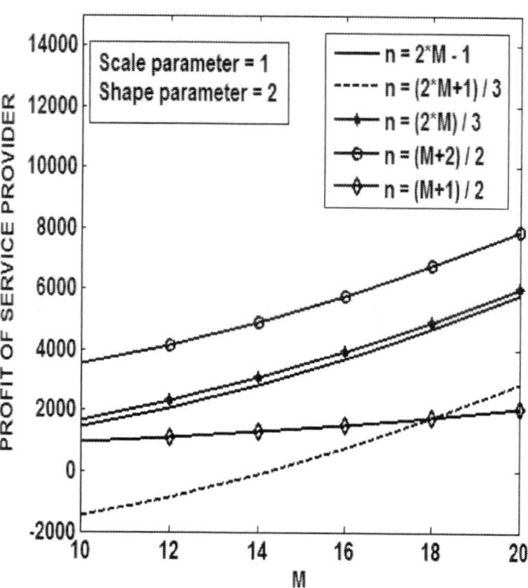

Figure 4. Profit of service provider using Weibull distribution.

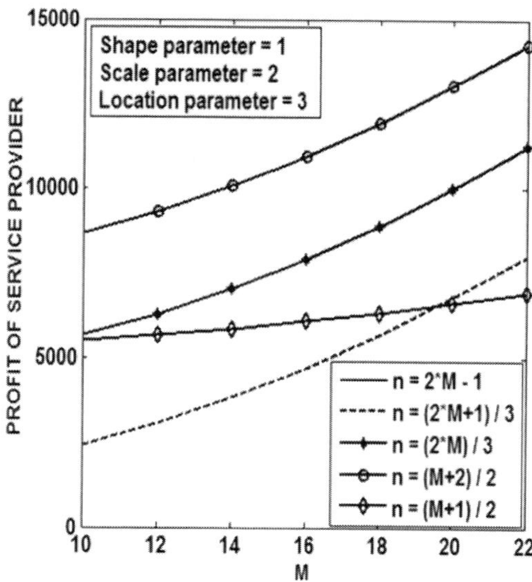

Figure 5. Profit of service provider using Pareto distribution.

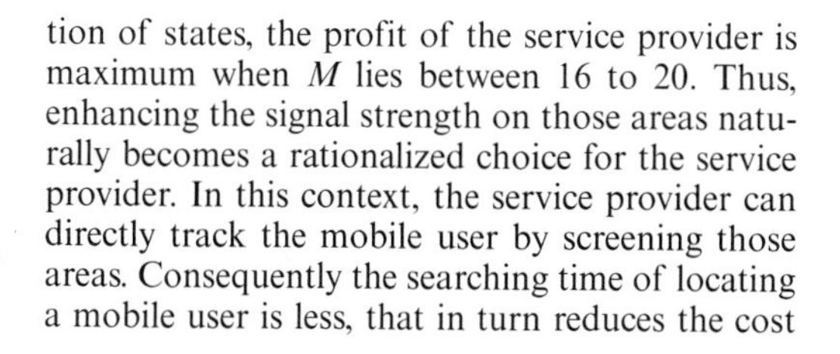

tion of states, the profit of the service provider is maximum when M lies between 16 to 20. Thus, enhancing the signal strength on those areas naturally becomes a rationalized choice for the service provider. In this context, the service provider can directly track the mobile user by screening those areas. Consequently the searching time of locating a mobile user is less, that in turn reduces the cost of repetition updates of the mobile user, thereby reducing the cost of overall LM.

5 CONCLUSION AND FUTURE SCOPE

In this article, a new approach for profit analysis of the service provider in LM of PCS network has

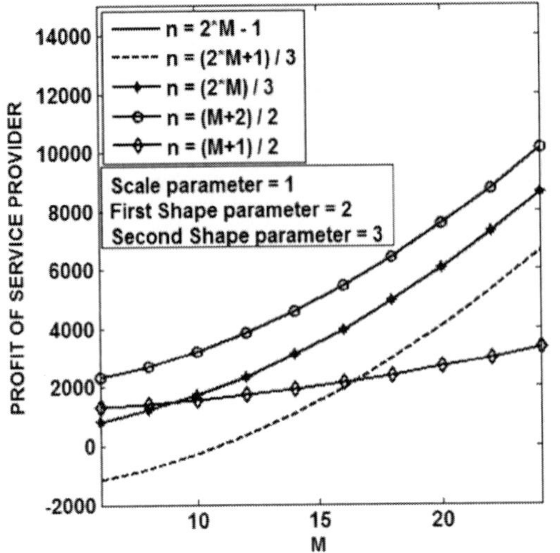

Figure 6. Profit of service provider using Burr distribution.

been proposed. The profit of service provider is analyzed by using the following afore-mentioned probability distributions. This study highlights an optimal choice of the number of states and from the simulation section it can also be inferred that the state number selection does not alter for different distribution of states. Thus, overall it may be concluded that, this article contributes to provide an idea regarding the number of states that may be bounded within a specific limit to maximize the profit of a service provider in a PCS network as well as reducing the cost of LM.

REFERENCES

Akyildiz I. F. Wang W. and Lee Y-J., Aug 2001, Location Management in 3G/4GWireless Systems" in *Proc. SPIE—Convergence Inf. ITCom Conf.*, Denver, CO, 1–13.

Chen I.R., Chen T-M. and. Lee C, 1998, "Performance Evaluation of Forwarding Strategies for Location".

Halawani S M, Ahmad Ab.R. bin and Khan M.Z, 2011, "A study for Issues of the Location Management In Mobile Networks", International Journal of Computer Science Issues, 8(1), 420–429.

Keshav T. R, 2006 "Location management in wireless cellular networks," Survey paper, CSE574S, University of Washington at St. Louis, Spring Available [Online].: http://www.cs.wustl.edu/~jain/cse574-06/cellular_location.htm.

Management in Mobile Networks, *The Computer Journal*, 41(4), 243–253.

Wang X., Lei X, Fan P., Qingyang R. H and Horng S-J., 2014, "Cost Analysis of Movement-Based Location Management in PCS Networks: An Embedded Markov Chain Approach".

Wang X., Fan P, Li J, and Pan Y, Nov 2008 "Modeling and Cost Analysis of Movement-Based Location Management for PCS Networks With HLR/VLR Architecture, General Location Area and Cell Residence Time Distributions", *IEEE Transactions on Vehicular Technology*, vol. 57(6), 3815–3831.

Zhang J., 2003, Location Management in Cellular Networks", *Handbook of Wireless Networks and Mobile Computing*, Chapter 2, John Wiley & Sons.

Computational Science and Engineering – Deyasi et al. (Eds)
© 2017 Taylor & Francis Group, ISBN 978-1-138-02983-5

Cyclostationary feature detection based FRESH filter in cognitive radio network

Sahadeb Shit & Srijibendu Bagchi
RCC Institute of Information Technology, Kolkata, India

ABSTRACT: Cognitive radio is a method where secondary user searches for a free band to utilize when licensed frequency band is not utilized. Spectrum sensing is the fundamental necessity of a cognitive radio that empowers to look for the free band and utilize accordingly. The expanded interest for portable correspondences and new remote applications raises the need to proficiently utilize the accessible range assets. This paper manages Cyclostationary based spectrum detecting in Cognitive Radios to empower unlicensed secondary users to craftily get to an authorized band. The alternative FRESH (Frequency Shift) filtering technique using knowledge of the signal cyclostationarity is used to detect the desired signal from the spectrum overlapping. Directions for improvements of these filters are given in this paper. The outcomes demonstrate that for signals which spectrally overlap, the versatile FRESH filter can perform exceptionally well while normal filters come up short.

1 INTRODUCTION

Cognitive Radio is a reconfigurable device for wireless communication that can enable unlicensed frequency band users to deploy licensed spectrum opportunistically by Dynamic Spectrum Access (Zhang and Wong, 1995). The co-channel interference problem is a common problem in multiuser communication systems (Haykin, 2005). This problem is caused by spectrally overlapped signals; one of which is desired signal and the other one is the interference. Conventional filters cannot give a solution to this problem. However, by using the cyclostationary properties of modulated signals, the adaptive Frequency Shift filtering techniques (FRESH) filters, (Gardner, 1986), can resolve this problem. The basic idea of FRESH filters depends on some properties of cyclostationary signals such that for a cyclic signal (Haykin, 1991), frequency shifted versions of it are correlated at various frequencies (Tian and Giannakis, 2006).

In this correspondence, we examine an alternative FRESH filtering technique to extract the desired signal from spectrally overlapped interference. Different filtering process have been proposed in some literature (Cabric et al., 2004). In practical communication systems, signals often experience spectral overlapping. In order to extract the desired signals from these interference, Gardner and his partners have proposed the Cyclic Wiener Filter (Gardner, 1993), that can isolate the desired signal components by utilizing the cyclostationarity of signals.

Consider a cyclic Wiener filter whose input $x(t)$ and output $\hat{y}(t)$ are correlated by

$$\hat{y}(t)=\int_{-}^{\infty} h(t,u)x(u)du \tag{1}$$

where $h(t, u)$ is the impulse response of the filter which can be expanded as

$$h(t,u)=\sum_{m=1}^{M} h_m\,(t-u)\exp\left(j2\pi\alpha_m u\right) \tag{2}$$

with α_m being the cyclic frequency of signals and $h_m(t-u)$ being a time invariant filter (Gardner, 1986). Using equation (2) we can write the output of the filter as

$$\hat{y}(t)=\sum_{m=1}^{M} h_m\,(t)\otimes\left[x(t)\exp\left(j2\pi\alpha_m\,(t)\right)\right] \tag{3}$$

It is proved that the Fourier transform of the optimum transfer function $h_m(t)$ must satisfiy the relation $s_{xx}^{\alpha}(f)H(f)=s_{yx}^{\alpha}(f)$ where $s_{xx}^{\alpha}(f)$ is the spectral autocorrelation density matrix (Roberts et al., 1991, Cabric 2008). The design of an optimum FRESH filter to extract the desired signal necessitates knowledge of the auto-spectral density matrix of the input as well as the cross-spectral density vector (Urkowitz, 1967, Gardner and Franks, 1975) between the input and the desired output.

1.1 *Implementation of FRESH filter*

Frequency Shift or FRESH filters are time varying filters which exploit the spectral coherence in cyclostationary signals to optimally estimate them. As the name suggests, a FRESH filter consists of many branches, each having a frequency shifter followed by a time invariant filter (Arslan and Yücek, 2007, Akyildiz et al., 2006) where each shift is equal to the cyclic frequency of the input signal.

Design a FRESH filter to detect a BPSK Signal with known cyclic frequency in AWGN, BPSK signal can be written as $x(n) = s(n)\exp(j2\pi f_c n)$ $s(n)$ is a random NRZ binary data sequence with full duty cycle square pulses of duration T_0. The carrier frequency f_c and baud rate $1/T$, are chosen as 30720 Hz and 3200 Hz respectively. At the receiver, the signal $r(n)$ is the sampled value

$$r(n) = x(n) + w(n) \tag{4}$$

$$x_{\alpha_p}(n) = r(n).\{\exp(j2\pi\alpha_p nT_s)\}_{n=1:M_p} \tag{5}$$

$$x^*_{-\beta_q}(n) = r^*(n).\{\exp(j2\pi\beta_q nT_s)\}_{n=1:N_p} \tag{6}$$

where α_p and β_q are cyclic frequencies of the input signal $a_p(n)$ and $b_p(n)$ the weight vectors of the FIR filters consider the concatenated weight vector $w(n)$ equal to $\{a_1(n)......a_m(n), b_1(n)......b_m(n)\}$

The output can written as

$$y(n) = \sum_{p=1}^{M} a_p(n) \otimes \mathrm{X}_{\alpha_p}(\mathrm{n}) + \sum_{q=1}^{N} b_q(n) \otimes x^*_{-\beta_q}(\mathrm{n}) \tag{7}$$

Then the weight update equations can be written as follows

$$\in(n) = r(n) - y(n) \tag{8}$$

$$w(n+1) = w(n) + \mu\in^*(n)x(n) \tag{9}$$

Here μ is convergence factor, (.) denotes element-wise product. Sampled at a sufficiently high rate to avoid aliasing at cyclic frequencies close to $2f_c$. This signal is then passed through a FRESH filter as follow in figure. Three cyclic frequencies of the BPSK signal $x(n)$ are utilized in the FRESH filter $\alpha_1 = 1/T_0, \alpha_2 = 2f_c, \alpha_3 = 2f_c + 1/T_0$,. The sampling frequency f_s is chosen to be 32 times the baud rate.

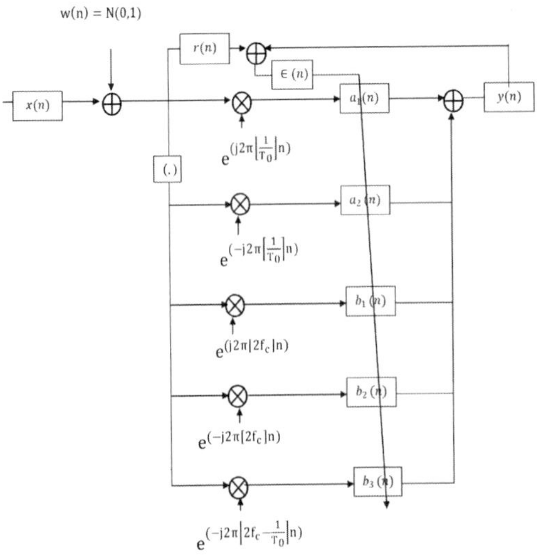

Figure 1. FRESH filter with five branches for estimating a BPSK signal.

2 SIMULATION RESULTS

2.1 *Number of iteration & error (t (n)) number of iteration* = 50, μ = 3*10^{-3}

Weight vectors are

$a_1(1) = 1000 + 52699*j$;

$a_2(1) = 2000 + 5888*j$;

$b_1(1) = 4000 + 2987*j$;

$b_2(1) = 6689 + 2569*j$;

$b_3(1) = 4569 + 2598*j$;

It is observed that FRESH filter aided cyclostationary sensing gives significantly better detection performance than all other conventional filter techniques in the low SNR. As the quantity of iteration available increases than the quantity of mean square error decreases, the detection performance improves concurrently.

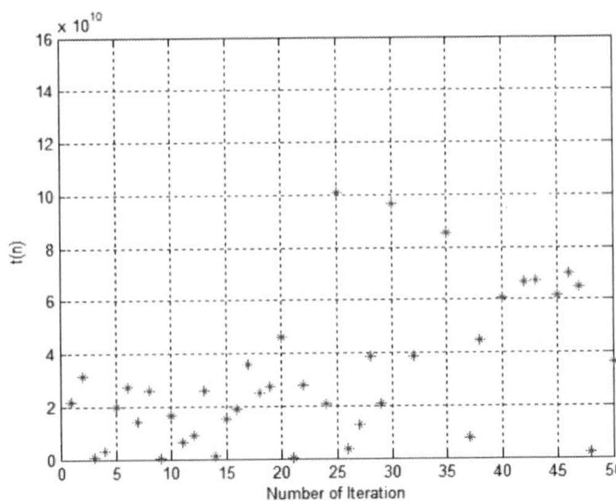

Figure 2. Variation of error with no of iteration n = 50.

2.2 *Number of iteration & error (t (n)) number of iteration* = 100, $\mu = 3*10^{-4}$

Weight vectors are

$a_1(1) = 1300 + 55999*j;$

$a_2(1) = 200 + 1886*j;$

$b_1(1) = 6000 + 4587*j;$

$b_2(1) = 6689+1569*j;$

$b_3(1) = 3669 + 2556*j;$

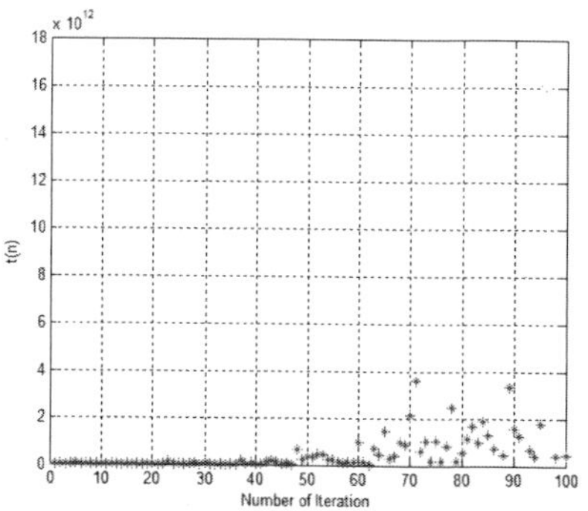

Figure 3. Variation of error with no of iteration n = 100.

2.3 *Number of iteration & error (t (n)) number of iteration* = 500, $\mu = 3*10^{-5}$

Weight vectors are

$a_1(1) = 4000 + 5874*j;$

$a_2(1) = 2560 + 1866*j;$

$b_1(1) = 6890 + 5687*j;$

$b_2(1) = 6689 + 1569*j;$

$b_3(1) = 3658 + 4576*j;$

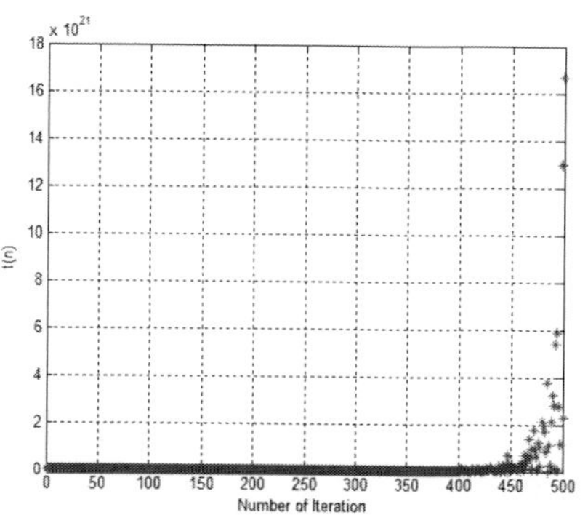

Figure 4. Variation of error with no of iteration n = 500.

2.4 *Number of iteration & error (t (n)) number of iteration* = 9000, $\mu = 4*10^{-3}$

Weight vectors are

$a_1(1) = 5258 + 5574*j;$

$a_2(1) = 2369 + 4663*j;$

$b_1(1) = 5687 + 5984*j;$

$b_2(1) = 6384 + 4569*j;$

$b_3(1) = 5683 + 6662*j;$

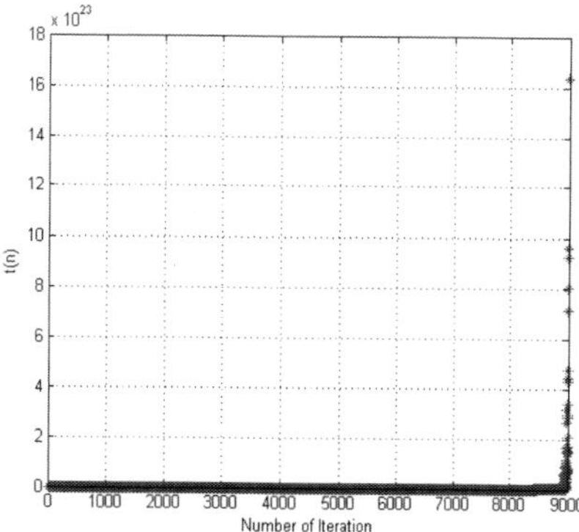

Figure 5. Variation of error with no of iteration n = 9000.

3 CONCLUSION

Design cyclostationary spectrum sensing using FRESH filter and find FRESH filter detect a cyclostationary signal. So it can conclude that cyclostationary spectrum sensing using FRESH filter play a vital role in spectrum sensing technique and a FRESH filter which can detect a cyclostationary signal will be implemented.

REFERENCES

Akyildiz I.F., Lee W. Y., Vuran M. C., and Mohanty S. 2006. NeXt generation dynamic spectrum access/cognitive radio wireless networks: a survey. Comput. Networks 50(13): 2127–2159.

Arslan H, and Yücek T. 2007. Spectrum Sensing for Cognitive Radio Applications. Cognitive Radio, Software Defined Radio, and Adaptive Wireless Systems Springer 3: 263–289.

Cabric D, Mishra S., and Brodersen R. 2004. Implementation issues in spectrum sensing for cognitive radios. Proc. Asilomar Conf. on Signals, Systems and Computers 1: 772–776.

Gardner W. A. 1993, "Cyclic Wiener filtering: Theory and method," IEEE Trans. Commun., 41, 151–163.

Gardner W.A. 1986, "The spectral correlation theory of cyclostationary time series", Signal Processing—July 1986.

Gardner W. A. and Franks L. E. 1975, "Characterization of cyclostationary random signal processes," IEEE Trans. Inform. Theory, IT-21, pp. 4–14.

Haykin S, 2005. Cognitive radio: brain-empowered wireless communications. IEEE Journal on Selected Areas in Communications 23(2): 201–220.

Haykin S. 1991, Adaptive Filter Theory, 2nd ed. New York: Wiley.

Roberts R., Brown W. A., and Loomis H. 1991, "Computationally Efficient Algorithms for Cyclic Spectral Analysis, IEEE SP Magazine 1991.

Tian Z. and Giannakis G.B. 2006. A Wavelet Approach to Wideband Spectrum Sensing for Cognitive Radios. 1st International Conference on Cognitive Radio Oriented Wireless Networks and Communications: 1–5.

Urkowitz H., 1967. Energy detection of unknown deterministic signals. Proceedings of the IEEE 55(4): 523–531.

Zhang J and Wong K. M, 1995, "A new kind of adaptive frequency shift filter," in Proc. ICASSP, 1995, 2, 913–916.

Computational Science and Engineering – Deyasi et al. (Eds)
© 2017 Taylor & Francis Group, ISBN 978-1-138-02983-5

Remote sensing image retrieval system based on content based image segmentation

Ripan Roy, Sabir Ahamed & Sagnik Chattopadhay
Department of ECE, Techno India College of Technology, Kolkata, India

Soumen Santra
Department of MCA, Techno India College of Technology, Kolkata, India

Kalyani Mali
Department of CSE, University of Kalyani, Kalyani, Nadia, India

ABSTRACT: Information retrievals were not enough as now a day. Image retrievals are also gaining its popularity. Image retrievals are being in use but only text based was not enough and only to satisfy real life image sensing image based problems. Content based are introduced to make the retrievals more accurate, efficient, optimize, feasible and satisfactory for images. In this paper, content based image retrievals are disused, analyzed and shown. Here it is shown how it can be retrieved and what the techniques behind content based image retrieval are.

1 INTRODUCTION

The image retrievals has become the part and parcel of our daily life. The need is everywhere—Criminal records, Medical Zones, Satellite launch and even for our official works also. [Jun Yue et al]

Each and every second, the number of images are increasing, thus for them database storage is also increasing. So for searching or retrieval images from such enormous collections of images (big data) can be achieved by Content Based Image Retrievals System (CBIR).

CBIR basically uses the technique of finding the desired images with help of its visual feature. This visual feature may include color, shape, texture etc. [Markus Stricker et al., Jun Yue et al.]

Before, text-based image retrievals were in practiced which finds the desired image by the entered metadata. But, however, from the enormous image collections, the output was also very large and inaccurate. Thus, the user had to search manually from the output images. So, here extra manual labor introduced which ultimately hampers work and time.

In CBIR no such factors are there to be taken in concern. The main characteristics of CBIR are to show images appropriately and accurately. It retrieves images by the content-image itself instead any user input keywords, using color features. Color feature is the main feature of any particular image which defines any image completely. [en.wikipedia.org/wiki/Content-based_image_retrieval] It includes three fundamentals on which CBIR is totally based on: extraction of visual content, indexing and retrieval designing. According to the RGB model, the image pixels can be processed by the methods like Averaging and Histograms to get the color content of an image.

There are several image retrievals systems which uses different techniques to retrieve the searched images:

1.1 *Retrieval featuring color*

The images which is stored every day in the databases are analyzed with different color featuring techniques and the data are saved. When the user search for the image by uploading a sample image (say) or may enter the color proportions (i.e. 25% red, 46% green) or its color feature is extracted, calculated and then compared with the saved values. Thus the output images with closed results are retrieved. [W. Niblack, Markus Stricker et al.]

1.2 *Retrieval featuring structure*

It happens many time, color of many images is similar but in close observation the image is actually different. The structure or texture has ability to make distinguish in these scenarios. Texture similarity of an image can be done best by comparing image contour values. It is also known as second

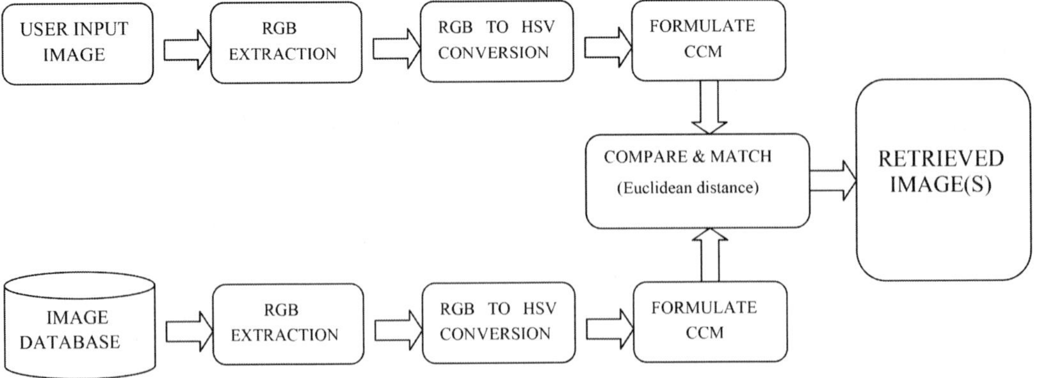

Figure 1. CBIR using color feature extraction.

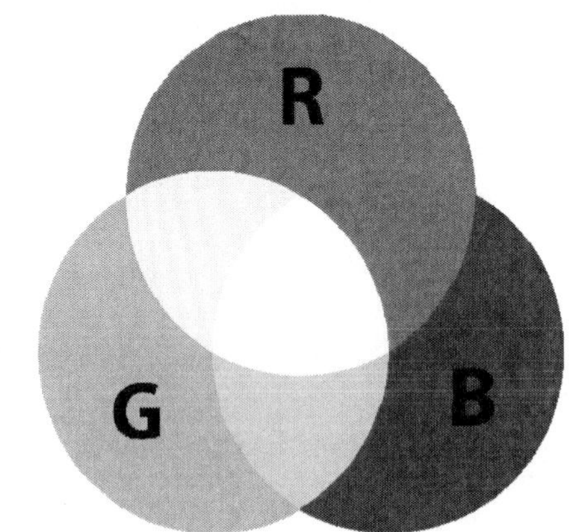

Figure 2. RGB Model of colors.

order statistics calculated from query and stored images. [Krishna Kant Singh et al].

Thus, from all above it is clear that these methods are able to calculate the relative resolution from selected pixels from an image and also find out the image texture properties like degree of coarseness, contrastness etc. [en.wikipedia.org/wiki/Content-based_image_retrieval] Similarly, for an image, not only texture based but, also color based or content based or both can be implemented.

1.3 *Retrieval featuring other factors*

The other feature may include with the help of the positioning of an image, by its shape and other methods are still in research. In primitive level, the image retrieval by its shape feature is most obvious technique. In an image, the natural objects are defined by its shape. When user feeds the query image, several shape extracting features are performed and simultaneously same is performed on the image of stored database. After closed comparison between those two image results are given. Shape matching for three dimensional images which are more challenging issue and also possible using CBIR system. Finding image by its location access will definitely contribute towards the Geographical Information Systems (GIS) which is one of the elegant research work for many years till now.[en.wikipedia.org/wiki/Content-based_image_retrieval] Figure 2 [thestudioexec.com] explains the RGB Model of colors. The benefit of all these techniques may be that they can describe an image at varying levels of image details. This is mainly found in natural or real life images where objects of interest may appear in variety of guises. [Jun Yue et al.]

2 WORK TO BE DONE

In this paper, it has been discussed about color feature extraction of an image. Extracting of color feature lead to minimum error as it is robust to background complication, viewing angle, translation and independent of image size. The main benefit of color feature over texture and shape is that the deformable objects and substances can be identified by mass nouns. It provides an intelligent and automatic solution.

Any image is composed of set of pixels. These pixels is composed of intensity and color. These colors can be represented by different color models such as Red-Green-Blue (RGB) and Hue-Saturation-Value (HSV).[Markus Stricker et al.] Here Figure 2 explains the flow chart of CBIR system. We can use any version of database software for storing the images. Here we use the JAVA version jdl 1.7.0 with NetBeans IDE 8.0 for implementing the CBIR system. Several java packages and their corresponding classes are used here.

In Figure 3 [rheumtutor.com] shown all variation of RGB Model.

Figure 3. RGB description in an image.

Table 1. RGB color codes.

Red (r)	Green (g)	Blue (b)	RGB Value	Color
255	0	0	255	RED
0	255	0	65280	GREEN
0	0	255	16711680	BLUE
0	255	255	16776960	CYAN
255	0	255	16711935	MAGENTA
255	255	0	65535	YELLOW
255	255	255	16777215	WHITE
128	128	128	8421504	GRAY
0	0	0	0	BLACK

*Here are only some color values.

Table 2. Extracting elements from RGB values.

RGB-Value	Hexadecimal-Value	(r)	(g)	(b)
6579300	646464	100	100	100
16422450	FA9632	50	150	250
32896	008080	128	128	0

2.1 *RGB color feature*

In an image, each pixel consists red, green and blue value ranging from 0 to 255 giving total 16777216 different colors. These colors are also known as additive primaries as any desired colors. These can be produced by adding others to these colors. [https://en.wikipeia.org/wiki/RGB_color_model]

Another feature of color extraction is that it closely relates to the same way as our eyes perceive color. This uses Cartesian coordinate systems.

The RGB values can be calculated by:

RGB = r + (g*256) + (b*256*256)

In this algorithm we need individual RGB values. Thus individual RGB can be calculated by converting the RGB values to Hexadecimal values.

If there are less than six digit hexadecimal numbers then fill the blank column with zero.

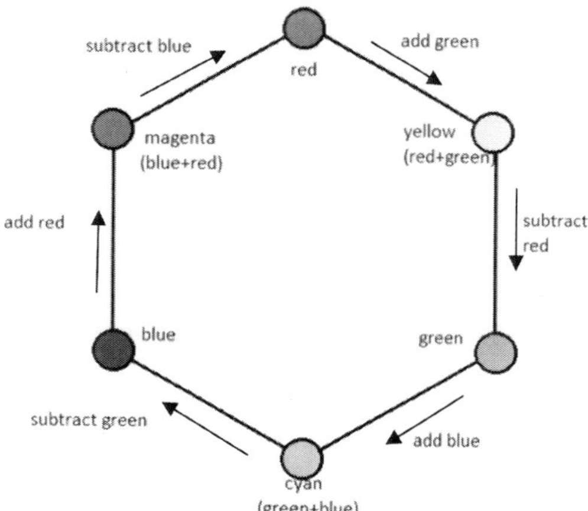

Figure 4. The HSV color cycle.

2.2 *Hue-Saturation-Value (HSV) color feature*

In HSV color feature, it defines the colors of the image in terms of shades and brightness or luminance.

Basically Hue, in terms of spectrum colors, represents dominant wavelength in light. Figure 4 expressed from 0° to 360° in which hues of red starts from 0° yellow starts from 60°, green starts 120°, cyan starts from 180°, blue starts at 240° and magenta starts at 300°.

Saturation represents the dominance of hue in color or intensity of color. Degree of purity of color can be expressed from this. Vivid is the high saturated color whereas muted is the low saturated color and grey image has no saturation. [https://en.wikipedia.org/wiki/HSL_and_HSV]

Value describes a relative lightness or darkness of a color.

Here are the formulas for Hue, Saturation and Value to be used in the algorithm:

$$H = \cos^{-1}\left(\frac{\frac{1}{2}\{(r-g)+(r-b)\}}{\sqrt{(r-g)^2+(r-b)*(g-b)}}\right)$$

$$S = 1 - \frac{3}{(r+g+b)}\{(r,g,b)_{\min}\}$$

$$V = \frac{1}{3}(r+g+b)$$

2.3 *Color conversion*

The triangle (Figure 5) has shown the **RGB** to HSV color triangle whose vertices are denoted by Red (r), Green (g) and Blue (b). The point **C** is the hue point which is formed by the angle between the line connecting **C** to the centre of the triangle and the line connecting RED point to the centre of the triangle. [http://www.rapidtables.com/convert/color/rgb-to-hsv.htm]Saturation (of point **C**) is the distance between the point **C** and centre of the triangle. The value (intensity) (of point **C**) is represented as length of the line bisecting the triangle and passing through centre and also getting the greyscale points too.

2.4 *Color Co-occurrence Matrix*

Color Co-occurrence Matrix (CCM) technique, used for the texture analysis. According to Robert M Haralick, who first introduced this technique, measures the texture of the image. It is basically a matrix which is defined over an image to be distribution of co-occurring values at a given offset. Here we take the Hue, Saturation and Values of specified pixel.

CCM made in use when Query Image is submitted by the user and the Hue Value, Saturation value and Value of the pixels are taken and then CCM is used.

The CCM is used by this formulae:

$$CCM = (9*H) + (3*S) + (V)$$

2.5 *Euclidean Distance*

Euclidean Distance plays the important role in this particular algorithm as it compares as well as measures the similarity between two feature vectors (Here F1 is the feature vector of the input image/query image and F2 is the feature vector of the database image.)

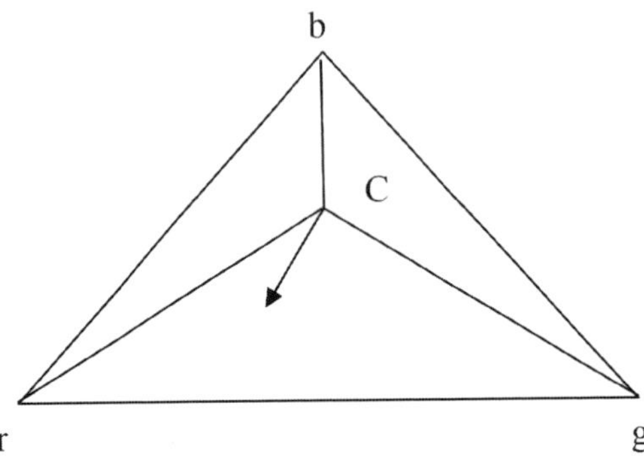

Figure 5. RGB to HSV conversion.

The formulae to measure the Euclidean Distance is:

$$E_{distance} = \sqrt{\sum_{i=1}^{n}\left(F1_i - F2_i\right)^2}$$

3 EXPERIMENTAL RESULTS

At First query image inserted from which r, g and b values are converted to desired Hue, Saturation and Value and necessary co-color occurrence matrix is obtained and the Euclidean Distance sorts the results are shown below:

4 FUTURE

There are still many points which have to be taken care and modify. The performance and efficiency of the content based image retrieval techniques may be increased by integrating it with texture and shape feature also. [Yi Zhuang et al.].

More than one feature shall definitely make the search more and more accurate results. The system designing with such combined techniques are in process with several research works. [Michal Haindl et al.] More feature extractions are

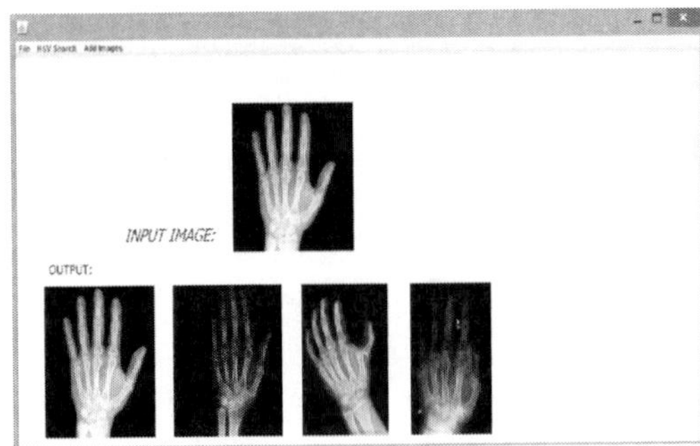

Figure 6. Output for medical images.

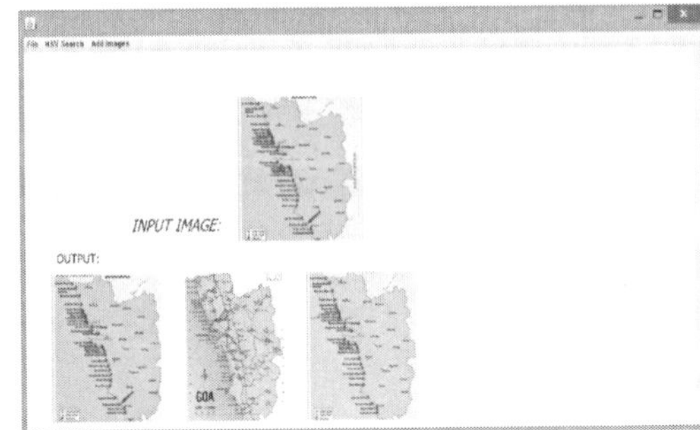

Figure 7. Output for map images.

still to be discussed if implement like shape and structure combination and also text and the content both can be search at the same time for better precision.

5 CONCLUSION

The main aim of this paper is to show the functionality of the content based image retrieval systems. Still most of the systems use text based image retrievals. Content based Image retrievals has been successful in the field of robust retrievals with accurate and appropriate results. The output search results are less in numbers such that the user don't have to put extra manual work to search the image. The Euclidean Distance helps to compare the results of the input query image and the database images. The algorithm used here improves quality results, reduces complexity time and increases the user interaction as well.

REFERENCES

Dana E. Ilea n, Paul F. Whelan Image segmentation based on integration of color—texture descriptors—A review 0031-3203/2011 Elsevier Ltd. doi:10.1016/j.patcog.2011.03.005.

Jun Yue, Zhenbo Li, Lu Liu, Zetian Fu, Content-based image retrieval using color and texture fused features, 08957177, 2010 Elsevier Ltd. All rights reserved. doi:10.1016/j.mcm.2010.11.044

Krishna Kant Singh, Akansha Singh A study of Image Segmentation Algorithms for Different Types of Images IJCSI International Journal of Computer Science Issues, Vol. 7, Issue 5, September 2010 ISSN (Online): 1694–0784 ISSN (Print): 1694–0814

Markus Stricker & Markus Orengo (1995), "Similarity of Color Images", Proceedings/Conference Na Proceedings of SPIE Storage and Retrieval for Image and Video Databases

Michal Haindl and Stanislav Mikes, A competetion in Unsupervised Color Image Segmentation, Patter Recognition, http://dx.doi.org/10.1016/j.patcog.2016.03.003.

Niblack W. (1993), "The QBIC Project: Querying Images by Content using Color, Texture and Shape", Proceedings of SPIE.

Yi Zhuang, Nan Jiang, Zhiang Wu, Qing Li, Dickson K.W. Chiu, Hua Hu, Efficient and robust large medical image retrieval in mobile cloud computing environment 0020-0255/2013 Elsevier Inc. http://dx.doi.org/10.1016/j.ins.2013.10.013

https://en.wikipedia.org/wiki/RGB_color_model

https://en.wikipedia.org/wiki/HSL_and_HSV

http://www.rapidtables.com/convert/color/rgb-to-hsv.htm

https://en.wikipedia.org/wiki/Content-based_image_retrieval

Images Sources:

http://www.telegraph.co.uk

http://www.thestudioexec.com

http://mapsofindia.com

http://goindia tourism.com

http://rheumtutor.com

Computational Science and Engineering – Deyasi et al. (Eds)
© 2017 Taylor & Francis Group, ISBN 978-1-138-02983-5

Expression detection from face images

Samiran Singha & Ranjan Jana
RCC Institute of Information Technology, Kolkata, West Bengal, India

ABSTRACT: Facial expression can convey non-verbal cues in face to face communication. It is the process to recognize the speaker's emotion to predict the intention of the speaker. Automatic facial expression detection can improve human-machine interaction. There are several methodologies introduced for detection of facial expression. Image processing is a very interesting part in computer science when it comes to human face recognition. Human gestures are mainly identified by observing movements of facial object like eye, nose, mouth, eyebrows. It is still a challenge to develop a digital system which can reliably detect an expression of a human. In this paper, a simple method is proposed to detect expressions from facial images. Face features are extracted from face image to classify expression using fuzzy c-means clustering algorithm. The obtained results are significant and remarkable.

1 INTRODUCTION

Every facial expression carries an emotional content with it. Humans communicate effectively and are responsive to each other's emotional states. Computers must also gain this ability. Facial expression, consistency movement and physiological reactions are the basic units of non-verbal communication. Expression recognition plays a significant role in recognizing one's affect and in turn helps in building meaningful and responsive Human Computer Interfaces. By observing face, one can decide whether a man is serious, happy, thinking, sad, feeling pain and so on. Recognizing the facial expression of a man can help in many of the sphere like in the field of medical science where a doctor can be alerted when a patient is in severe pain. It helps in taking prompt action at that time. Facial expression recognition has attracted increasing attention in visual sensation, pattern recognition, and human-computer interaction inquiry communities. Automatic recognition of facial expression therefore forms the essence of various next genesis computer science tools including affective computing engineering, intelligent tutoring arrangement, patient profiled individual wellness monitoring systems, etc. This paper proceeds with following sections. Section 2 reflects the previous work. Section 3 includes implementation details for this work. Experimented results are depicted in section 4. Finally, section 5 contains conclusions.

2 PREVIOUS WORKS

Facial expression detection can improve human-machine interaction. There are several methodologies introduced for detection of facial expression. Raheja et al. propose separating emotional states from non-emotional states are done using back propagation neural network. Saudagare et al. propose another technique for facial expression detection using neural network which is able to recognize a large variety of facial action units. Emotion recognition approach can be divided into two groups based on appearance based feature and geometry-based features is proposed by Valstar et al. Human face has several important regions such as, lip, eye, nose and eyebrows. Each of these regions has a unique shape and a unique pattern. So, every expression provides unique features. Using back propagation method the extracted features have been classified to detect facial expression is proposed by Shukla et al. Automatic emotion recognition can progress the objectivity and efficiency of emotion-related studies in education, psychology and psychiatry. Vasani et al. propose one method to detect emotional facial expressions occurred in a realistic human conversation setting. Jana et al. propose that detection of face regions from face image is one important part for facial expression detection. Face regions have been detected using the technique described by Viola and Jones. Automatic face detection is not an easy task, because face patterns can have significantly variable image due to many factors such as hairsbreadth styles and glasses. So, Su et al. propose a SOM-based automatic facial expression detection. Kharatkar et al. propose a face detection algorithm using multi-layer perceptron to implement for classification of seven reflection including six basic facial expressions (i.e., happiness, sorrowfulness, smile, surprise, fear, and disgust). Different light atmospheric condition and posture changes make extreme variations in the face image. Therefore, Bhatt et al. proposed one technique using possibility fuzzy c-means for facial expression detections. Good feature extraction is the most crucial part of facial expression detection. Jana et al. propose one efficient technique for good feature extraction from face image and fuzzy c-means clustering technique classification.

3 IMPLEMENTATION

Several different approaches have been proposed to solve the problem of facial expression detection. Here a new approach has been proposed to implement facial expression detection. The following four steps: preprocessing, feature extraction, training, and testing have been implemented for this detection.

3.1 *Preprocessing*

Initially the sample images of front face as shown in Fig. 1 are taken through digital camera with several gestures. Now only the face portion from the image is cropped using Viola-Jones algorithm technique as shown in Fig. 2.

3.2 *Feature extraction*

The most important regions of face like forehead, left eye, right eye, left cheek, right cheek, mouth and nose are detected as shown in Fig. 3. After performing these operations, the extracted features are converted to gray image to perform canny edge calculation. Edges are significant local changes of intensity in an image. Color face image is converted into gray scale image to detect edges on face image as shown in Fig. 4. Gray scale image is processed into binary image with edges using canny edge detection technique as shown in Fig. 5, where edges consists of a set of white pixels.

Figure 1. Original face image.

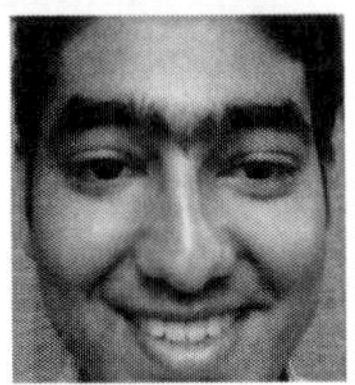

Figure 2. Cropped face image.

After that five features F1 to F5 are calculated depends of numbers of white pixels on individual face regions as shown in Table 1.

F1 = Number of white pixels of forehead area

F2 = Number of white pixels of nose area

F3 = Number of white pixels of left Cheek

F4 = Number of white pixels of right Cheek area

F5 = Number of white pixels of eyebrows area

3.3 *Training*

A training database is made which includes the five face features F1 to F5 of known expression images of smile, anger, disgust, and fear. Total 40 images of 4 different facial expressions of smile, anger, disgust, and fear are used as training dataset.

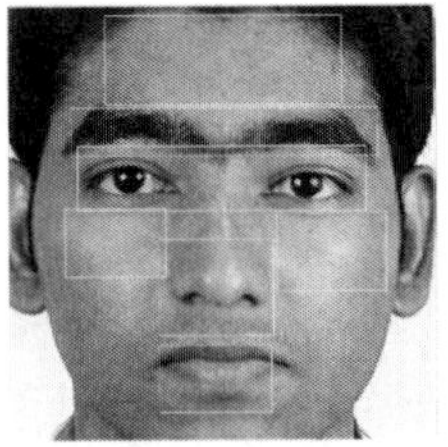

Figure 3. Detection of face regions.

Figure 4. Gray level face image.

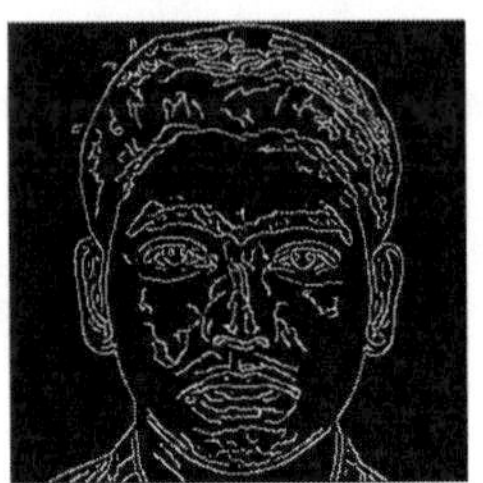

Figure 5. Binary image with edges.

Table 1. Features of face images.

Image	Type	No of White Pixels				
		F1	F2	F3	F4	F5
	Smile	1008	828	364	186	698
	Anger	376	484	292	369	426
	Disgust	542	589	197	182	606
	Fear	425	490	280	220	390
	Unknown	1008	759	327	277	685

Table 2. Result of FCM with clustering values.

Image	Type	FCM Membership Value			
		Cluster 1	Cluster 2	Cluster 3	Cluster 4
	Smile	0.00	0.00	0.99	0.01
	Anger	1.00	0.00	0.00	0.00
	Disgust	0.00	0.00	0.00	1.00
	Fear	0.00	1.00	0.00	0.00
	Unknown	0.00	0.00	0.99	0.01

Table 3. Experimented result.

Face type	No. of image	Correct	Correctness
DISGUST	15	12	80%
FEAR	15	11	73.3%
SMILE	15	13	86.7%
ANGER	15	12	80%
Total	60	48	80%

3.4 *Testing*

The FCM Clustering is applied on features F1, F2, F3, F4, F5 to get the maximum membership value. This will help to detect the emotional behavior of the unknown face image. The database is set for 4 kind of emotional behavior. So the total cluster is 4. The maximum member ship value is highlighted for each entry to detect the expression type of a cluster as shown in Table 2.

4 EXPERIMENTAL RESULT

The experiment has been done on 60 images of four different gestures like smile, fear, disgust and surprise. There are 15 face images for disgust, 15 face images for fear, 15 face images for smile and 15 face images for surprise. The experimented results are shown in Table 3. The overall accuracy is 80%, which is remarkable.

5 CONCLUSIONS

In this system initially the human face image is taken as input. Then face is cropped and necessary facial regions are detected using the viola-jones algorithm. After that face image is converted to gray scale image to perform edge detection using canny edge detection algorithm. Total five features F1 to F5 are extracted using total number of white pixels present on those facial regions. After that fuzzy c-means clustering has been applied on those features to detect the emotional behavior of human face. Fifteen face images of four different expressions are tested. So, out of sixty images forty eight images are correctly classified. It is seen that the performance of the system for all gestures are not same, some of them are found with less expectation. Hence the system is not fully optimized and further improvements are necessary. There are several areas in which the system can be improved. Extracting more features can improve the performance of the system.

REFERENCES

Bhatt, A. 2014. Possibility Fuzzy C-Means Clustering for Expression Invariant Face Recognition. *International Journal on Cybernetics & Informatics* 3(2). April 2014: 35–45.

Jana, R. & Datta, D. & Saha R. 2013. Age Group Estimation using Face Features. *International Journal of Engineering and Innovative Technology* 3(2). August 2013: 130–134.

Jana, R. & Datta, D. & Saha, R. 2014. Age Estimation from Face Image using Wrinkle Features. *International Conference on Information and Communication Technologies).* Kochi. India. December 2014: 1754–1761.

Kharatkar, A. & Shekokar, S. & Sable, N. & Lokhande, S. 2014. Face Expression and Analysis. *International Journal of Advanced Research in Computer Science and Software Engineering* 4(3). March 2014: 1302–1309.

Raheja, L. J. & Kumar, U. 2010. Human Facial Expression Detection from Detected in Captured Image using Back Propagation Neural Network. *International Journal of Computer Science & Information Technology* 2(1). February 2010: 116–123.

Saudagare, V. P. & Chaudhari, S. D. 2012. Facial Expression Recognition using Neural Network—An Overview. *International Journal of Soft Computing and Engineering* 2(1). March 2012: 224–227.

Shukla, N. & Kumar, A. 2013. Using Back-Propagation Recognition of Facial Expression. *Journal of Environmental Science, Computer Science and Engineering & Technology* 2(1). December 2012—February 2013: 39–45.

Su, C. M. & Yang, K. C. & Lin, C. S. & Huang, Y. D. & Hsieh, Z. Y. & Wang, C. P. 2013. An SOM-based Automatic Facial Expression Recognition System. *International Journal on Soft Computing, Artificial Intelligence and Applications* 2(4). August 2013: 45–57.

Valstar, F. M. & Mehu, M. & Jiang, B. & Pantic, M. & Scherer, K. 2012. Meta-Analysis of the First Facial Expression Recognition Challenge. *IEEE Transactions on Systems, Man, and Cybernetics—Part B: Cypernetics* 42(4). August 2012: 966–979.

Vasani, B. G. & Senjaliya, S. R. & Kathiriya, V. P. & Thesiya, J. A. & Joshi, H. H. 2013. Human Emotional State Recognition Using Facial Expression Detection. *International Journal of Engineering and Science* 2(2). January 2013: 42–44.

Computational Science and Engineering – Deyasi et al. (Eds)
© 2017 Taylor & Francis Group, ISBN 978-1-138-02983-5

Trajectory-based cooperative relaying Dense Wireless Sensor Networks (DWSNs) for better energy-efficiency

D. Saha
Department of ECE, RCC Institute of Information Technology (RCCIIT), Kolkata, India

ABSTRACT: For Dense Wireless Sensor Networks (DWSNs), this work proposes to harness micro-level wireless broadcast diversity using multiple cooperative relays within each hop of Trajectory Based Forwarding (TBF) – a popular source routing technique, where the source node, after determining the current best trajectory, embeds it in the packet header so that the forwarding nodes need not calculate the next hop. In the proposed protocol called Cooperative relay-based TBF (CBTF), although every forwarder knows the next hop from the packet header, it does not transmit directly to its designated neighbor. Instead it lowers its transmission power so as to send the packet to some intermediate relay nodes, lying close to the trajectory, which, in turn, transmit concurrently the packet to the next forwarder. This forces cooperative relaying between two adjacent forwarders, which avoid direct communication between them. The paper shows analytically that the proposed relay-based method in CTBF is more energy-efficient than the direct forwarding used in TBF.

1 INTRODUCTION

Dense Wireless Sensor Networks (DWSNs) are becoming commonplace now-a-days due to their numerous applications in factory floors, smart homes, healthcare and smart cities (Boukerche & Darehshoorzadeh, 2014). Typically, low power sensor nodes with limited communication ranges constitute DWSNs, whose energy-efficiency is a major concern for their intended longer lifetime. DWSNs typically employ source-routing techniques, such as Trajectory Based Forwarding (TBF) scheme (Niculescu & Nath, 2003), because they are easily scalable with the density of the network. Recent proposals (Boukerche & Darehshoorzadeh, 2014) also try to in still robustness and reliability into TBF by added opportunism to the original protocol so that packets are forwarded along multiple extra routes (Coutinho, Boukerche, Vieira, & Loureiro, 2016). But this increases delay as well as complexity.

This paper offers an alternate proposal here, which uses cooperative wireless broadcast diversity at the micro-level by utilizing some selected relay nodes within each hop of TBF. As an example, we have shown the use of three intermediate (aka in-hop) relays within each hop of the trajectory T<$\cdots v_{i-1}, v_i, v_{i+1}, \cdots\cdots$> in Fig. 1. Each forwarding node v_i sends the packet to all the three intermediate relays, which, in turn, cooperatively forward it to the next forwarder v_{i+1}. Obviously, this reduces the transmission power of the forwarder nodes by some fraction. Additionally, the relays coordinate their transmission in such a way that the received signals at v_{i+1} from the relays are coherently combined (i.e., signals from all relays arrive at v_{i+1} in phase) with proper coordination (Khandani, Abounadi, Modiano, & Zheng, 2007). Consequently, even if we employ multiple relays, their transmission energies do not add up commensurately, giving us some gain in overall energy efficiency. This work shows that the proposal of using relays within each hop reduces the energy consumption considerably. Moreover, this helps the DWSN combat the negative impact of fading and other channel degradations prevalent in indoor environments. Obviously, participation of relays in forwarding the data brings in local redundancy at the hop level, thereby improving the reliability to some extent at the glo-

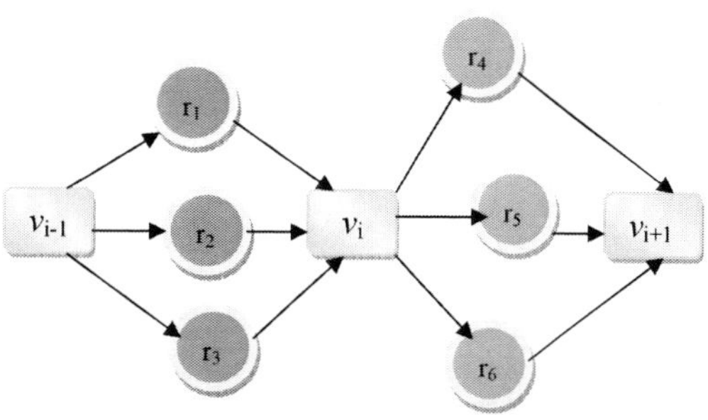

Figure 1. Use of in-hop relays between every pair of forwarders in a trajectory T<$\cdots v_{i-1}, v_i, v_{i+1}, \cdots\cdots$>.

bal (i.e., trajectory) level. Use of in-hop relays is appropriate for DWSNs because the higher the density, the higher the likelihood of finding more relays within each hop.

The paper is organized as follows. Section II briefly presents the background and the model used. Section III analyzes CTBF mathematically and presents relevant comparative results. Section IV concludes the paper.

2 BACKGROUND AND MODEL

This work considers an ad-hoc DWSN with a distributed topology. The sensor nodes are mostly active having the capability to sense and route data. A synchronization algorithm (Xie, Cui, & Lao, 2006) maintains connectivity among the active nodes, while effecting a more-or-less uniform energy distribution among them. If a node loses sensing capability due to some malfunction or damage so that it can take part in communication only, we consider them as passive nodes in the DWSN.

Source routing-based TBF or its extensions like Vector Based Forwarding (VBF) (Xie, Cui, & Lao, 2006) are popular routing techniques for such DWSNs. In source routing, the source node uses a suitable method (Catanuto, Toumpis, & Morabito, 2007) to pre-calculate the complete route, called trajectory T, consisting of only active nodes, and puts it in the packet header. Pre selection of an optimal T (known as source routing) is preferable to computing the next hop locally, based on available choices at every node (known as destination routing), because the latter approach increases delay as well as entails in energy depletion due to repetitive computations.

Assuming all hop distances to be nearly equal in a trajectory T whose complete length is L, the hop count in T turns out to be approximately $^{L}/_{x}$(as $L>>x$). Consequently, the total energy E_{total} expended in routing a packet over T is given by:

$$E_{total} = E_{hop} \times (^{L}/_{x}) \tag{1}$$

where E_{hop} is the energy spent in each hop. Since we have considered the hop distances to be equal, E_{hop} is also the same for every hop.

Let $V = \{v_1, v_2, \ldots\ldots, v_N\}$ be the set of active nodes and $P = \{p_1, p_2, \ldots\ldots, p_K\}$ be the set of passive nodes in DWSN. Thus, the potential set of relay nodes $R = \{r_1, r_2, \ldots\ldots, r_M\} = \{V \cup P\}$, M = N+K. Note that when a source node calculates the optimal T, it takes into account the members from V only. For example, Figure 2 considers the trajectory $T< \cdots v_{i-1}, v_i, v_{i+1}, \cdots >$ that consists of elements of V only. Let $\Psi(t_i)$ denote the maximum transmission range

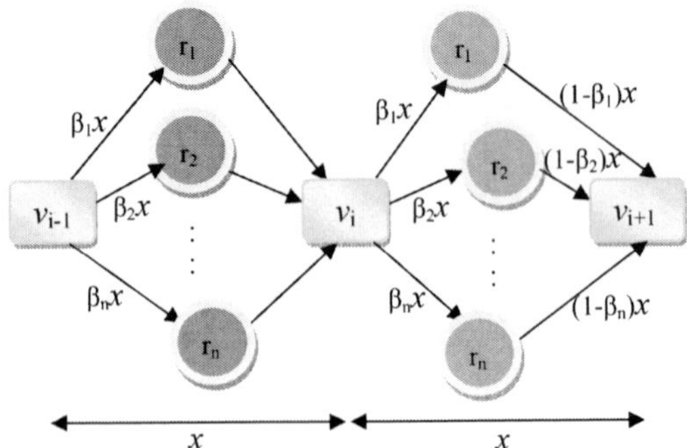

Figure 2. Illustration of n number of in-hop relays within each single hop.

of a particular node t_i, ($t_i \in$ R); $\Psi(t_i)$ is a function of the remaining battery power of t_i. The energy model for transmission at a distance z by any t_i is given by (Catanuto, Toumpis, & Morabito, 2007):

$$E(z) = az^b + c_{TR} \tag{2}$$

where a is the received signal energy at a distance z from $t_i(z \le \Psi(t_i))$, b is the path loss exponent ($2 \le b \le 6$) that depends on the wireless medium around t_i (Niculescu & Nath, 2003), and c_{TR} is the energy spent in transmitter and receiver electronics. Here, $c_{TR} = c_T + c_R$, where c_T and c_R are the energy consumed in transmitter electronics and the receiver electronics, respectively. Compared to az^b and c_{TR}, we consider the processing energy to be negligible; so we ignore it. Thus, for TBF with average hop distance x, we can derive from (2) the energy needed per hop as:

$$E_{hop}^{TBF} = E(x) = ax^b + c_{TR} \tag{3}$$

Substituting the above expression in (1), we obtain the total trajectory energy in TBF as:

$$E_{total}^{TBF} = E_{hop}^{TBF} \times \frac{L}{x} = (ax^b + c_{TR}) \times \frac{L}{x} \tag{4}$$

We next calculate the energy required in the proposed *cooperative* relay-based *TBF* approach (called CTBF in this work), and compare it with the above energy needed in original TBF approach.

3 COMPARATIVE ANALYSIS AND RESULTS

We concentrate on a single hop of CTBF first and find out the energy required. As discussed earlier,

CTBF enhances the TBF approach by inserting n number of in-hop relay nodes (say $r_1, r_2, \ldots, r_n$ where $n \leq M$) in between every adjacent pair of forwarders (say v_i and v_{i+1}) such that each r_j ($1 \leq j \leq n$) accepts packet from v_i and relays the same to v_{i+1} (Fig. 2).

Every intermediate node cannot become a relay node because, to be are lay node, r_j must satisfy the following criteria: (i) $r_j \in R$, (ii) its distance from v_i should be about $\beta_j x$, where $0 \leq \beta_j \leq 1$, (iii) its distance from v_{i+1} should be about $(1-\beta_j)x$, and (iv) $(1-\beta_j)\, x \leq \Psi(r_j)$ and $\beta_j x \leq \Psi(v_i)$. Conditions (ii) and (iii) restrict the relay nodes to lie in between v_i and v_{i+1} around T. We may find a suitable β_j in order to get sufficient number of *relays* within each hop. Note that, since transmission by v_i is broadcast in nature, all relay nodes within the range $\Psi(v_i)$ hears it; but most of them cannot participate in cooperative relaying because they do not satisfy the aforementioned conditions (ii) and (iii).

Thus, hop energy for CTBF comprises energy required for two sub-hop communications: (i) wireless broadcast from forwarder v_i to set of n relays $\{r_1, r_2, \ldots, r_n\}$, and (ii) synchronized additive many-to-one communication from n relays $\{r_1, r_2, \ldots, r_n\}$ to next forwarder v_{i+1} (Khandani, Abounadi, Modiano, & Zheng, 2007). If $E_{i,n}$ indicates the energy for the first sub-hop communication, and $E_{n,i+1}$ represents that for the second part, we may write:

$$E_{hop}^{CTBF} = E_{i,n} + E_{n,i+1} \tag{5}$$

For the sake of brevity, we present here only the final results as derived in (Khandani, Abounadi, Modiano, & Zheng, 2007) for the calculation of each sub-hop energy component.

$$E_{i,n} = a \times \left[\max\left((\beta_1 x)^b, (\beta_2 x)^b . (\beta_n x)^b\right)\right] + (c_T + nc_R) \tag{6}$$

and

$$E_{n,i+1} = a \times \left(\frac{1}{\sum_{i=1}^{n} \frac{1}{((1-\beta_i)x)^b}} \right) + (nc_T + c_R) \tag{7}$$

If the expected value of the distances of the n relays from v_i be βx, then the expected value of the distances of the n relays from v_{i+1} becomes $(1-\beta)x$. We may then simplify (6) and (7) as (Khandani, Abounadi, Modiano, & Zheng, 2007):

$$E_{i,n} = a(\beta x)^b + (c_T + nc_R) \tag{8}$$

and

$$E_{n,i+1} = a \times \left(\frac{1}{\frac{n}{((1-\beta)x)^b}} \right) + (nc_T + c_R) \tag{9}$$

Putting the above two expressions in (5), we arrive at:

$$E_{hop}^{CTBF} = a(\beta x)^b + a\frac{((1-\beta)x)^b}{n} + (n+1)c_{TR} \tag{10}$$

Substituting the above expression in (1), we obtain the total trajectory energy in CTBF as:

$$E_{total}^{CTBF} = E_{hop}^{CTBF} \times \frac{L}{x}$$
$$= \left[a(\beta x)^b + a\frac{((1-\beta)x)^b}{n} + (n+1)c_{TR} \right] \times \frac{L}{x} \tag{11}$$

An interesting observation is that, when $\beta = 1$ and $n \to 0$, equation (11) reduces to equation (4), which implies that CTBF is a general case of TBF. When $\beta = 0$ and $n \geq 1$, CTBF reduces to opportunistic routing (Boukerche & Darehshoorzadeh, 2014).

Since the number of hops is identical for both TBF and CTBF, it is sufficient to compare their hop energies (Fig. 3) using equations (3) and (10), respectively, in order to get an idea about overall energy consumption. To evaluate equations (3) and (10), we initially assume $a = 1.0$ pJ, $b = 3$, $c_{TR} = 0.04$ pJ, $\beta = 0.5$ and $n = 3$ (Boukerche & Darehshoorzadeh, 2014). It is evident from Fig. 3 that, as hop distance x increases, the gap between hop energies for TBF and CTBF increases considerably. Except-

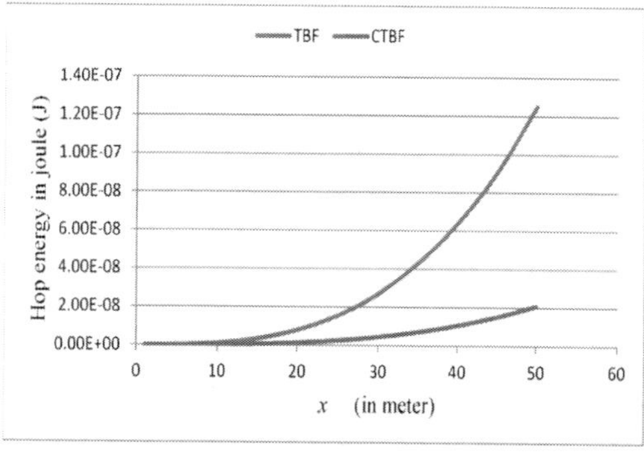

Figure 3. Comparison of single hop energy between TBF and CTBF ($a = 1.0$ pJ, $b = 3$, $c_T = 0.3$ pJ, $c_R = 0.1$ pJ, $n = 3$, $\beta = 0.5$).

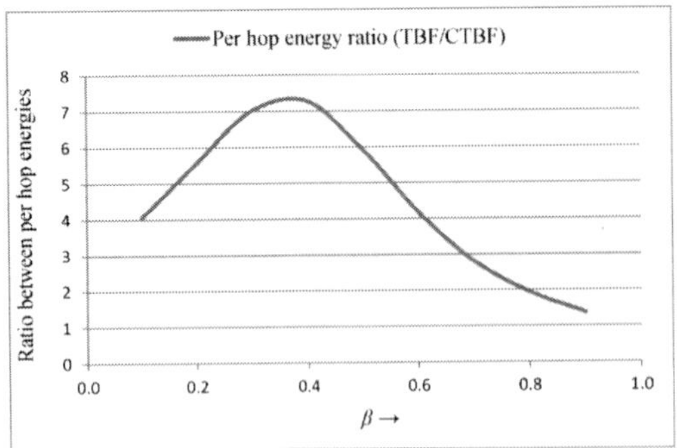

Figure 4. Ratio of hop energies versus β (a = 1.0 pJ, b = 3, x = 10 m, c_T = 0.3 pJ, c_R = 0.1 pJ, n = 3).

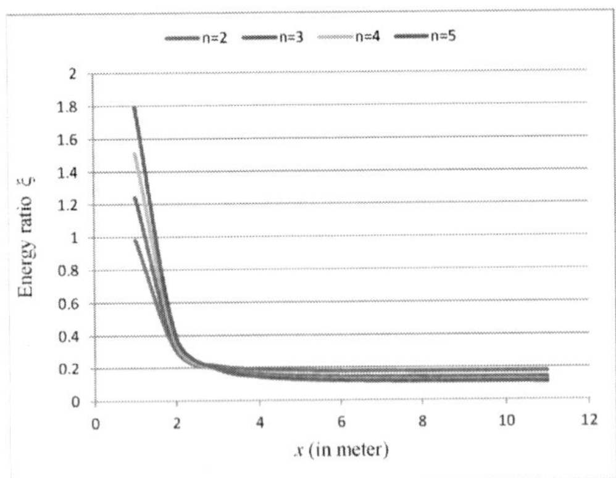

Figure 5. Energy ratio ξ versus hop distance for different values of n(a = 1.0 pJ, b = 3, β = 0.4, c_T = 0.3 pJ, c_R = 0.1 pJ).

ing for very small values of x (such as 1 m), hop energy in CTBF is always less than that in TBF. This clearly shows the superiority of CTBF over TBF in terms of energy-efficiency.

To compare the energy values further, Fig. 4 plots the ratio of equation (3) to equation (10) against β. It clearly indicates that hop energy in CTBF is always less than that in TBF for all values of β, which further substantiates our claim that CTBF is better than TBF in respect of energy usage. Mathematically speaking, hop energy for TBF does not depend on β; it is constant (as x is fixed at 10 m). The hop energy for CTBF, however, varies with β for the same x. Hence, their ratio in Fig. 4 takes a convex shape to imply that there is an optimal value of β around 0.4 (for the given setting). So, henceforth, future plots will use β = 0.4, whenever x = 10 m.

Let us now introduce a comparative measure of energy. We define ξ as the ratio of total CTBF energy to total TBF energy.

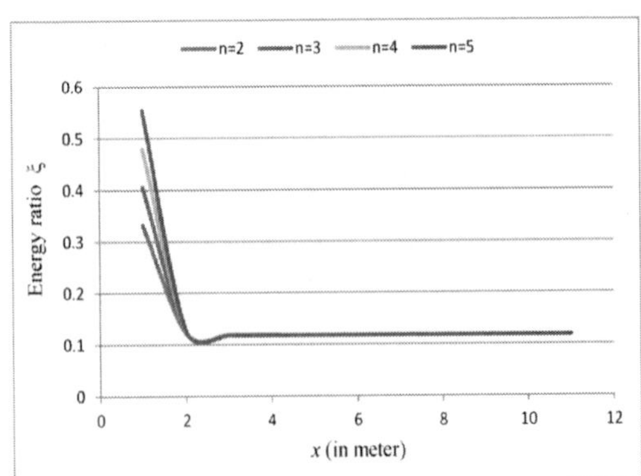

Figure 6. Energy ratio ξ versus hop distance for different values of n in indoor environment with higher received signal strength (a = 5.0 pJ, b = 6, β = 0.7, c_T = 0.3 pJ, c_R = 0.1 pJ).

$$\xi = E_{total}^{CTBF} \Big/ E_{total}^{TBF} = \left[\frac{ax^b + c_{TR}}{a(\beta x)^b + a\left(\frac{((1-\beta)x)^b}{n}\right) + (n+1)c_{TR}} \right] \tag{12}$$

Obviously, the smaller the value of ξ, the better the energy-efficiency. The plot of ξ against hop distance x for various values of n in Fig. 5 helps designers judge the impact of the number of relays on energy values. Fig. 5 shows that ξ saturates quickly as soon as x crosses 2 m, which is a small distance.

Next, in order to check the behavior indoor, we set b = 6 and a = 5 pJ because interferences are more indoor and path loss is also more (Catanuto, Toumpis, & Morabito, 2007). The plot of ξ against hop distance x for various values of n in Fig. 6 reveals a similar pattern as that is observed in Fig. 5. So, for all practical purposes, we can conclude that 2 or 3 relays are enough for CTBF. For a moderately dense network, finding 2 relays per hop is not a difficult proposition at all. Figure 5 also indicates that if the hop distance is very small—of the order of a few meters – it is not wise to go for CTBF because ξ is more than unity in that case.

4 CONCLUSION

Recently, DWSNs have emerged as cost-effective infrastructure to support large-scale, dense, networked systems such as Internet of Things. The sensor nodes sense relevant data from environment and then forward the aggregated data to the sink or base station using some routing method. Source routing like TBF is very common for DWSNs to limit energy consumption. Along that line, this paper shows that CTBF utilizes cooperative wireless diversity in-hop to increases energy-efficiency further. This minor modification is useful in increasing network lifetime in WSN applications utilizing low-power relays (Boukerche & Darehshoorzadeh, 2014).

Moreover, CTBF increases per-hop reliability too because, due to redundancy, even if some relays are inoperative, others will continue with relaying, and the receiving node will not find it difficult to recover the packet. Finally, in CTBF, every hop is effectively point-to-point by virtue of cascading a point-to-multipoint communication with a matching multipoint-to-point communication. This restricts the fan-out to in-hop relays only, keeping the flooding of packets in check.

REFERENCES

Boukerche, A., & Darehshoorzadeh, A. (2014). Opportunistic Routing in Wireless Networks: Models, Algorithms, and Classifications. *ACM Comp. Surveys, vol. 47, no. 2*, 22:1–22:36.

Catanuto, R., Toumpis, S., & Morabito, G. (2007). Opti{c,m}al: Optical/Optimal Routing in Massively Dense Wireless Networks. *Proc. IEEE INFOCOM 2007*. IEEE.

Coutinho, R. W., Boukerche, A., Vieira, L. F., & Loureiro, A. A. (2016). Design Guidelines for Opportunistic Routing in Underwater Networks. *IEEE Communications Magazine, vol. 54, no. 2*, 40–48.

Khandani, A. E., Abounadi, J., Modiano, E., & Zheng, L. (2007). Cooperative Routing in Static Wireless Networks. *IEEE Trans Commun*, 2185–2192.

Niculescu, D., & Nath, B. (2003). Trajectory based forwarding and its applications. *Proc. ACM Mobicom 2003* (pp. 260–272). ACM.

Xie, P., Cui, J.-H., & Lao, L. (2006). VBF: Vector-Based Forwarding Protocol for Underwater Sensor Networks. *Proc. 5th Int'l. IFIP-TC6 Networking Conf* (pp. 1216–1221). IFIP.

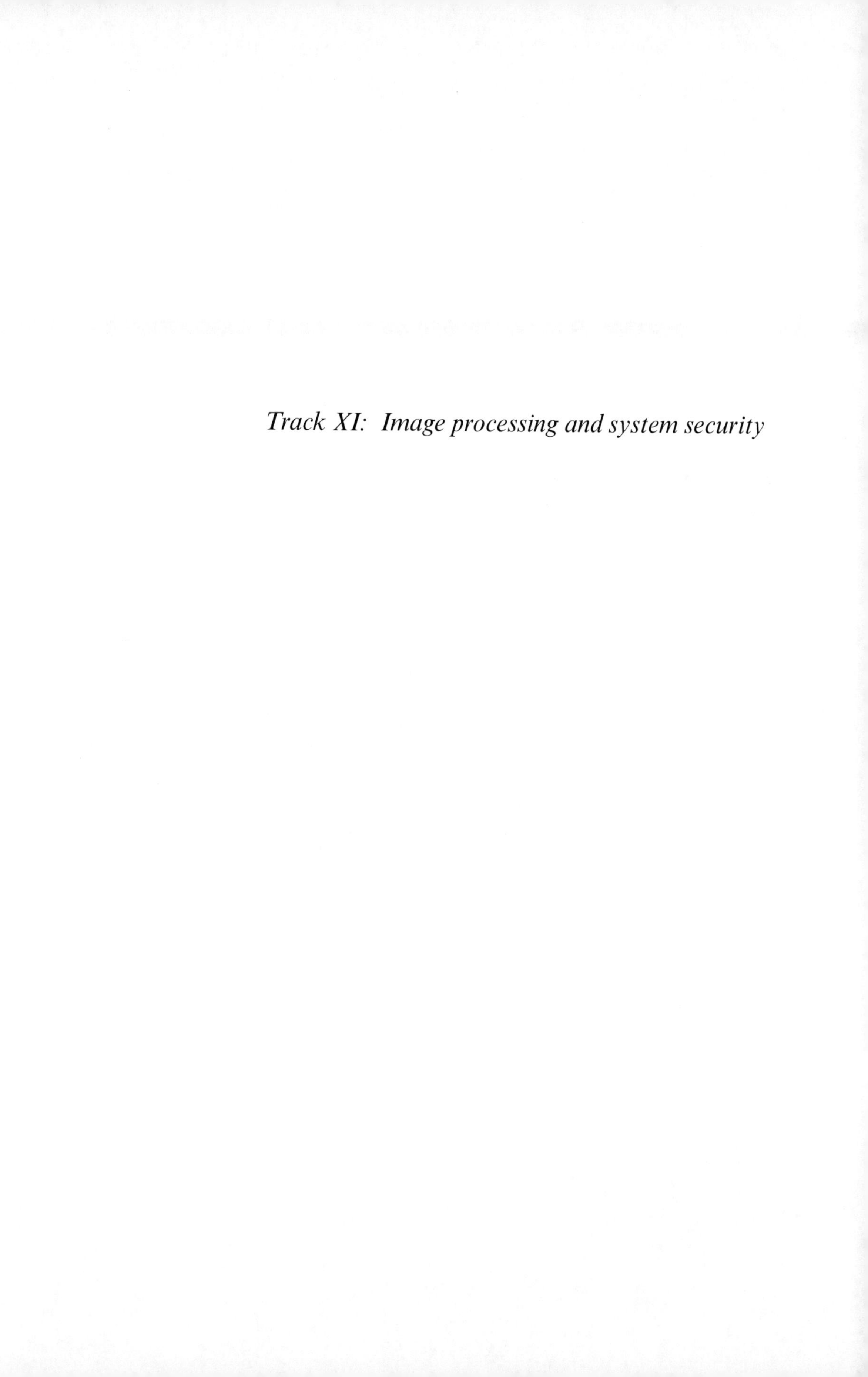

Track XI: Image processing and system security

Computational Science and Engineering – Deyasi et al. (Eds)
© 2017 Taylor & Francis Group, ISBN 978-1-138-02983-5

Plant flower recognition using image texture analysis

Soumen Mukherjee, Arup Kumar Bhattacharjee, Paromita Banerjee,
Suman Talukdar & Sudipta Paul
RCC Institute of Information Technology (RCCIIT), Kolkata, India

ABSTRACT: This paper is based on a small application which is used to recognize an unknown flower species. The application gathers images of different types of flowers to build the repository and test the unknown image for identification. The whole process involves various steps like accession, preprocessing, training and testing. The whole process is done using 3 varieties of flowers with around 90 images and with different image features like morphological features, GLCM and color features.

1 INTRODUCTION

Recognition of flowers using soft computing tools is very challenging topics because of large variety of flower species are available (Jain 1989). It is difficult to extract different types of features of all the flowers that are present in the world. Computer aided flower identification is a very useful tools for plant species identification aspect (Leaf and Flower Recognition System). To handle such volumes of information of the flowers, development of a quick and efficient classification method is very important so that it can be used in different aspects such as educational fields, and in various medicinal purposes etc. A number of works is still undergoing in this domain. So in that respect we propose to prepare a plant flower identification application using soft computing tools. Creating the database of flower images involves steps like gathering of the different types of flowers image followed by few basic level of preprocessing for standardization. The known samples are tested with preprocessed known images and results are generated.

2 OUR WORK

The proposed application takes an unknown image from the users and compares the image with different flower's image kept in the repository. This comparison is consist of 4 steps. The first step is gathering of the different types of flowers image. This step is called accession. In image accession the objective is taking an image of unknown flower by a digital camera and the backgrounds of all the images are white.

The 2ndstep is image preprocessing. In image-preprocessing there are several preprocessing technique which helps in features extraction of an image. Different image preprocessing techniques are 1) Resize 2) RGB to gray conversion 3) Gray to binary conversion. The 3rdstep is image recognition where the system extract the texture feature using Gray Level Co-Occurrence Matrix (GLCM) and the in final step the system display the results.

3 NEURAL NETWORK

Neural networks are consisting of one or more layers of interconnected nodes. These individual nodes are known as perceptrons. In a Multi Layered Perceptron (MLP) perceptrons are arranged into layers and layers are connected with other another. In the MLP there are three types of layers namely, the input layer, hidden layer(s), and the output layer. The objective of the neural network is to minimize some measure of error. In Neural networks, samples with known features are used for training purposes. Usually training data set includes a number of cases, each containing values for a range of input and output variables and generate different classes. This operation requires few decisions like: which samples to use, and how many (and which) samples to gather. After training, the purpose of the network is to assign each test (unknown) sample to one of the number of class generated using training process (classification).

4 METHODOLOGY

4.1 *Image accession*

The recognition process start with collecting different type of flowers and putting all of them on white background. Clicking all those flowers by digital camera from a same distance and storing all these images into the system database (Figure 1).

4.2 *Image processing*

This is the most important part in recognizing a flower image. This module has 3 sub modules.

4.2.1 *Resize the flower image*

In this module all flower images are resized in a same size. It resizes the length and width of a flower (Figure 2).

4.2.2 *RGB to gray conversion*

After resizing of each flower image this module convert the flower image to gray scale color from the RGB color image (Figure 3).

4.2.3 *Grey to binary*

This module convert the gray color image to binary image (Figure 4).

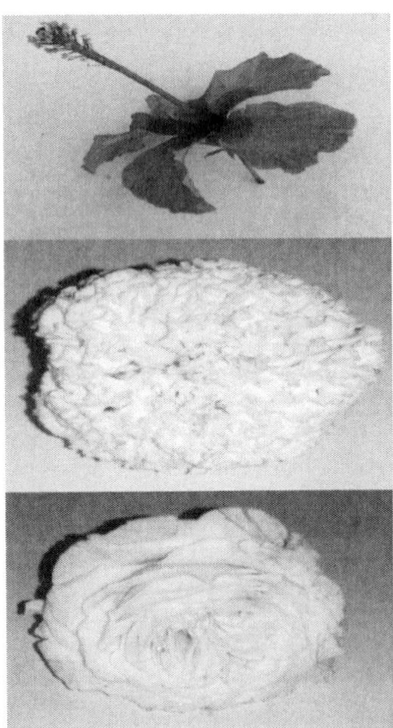

Figure 1. Images with white background.

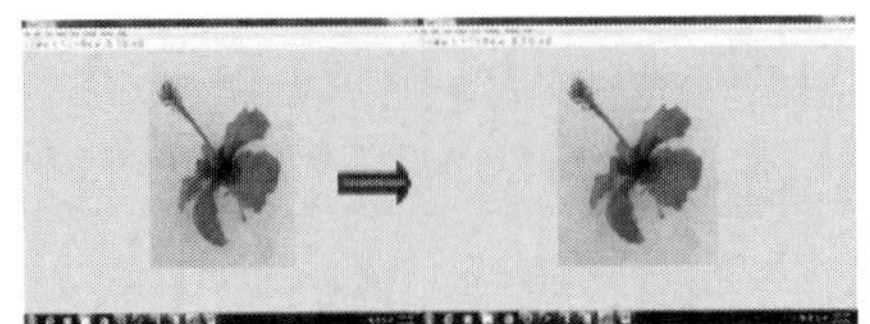

Figure 2. After applying the resize operations.

4.3 *Image recognition*

In this phase some of the features of a flower image are extracted.

This sub module extracts 3 main features of a flower. The features have the following details.

4.3.1 *Area*

It finds the area of a particular image and area of different image by using the MATLAB functions.

4.3.2 *GLCM*

The GLCM (Gray-Level Co-occurrence Matrix) actually contains information of the frequency of occurrence of two neighboring pixel combination in an image. Although 22 features can be derived from GLCM, usually only 5 are considered as parameters of importance: Contrast, Homogeneity, Dissimilarity, Energy and Entropy. By calculating this five texture features it is possible to see how they behave for different textures.

4.3.3 *Color feature*

This sub module find the color feature of different flower image.

5 RESULT

In our system we have used Percpeptron (a type of neural network) to classify two different images from a set of 90 images. In this below graph we can see that one line is dividing the graph into 2 portions, the line in the following Fig. 5 is an example of the classification phase, in which two different type of flower are classified. The bubble is representing the Marigold and the positive sign is representing the China rose

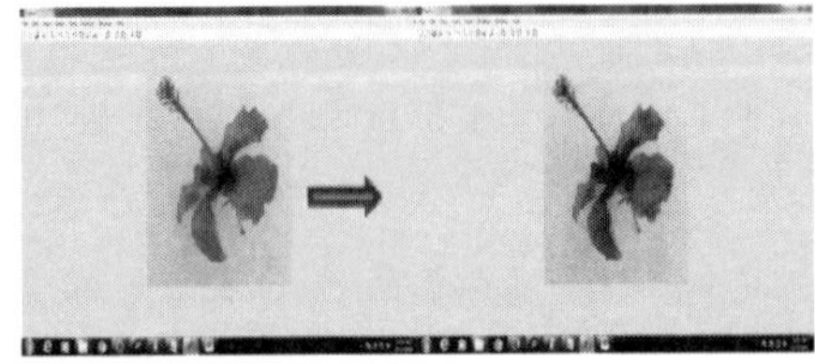

Figure 3. After performing the black and white conversion operations.

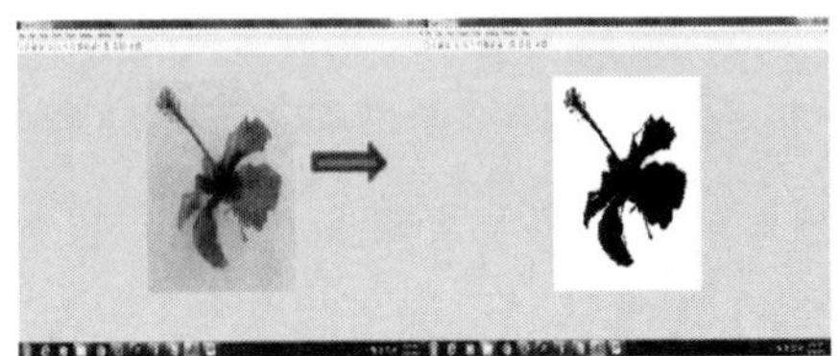

Figure 4. Binary conversion.

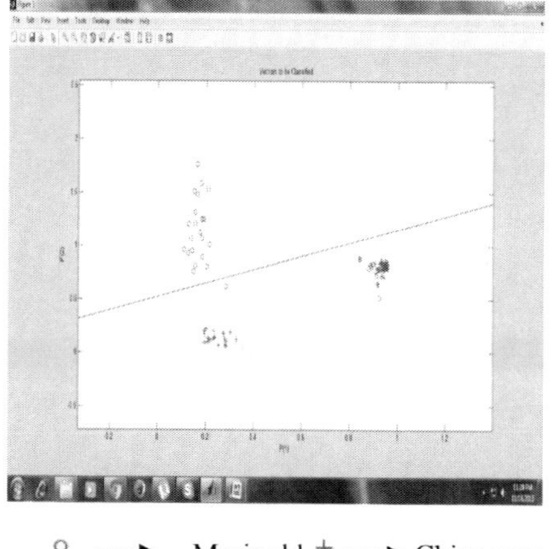

o → Marigold + → China rose

Figure 5. Graphical representation of GLCM of Marigold vs Chinarose.

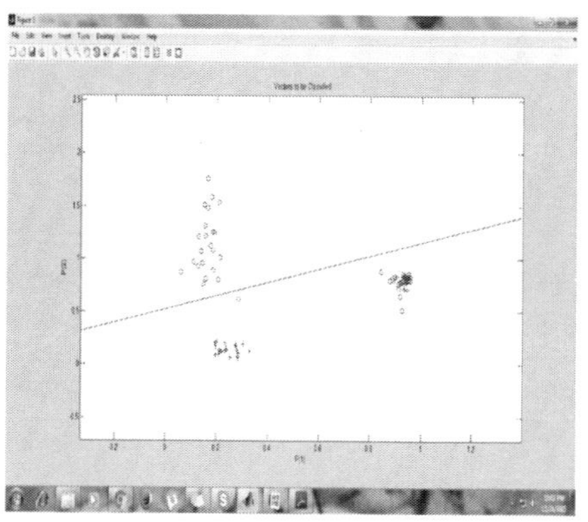

o → Mariegold flower classified by perceptron

Figure 6. Graphical representation of testing set.

The next graph (Fig. 6) is the example of the testing sets which classifies and recognizes the type of the flower. In this graph we can see that one red bubble (Fig. 6), which representing that the flower type is Marigold.

6 FUTURE SCOPE

The system is an initial stage and it comes with certain limitation like it requires taking a clear picture of a flower without trimming off a flower. As well as percentage of recognize is very small since extraction of all the features of a flower is very difficult. It needs more time to train a huge amount of flower data set used for recognizing a flower.

7 CONCLUSION

The system is able to fulfill the research objective by extracting four main features of a flower and recognizing it. The system trains three kinds of flowers with 90 images in the system database. The system tests its performance on three kinds of flowers with 60 images for a training data set. Moreover the system tests two kinds of flowers within 30 images for an un-training data set. The precision rate is 75.19 percent for the training dataset and 76 percent for the un-training data set.

REFERENCES

Amadasun, M. and King, R. (1989) 'Textural features Corresponding to textural properties', IEEE Transactions on Systems, Man, and Cybernetics, vol. 19, no. 5, pp. 1264–1274.

Clausi D A. An analysis of co-occurrence texture statistics as a function of grey level quantization. Canadian Journal of Remote Sensing. 2002. Vol. 28(1). P. 45–62

Devendran V et. al., "Texture based Scene Categorization using Artificial Neural Networks and Support Vector Machines: A Comparative Study," ICGST-GVIP, Vol. 8, Issue IV, pp. 45–52, December 2008

Gonzalez, R. and Woods, R. (2002). Digital Image Processing. 3rd Edn., Prentice Hal Publications, pp. 50–51

Haralick R. M., Shanmugam K., Dinstein I. Textural Features of Image Classification. IEEE Transactions on Systems, Man and Cybernetics. 1973. Vol. 3(6). P. 610–621

Haralick, R.M. (1979) 'Statistical and Structural Approaches to Texture', Proceedings of the IEEE, Vol. 67, No. 5, pp. 786–804.

Haralick, R.M., Shanmugan, K.. and Dinstein, I. (1973) 'Textural Features for Image Classification', IEEE Tr. on Systems, Man, and Cybernetics, Vol SMC-3, No. 6, pp. 610–621.

Jain Anil K. (1989) Fundamentals of Digital Image Processing: Prentice Hall.

Jensen. 1996. *Introductory Digital Image Processing.* (Upper Saddle River, NJ: Prentice Hall). Ed. 2. Ch. 2 (60–61), Ch. 3, and Ch. 4

Jong Kook Kim, JeongMi Park, KounSik Song and Hyun Wook Park "Texture Analysis and Artificial Neural Network for Detection of Clustered Micro calcifications on Mammograms" IEEE, pp.199–206, 1997.

Leaf and Flower Recognition System (e-Botanist)

Torheim G., Godtliebsen F., Axelson D., Kvistad K. A., Haraldseth, O. and Rinck P. A., "Feature extraction and classification of dynamic contrast-enhanced T2*-weighted breast image data," IEEE Transactions on Medical Imaging, vol. 20, pp. 1293–301, 2001.

Computational Science and Engineering – Deyasi et al. (Eds)
© 2017 Taylor & Francis Group, ISBN 978-1-138-02983-5

Plant leaf image recognition and classification using perceptron

Soumen Mukherjee, Arup Kumar Bhattacharjee, Manas Ghosh, Trishita Ganguly, Rick Punyadyuti Sinha & Swarup Mandal
RCC Institute of Information Technology, Kolkata, India

ABSTRACT: It is a challenging task for any lay man to analyze plant leaf images because of the many minute variations that exist in them, hence the need for a large data set for analysis. It is a difficult task to develop an automated recognition system which is able to process large information and provide correct estimations and figures. The result of the recognition process categorizes each individual plant sample to a descending series of related plants in terms of their characteristics. It is very time consuming. The focus of the computerized plant recognition system is on the detection and extraction of stable features of the plants. The information extracted from leaf veins, thus, play an important role in the process of identification. The goal of this work is to develop a system where the user can input a picture of any unknown plant leaf and the system classify the plant species in question and display sample images of the nearby matches. Artificial neural networks have applied to this type of problems in pattern recognition, classification and image analysis.

1 INTRODUCTION

Plants play an integral role in the sustenance of life on Earth. They provide us with certain important factors that affect our survival in the form of oxygen, food, medicine, fuel and much more [Chaki, Parekh]. It is imperative to have a proper understanding of plants to increase productivity in agriculture and make it sustainable. The human population is ever increasing, and coupled with a varying climate, there is a constant threat to the ecosystems. Therefore, it is very much essential to be able to identify new or rare species of plants and to observe their manifold characteristics and geographical extent as part of wider biodiversity projects [Chaki, Parekh]. The most popular choice in plant classification is done based on leaf images and their proper identification. Sampling leaves and photographing them is quite easy and cost effective. We can then transfer those images into a computer and then run various image processing techniques on them to eventually recognize the plant using machine learning techniques.

There are lots of kinds of plant species that exist on Earth [Chaki, Parekh]. Plants play an important role in human life and lives of other species on the planet. Due to many serious reasons including global warming, deforestation etc. these plant species are on the decline. There is an imperative need now, hence, to save these plant species from the point of extinction. Fortunately, Botanists are recognizing this need to save plants and are devising new methods to protect them [Chaki, Parekh]. In a same fashion, there is a need to identify a plant by its category to help these agriculturists and farmers to recognize a plant for what it is in order to help preserve it. To help them in their venture, several researches are underway to help them classify plants on just looking at a photo of a leaf or a flower of a particular plant. Here, we deal with the leaf identification only [Gupta].

Automatic plant leaf identification can be useful in a wide variety of applications, such as environmental protection, plant resource survey, as well as for educational purposes. Creating a plant database is an important step towards environmental conservation. Certain features of advanced Information Technology, Image Processing and Plant Leaf database creation helps make this process much more efficient and accurate. Designing a convenient and efficient identification system is necessary and important because it can facilitate fast classification of plants and understanding and managing them [Hsiao].

This project aims at developing a software system that, when fed with an image of a plant leaf, will be able to identify the name of the plant with maximum accuracy. The use of Information Technology along with various Image Processing techniques and database creation will enable us

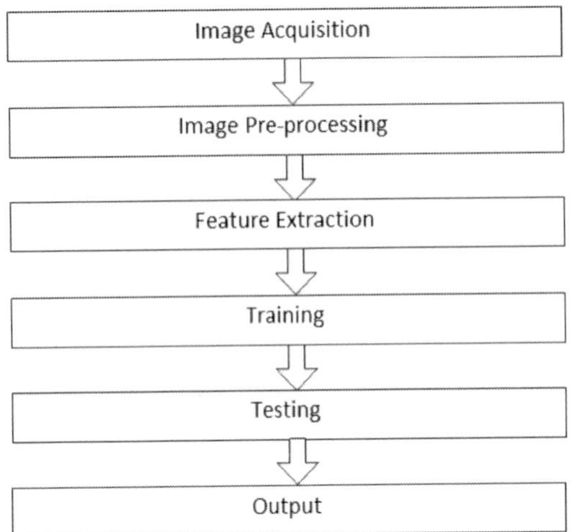

Figure 1. Flowchart for the proposed system of leaf recognition [Gupta].

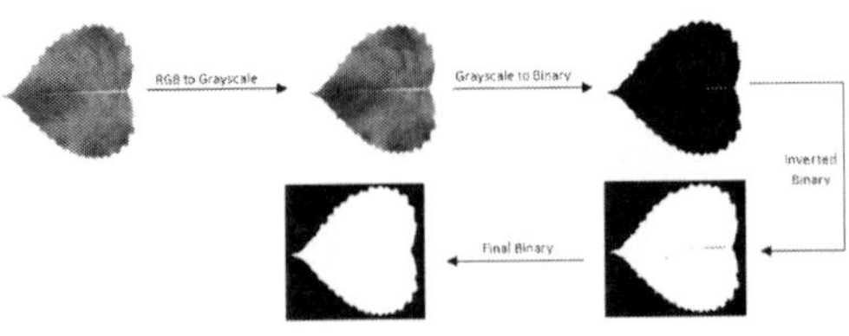

Figure 2. Image pre-processing stages.

to achieve an outcome that will provide us with a highly accurate result, that we hope can be up to the standards required for its practical application [Hsiao].

2 PROPOSED SYSTEM

In this paper system implements a plant leaf recognition algorithm by using easy to extract features like shape and texture features [Kebapci, Mzoughi, Mouine]. The learning algorithm used here is Perceptron which is a binary classifier. The following diagram shows a Flowchart of the proposed technique.

For this project we have used a dataset named Flavia, downloaded from http://flavia.sourceforge.net/.

3 IMAGE PRE-PROCESSING

The leaf image pre-processing means to the primary processing of input image to correct the geometric distortions calibrate the data and eliminate the noise and error that are present in the data.

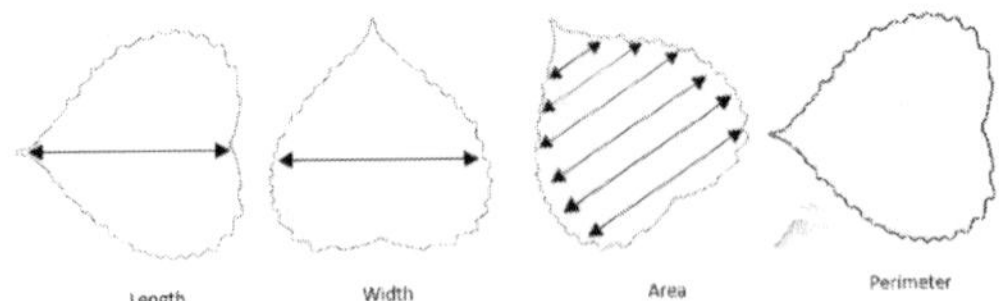

Figure 3. Shape features.

4 FEATURE EXTRACTION

4.1 *Shape features*

- Leaf Area: It is the number of pixels of binary value 1 on smoothed leaf image.
- Leaf Perimeter: It is the number of pixels along the closed contour or border of the leaf.
- Physiological Length: It is the number of pixels present in the major axis of the leaf.
- Physiological Width: It is the number of pixels present in the minor axis of the leaf.[Satti]

4.2 *Texture features*

- Contrast: This feature returns a measure of the intensity contrast between a pixel and its neighbor pixels over the whole image.

$$\sum_{i,j} |i-j|^2 p(i,j)$$

- Correlation: This feature returns a measure of how correlated a pixel is to its neighbor pixels over the whole image.

$$\sum_{i,j} \frac{(i-\mu i)(j-\mu j)\, p(i,j)}{\sigma_i \sigma_j}$$

- Energy: This feature returns the sum of squared elements in the GLCM gray-level co-occurrence matrix.

$$\sum_{i,j} p(i,j)^2$$

- Homogeneity: This feature returns a value that measures the closeness of the distribution of elements in the GLCM to the GLCM diagonal [Kulkarni]

$$\sum_{i,j} \frac{p(i,j)}{1+|i-j|}$$

5 EXPERIMENTAL RESULTS

Table 1. Physiological features of Chinese horse chestnut.

Serial no.	Area	Perimeter	Length	Width
1.	34513	907.394011026843	344.531009341795	128.490019095019
2.	36801	926.690475583125	341.423811911017	138.107124317821
3.	40497	961.945309579064	358.563738148172	144.875308447879
4.	49046	1095.82756057297	405.983893349434	154.717351582182
5.	51337	1097.14126907196	396.148075117630	166.245327427782
6.	47720	1067.14126907196	401.547060695648	152.071968855282
7.	48792	1044.17070632348	375.865278925007	166.396746478218
8.	49854	1075.24177413535	391.299297040993	163.949497010983
9.	44174	1018.12907576314	374.638313269977	151.191530001936
10.	45112	1030.87424176720	372.152734954420	156.384409415273

Table 2. Physiological features of Chinese Redbud.

Serial no.	Area	Perimeter	Length	Width
1.	58939	964.472221513645	285.740196099259	267.979334173589
2.	54023	938.028570699749	276.748390593410	254.424157326613
3.	56905	929.259018078049	279.877433459147	262.059819299049
4.	55974	967.584919885853	276.321001926541	265.927475147931
5.	56050	943.986940139406	275.867853436067	265.842642119292
6.	54665	922.572726577033	273.589159004449	258.416025740081
7.	49896	886.028570699749	261.273600012155	249.259000525760
8.	56500	957.986940139407	272.670390281482	268.939365800934
9.	46936	856.874241767198	262.998547910948	231.300067593887
10.	57618	929.401153701779	288.971656724895	257.476007040440

6 COMPARISON WITH OTHER SIMILAR SYTEMS

Table 3. Comparison with accuracy of other similar systems.

Author	Paper	Accuracy (in%)
Stephen Gang Wu, Forrest Sheng Bao, Eric You Xu, Yu-Xuan Wang, Yi-Fan Chang and Qiao-Liang Xiang	A Leaf Recognition Algorithm for Plant Classification Using Probabilistic Neural Network	90.312
X. Gu, J.-X. Du, and X.F. Wang	Leaf recognition based on the combination of wavelet transform and Gaussian Interpolation	93
Y. Ye, C. Chen, C.-T. Li, H. Fu, and Z. Chi	A computerized plant species recognition system	71
M. Khalil and M. Bayoumi	A dyadic wavelet an e invariant function for 2D shape recognition	92
Yuan Tian	Multiple Classifier Combination For Recognition of Wheat Leaf Diseases Intelligent Automation and Soft Computing	94
Abdul Kadir	Experiments Of Zernike Moments For Leaf Identification, Journal of Theoretical and Applied Information Technology	93.44
X.-F. Wang, J.-X. Du, and G.-J. Zhang	Recognition of leaf images based on shape features using a hypersphere classifier	92
A.H. Kulkarni, Dr. H.M.Rai, Dr. K.A. Jahagirdar, P.S. Upparamani	A Leaf Recognition Technique for Plant Classification Using RBPNN and Zernike Moments	93.82
Sapna Sharma, Dr.Chitvan Gupta	Recognition of Plant Species based on leaf images using Multilayer Feed Forward Neural Network	91.13
Proposed	Plant Leaf Identification and Recognition using Soft Computing Tools	100*

*Using binary classifier perceptron consisting of 2 types of leaves with 50 images as training set and 10 images as test set per leaf type.

7 FUTURE PROSPECTS

Adding more features would increase the efficiency and accuracy that we can expect from the finished system.

There are several areas where there is scope of improving further.

- Finding and implementing more features that offer increased accuracy and reliability.
- Implementing other techniques and algorithms and combining their best features to this method to improve efficiency and accuracy.
- Increasing the size of the database increases the range that this system can work upon.

8 CONCLUSION

A simple method of plant leaf recognition is presented in this project. Achieved results are encouraging and suggest the adequacy of the selected features. This proposed system is based on some previous works on the same topic, and those algorithms along with this one will help progress in the field of plant leaf identification and classification. This work studies an image classification process based on a soft computing approach to handle identification of plant leaves of different types.

Experimental results of this proposed system clearly show that this method can indeed differentiate the unknown plant leaf types. So, it is safe to say that this technique helps us get a very simple and yet effective way of solving our problem of identification and classification of plant leaves.

REFERENCES

Chaki, Jyotismita & Parekh, Ranjan. Plant Leaf Recognition using Gabor Filter. *International Journal of Computer Applications* (0975–8887) Volume 56–No. 10, October 2012.

Gupta, Dr. Chitvan & Sharma, Sapna. Recognition of Plant Species based on leaf images using Multilayer Feed Forward Neural Network. *International Journal of Innovative Research in Advanced Engineering (IJIRAE)* ISSN: 2349–2163 Issue 6, Volume 2 (June 2015).

Hsiao, Jou-Ken, Kang, Li-Wei, Chang, Ching-Long & Chen, Chao-Yung Hsuand Chia-Yen. 2014. Learning Sparse Representation for Leaf Image Recognition. *ICCE-Taiwan.*

Hsiao, Jou-Ken, Kang, Li-Wei, Chang, Ching-Long & Lin, Chih-Yang. Comparative Study of Leaf Image Recognition with a Novel Learning-based Approach. *Science and Information Conference 2014 August 27–29, 2014 | London, UK.*

Kebapci, H., B. Yanikoglu, & G. Unal. Plant image retrieval using color, shape and texture features. *The Computer J.*, vol. 54, Sept. 2011.

Kulkarni, A.H., Rai, Dr. H. M., Jahagirdar, Dr. K. A. & Upparamani, P. S. A Leaf Recognition Technique for Plant Classification Using RBPNN and Zernike Moments. *International Journal of Advanced Research in Computer and Communication Engineering* Vol. 2, Issue 1, January 2013.

Mouine S. et al. Advanced shape context for plant species identification using leaf image retrieval. *Proc. ICMR*, June 2012.

Mzoughi O. et al. Advanced tree species identification using multiple leaf parts image queries. *Proc. ICIP*, Sept. 2013, pp. 3967–3971.

Satti, Vijay, Satya, Anshul & Sharma, Shanu. An Automatic Leaf Recognition System For Plant Identification Using Machine Vision Technology. *International Journal of Engineering Science and Technology (IJEST)* ISSN: 0975–5462, Vol. 5 No.04 April 2013.

Computational Science and Engineering – Deyasi et al. (Eds)
© 2017 Taylor & Francis Group, ISBN 978-1-138-02983-5

A new random symmetric key generator: AASN algorithm

Asoke Nath, Aashijit Mukhopadhyay, Somnath Saha & Naved Ahmed Tagala
St. Xavier's College (Autonomous), Kolkata, India

ABSTRACT: In this paper, the authors have introduced a new method for transferring encryption key between sender and receiver. In symmetric key cryptography it is difficult to send any key over any insecure channel as any moment the intruder may hack the key. In the present paper the authors have designed a method where any trivial secret key can be pre-decided between the sender and the receiver. This key is then embedded with the timestamp extracted from the machine in a random manner. The modified key can be used to encrypt any plain text. After the encryption is over the same time stamp will be embedded in the cipher text in random manner. The receiver has to first detect the time stamp and that is to be clubbed with predefined key for decryption purpose. The advantage of this method is that even if someone knows original key but he/she will fail to decrypt the cipher text as he/she has to know idea about the exact time stamp. The present method studied at different time span and the results found are quite satisfactory.

Keywords: symmetric key, timestamp, logarithmic spiral, prime numbers, brute force attack

1 INTRODUCTION

The greatest challenges in the Symmetric key cryptography is key exchange mechanism between sender and receiver. It has been proved that after getting hold of the key, the cipher text will be very easy to be broken. So, in this paper the authors have developed a new Key Generation Algorithm based on the One Time Pad concept and the time stamping concepts called AASN (Asoke, Aashijit, Somnath, Naved) algorithm.

The One Time Pad taken from the User is intermixed with the system time in milliseconds (Time elapsed after 1970). The total message then is broken into bits and stored in a two dimensional binary array. According to few random bytes extracted from the time stamp we generate two values namely 'a' and 'b', such that $0 < a < 0.9$ and $0 < b < 0.5$. Now according to the different values of 'a' and 'b' a logarithmic spiral is generated (as shown in the following figures) and store the points in a list. Now at the initial position a perturbation is generated and make it travel through the total bit pattern following the path as mentioned by the spiral. So, one has to add the first bit with the second bit following rules of binary addition and replace the previous bit with the added bit. The overflowed bit is used with the next bit in the path. The end overflowing bit is ignored. Finally, a disrupted array of binary digits is generated.

After that one has to generate the numerical constant π up to a very high accuracy using any standard π generation techniques available. High order prime numbers are being extracted from a random number calculated using the time stamp. A specific number of extracted digits are used to extract the particular position from the disrupted bit array and they provide 16-bit numbers which can be used for any form of encryption.

Different shuffling operations are used in order to shuffle a single dimensional array of integer numbers which again can be used for encryption in any transportation techniques.

Finally, the time stamp used for encryption is broken into bit stream and embedded into the cipher text in a random fashion. Thus, transfer of the key will be totally safe through the insecure channel.

1.1 *Generated figures of Logarithmic Spiral for different values of 'a' and 'b'*

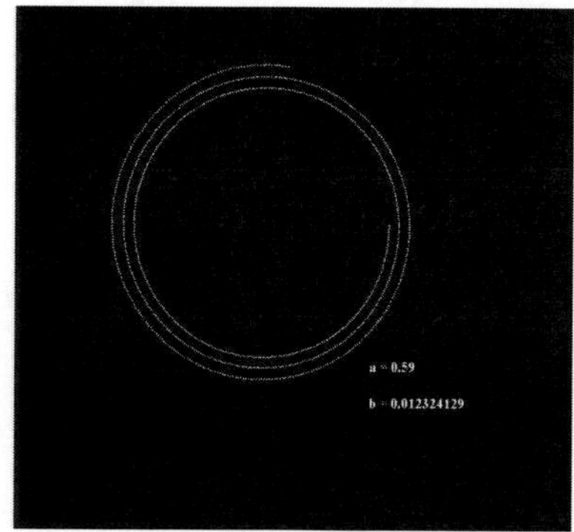

Figure 1. Generated Figure-1.

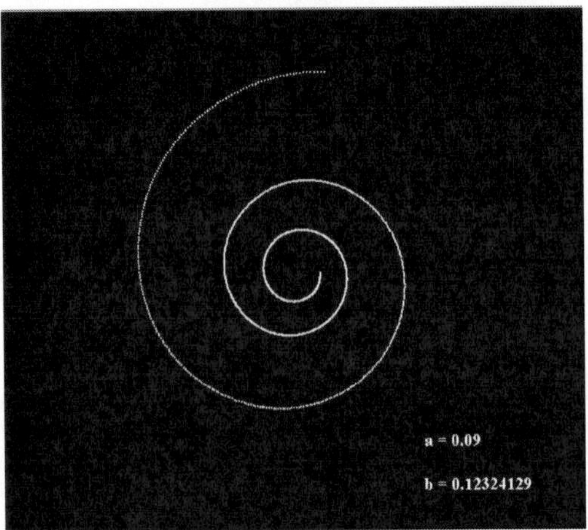

Figure 2. Generated Figure-2.

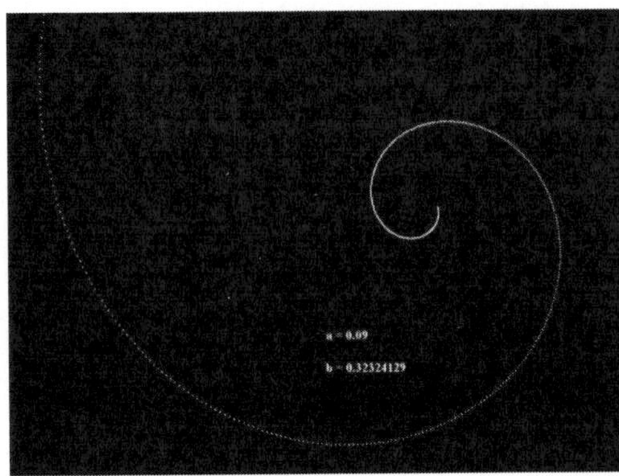

Figure 3. Generated Figure-3.

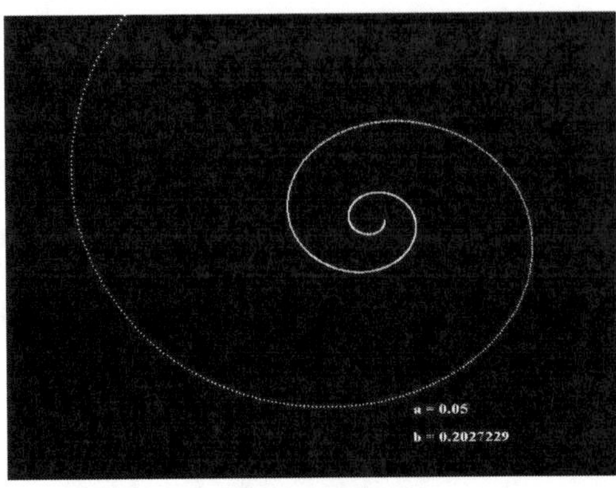

Figure 4. Generated Figure-4.

2 ALGORITHM

2.1 *Governing equation*

Logarithmic Spiral:

$$x(t) = r(t)\cos(t) = ae^{bt}\cos(t)$$
$$y(t) = r(t)\sin(t) = ae^{bt}\sin(t)$$

2.2 *Block model of the process*

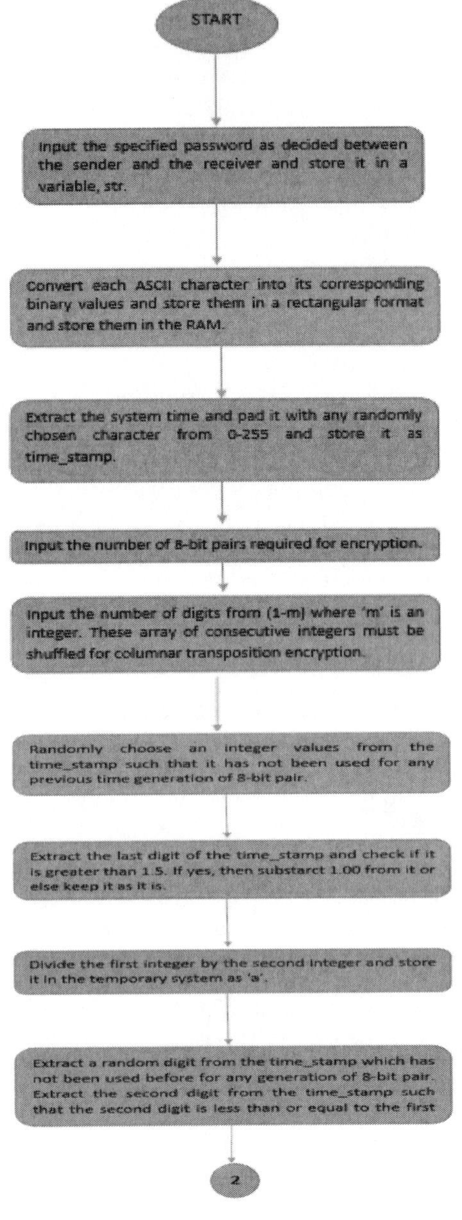

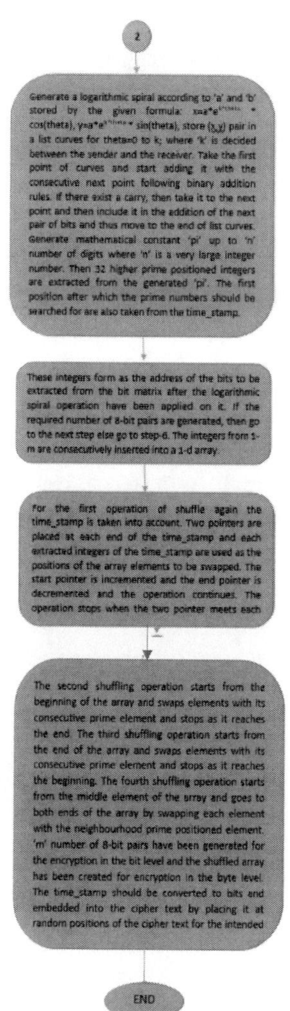

3 RESULTS

Test Case 1. (When the same secret keys have been used for a number of times).

Secret Key Entered	16-bit Integer numbers produced as required	Combinations for Columnar Transposition
Enter The Password:- *abcd*	1000000000000000 0000000000000010 0000000000000000	3 9 2 1 6 5 7 8 4 10
Enter The Password:- *abcd*	1000101000100000 0111001010010000 0000000010000001	7 2 1 9 6 5 8 4 3 10
Enter The Password:- *abcd*	1001010000000010 0000000000010000 0010000001001001	1 7 2 10 6 8 3 4 9 5
Enter The Password:- *abcd*	0100010000001000 0001000000110010 0000000000001000	4 7 2 8 5 1 6 3 9 10

Test Case 2. (Any trivial Alpha-Numeric can be used as a Secret key to generate randomness in the result).

Secret Key Entered	16-bit Integer numbers produced as required	Combinations for Columnar Transposition
Enter The Password:- *Cryptography*	0000111000000001 1000001000010001 1100100000000000	6 7 1 3 2 9 4 8 5 10
Enter The Password:- *MadaM*	0000000100000100 0111011100000000 1000000000000000	5 3 1 10 6 4 9 8 7 2
Enter The Password:- *111111*	0000001000011000 0000001000000000 1010000000011100	3 7 1 5 6 10 4 8 9 2
Enter The Password:- *3*	0010010001100010 0001100001010000 0001100000010010	10 3 1 5 6 8 9 4 7 2

4 CONCLUSIONS AND DISCUSSIONS

The Secret Key entered by the user is pre-decided between a particular user-receiver pair. This secret key is embedded along with the timestamp extracted from the system. This time stamp is also embedded into the cipher text during transmission. So, the key is not being transferred through any channel, only the time stamp embedded into cipher text in a random manner is transferred through an insecure channel. This helps to increase the secrecy of the transferred bits.

The Key generation technique has been successful in generating random keys using the same one time pad at different time. It also gives a new technique to transfer a part of the key (that is the time stamp) through a very insecure channel. It has also proved itself in the above produced results that it is totally free from Brute Force attacks and other statistical attacks. Thus it can be used by any Symmetric Key Encryption technique to encrypt very valuable data.

The use of time stamp will be replaced by some random number generation technique in the next work.

REFERENCES

Cryptosystems on International Journal of Computer Science and Business Informatics.

Cryptography Engineering: Design Principles and Practical Applications by Niels Ferguson, Bruce Schneier, Tadayoshi Kohno.

Dragan Vidakovic and Dusko Parezanovic, "Generating Keys in Elliptic Curve".

Rizwan Haider, Bhavana Srivastava and Umar Badr Shafeeque, "Multilingual Key Generator", COTII-2013, ISBN: 978-81-924738-5-7, Vol. 1, 2013.

Rizwan Haider and Umar Badr Shafeeque, "Enhanced Multilingual Key Generator" on International Journal of Advancements in Technology.

William Stallings, Cryptography and Network Security, fourth Edition, Pearson Education, ISBN-978-81-775-8774-6.

Computational Science and Engineering – Deyasi et al. (Eds)

A stream cipher based symmetric key cryptographic technique using ASCII value of number and position of 0's and 1's

Sarbajit Manna
Ramakrishna Mission Vidyamandira, Howrah, India

Saurabh Dutta
Dr. B.C. Roy Engineering College, Fuljhore, Durgapur, India

ABSTRACT: This research work proposes a symmetric key encipherment technique based on stream cipher where the binary presentation of the ASCII corresponding to each and every character of the plain text is scanned to find out the number of 0's and 1's and also their positions. This technique uses a 96-bit symmetric key consisting of number of 0's and 1's and their positions. A large number of files (.txt, .sys, .com, .dll, .exe etc) of different sizes can use this technique effectively. The symmetric key is created for each plaintext character dynamically so that the security of the symmetric key technique can be enhanced. An 8 X 12 matrix is used in encryption as well as in decryption to create a key value. This key value is used in both encryption as well as in decryption to make the plaintext file more secure. The receiver, after getting the ciphertext and the symmetric key, can easily calculate the key value and decrypt the ciphertext using a reversible operation.

Keywords: stream cipher, symmetric key, ASCII value, plain text, cipher text

1 INTRODUCTION

As Internet is the most essential means of data communication today, the need for information security is increasing every day. So, every day the need to have more complex cryptographic technique is increasing. Some algorithms use symmetric key technique, some use asymmetric key technique, some use a combination of the two (Mukhopadhyay *et. al.*, Kahate, Stallings 2002).

In this paper, every plaintext character is encrypted using a key value generated from a dynamic 96-bit symmetric key and the corresponding ciphertext character is generated. Thus the entire plaintext file is converted to a ciphertext file and the ciphertext file along with the symmetric key is transported over the Internet to the receiver and the ciphertext file is converted back to the plaintext file by the receiver applying the reverse procedure. The technique has the following four steps (Manna *et. al.*, 2014, Das *et. al.*,).

1.1 *Symmetric key generation*

Plain Text → Symmetric Key

1.2 *Symmetric key value generation*

Symmetric Key → Symmetric Key Value

1.3 *Encipherment mechanism*

XOR
Plain Text ←↓→ Symmetric Key Value → Cipher Text

1.4 *Decipherment mechanism*

XOR
Cipher Text ←↓→ Symmetric Key Value → Plain Text

Section 2 describes the encipherment and decipherment techniques. The experimental result and analysis are shown in section 3. Conclusion is drawn in section 4. Finally, future scope of work is represented in section 5.

2 PROPOSED ALGORITHM

The proposed algorithm is based on the ASCII value of number and position of 0's and 1's corresponding to each and every plain text character. Section 2.1 and 2.2 describes the details of the encipherment mechanism and the decipherment mechanism.

2.1 *Encipherment mechanism*

The encipherment mechanism is described from section 2.1.1 to section 2.1.5.

2.1.1 *Binary bit stream formation*

Each plain text character of the input file is read one by one and is converted into its 8-bit binary bit stream corresponding to its ASCII value.

Let, 'F' (ASCII value 70) be the first plain text character. The 8-bit binary bit stream of 'F' will be 01000110 [7].

2.1.2 *Symmetric key generation*

The symmetric key is having 12 blocks of size 8 bits each. So the total size of the symmetric key is 96 bits. The 1st block and the 3rd block represent number of 0's (Even/Odd) and number of 1's (Even/Odd) respectively. If it is Even/Odd, then ASCII value of 'E'/ 'O' is stored in the required block. The 2nd block and the 4th block represent number of 0's and number of 1's respectively, though the ASCII values of those numbers are stored in the required block. The 5th and the 6th block represent ASCII value of the position of first and position of last zero respectively. The 7th and the 8th block represent ASCII value of the position of first and position of last one respectively. Remaining blocks represent position of remaining 0's followed by position of remaining 1's respectively. Position starts from 1 from the right hand side of the 8-bit binary presentation of the plaintext character's ASCII value.

So, for plaintext character 'F', the symmetric key block-1 to block-12 contains binary values of 79 (O_{ASCII}), 53 (5_{ASCII}), 79 (O_{ASCII}), 51 (3_{ASCII}), 49 (1_{ASCII}), 56 (8_{ASCII}), 50 (2_{ASCII}), 55 (7_{ASCII}), 52 (4_{ASCII}), 53 (5_{ASCII}), 54 (6_{ASCII}), 51 (3_{ASCII}) respectively [7].

2.1.3 *Symmetric key value generation*

Twelve 8-bit blocks starting from the first block are stored in an 8 X 12 matrix starting from the first column in top-down fashion. Then the key value is formed by choosing the elements having same row and column index from the top row.

So, for plaintext character 'F', the key value is 00010011.

2.1.4 *Cipher Text (CT) generation*

Using modulo-2 addition between the binary bit stream of plaintext character's ASCII value and the symmetric key value, the cipher text character is generated.

So, for plaintext character 'F', the cipher text becomes 010000110_2 + 00010011_2 = 01010101_2 which is U (85_{ASCII}) [7].

2.2 *Decipherment mechanism*

Section 2.2.1 to section 2.2.3 describes the decipherment mechanism.

2.1.1 *Binary bit stream formation from the Cipher Text (CT)*

The receiver forms the binary bit stream corresponding to each ciphertext character in the entire encrypted file from their ASCII values after receiving it from the sender.

So, ASCII value of cipher text 'U' is 85_{10} = 01010101_2 in binary.

2.1.2 *Symmetric key value formation*

The receiver generates the symmetric key value by applying the same procedure as the sender.

The receiver generates the key value 00010011_2 from the 96-bit symmetric key.

2.1.3 *Plain Text (PT) generation*

Using modulo-2 addition between the binary bit stream of cipher text character's ASCII value and the symmetric key value, plain text character is generated.

So, for cipher text character 'U', the plain text becomes 01010101_2 $+_2 00010011_2$ = 01000110_2 which is F (70_{ASCII}).

3 RESULT AND ANALYSIS

Results of five different categories of files namely .EXE, .DLL, .COM, .SYS, .TXT respectively is presented in this section. Ten different files in size and content are analyzed for this section. The algorithm has been implemented and tested using C language on a computer with Intel Corei5 2.66 GHz processor having 4 GB RAM.

To test the non-homogeneity between the source and the encrypted files Pearson's chi-squared test has been done. It signifies that whether the observations onto encrypted files are in good agreement with a hypothetical distribution. The Chi square distribution has been done with (256–1) = 255 degrees of freedom, 256 is the total number of classes of possible characters in the source file as well as in the encrypted file. If the observed value exceeds the tabulated value, then the null hypothesis is rejected (Dutta 2004).

Following equation defines the "Pearson's chi-squared" or the "Goodness-of-fit chi-square":

$$X^2 = \Sigma \{(f_0 - f_e)^2 / f_e\} \quad (1)$$

Here f_0 and f_e respectively mean that the frequency of a character in the encrypted file and the frequency of the same character in the corresponding source file. The chi-square values are calculated using this equation for a source file and the corresponding encrypted file (Dutta 2004).

3.1 *Result for .EXE files*

Table 1. Result analysis parameters for .EXE files.

Source/ Target File Size (KB)	Encryption Time (Sec)	Decryption Time (Sec)	Chi Square Value	Degree of Freedom
12.00	0.63	0.01	2304983	209
11.50	0.61	0.02	621545	252
11.00	0.60	0.02	437265	251
8.00	0.42	0.01	691858	253
5.50	0.29	0.01	2199725	241
7.50	0.41	0.01	358776	248
18.00	0.95	0.03	366129	255
15.00	0.81	0.03	520040	254
21.30	1.27	0.04	475822	255
12.00	0.62	0.02	2181512	227

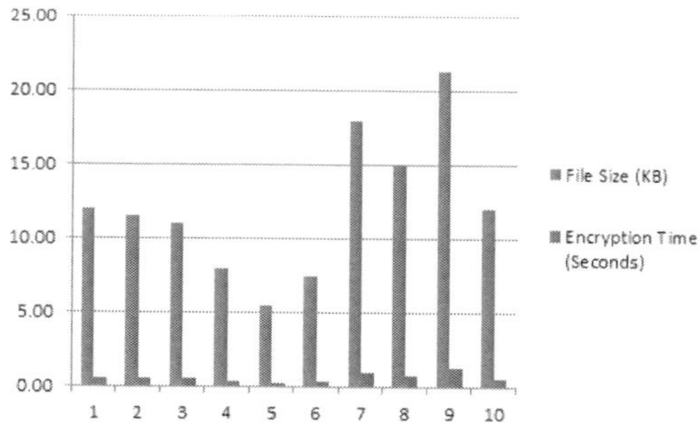

Figure 1. Column chart representing Source file size versus encryption time (for .EXE files).

3.1 *Result for .DLL files*

Table 2. Result analysis parameters for .DLL files.

Source/ Target File Size (KB)	Encryption Time (Sec)	Decryption Time (Sec)	Chi Square Value	Degree of Freedom
17.00	0.91	0.03	653156	255
17.00	0.43	0.02	12748792	195
13.00	0.67	0.02	8410593	251
29.50	1.53	0.03	1904674	255
34.00	2.12	0.04	2595540	255
11.00	0.59	0.02	567651	253
16.50	0.86	0.02	6520249	249
10.50	0.55	0.02	951156	254
15.50	0.84	0.02	578876	255
13.00	0.70	0.02	2754797	252

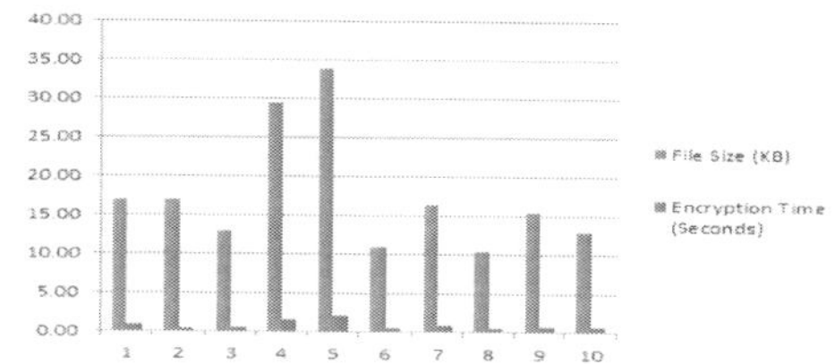

Figure 2. Column chart representing Source file size versus encryption time (for .DLL files).

3.2 *Result for .COM files*

Table 3. Result analysis parameters for .COM files.

Source/ Target File Size (KB)	Encryption Time (Sec)	Decryption Time (Sec)	Chi Square Value	Degree of Freedom
7.50	0.40	0.01	355776	250
9.00	0.47	0.02	372969	253
7.00	0.39	0.01	545229	244
68.20	4.45	0.07	1281425	255
29.00	1.74	0.03	1697870	254
25.50	1.39	0.04	1499457	255
19.20	1.00	0.03	511186	255
6.85	0.38	0.01	320624	255
14.30	0.77	0.03	359580	255
1.10	0.07	0.01	26912	185

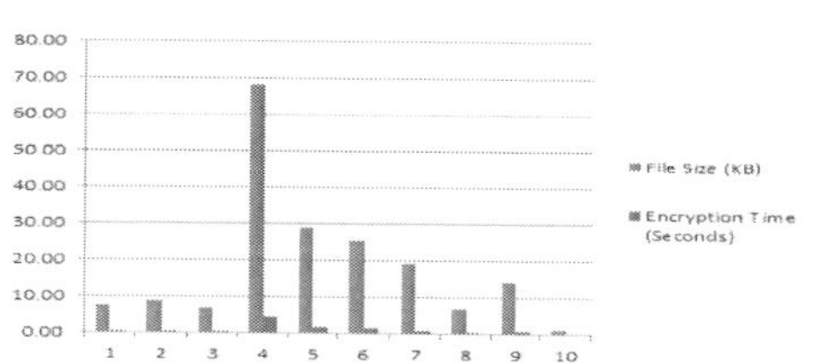

Figure 3. Column chart representing Source file size versus encryption time (for .COM files).

3.4 *Result for .SYS files*

Table 4. Result analysis parameters for .SYS files.

Source/ Target File Size (KB)	Encryption Time (Sec)	Decryption Time (Sec)	Chi S quare Value	Degree of Freedom
10.00	0.55	0.02	1001301	254
11.00	0.58	0.02	1363951	253
15.50	0.85	0.02	1344484	255
10.00	0.52	0.01	1153160	252
10.00	0.52	0.02	895640	254
8.81	0.47	0.01	417656	253
4.12	0.22	0.01	4028556	240
16.00	0.85	0.02	9729828	249
4.65	0.25	0.01	105858	250
17.20	0.93	0.03	4580947	254

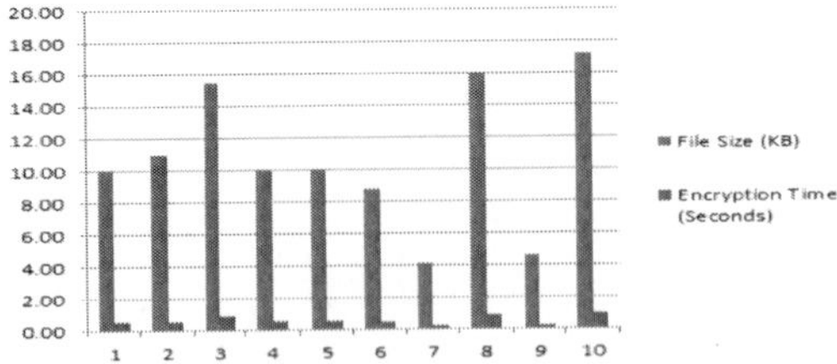

Figure 4. Column chart representing Source file size versus encryption time (for .SYS files).

3.5 *Result for .TXT files*

Table 5. Result analysis parameters for .TXT files.

Source/ Target File Size (KB)	Encryption Time (Sec)	Decryption Time (Sec)	Chi Square Value	Degree of Freedom
1.30	0.08	0.01	2438	89
5.29	0.29	0.01	1110951	92
6.30	0.35	0.01	21442	94
0.13	0.01	0.01	132	45
0.81	0.06	0.01	83	88
22.40	1.20	0.03	1583346	93
2.15	0.12	0.01	118910	86
2.44	0.14	0.01	3901	85
31.80	2.05	1.97	1016704	83
18.60	1.02	0.02	4355912	99

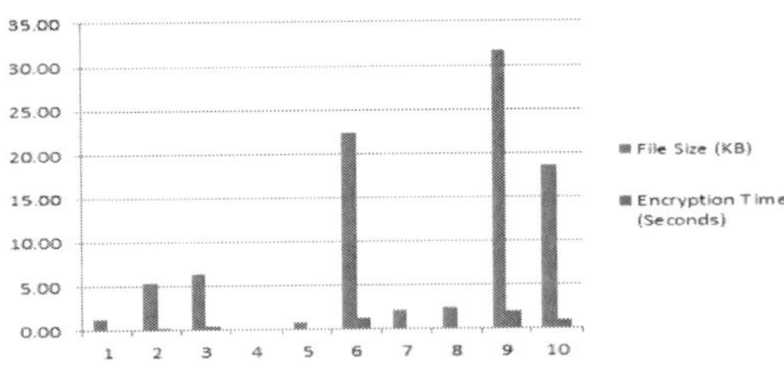

Figure 5. Column chart representing Source file size versus encryption time (for .TXT files).

4 CONCLUSION

This technique has been developed to give an extra edge to the security. The symmetric key is generated from the plain text dynamically. The symmetric key as well as the symmetric key value change automatically with the change in the plain text. The symmetric key size is 96 bits for each plain text character. So, the total size of the symmetric key is 96 times the number of plain text character present in the plain text file. The key is made so large to prevent the brute force attack. This technique can be implemented using any high level language but encryption and decryption time can be higher for the languages which uses virtual machines to execute the program. The encrypted and the decrypted file size is same as the source file size. So, extra memory space is not required to store the encrypted and decrypted files. The execution time is dependent on the source file size as well as the platform on which the program is being run. Not only addition modulo-2 operation but also any reversible operations can be applied on the plain text file.

5 FUTURE SCOPE OF WORK

This algorithm will work much better if randomness is included. Randomness can be included in the algorithm if the bits of the key value are chosen randomly each time from the 8 X 12 matrix.

REFERENCES

Das. S., Dutta. S., (2013) An Approach of Bit-level Private-key Encryption Scheme based on Armstrong Number and Perfect Number in Selective Mode, Bulletin & Engineering Science, 5(1).

Dutta, S., (2004) An Approach towards Development of Efficient Encryption Technique, A thesis submitted to the University of North Bengal for the Degree of Ph.D.

Kahate, A., Cryptography and Network security, Tata McGraw-Hill Publishing Company Limited.

Manna. S., Dutta, S. (2014) A Stream Cipher based Bit-Level Symmetric Key Cryptographic Technique using Chen Prime Number, International Journal of Computer Applications, 107(12), 18–22.

Mukhopadhyay, D., Forouzan, B. A., Cryptography and Network Security, Tata McGraw-Hill Publishing Company Limited, 2nd Edition.

Stallings, W., (2002) Cryptography and Network security: Principles and Practice (Second Edition), Pearson Education Asia, Sixth Indian Reprint http://www.asciitable.com/.

Computational Science and Engineering – Deyasi et al. (Eds)
© 2017 Taylor & Francis Group, ISBN 978-1-138-02983-5

Microcontroller based automated life savior—Medisûr

Soumick Chatterjee, Pramod George Jose, Priyanka Basak, Ambreen Athar, Bindhya Aravind, Romit S. Beed & Rana Biswas
Department of Computer Science, St. Xavier's College (Autonomous), Kolkata, India

ABSTRACT: With the course of progress in the field of medicine, most of the patients' lives can be saved. Only thing required is the proper attention in proper time. Our wearable solution tries to solve this issue by taking the patients vitals and transmitting them to the server for live monitoring using mobile app along with patient's current location. In case of an emergency, that is if any vitals show any abnormalities, a SMS is sent to the caregiver of the patient with the patient's location so that he can reach there on time.

1 INTRODUCTION

1.1 *Objective*

The 21st century has witnessed a major paradigm shift in diagnosis technology in the field of medicine. Embedded technology has been effectively used for successful diagnosis, leading to more accurate treatments. High speed calculations, more complex testing methodologies, more accurate and reliable results have forever enhanced diagnosis process.

Today's world is becoming increasingly dependent on heuristic data; medical treatment is no exception. Large volume of data is generated about patients, symptoms and diagnosis. In the previous decade, doctors would have recorded certain basic information about the patients in a prescription. In contrast, these days he will have a long stream of digital data, from minute details about the patient to high-definition medical images.

Since the world is going green, this data is needs to be made available in digital format. This in turn has broadened the application of informatics to improve health care and has also contributed in medical research. Today, informatics is being applied at every stage of health care from basic research to care delivery and includes many specializations such as bioinformatics, medical informatics, and biomedical informatics, and is often referred to as "in silico" research. Informatics specializes on introducing better means of using technology to process information.

The medical field has developed sufficiently to ensure critical patients in majority of the cases, can survive cardiac arrest or similar issues. The person can return to normalcy, with proper diagnosis and medication. But the treatment has to be given at the right time, which is during the initial stage of an attack. This becomes difficult to achieve when the person is alone, or when the patient is a senior citizen. Medisûr tries solves the problem by being an automated companion who watches the patient's body vitals and calls for help when in danger.

The Medisûr gadget monitors the body vitals and updates the data in a database. Caretaker of the patient can live monitor this information. In case, if the vitals are in the danger zone, an alert message is sent to the caretaker. So, the patient receives proper treatment on time.

1.2 *Working principal*

The electronic gadget under consideration consists of a wearable device and a receiving kit. The wearable device is supposed to detect and get the body vitals of the person wearing it, post which it sends the same to the receiving kit. The wearable band has sensors to serve this purpose and an emergency button in case of emergency. The receiving kit has three tier functionality. The first to track the current location of the person wearing it, second, it sends the collected information to a centralized server for live monitoring and also for future reference and lastly, to check whether the received vitals are within the normal range or not. If not, then it sends a SMS to the pre-configured number with the current location along with the current vitals for immediate assistance. Another scenario could be, if the patient feels any discomfiture, feels dizzy that can never be detected by any sensor. In that case, the patient presses the emergency button, then also a similar SMS will be sent to the pre fed mobile number stating the person requires attention. This being an embedded system, it has related software modules. There is a central server, where the current vitals & location are sent from the receiving kit. Server also stores daily data for future reference, from which a caregiver can get an idea about how the patient's health was in past 6 months. Again, there is a Multi-platform mobile App. The app fetches the latest vitals and location from the server and shows it on screen, so that caregiver can check anytime how his loved one is doing and where he is currently.

2 COMPONENTS

2.1 *A Tmega 328*

ATmega328 is an 8-bit AVR RISC-based microcontroller by Atmel which is used in the wearable gadget, to collect sensor information, then encrypt it and send it to the receiving kit.

2.2 *Arduino Uno*

Arduino Uno is a versatile microcontroller board based on the ATmega328 which is used in the receiving kit to receive sensor data and location from the wearable gadget and then it will send the data to the server. If required, this will also help in sending SMS. This can be replaced with just another ATmega328 microcontroller instead of this complete board.

2.3 *LM35 temperature sensor*

LM35 series are precision integrated-circuit temperature devices which gives an output voltage linearly proportional to the Centigrade temperature it receives. This LM35 Sensor is used to detect body temperature in the prototype.

2.4 *M212 Pulse Sensor*

Pulse Sensor which is used in this implementation is a photoplethysmograph, which is a well-known medical device used for non-invasive heart rate monitoring. Sometimes, it measures blood-oxygen levels (SpO2), sometimes they don't. The heart pulse signal that comes out of a photoplethysmograph is an analog fluctuation in voltage, and it has a predictable wave shape which is shown in Figure 1.

The depiction of the pulse wave is called a photoplethysmogram, or PPG. The goal is to find successive moments of instantaneous heart beat and measure the time between them, called the Inter Beat Interval (IBI), which can be archived by following the predictable shape and pattern of the PPG wave. In this implementation, it is used to measure the pulse in the wearable gadget and send it to the ATmega328 for further processing.

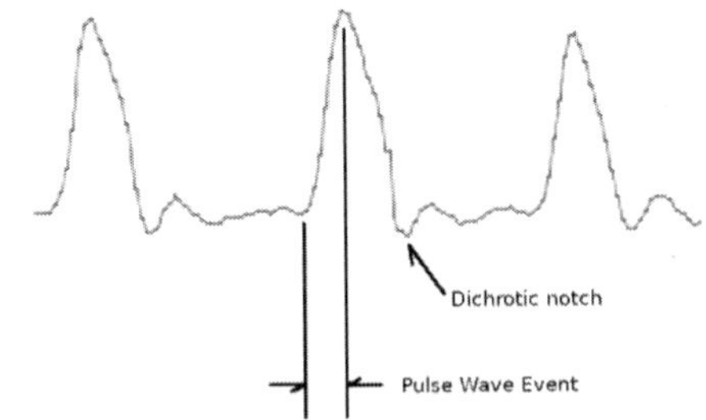

Figure 1. Pulse wave.

2.5 *434MHz Radio Frequency Receiver (Rx) & Transmitter (Tx)*

RF transmitter and receiver module is used to transmit and receive the radio frequency for the wireless communication which is used to send sensor information from the wearable gadget to the receiving kit.

2.6 *NEO6MV2 GPS module*

The NEO-6 module series is a family of GPS receivers which consist of high performance u-blox 6 positioning engine, giving it receivers excellent navigation performance even in the most challenging environments. This is used to triangulate location of the user and send it to the Arduino for further processing as required.

2.7 *SIM300 GSM module*

A GSM Modem is a device that modulates and demodulates the GSM signals and in this particular case 2G signals. The modem which is used in this implementation is a SIMCOM SIM300 which is a Tri-band GSM/GPRS Modem. The TTL interface present in it allows to be directly interfaced with the Arduino present in the receiving kit, using which, the device is connecting to the designated web service and sending sensor plus location information. Also this GSM Module is used to send SMS when required.

2.8 *AES 256bit encryption*

The Advanced Encryption Standard (AES), also known as Rijndael, is a specification for the encryption of electronic data. In this implementation AES with 256 bits Key Size is used for encrypting data while sending from wearable gadget to the receiving kit and from receiving kit to the server.

2.9 *Microsoft Azure virtual machine*

Microsoft Azure is a cloud computing platform and infrastructure created by Microsoft for building, deploying, and managing applications and services through a global network of Microsoft-managed data centers. This is used to host the Web Service and also to host Microsoft SQL Server in which the data received from receiving kit is stored for further use. Mobile App can access this to show the live information and also can be used to track patient history.

3 WORKING MECHANISM

3.1 *Module 1: Wearable gadget*

First module is a wearable gadget in the form of a wrist band. This wearable gadget consists of many bio sensors. For this implementation, a LM-35 is used as the temperature sensor. For sensing the pulse, M212 heart rate sensor is used. Both the sensors are placed on the inner side of the band, placed strategically so that it directly touches the skin. Pulse sensor should be placed in such a way that it can detect the pulse. Sensors are connected to a circuit board, where ATmega328 Microcontroller is placed. Data collected by the sensors are periodically sent to the microcontroller. A string is constructed by concatenating the sensor values. If the user feels unwell, that can't be detected by any sensor. For that, there is an Emergency Button. If the button is pressed then the value of the switch is one, or else zero. This value is also concatenated with the sensor values. For example, if the switched is pressed along with that, current body temperature is 98°F and the current pulse rate is 77 bpm, then the string will be constructed as - 1,98,77 This string is now transmitted from the wrist band to the second module, which is the receiving kit, explained later. In this implementation, this transmission is done using 434Mhz RF-Transmitter. Here a problem may creep up, if two or more radio device with the same frequency is used in close range, it may interfere with each other. To overcome this, a unique prefix is added to the above string. In the prototype implementation, the prefix—"SXCMS" have been added. So, the string becomes—SXCMS:1,98,77 But this isn't enough. As wireless technology is used here, security is a concern. So, the above string is encrypted using AES 256bit encryption before transmitting using the unique device ID as the key. For the prototype, device ID is MS001. So, the final encrypted string that is to be transmitted from the wearable gadget becomes:—Ql05 Ng7kM1M9PPY36BO/ LbiLyJsVpDSp4hU4tVz9 Nw =.

3.2 *Module 2: Receiving kit*

Second module of this implementation is the receiving kit. In the receiving kit a 434Mhz RF-Receiver is connected to the Arduino, which receives the encrypted string from the wrist band and sends it to the Arduino Uno microcontroller board, which first decrypts it using the same AES scheme with the device ID as the key. This device ID is unique for each Wrist Band & Receiving Kit combo. So, the receiving kit has the same key which the wrist band used to encrypt the string. So, in this prototype MS001 is used for decryption and the plain text SXCMS: 1,98,77 is obtained. First, the string is validated to check whether it is coming from the correct device. If the string is starting with SXCMS then it is understood that it is coming from the correct device. After this, the string is broken to extract the parameter values of switch, hear rate and temperature. In this implementation, NEO6MV2 GPS module is used for triangulating the location of the device. Utmost priority is given to the switch. If the switch value is 1, irrespective of the sensor values, a SMS is sent using the SIM300 module present, to the care giver stating that person wearing this device needs his attention along with the exact location and sensor information. Apart from this, using the same SIM300 module, a web service is called, which is hosted in the Microsoft Azure server, by sending a simple HTTP Get Request. While sending the request, the device ID is also sent to identify the patient uniquely. So, the plain text that is going to be sent via the get request is MS001,98,77, Park Street where MS001 is the device ID, 98°F is the temperature, 77 is the pulse rate and Park Street is the triangulated location. For security reasons, encrypting this string is required and this is done using AES 256bit encryption with a global encryption key, in this case MedS, by which the cipher text is obtained as vLbOEhcctSgLS8 W66U6M6QdRnL52kIjcD-N9ONNuHoWI = which is sent via the query string parameter. This is sent to the server at a regular interval, say every minute. Besides sending this information to the server, it is also checked if there is any anomalies in the pulse rate or temperature. Say, if the temperature goes beyond 99°F or pulse rate goes below 40 bpm, then an SOS message is sent to the care giver with the current sensor values plus the triangulated location stating that the person wearing this device may require medical help immediately.

3.3 *Software modules*

An embedded system is incomplete without software modules First soft-module is the web service which receives the cipher text from each of the Medisûr device, and decrypts it using the global key MedS. From the decrypted text, the device is first identified from the device ID. Consequently the individual sensor data are extracted, and then lastly the location.

Furthermore, this information is updated in the database, in the currentPatientData table. In every six hours, this information is stored in patient Histry table as well. Both web service and database are stored in the Microsoft Azure virtual machine, so as to access it centrally.

The last and one of the most vital part of our solution pertains to a cross platform mobile app. In the mobile app, a caregiver can add multiple persons whom he/she is looking after. When a person is selected, mobile app calls a WCF service, which is hosted in the same server. This provides latest data from the currentPatientData table. So, a caregiver can check the live vitals and location of patients, updated every minute. Using the app, the history of past 6 months can be checked, so that he/she can monitor his patient more efficiently.

4 RESULTS AND DISCUSSION

The Authors have implemented this concept by creating a minimalistic prototype consisting most of the components except the GPS Module, which is shown in Figure 2.

The outputs observed were close to the desired ones, but can be fine-tuned further. It was detecting heart-beat accurately, for body temperature, other sensors would give better results.

This prototype is quite bulky, and not very easy to carry. But this can be made quite compact by using latest industry based fabrication techniques.

For more accurate results, some of the components can be replaced with some superior ones. Like custom made body temperature sensor based on DS1624, DS18B20 could be better for precision while sensing the body temperature. For the transmission of sensor values from the wrist band to the receiving kit, other transmission modules such as NRF24 L01 or Zigbee can also be used to improve the transmission. For the GPRS Based communication, it would be better to use SIM900 over SIM300.

Security is a major concern these days. While transmitting data from the receiving kit to the server, in this implementation, AES is used with a global key which is not that secure, as if someone gets hold of it, he can access such vital confidential information including locations. For that, any other means of security can be provided such as RSA.

This implementation is very scalable as more sensors can be added to this for collecting other vitals from the patient.

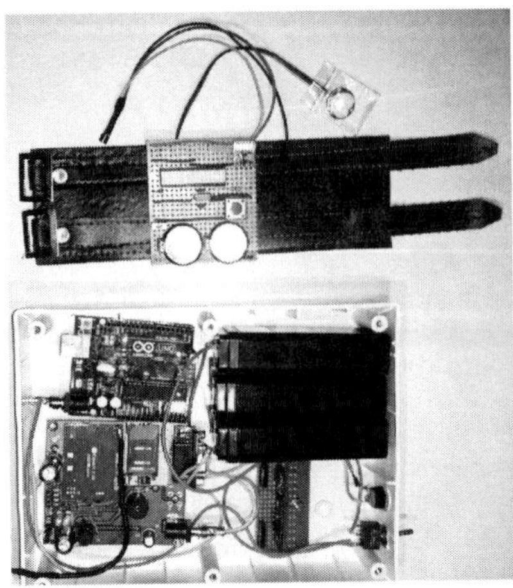

Figure 2. First Working Prototype.

5 CONCLUSION

This device is not a medical device to help in treatment but it provides the caregiver to monitor the health condition and also to get alert when in need. Location tracking gives this solution that very required edge so that when a person is in trouble, caregiver can also know where that person is.

REFERENCES

Castro, Daniel 2009. The Role of Information Technology in Medical Research. Atlanta Conference on Science, Technology and Innovation Policy.

Chandra, Rupa; Rathee, Asha; Verma, Pooja & Chougule, Archana 2014. GSM Based Health Monitoring System. IRF International Conference.

Coyle, G; Boydell, L & Brown, L 1995. Home Telecare for The Elderly. Journal of Telemedicine and Telecare 1: 183–185.

Schneier, Bruce; Kelsey, John; Whiting, Doug; Wagner, David; Hall, Chris; Ferguson, Niels; Kohno, Tadayoshi; et al. 2000. The Twofish Team's Final Comments on AES Selection.

Shaikh, Rubina. A. 2012. Real Time Health Monitoring System of Remote Patient Using Arm7. International Journal of Instrumentation, Control and Automation 1 (3,4): 102–105.

Sowmya, G. & Sandeep, B.L. 2016. Remote Health-Care Monitoring System Using Arduino Board over Distributed Ubiquitous Environment. International Journal of Advanced Research in Computer and Communication Engineering 5 (4): 816–819.

Tahat, Ashraf A. 2009. Body Temperature and Electrocardiogram Monitoring Using an SMS-Based Telemedicine System. 4th International Symposium on Wireless Pervasive Computing.

Verma, Sanjay & Gupta, Namit 2012. Microcontroller-based Wireless Heart Rate Telemonitor for Home Care. IOSR Journal of Engineering 2 (7): 25–31.

Yelton, Jeff 2009. The Advancement of Medical Technology in the 21st Century. WiredFox Technologies White Paper.

Zhang, Xiaoyu; Jiang, Hanjun; Zhang, Lingwei; Zhang, Chun; Wang, Zhihua & Chen, Xinkai 2009. An Energy-Efficient ASIC for Wireless Body Sensor Networks in Medical Applications. IEEE Transactions on Biomedical Circuits and Systems 4 (1): 11–18

2001. Announcing the Advanced Encryption Standard (AES). Federal Information Processing Standards Publication 197.

Computational Science and Engineering – Deyasi et al. (Eds)

Entropy and similarity measurement of intuitionist multi-fuzzy sets and their application in decision making

Amalendu Si
Department of Computer Science and Engineering, MIT, Bankura, India

Anup Kumar Kolya
Department of Computer Science and Engineering, RCCIIT, Kolkata, India

ABSTRACT: In this paper, we propose a method to measure entropy and similarity for Intuitionist Multifuzzy Set (IMFS). Initially, we construct intuitionistic multifuzzy matrix from decision maker with respect to the objective criteria. Then, we construct aggregated intuitionistic multi fuzzy matrix using the Intuitionistic Multifuzzy Weighted Averaging (IMFWA) operator from intuitionistic multifuzzy matrix and decision maker weight. Decision maker weights are calculated by the proposed entropy base weight estimation method. Finally, similarity measurement method is applied to solve the problem on pattern reorganization, multi criteria decision making, medical diagnosis etc.

Keywords: Group Decision Making (GDM), Multi Group Decision Making (MGDM), Intuitionistic Fuzzy Vector (IFV), Intuitionistic Multi-Fuzzy Weighted Arithmetic (IMFWA), Intuitionistic Multi Fuzzy Matrix (IMFM), entropy, similarity

1 INTRODUCTION

Intuitionist fuzzy sets were proposed by Atanassav [1] in 1999 and later extended as a generalized form of Zadeh's fuzzy set theory [2]. Recently, IFS is treated as an active research field to solve the multi criteria decision making problem in different domain of uncertain information like marketing, medical diagnosis, and in Group Decision Making (GDM) problem [3–5]. Entropy and similarity measure are two important parts in the fuzzy set theory field. Different researcher investigated those two things in fuzzy set theory in different way of different perspective points. First time in 2001, Burillo and Bustince [6] introduced the concept of entropy to measure the degree of intuitionism of an IVFS or IFS. Nonprobabilistic type entropy measure with a geometric interpretation of IFSs introduced E. Szmidt, J. Kacprzyk [7]. Introduced Entropy of intuitionistic fuzzy set based on similarity measurement between IFSs is described by W. Zeng, F. Yu, X. Yu, H. Chen and S. Wu [8]. Similarity measurement of IFS is used to estimate the degree of similarity between two IFSs. Szimidt and Kacprzyk [9] defined a similarity measurement using a distance measurement which involves both similarity and dissimilarity. Expanding upon this work, Szimidt and Kacprzyk in [10] considered a family of similarity measurements and compared with some existing similarity measurements. IFS or IVFS can be used as a proper tool for representing both membership and non membership of an element to an object. But there is situation where each element has different membership values. In such situation an Intuitionistic Multi Fuzzy Set (IMFS) is more appropriate. Here, IMFS is presented as a tool for representing this type situation linguistically. Authors combined the concept of IFS and MFS produced IMFS and applied this to group MCDM problem in [12]. In-group, MCDM a number of decision maker gave valuable opinions based on their observation with a set of attribute values. Then, multiple individual opinions are sum up as a collective opinion. The authors [12] proposed a suitable algorithm to find out the final outcome, at first constructed Intuitionistic Multi Fuzzy Matrix (IMFM) then constructed the aggregated intuitionistic multi fuzzy matrix which reflects the combined opinions of all decision maker using newly proposed Intuitionistic Malty-Fuzzy Weighted Averaging (IMFWA) operator and arbitrary attribute weight. Attribute weight play an important and vital role in decision making procedure, because the final ranking order may be changed due to change of attribute weight [11]. Group multi-criteria decision making was proposed using IMFS with arbitrary attribute weight information in this literature [12]. In this paper, we propose a new group multi-criteria decision making technique based on similarity measure using entropy based attribute weight under IMFS domain, which is important in this context. The rest of this paper is organized into two parts. Firstly, we give the definition of IMFS entropy and propose a set of IMFS entropies with some importance perquisites; secondly we propose a new MCDM method based on similarity measure using entropy based attributes weights to deal with MCDM problems with unknown attribute weights information under IMFs environment.

2 INTUITIONISTIC FUZZY SET

In this section, we briefly review about IFS, its operation and comparison. Definition of IFS and IFV with Necessary Conditions Atanassov [1] give the concept of intuitionistic fuzzy sets in 1965. Consider a set E be a universal set. An intuitionistic fuzzy sets or IFS in E is an element having the form $A=\{\langle x,\mu_A(x),\nu_A(x)\rangle | x\in E\}$ where the function $\mu_A : E\rightarrow[0,1]$ and $\nu_A : E\rightarrow[0,1]$. denoted the degree of membership and the degree of non membership respectively of the element x from E to set A. For any element $x \in$ E satisfy the condition $x\in E, 0\leq\mu_A(x)+\nu_A(x)\leq 1$.

The function $\pi_A : E\rightarrow[0,1]$ specified by $\pi_A(x)=1-\mu_A(x)-\nu_A(x), x\in E$ defines the degree of uncertainty of membership of the element x to set A called hesitant function. For a fixed $x \in$ E, an object $\{\mu_A(x),\nu_A(x)\}$ is usually called Intuitionistic Fuzzy Value (IFV) or Intuitionist Fuzzy Number (IFN).

2.1 *Basic operation on Intuitionistic Fuzzy Sets*

Addition $\oplus$ and Multiplication $\otimes$ of IFSs

The operation of addition $\oplus$ and multiplication $\otimes$ on intuitionistic fuzzy values are defined by Atanassov [16] as follows: Let $A=(\mu_A(x),\nu_A(x))$ and $B=(\mu_B(x),\nu_B(x))$ be two IFVs, then

$$A\oplus B=(\mu_A+\mu_B-\mu_A.\mu_B,\nu_A.\nu_B)$$

$$A\otimes B=(\mu_A.\mu_B,\nu_A+\nu_B-\nu_A.\nu_B)$$

These operations are constructed in such a way that they generate IFV, since it is proved that $0\leq\mu_A+\mu_B-\mu_A.\mu_B+\nu_A.\nu_B\leq 1$ and $0\leq\mu_A.\mu_B+\nu_A+\nu_B-\nu_A.\nu_B\leq 1$.

Each ordinary fuzzy set A may be written as: $A=\{<x,\mu_A(x),1-\mu_A(x)>|x\in X\}$.

If $A,B\in IFSs$, then the following operations can be found in

1. $A\subseteq B$ iff $(\forall x\in X)(\mu_A(x)\leq\mu_B(x)$ $\&\nu_A(x)\geq\nu_B(x))$;
2. $A=B$ iff $(\forall x\in X)(\mu_A(x)=\mu_B(x)$ $\&\ \nu_A(x)=\nu_B(x))$;
3. $A^C=\{\langle x,\nu_A(x),\mu_A(x)\rangle | x\in X\}$.

Multiplication $\otimes$ of IFS with real values

The addition and multiplication of IFSs with real values are evaluated using expressions (1) and (2) in [13].

$$n_A=\{1-(1-\mu_A)^n,\ \nu_A^n\}\ \&\ A^n=\{\mu_A^n,1-(1-\nu_A)^n\},$$

where n is any integer.

These aforementioned operations produce IFVs which depends not only for integer n but also for all real value $\lambda>0$ i.e.,

$$\lambda_A=\{1-(1-\mu_A)^\lambda,\ \nu_A^\lambda\}, A^\lambda=\{\mu_A^\lambda,1-(1-\nu_A)^\lambda\}$$

Intuitionistic weighted arithmetic mean: the IWAM can be found using expression (1) and (3) as follows:

$$IWAM=w_iA_i\oplus w_2A_2\oplus\ldots\ldots\oplus w_nA_n$$

$$=\left\{1-\prod_{i=1}^{n}\left(1-\mu_{A_i}\right)^{w_i},\prod_{i=1}^{n}\nu_{A_i}^{w_i}\right\}\quad \text{Where } 0\leq w_i\leq 1$$

and $\sum_{i=1}^{n} w_i=1$.

IFVs is determined using this aggregating operator which is the most popular in the solution of MCDM problem in the intuitionistic fuzzy setting.

3 INTUITIONISTIC MULTI FUZZY SET

In this section briefly describes multi-fuzzy set, intuitionistic multi-fuzzy set and a few operations on IMFS including IMFWA operator and distance measurements.

3.1 *MFS and IMFS*

Let U be a universal set and k be a positive integer. A multi-fuzzy set A over U is a set of ordered sequence

$$A=\left\{x/\left(\mu_A^1(x),\mu_A^2(x)\ldots\ldots\ldots\mu_A^i(x)\ldots\ldots\mu_A^k(x)\right):x\in U\right\}$$

where $\mu_A^i\in p(U)$. $i=1,2,\ldots\ldots\ldots k$.

The function $\mu_A = (\mu_A^1 \mu_A^2 ....\mu_A^i)$ is called a multi membership function of multi fuzzy set A and k is called the dimension of multi-fuzzy set A.

An intuitionistic multi-fuzzy set A is defined by $A = \left\{ \begin{matrix} x/\left(\mu_A^1(x), \mu_A^2(x) .... \mu_A^k(x)\right), \\ \left(\nu_A^1(x), \nu_A^2(x) .... \nu_A^k(x)\right): x \in U \end{matrix} \right\}$ where $0 < \mu_A^i(x) + \nu_A^i(x) < 1, \pi_X^2 = 1 - \mu_A^i(x) + \nu_A^i(x)$ for every $x \in U$ and $i = 1,2,..........k$ where k is appositive integer and U is an universal set.

For a fixed $x \in U$, an object $\{\mu_A^i(x), \nu_A^i(x)\}$ is usually called an Intuitionistic Multi-Fuzzy Value (IMFV) or Intuitionistic Multi-Fuzzy Number (IMFN) where $i = 1,2,..........k$ and $k > 0$.

3.2 *Some useful operations on IMFS*

For any two intuitionistic multi-fuzzy sets A and B of dimension k, the following operations can be defined, Let two IMFS are

$$A = \left\{ \begin{matrix} x/\left(\mu_A^1(x), \mu_A^2(x) .... \mu_A^k(x)\right), \\ \left(\nu_A^1(x), \nu_A^2(x) .... \nu_A^k(x)\right): x \in U \end{matrix} \right\}$$

$$B = \left\{ \begin{matrix} x/\left(\mu_B^1(x), \mu_B^2(x) .... \mu_B^k(x)\right), \\ \left(\nu_B^1(x), \nu_B^2(x) .... \nu_B^k(x)\right): x \in U \end{matrix} \right\}$$

Addition of IMFS (A ⊕ B)

In A ⊕ B, the membership and non-membership values are obtained as follows

$$\mu_{A \oplus B}^j(x) = (\mu_A^j(x) + \mu_B^j(x) - \mu_A^j(x).\mu_B^j(x)),$$
$$\nu_{A \oplus B}^j(x) = \nu_A^j(x).\nu_B^j(x)$$
$$j = 1,2,..........k \; x \in U$$

Multiplication of IMFS (A ⊗ B)

In A ⊕ B, the membership and non membership values are obtained as follows

$$\mu_{A \otimes B}^j(x) = \mu_A^j(x).\mu_B^j(x)$$
$$\nu_{A \oplus B}^j(x) = \nu_A^j(x) + \nu_B^j(x) - \nu_A^j(x).\nu_B^j(x)$$
$$j = 1,2..........k, x \in U$$

Example 1 let A and B be two intuitionistic multi-fuzzy sets of dimension 3 defined as follows:

A = {(0.7, 0.6, 0.5), (0.1, 0.2, 0.4)}

B = {(0.4, 0.4, 0.3), (0.3, 0.4, 0.6)}

The addition A ⊕ B and multiplication A ⊗ B of A and B can be obtained by

A ⊕ B = {(0.82, 0.76, 0.65), (0.03, 0.08, 0.24)}
A ⊗ B = {(0.28, 0.24, 0.45), (0.37, 0.52, 0.76)}

Multiplication of IMFS A with real value

Multiplication of IMFS A of dimension k with real value $\lambda > 0$ can be defined as

$$\lambda A = \left\{1 - \left(1 - \mu_A^i\right)^\lambda, \left(\nu_A^i\right)^\lambda\right\}, A^\lambda = \left\{\left(\mu_A^i\right)^\lambda, \left(1 - \left(1 - \nu_A^i\right)\right)^\lambda\right\}$$
$$j = 1,2..........k, x \in U$$

Example Let A be an intuitionistic multi-fuzzy set of dimension 3 where A is defined as follows and $\lambda = 0.5$

A = {(0.7,0.6,0.5),(0.1,0.2,0.4)}

λA = {(0.45,0.37,0.29),(0.32,0.45,0.63)},

A^λ = {(0.84,0.77,0.71),(0.05,0.11,0.23)}

3.3 *IMFWA operator*

Das, Kar and Kar [12] introduced the intuitionistic multi-fuzzy weighted arithmetic operator can be defined by

$$\text{IMFWA} = w_1 A_1 \oplus w_2 A_2 \oplus ...... \oplus w_n A_n =$$
$$\left\{1 - \prod_{i=1}^{n}\prod_{i=1}^{k}\left(1 - \mu_{A_i}^j\right)^{w_i}, \prod_{i=1}^{n}\prod_{j=1}^{k}\left(\nu_{A_i}^j\right)^{w_i}, \prod_{i=1}^{n}\prod_{i=1}^{k}\left(1 - \mu_{A_i}^j\right)^{w_i} - \prod_{i=1}^{n}\prod_{j=1}^{k}\left(\nu_{A_i}^j\right)^{w_i}\right\}$$

where $A_1, A_2 ....A_n$ are IMFS of dimension k and $w_1, w_2 ....w_n$ are their weights where $\Sigma_{i=1}^n w_i = 1$ and $w_i \in [0,1]$.

Example Let A_1 and A_2 be two intuitionistic multi-fuzzy sets of dimension 3 and weight $w_1 = 0.4$ and $w_2 = 0.6$ are assigned to them. Let

A_1 = {(0.7, 0.6, 0.5), (0.1, 0.2, 0.4)}

A_2 = {(0.4, 0.4, 0.3), (0.3, 0.4, 0.6)}

The aggregated IMFWA is computed as

$$\text{IMFWA (A1,A2)} = w_1 A_1 \oplus w_2 A_2 =$$
$$1 - \left\{(1-0.7)^{0.4}(1-0.6)^{0.4}(1-0.5)^{0.4}(1-0.4)^{0.6}\right\},$$
$$\left\{(0.1)^{0.4}(0.2)^{0.4}(0.4)^{0.4}(0.3)^{0.6}(0.4)^{0.6}(0.6)^{0.6}\right\}$$
$$= \{0.95, 0.09\}$$

Definition of score function S(x), $x \in U$ and accuracy function H(x), $x \in U$ for IMFSs

If x is an IMFV where $x = \{(\mu^1{}_X, \mu_X^2 ....\mu_X^k), (\nu_X^1, \nu_X^2 ....\nu_X^k)\}$, the score function can be defined as S(x) = $\Sigma_j \{\mu^j(x) - \nu^j(x)\}, j = 1,2,......k$

The accuracy function can be defined as $H(x)=\sum_{j}\{\mu^{j}(x)+\nu^{j}(x)\}, j=1,2,......k$

Detail utilization of score and accuracy function in [12].

The definition of distance function d (A, B) between two IMFS A and B as

$$d(A,B)=\frac{1}{2n}\left(\sum_{j=1}^{n}\sum_{i=1}^{k}\left(\left(\mu^{i}_{A_j}-\mu^{i}_{B_j}\right)^2+\left(\nu^{i}_{A_j}-\nu^{i}_{B_j}\right)^2+\left(\pi^{i}_{A_j}-\pi^{i}_{B_j}\right)^2\right)\right)^{\frac{1}{2}}$$

3.4 *Presentation of IMFS to IFA conversion*

An intuitionistic multi-fuzzy set A = $\{x/(\mu^1_A(x), \mu^2_A(x)..........\mu^k_A(x)),(\nu^1_A(x),\nu^2_A(x)..........\nu^k_A(x)): x\in U\}$ can be degenerate to an IFS A^G where $A^G=\{x/\mu^G_A(x),\nu^G_A(x) \in U\}$. $\mu^G_A(x)=\frac{1}{k}\sum_{I=1}^{K}\mu^i_A$, $\nu^G_A(x)=\frac{1}{k}\sum_{I=1}^{K}\nu^i_A$ and $0\le\mu^G_A(x)+\nu^G_A(x)\le 1, \pi^G_A(x)=$ for every $x\in U$ and $i=1,2,..........k$ where k is appositive integer and U is a universal set.

Example. Let A be an IMFS of dimension 3. If A = {(0.7,0.6,0.5),(0.1,0.2,0.4), then IFS A^G can be expressed as A^G = {(0.7,0.6,0.5)/3,(0.1,0.2,0.4)/3} = (0.6,0.2).

An intuitionistic multi-fuzzy set $A=\{x/(\mu^1_A(x), \mu^2_A(x)..........\mu^k_A(x)),(\nu^1_A(x),\nu^2_A(x)..........\nu^k_A(x)): x\in U\}$ can be degenerate to an IFS A^G where A^G = $\{x/\mu^G_A(x),\nu^G_A(x)\in U\}$. $\mu^G_A(x)=\max\{\mu^1_A(x),\mu^2_A(x)..........\mu^k_A(x)\}$, $\nu^G_A(x)=\min\{\nu^1_A(x),\nu^2_A(x)..........\nu^k_A(x)\}$ and $0\le\mu^G_A(x)+\nu^G_A(x)\le 1, \pi^G_A(x)=1-\mu^G_A(x)-\nu^G_A(x)$, for every $x\in U$ and $i=1,2,..........k$ where k is appositive integer and U is a universal set.

Example. Let A be an IMFS of dimension 3. If A = {(0.7,0.6,0.5),(0.2,0.1,0.4)}, then IFS A^G can be expressed as A^G = {max(0.7,0.6,0.5), min(0.1,0.2,0.4)} = (0.7,0.1).

4 ENTROPY ON IMFS

The Entropy of a fuzzy set describes the degree of fuzziness for the fuzzy set. Zadeh [1] introduce the entropy of fuzzy set first in 1965 and in 1996 Burillo and Bustince [6] proposed the intuitionistic fuzzy entropy. We defined the entropy on IMFS based on the work [13] and [6].

A real function $E: IMFS(X)\to[0,1]$ is called the entropy of IMFS(X), if E satisfies the following properties:

(E1) E(A) = 0 iff A is a crisp set;

(E2) E(A) = 1 if $\mu_A(x)=\nu_A(x)$ where j = 1,2,..... k for all $x\in X$;

(E3) E(A) ≤ E(B) if A is less fuzzy than B, i.e., $\mu^j_A(x)\le\mu^j_B(x)$ & $\nu^j_A(x)\ge\nu^j_B(x)$ for $\mu^j_B(x)\le\nu^j_B(x)$ for j = 1,2......k Or $\mu^j_A(x)\ge\mu^j_B(x)$ & $\nu^j_A(x)\le\nu^j_B(x)$ for $\mu^j_B(x)\ge\nu^j_B(x)$ for j = 1,2..... k (E4) E(A) = E(A^c)

When X = $\{x_1,x_2,x_3............x_n\}$

$$A=\{\langle x/((\mu^1_A(x),\mu^2_A(x),.......,\mu^k_A(x)),(\nu^1_A(x),\nu^2_A(x),.......\nu^k_A(x))):x\in U\}$$

where $0\le\mu^i_A(x)+\nu^i_A\le 1$

$\pi^i_A(x)=1-(\mu^j_A(x)+\nu^j_A(x))$ for every $x\in$ U & i = 1,2,.... k where k is a positive integer & U is a universal set.

$$\text{Then we have } E^1(A)=\frac{\sum_{i=1}^{k}\sum_{j=1}^{k}(\mu^j_{A_i}(x)\wedge\nu^j_{A_i}(x))}{\sum_{i=1}^{k}\sum_{j=1}^{k}(\mu^j_{A_i}(x)\vee\nu^j_{A_i}(x))} \quad (1)$$

is the entropy on IMFS(X).

Example 1. Let A be two intuitionistic multi-fuzzy sets of dimension 3 defined as follows:

A = {(0.7, 0.6, 0.5),(0.1,0.2,0.4)}

Entropy of set A can be calculated as $E(A)=\frac{0.1+0.2+0.4}{0.7+0.6+0.5}=\frac{0.7}{1.8}=\frac{7}{18}$

4.1 *Fuzzy cross entropy*

For two IMFSs A and B is called intuitionistic multi fuzzy

$$I(A,B)=\sum_{i=1}^{n}\left[\mu^i_A(x)\ln\frac{\mu^i_A(x)}{\frac{1}{2}\left[\mu^i_A(x)+\mu^i_B(x)\right]}+\nu^i_A(x)\ln\frac{\nu^i_A(x)}{\frac{1}{2}\left[\nu^i_A(x)+\nu^i_B(x)\right]}+\pi^i_A(x)\ln\frac{\pi^i_A(x)}{\frac{1}{2}\left[\pi^i_A(x)+\pi^i_B(x)\right]}\right]$$

cross entropy between A and B

I(A, B) is not symmetric with respect to its arguments, then we can get the symmetric cross entropy measure of IMFS A and B: D(A, B) = 1/2[I(A, B)+I(B, A)]

5 SIMILARITY ON IMFS

Similarity measure is a term that measures the degree of similarity between IFSs. As an important content in fuzzy mathematics, similarity measures between IFSs have gained much attention for their wide applications in real world, such as pattern recognition, machine learning, decision making and market prediction.

A real function $S: IFSs(X)\times IFSs(X)\rightarrow[0,1]$ is called the similarity measure on IMFS(X), fundamentally it is useful in various domain such as decision making, pattern recognition and machine learning. if S satisfies the following properties:

(S1) $S(A, A^c) = 0$ if A is a crisp set;
(S2) $S(A,B)=1 \Leftrightarrow A=B$;
(S3) $S(A, B) = S(B, A)$
(S4) For all $A,B,C\in IMFS(X)$,
If $A\leq B\leq C$, Then $S(A,C)\leq S(A,B)$ & $S(A,C)\leq S(B,C)$. When $X=\{x_1,x_2,x_3............x_n\}$,

$$A=\left\{\langle x/\left((\mu_A^1(x),\mu_A^2(x),........,\mu_A^k(x)),(\nu_A^1(x),\nu_A^2(x),........\nu_A^k(x))\right)\right\rangle : x\in U\}$$

$$B=\left\{\langle x/\left((\mu_B^1(x),\mu_B^2(x),........,\mu_B^k(x)),(\nu_B^1(x),\nu_B^2(x),........\nu_B^k(x))\right)\right\rangle : x\in U\}$$

Then we have:

$$S^1(A,B) = 1-\frac{1}{n^2}\sum_{i=1}^{n}\sum_{j=1}^{k} max\left\{\left|\mu_{A_i}^j(x)-\mu_{B_i}^j(x)\right|\left|\nu_{A_i}^j(x)-\nu_{B_i}^j(x)\right|\right\}$$

(2) is the similarity measure on IMFS(X).

Example 2. Let A be two intuitionistic multi-fuzzy sets of dimension 3 defined as follows: A = {(0.7,0.6,0.5), (0.1,0.2,0.4)}, B = {(0.4,0.4, 0.3), (0.3,0.4,0.6)}

Similarity of set A and B can be calculated as: $S(A,B)=1-\frac{1}{9}(0.3+0.2+0.2)=1-\frac{1}{9}\times 0.7 =$ 0.93 let X = {x_1, x_2, x_3,................, x_n} and A, B∈IMFSs(X). The similarity between A and B can be define as follows

$$S(A,B) = 1-\left[\frac{1}{3n^2}\sum_{i=1}^{n}\sum_{j=1}^{k}\left(\begin{array}{l}\left|\mu_{A_i}^j(x)-\mu_{B_i}^j(x)\right|^\lambda\\+\left|\nu_{A_i}^j(x)-\nu_{B_i}^j(x)\right|^\lambda\\+\left|\pi_{A_i}^j(x)-\pi_{B_i}^j(x)\right|^\lambda\end{array}\right)\right]^{\frac{1}{\lambda}} \quad \lambda\geq 1$$

6 DEVELOPMENT ALGORITHM

Step 1: Intuitionist multi-fuzzy matrixes are developed by the information given the experts. Consider that the rating of the object x with respect to the jth criteria c_j given the lth decision maker expressed in IMFS as r_{li} = {($\mu^{lj}_1(x)$, $\mu^{lj}_2(x)$,..... $\mu^{lj}_k(x)$), ($\nu^{lj}_1(x)$, $\nu^{lj}_2(x)$,..... $\nu^{lj}_k(x)$),($\pi^{lj}_1(x)$, $\pi^{lj}_2(x)$,..... $\pi^{lj}_k(x)$): x∈U} where k is the dimension of IMFS. N and L are the number of criteria and decision maker respectively. Multi-criteria group decision making problem represent using IMFS by the following matrix.

$$R=(r_{ij})_{lXn}=\begin{pmatrix} r_{11} & r_{12}\cdots\cdots & r_{1n}\\ r_{21} & r_{22}\cdots\cdots & r_{2n}\\ r_{l1} & r_{l2}\cdots\cdots & r_{ln}\end{pmatrix} \tag{3}$$

Step 2: let R be the decision matrix. Then we construct the entropy matrix $E = (E_{ij})_{l\,X\,n}$ from the IMFS decision matrix R. Where $E_{ln} = M_{ln}+N_{ln}+H_{ln}$ where

$$M_{ln}=m_{ln}-\sum_{i=1}^{k}(m_{ln}-\mu_j^{ln}),\ m_{ln}=\max(\mu^{lj}_1(x),\mu^{lj}_2(x),.....\ \mu^{lj}_k(x))$$

$$N_{ln}=s_{ln}-\sum_{i=1}^{k}(s_{ln}-\nu_j^{ln}),\ s_{ln}=\max(\nu^{lj}_1(x),\nu^{lj}_2(x),.....\ \nu^{lj}_k(x))$$

$$H_{ln}=h_{ln}-\sum_{i=1}^{k}(h_{ln}-\pi_j^{ln})\ h_{ln}=\max(\pi^{lj}_1(x),\pi^{lj}_2(x),.....\ \pi^{lj}_k(x))$$

$$E=\begin{pmatrix} E_{11} & \cdots & E_{1k}\\ \vdots & \ddots & \vdots\\ E_{n1} & \cdots & E_{nk}\end{pmatrix} \tag{4}$$

Then normalize the entropy matrix E as follows

$$\overline{E}=\begin{pmatrix} \overline{E_{11}} & \cdots & \overline{E_{1k}}\\ \vdots & \ddots & \vdots\\ \overline{E_{n1}} & \cdots & \overline{E_{nk}}\end{pmatrix} \tag{5}$$

$$\overline{E_{ij}}=E_{ij}\big/\max\{\mu_{j1},\mu_{j2}\,.........\,\mu_{jl}\} \tag{6}$$

where j = 1,2,3.... n;

Finally, the attribute weight w_j (j = 1, 2 ....., n) is calculate

$$w_j=z_j/b_j \tag{7}$$

where $z_j=\sum_{i=1}^{n}E_{ji}$ and $b_j=\sum_{i=1}^{l}z_j$

Step 3: Based on attribute weight in step can be considered as a weight of decision maker $d_l(l \in L)$ which can be expressed as $w_j \in |0,1|$ $i \in L$ where $\sum_{i=1}^{l} w_i = 1$.

Step 4: An aggregated intuitionistic multi fuzzy is constructed based on the opinion of the student performance. Let R = $(r_{ij})_{l \times n}$ be the intuitionistic multi fuzzy matrix of decision maker. In group decision making process all individual decision makers' opinion is focused in to a group opinion to construct an aggregated intuitionistic multi fuzzy matrix r_{ij} with the help of an IMFWA operator

$$r_{ij} = \text{IMFWA}\left(r_{ij}^1, r_{ij}^2, \ldots\ldots r_{ij}^l\right) = \omega_1 r_{1j}^1 \oplus \omega_2 r_{2j}^2 \oplus \omega_3 r_{3j}^3 \ldots\ldots \oplus \omega_l r_{lj}^l$$
$$= \left\{ 1 - \prod_{j=1}^{l}\prod_{i=1}^{k}(1-\mu_i^j)^{\omega_j}, \prod_{j=1}^{l}\prod_{i=1}^{k}(v_i^j)^{\omega_j}, \prod_{j=1}^{l}\prod_{i=1}^{k}(1-\mu_i^j)^{\omega_j} - \prod_{j=1}^{l}\prod_{i=1}^{k}(v_i^j)^{\omega_j} \right\} \tag{8}$$

the aggregated intuitionistic multi fuzzy matrix can be defined as $R = (r_1, r_2, r_3 \ldots\ldots r_n)$ where

$$r_i = \left\{ \left(\mu_1^j, \mu_2^j \ldots\ldots \mu_k^j\right), \left(v_1^j, v_2^j \ldots\ldots\ldots\ldots v_k^j\right), \left(\pi_1^j, \pi_2^j \ldots\ldots\ldots\ldots \pi_k^j\right) \right\} \tag{9}$$

Step 5: calculate the similarity between the aggregated intuitionistic multi fuzzy matrix R and Section wise cut-off sample matrix S using equation (2) denoted $s = (s_1, s_2, s_3 \ldots\ldots\ldots s_k)$

Step 6: select largest one denoted by S_L among the all value of S. then the lth option is the best choice.

6.1 *Case study*

In an examination system, there are five sections Language1, Language2, General Science, Mathematics and Indian History each sections have three subsections. The examination system four phases are screening, preliminary, intermediate and final. The actual result depends on all those phase according to the result of the examination represent for each section with three subsections correct option(μ), wrong option(ν) and on option(π).

Let D be the category of the result outstanding, excellent, Good and Average express by D = {O, E, A, B} and S be the set of five section of examination is given by S = {s_1, s_2, s_3, s_4}. Four phase of examination are denoted by E = {e_1, e_2, e_3, e_4} appeared by students and according to their performance classify the standard.

Step 1: intuitionistic multi fuzzy matrix R for four phase's examination system are constructed depending on the student performance. Table 1 show the intuitionistic multi fuzzy decision R for the student performance.

Step 2: calculate the attribute weights using Entropy equation (1) as follows w = (0.31, 0.25, 0.26, 0.2)

Step 3: An aggregated intuitionistic multi fuzzy matrix is constructed based on student perform-

Table 1. Section wise student performance.

Exam/Score	Language1	Language2	General Science	Mathematics	Indian History
Phase 1	(0.3,0.7,0.5) (0.2,0.1,0.4) (0.5,0.2,0.1)	(0.4,0.3,0.4) (0.3,0.6,0.4) (0.3, 0.1,0.2)	(0.1,0.2,0.0) (0.7,0.7,0.8) (0.2,0.1,0.2)	(0.5,0.6,0.7) (0.4,0.3,0.2) (0.1,0.1,0.1)	(0.4,0.3,0.4) (0.6,0.4,0.4) (0.0,0.3,0.2)
Phase 2	(0.5,0.6,0.4) (0.2,0.2,0.5) (0.3,0.2,0.1)	(0.3, 0.5,0.3) (0.3,0.2,0.5) (0.4,0.3,0.2)	(0.4,0.6,0.6) (0.2,0.3,0.2) (0.2,0.1,0.2)	(0.5,0.2,0.8) (0.3,0.6,0.1) (0.2,0.2,0.1)	(0.4,0.7,0.8) (0.4,0.0,0.1) (0.2,0.3,0.1)
Phase 3	(0.3,0.4,0. 7) (0.5,0.5,0.3) (0.2,0.1,0.0)	(0.3,0.8,0. 7) (0.6,0.1,0.2) (0.1,0.1,0.1)	(0.4,0.1,0.2) (0.3,0.5,0.8) (0.3,0.4,0.0)	(0.3,0.4,0.4) (0.4,0.5,0.2) (0.3,0.1,0.4)	(0.1,0.3,0.7) (0.7,0.3,0.3) (0.2,0.4,0.0)
Phase 4	(0.3,0.7,0.5) (0.2,0.1,0.4) (0.5,0.2,0.1)	(0.3,0.7,0.5) (0.2,0.1,0.4) (0.5,0.2,0.1)	(0.3,0.7,0.5) (0.2,0.1,0.4) (0.5,0.2,0.1)	(0.3,0.7,0.5) (0.2,0.1,0.4) (0.5,0.2,0.1)	(0.3,0.7,0.5) (0.2,0.1,0.4) (0.5,0.2,0.1)

Table 2. Section wise performance after aggregation.

Section	Language1	Language2	General Science	Mathematics	Indian History
Score	(0.36,0.55,0.55) (0.75,0.83,0.61) (0.61,0.73,0.93)	(0.41,0.55,0.48) (0.65,0.72,0.64) (0.76,0.84,0.85)	(0.33,0.33,0.32) (0.64,0.48,0.51) (0.69,0.84,0.81)	(0.49,0.42,0.64) (0.65,0.60,0.84) (0.83,0.82,0.8)	(0.3,0.44,0.61) (0.42,1.0,0.76) (0.88,0.44,0.85)

Table 3. Section wise cutoff.

Phase/Cut of Marks	Language 1	Language 2	General Science	Mathematics	Indian History
Phase 1	(0.8,0.1,0.1)	(0.2,0.7,0.1)	(0.3,0.5,0.2)	(0.5,0.3,0.2)	(0.5,0.4,0.1)
Phase 2	(0.7,0.2,0.1)	(0.5,0.1,0.4)	(0.4,0.2,0.4)	(0.4,0.1,0.5)	(0.1,0.2,0.7)
Phase 3	(0.5,0.3,0.2)	(0.3,0.5,0.2)	(0.8,0.1,0.1)	(0.3,0.2,0.5)	(0.4,0.4,0.2)
Phase 4	(0.1,0.7,0.2)	(0.3,0.6,0.1)	(0.5,0.1,0.4)	(0.2,0.6,0.2)	(0.3,0.3,0.4)

ance and attribute weight calculated in step 2. Table 2 show the corresponding aggregated intuitionistic multi fuzzy matrix.

Step 4: Obtain the similarity values between aggregated matrix of and cut off value matrix using equation (2) as follow s = (0.71, 0.76, 0.70, 0.68).

Step 5: Since $s_2 > s_1 > s_3 > s_4$, the most desirable alternative is s_2, i.e., the student is excellent category.

7 CONCLUSION

In this paper we extend the concept of IMFS with new axiomatic of IMFS entropy and Similarity measure, those are help full to merge the opinion of the expert into a single IMFM. Then we propose the entropy based attribute weight to solve the group multi criteria decision making problem. Furthermore, we present one numerical example are discussing to demonstrate the feasibility and utility of IMFS method. a future of this research work can enhance with number of objects.

REFERENCES

[1] Atanassov K., (1999) Intuitionistic fuzzy sets. Springer Physica, Heidelberg.

[2] Zadeh L. A., (1965) Fuzzy sets. Inf. Control. 8, 338–356.

[3] Atanassov K., Pasi G., Yager R., (2002) Intuitionistic fuzzy interpretations of multi-person multicriteria decision making. Proc. of 2002 First Intern. IEEE Symposium Intell. Syst. 1, 115–119.

[4] Atanassov K., Pasi G., Yager R., Atanassova V., (2003) Intuitionistic fuzzy graph interpretations of multi-person multi-criteria decision making. In: Proceedings of the Third Conference of the European Society for Fuzzy Logic and Technology EUSFLAT, Zittau, pp. 177–182.

[5] Pasi G., Atanassov K., Melo Pinto P., Yager R., Atanassov V., (2003) Multi-person multi-criteria decision making: intuitionistic fuzzy approach and generalized net model. In: Advanced design, production and management systems. Proceedings of the 10th ISPE International Conference on Concurrent Engineering, Madeira, pp. 1073–1078.

[6] Burillo P., Bustince H., (2001) Entropy on intuitionistic fuzzy sets and on interval-valued fuzzy sets, Fuzzy Sets and Systems 118 (3) 305–316.

[7] Szmidt E., Kacprzy k J., (2001) Entropy for intuitionistic fuzzy sets, Fuzzy Sets and Systems 118 (3), 467–477.

[8] Zeng W., Yu F., Yu X.,. Chen, Wu S H., (2009) "Entropy of intuitionistic fuzzy set based on similarity measure," International Journal of Innovative Computing, Information and Control, vol.5, no.12, pp. 4737–4744.

[9] Szmidt E., Kacprzy J., (2004) A concept of similarity for intuitionistic fuzzy sets and its application in group decision making, in: Proceedings of International Joint Conference on Neural Networks & IEEE International Conference on Fuzzy Systems, Budapest, Hungary, pp. 25–29.

[10] Szmidt E., Kacprzyk J., (2009) Analysis of similarity measures for Atanassov's intuitionistic fuzzy sets, in: Proceedings IFSA/EUSFLAT, pp. 1416–1421.

[11] Chen T., Li C., "Determining objective weights with intuitionistic fuzzy entropy measures: A comparative analysis,"

[12] Das S, Kar M B, Kar S, (2013) Group multi-criteria decision making using intuitionistic multi-fuzzy sets, Journal of Uncertainty Analysis and Applications

[13] Wei Cui-Ping, Pei Wang, Yu-Zhong Zhang, (2011) Entropy, similarity measure of interval-valued intuitionistic fuzzy sets and their applications, Information Sciences, vol.181, pp. 4273–4286.

[14] Dey S. K., Biswas R., Roy A.R., (2000) Some operations on intuitionistic fuzzy sets. Fuzzy Set. Syst. 114, 477–484.

[15] Xu Z., (2007) Intuitionistic preference relations and their application in group decision making. Inform. Sci. 177, 2363–2379.

[16] Atanassov K., (1994). New operations defined over the intuitionistic fuzzy sets. Fuzzy Set. Syst. 61, 137–142.

Computational Science and Engineering – Deyasi et al. (Eds)
© 2017 Taylor & Francis Group, ISBN 978-1-138-02983-5

Author index